Learning Materials in Biosciences

Learning Materials in Biosciences textbooks compactly and concisely discuss a specific biological, biomedical, biochemical, bioengineering or cell biologic topic. The textbooks in this series are based on lectures for upper-level undergraduates, master's and graduate students, presented and written by authoritative figures in the field at leading universities around the globe.

The titles are organized to guide the reader to a deeper understanding of the concepts covered.

Each textbook provides readers with fundamental insights into the subject and prepares them to independently pursue further thinking and research on the topic. Colored figures, step-by-step protocols and take-home messages offer an accessible approach to learning and understanding.

In addition to being designed to benefit students, Learning Materials textbooks represent a valuable tool for lecturers and teachers, helping them to prepare their own respective coursework.

Julijana Ivanisevic • Martin Giera
Editors

A Practical Guide to Metabolomics Applications in Health and Disease

From Samples to Insights into Metabolism

 Springer

Editors
Julijana Ivanisevic
Faculty of Biology & Medicine
University of Lausanne
Lausanne, Switzerland

Martin Giera
Center for Proteomics & Metabolomics
Leiden University Medical Center
Leiden, The Netherlands

ISSN 2509-6125   ISSN 2509-6133   (electronic)
Learning Materials in Biosciences
ISBN 978-3-031-44258-2   ISBN 978-3-031-44256-8   (eBook)
https://doi.org/10.1007/978-3-031-44256-8

This Springer imprint is published by the registered company Springer Nature Switzerland AG
The registered company address is: Gewerbestrasse 11, 6330 Cham, Switzerland

Paper in this product is recyclable.

Preface

The coronavirus pandemic was a serious challenge for our society. Lockdowns and restrictions affected all of us in different ways. Particularly, undergraduate students suddenly had to shift toward online lessons and courses, which severely impacted their (practical) training. However, this shift has also accelerated and led to permanent changes in our teaching systems with online content and learning becoming increasingly important. This is when the idea of a metabolomics book accompanied with online content and open-access datasets was born. Thereby, the main aim of this book is to facilitate online learning for undergraduate students in the field of metabolomics and to provide guidelines and materials for practical course or tutorial instructors.

As metabolomics has matured and become a routinely applied tool in (patho-) biochemistry, many excellent books cover the biological as well as chemical facets of this interdisciplinary analytical science. However, the content aimed at undergraduate students with a clear teaching focus and practical examples is lacking. With the presented book *A Practical Guide to Metabolomics Applications in Health and Disease: From Samples to Insights into Metabolism*, we aim to bridge this gap by providing the datasets and step-by-step protocols used for their treatment and embedded in the context of specific case studies to students and course supervisors to facilitate self-study and training, respectively.

Following a general introduction, the book first covers the hottest topics in the field, from the overlooked sample handling and preparation to highly debated data processing and metabolite annotation. The following two chapters present and discuss innovative strategies for data interpretation in the biochemically and physiologically relevant context. The presented protocols oriented on specific steps in metabolomics workflow are followed by the presentation of five stories told through metabolomics-derived data acquired in the context of specific physiological conditions (e.g., exercise and healthy aging) or disease (e.g., inflammation and cancer). Finally, the last three chapters present the case studies accomplished by the emerging mass spectrometry-assisted imaging technology, which is likely to pave the future of tissue and single-cell metabolomics. Practical examples from our laboratories with easy-to-follow instructions enable course instructors and students alike in designing educative training and working with "real-life" data.

Lausanne, Switzerland   Julijana Ivanisevic
Leiden, The Netherlands   Martin Giera

Contents

Part I

Introduction

Introduction

Martin Giera and Julijana Ivanisevic

1.1 Background and Definitions

The metabolome comprises the entity of all small molecules (metabolites) involved in energy metabolism (as fuels), maintenance of cell structure (as building blocks) and metabolic signaling, as messenger molecules. Metabolomics strives to identify, analyze, and quantify all metabolites of a given biological system. Consequently, metabolomics is defined as "the (quantitative) study of all small molecules (metabolites) of a biological system such as cells, tissues, or organisms." Metabolomics is significantly intertwined with analytical chemistry, biochemistry, and lately systems biology. Historically, metabolomics evolved from the centuries-old effort to describe and define the molecular fundaments of life. Ultimately it is a biochemical tool set to shed light on the molecular fundaments of metabolism. Importantly, metabolomics was only made possible by significant technological developments, most importantly mass spectrometry (MS) and nuclear magnetic resonance spectroscopy (NMR) [1]. Today, metabolomics can be divided into several subfields and categories. On one hand, technological characteristics define targeted and untargeted

M. Giera (✉)
LUMC, Leiden, The Netherlands

Metabolomics Unit, University of Lausanne, Lausanne, Switzerland
e-mail: m.a.giera@lumc.nl

J. Ivanisevic (✉)
LUMC, Leiden, The Netherlands

Metabolomics Unit, University of Lausanne, Lausanne, Switzerland

Lausanne, Switzerland
e-mail: julijana.ivanisevic@unil.ch

© The Author(s), under exclusive license to Springer Nature Switzerland AG 2023   3
J. Ivanisevic, M. Giera (eds.), *A Practical Guide to Metabolomics Applications in Health and Disease*, Learning Materials in Biosciences,
https://doi.org/10.1007/978-3-031-44256-8_1

metabolomics analysis. On the other hand, subfields, for example, lipidomics, computational metabolomics, plant metabolomics, and many more, have evolved in recent years. Targeted metabolomics refers to the (quantitative) analysis of large panels of known metabolites. In most cases, targeted methods make use of commercially available standard materials which are used to establish the analytical assays. Most targeted metabolomics panels are run by liquid chromatography tandem mass spectrometry (LC-MS/MS) or NMR with some applications also being developed for gas chromatography mass spectrometry (GC-MS) [2]. In turn, the observed chromatographic (or spectroscopic) signals are identified based on specific retention times (or chemical shifts in NMR) and characteristic mass spectrometric features (multiplicity) in comparison with genuine chemically characterized standard materials. In the case of untargeted metabolomics, compound identification takes place post-analysis. In other words, biological samples are most frequently analyzed using either LC-MS/MS, GC-MS, or NMR, and subsequently metabolite identification is initiated [3]. Several chapters of this book will give detailed instructions on how such data must be preprocessed and treated for successful metabolite identification and biological interpretation. Nevertheless, it is important to point out that untargeted metabolomics analysis usually results in hundreds to thousands of chromatographic signals for which mass spectrometric data has been obtained. While some of these data might be matched against existing databases (e.g., METLIN) or genuine synthetic standard materials, others are only putatively or ambiguously identified or remain unknown. Consequently this has led to the introduction of a classification system for the level of metabolite identification by the Metabolomics Standards Initiative (MSI) [4]. The original classification was comprised of four levels ranging from level 1 (unambiguously identified analyte) to level 4 (unidentified analyte). In this context, metabolic features should be distinguished from metabolites. A metabolic feature is basically a signal (or an ion) characterized by retention time and (tandem) mass spectrum (mass to charge or m/z ratio) which are recorded in untargeted metabolomics analysis. Additionally, the precise ion mobility (or collision cross section (CCS)) value of an ion reflecting its physicochemical properties can also be acquired nowadays (depending on the instrument setup). The process of metabolite annotation is the assignment of metabolite candidates to metabolic features, and a metabolite is an unambiguously identified chemical structure partaking in metabolism. Several important definitions are summarized in Box 1.1.

Box 1.1 Definitions of Common Terms in Metabolomics

Metabolite—a small molecule partaking in metabolism (roughly <1000 Da).

Metabolome—the entity of all metabolites.

Targeted metabolomics—(quantitative) targeted analysis of a panel of known metabolites implicated in a single or multiple pathways. Each metabolite is

(continued)

> **Box 1.1** (continued)
> characterized by its retention time (chemical shift) and mass spectrometric characteristics (multiplicity).
>
> *Untargeted metabolomics*—metabolite analysis without a priori hypothesis, requiring subsequent metabolite annotation.
>
> *Feature*—a signal of an ion characterized by its retention time and (tandem) mass spectrometric data recorded during untargeted metabolomics analysis. Each metabolite is represented by multiple features or ions corresponding to its isotopes, adducts, in-source fragments, multiply charged species, etc.
>
> *Annotation*—the process of systematically assigning metabolites to metabolic features.
>
> *MSI level*—confidence level in the annotation process.

1.2 Metabolomics in Biological Context

The elucidation of the DNA structure in 1953 and the subsequent publication of molecular biology's central dogma in 1958 set the stage for our present understanding of the information flow in biological systems [1]. It basically states that information flows from DNA (genome) to RNA (transcriptome) to protein (proteome), and when the proteome level has been reached, it cannot flow back, while this is possible between DNA and RNA. Remarkably, at this time, metabolites (metabolome) had not been considered a central part of information flow but pure biochemical bystanders. This view has dramatically changed during the last decennia [5]. For example, several important classes of signaling molecules, for example, fatty acyl esters of hydroxy fatty acids (FAHFA) [6], the eicosanoids or oxysterols, and other oxidized lipids, have been identified [7, 8]. Consequently, metabolites and the metabolome have become an integral part of our understanding of how information in biological systems is transferred and its relation to phenotypic observations (Fig. 1.1). Moreover, today the metabolome is considered as the phenotype at the molecular level, and it has become accepted that information can flow back to the above-indicated layers of biochemical organization. For example, posttranslational modifications represent a proteome-metabolome interaction; the same is true for post-transcriptional modifications (transcriptome—metabolome interaction) and epigenetics where the interaction between DNA and metabolites plays a significant role [5]. Even more so, it is becoming increasingly accepted that the metabolome (along with lipidome) is organelle- and cell type-specific. In other words, a specific cell type, for example, a macrophage or neutrophil or even subsets such as M1 or M2 macrophages, presents with a specific metabolome that is closely linked to their function and phenotype.

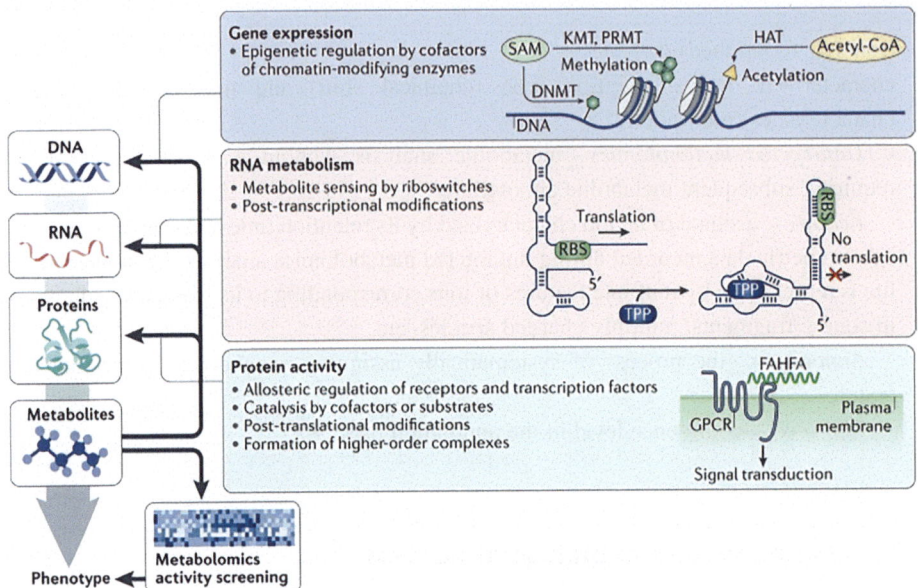

Fig. 1.1 Biological information flow and the role of metabolites, taken with permission from [5]

1.3 Analytical Technologies

As outlined above, metabolomics as a scientific field has largely been made possible by the rise of advanced analytical technologies. While Williams and co-workers were among the first to recognize the significance of individual biochemical signatures in the late 1940s, it was Dalgliesh et al. in 1966 who for the first time used gas chromatography coupled to mass spectrometry (GC/MS) to study urinary metabolites. Subsequently, Horning and colleagues introduced the term metabolic profiles and, together with Linus Pauling and Arthur Robinson, developed GC/MS methods to simultaneously monitor dozens of metabolites present in biological samples during the 1970s. However, the most significant cornerstone of metabolomics is electrospray ionization (ESI), which was mainly developed by John Fenn in the late 1980s. Today, a plethora of different analytical technologies are applied to deepen our understanding of the metabolome [1]. Below we will briefly describe these technologies, and an overview is given in Fig. 1.2.

1.3.1 Gas Chromatography: Mass Spectrometry (GC-MS)

GC-MS was among the first technologies applied in metabolomics analysis. Moreover, GC-MS is one of the first examples of online coupling a separation device (gas

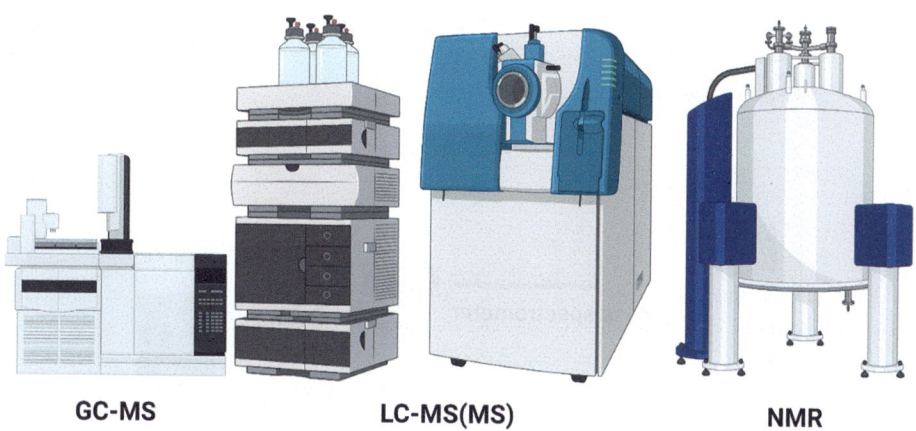

GC-MS   LC-MS(MS)   NMR

Fig. 1.2 Main analytical technologies used for metabolomics analysis (figure made with biorender. com)

chromatograph) with a detection device (mass spectrometer). Gas chromatography (GC) can be applied to volatile components and mainly relies on separation by individual boiling points. A GC consists of several components; a carrier gas (gas supply) usually helium or hydrogen is flushed through an inlet system (liner) that is connected to a separation column. Today, the most widely applied separation columns are capillary columns of 30 m length and 0.25 mm diameter. The most frequently applied stationary phase is poly(5% diphenyl/95% dimethyl siloxane). The column is located in a programmable oven. Nowadays, sample introduction is accomplished by autoinjection systems. Throughout the years sophisticated injection methods, for example, on-column injection, large-volume injections, and others, have been developed. However, split and splitless injections remain the most commonly applied techniques. Typically, 1–2 μL of a sample is injected into the hot (~300 °C) liner. The injection volumes in GC/MS are largely dependent on the liner volume, its temperature, the applied gas flow, and the sample solvent. A typical injection liner has roughly 1 mL volume. Overfilling the liner can lead to a backflush of injection solvent into the carrier gas lines, this might ultimately lead to increased background noise. It is important to keep in mind that water, for example, under standard GC/MS settings, has a 1000× higher gas volume when compared to its liquid volume. In other words, 1 μL water will become 1000 μL water vapor. Very useful GC calculators can be found here (https://www.agilent.com/en/support/gas-chromatography/gccalculators).

In the case of split injections, part of the sample is flushed away via a splitflow, and in the case of splitless injection, the entire sample will be transferred to the separation column. Subsequently the sample is focused on the head of the separation column; this is accomplished by keeping the starting temperature of the GC typically 5–10 °C below the boiling point of the employed solvent. Thereafter the programmable oven is ramped, and analytes

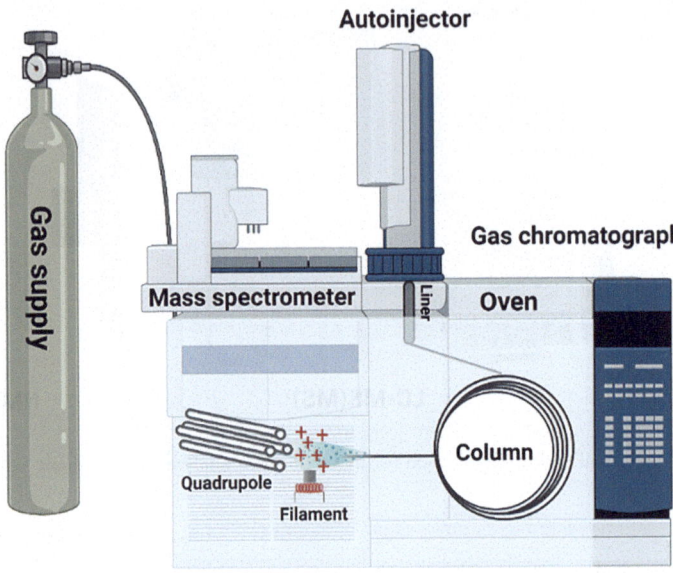

Fig. 1.3 The basic components of a GC-MS system (figure made with biorender.com)

are consecutively evaporated and transported through the column. In the case of GC-MS, the column outlet is introduced into the MS via a transfer line. The transfer line is basically a heated piece of capillary, ensuring that no analytes are trapped by a cold bridge between GC and MS. After entering the MS, ionization in most cases takes place in an electron ionization source (EI). Electrons expelled from a heated filament are accelerated into the ionization chamber. The kinetic energy of the so produced electrons can be controlled and is usually set to 70 eV. In the ionization chamber, neutral molecules, eluting from the GC column, are interacting with the accelerated electrons. When the accelerated electrons originating from the filament get into close proximity of a neutral molecule, an electron is being expelled, resulting in the formation of a positively charged molecule and its fragmentation. The so produced charged molecules and their fragments can now be filtered by a mass analyzer (usually a quadrupole) and quantified as electric current through means of an electron multiplier (Fig. 1.3).

The early success of GC-MS is related to the fact that coupling separation and detection technology is rather straightforward as the carrier gas automatically transports the separated molecules into the ionization chamber and can simply be pumped away. Moreover, for many decades, significant mass spectrometric libraries for compound identification only existed for EI-MS spectra. This is due to the fact that EI-MS spectra can be generated in a very reproducible way on a single-stage MS instrument. The main variable influencing the obtained EI-MS spectra is the applied ionization energy, and 70 eV has become the most widely accepted standard. Today the NIST GC-MS library contains more than 350,000 searchable spectra. Additionally, EI-MS spectra follow distinct fragmentation rules (not to

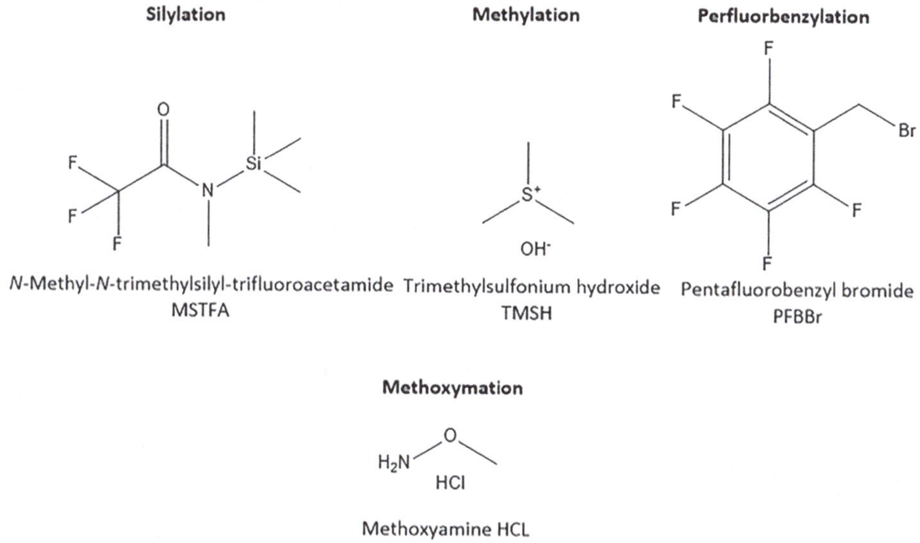

Silylation   **Methylation**   **Perfluorbenzylation**

N-Methyl-N-trimethylsilyl-trifluoroacetamide   Trimethylsulfonium hydroxide   Pentafluorobenzyl bromide
MSTFA   TMSH   PFBBr

Methoxymation

Methoxyamine HCL

Fig. 1.4 Common derivatization reagents used in GC/MS analysis

be discussed here) which to some extent allow for the identification of unknowns. However, besides these advantages, GC/MS does come with a couple of limitations. It is limited to vaporizable components, which also restricts its application to low molecular weight components (<750 Da). Moreover, EI is a very hard ionization technique that results in hundreds of fragment ions and very complex spectra, and last but not least, thermal stability is a prerequisite for GC/MS analysis. Throughout the years, researchers have developed strategies to address these limitations. In order to increase thermal stability of functional groups and to allow analysis of hard-to-vaporize molecules, numerous derivatization strategies for hydrophilic functional groups have been developed; examples include trimethylsilylation, methylation, acetylation, and many more (examples are shown in Fig. 1.4) [9]. A typical protocol for GC/MS-based metabolomics would combine methoximation and trimethylsilylation [10]. To allow for a milder ionization, so-called chemical ionization (CI) has been developed. CI makes use of a reagent gas, typically ammonia or methane. Within the ionization chamber, the number of reagent gas molecules by far exceeds the number of analyte molecules. In turn, the reagent gas will be ionized by the available free electrons whereafter a cascade of chemical steps leads to the ionization of the analyte molecules. In other words, the reagent gas works as an intermediate buffering the energies within the collision cell. Of note, unlike EI, CI can also produce negatively charged ions, a fact that has led to the development of electron capture negative ionization (ECNI). Particularly in combination with highly electrophilic derivatization, for example, pentafluorobenzyl bromide, GC-ECNI-MS is a very selective and sensitive approach for the analysis of carboxylic acids [11].

Box 2 Learning Objectives and Training GC/MS

GC-MS make yourself familiar with the components of a GC/MS system.

GC-MS study the basic principles underlying separation and detection in GC-MS analysis.

EI-MS familiarize yourself with the common principles of EI-MS. Exercise reading and interpreting EI-MS spectra. A demo version of the NIST database including structures and spectra can be found here: https://chemdata.nist.gov/dokuwiki/doku.php?id=chemdata:nist17

Derivatization what are the basic characteristics of derivatization reagents? Draw the derivatization reactions for several examples, e.g., trimethylsilylation with MSTFA, derivatization with PFBBr, methoximation, and silylation. Generate an overview of functional groups and suitable derivatization reagents. How does methylation with TMSH exactly work? Why is methoximation and trimethylsilylation combined for metabolomics analysis? Does the order of derivatization steps matter?

Example try to explain the major fragments in the spectrum of cholesterol-TMS. Compare the spectrum of cholesterol with its double-bond isomers. Can such chemically similar components be distinguished by EI-MS analysis? An exemplary data set of sterol isomers can be found in the supplementary material here [2].

1.3.2 Liquid Chromatography Tandem Mass Spectrometry (LC-MS/MS)

As outlined above GC-MS was the first hyphenated technique applied in metabolomics research. However, its limitations hampered its application particularly for larger as well as hydrophilic molecules. The solution to this issue was the development of electrospray ionization (ESI) and liquid chromatography mass spectrometry (LC-MS). Liquid chromatography (LC) is a partition-based separation technology. Partition of molecules takes place between a stationary phase and a liquid eluent. For a detailed description of the fundamental principles of LC, please refer to dedicated literature. A schematic of a LC-MS system is shown in Fig. 1.5. High-pressure LC pumps deliver a gradient of a weak and a strong eluent to a stationary phase which is connected to the MS detector. Modern LC systems usually operate at pressures between 200 and 800 bar. The most frequently applied eluent combinations in LC-MS analysis are methanol or acetonitrile as organic modifiers and water with 0.1% formic acid or MS-compatible buffer salts as aqueous phase. Given the fact that metabolomics analysis of bodily fluids or cells deals with hundreds to thousands of metabolites, it is understandable that superior chromatographic performance is desired. Due to the fact that LC separation is influenced by specific factors, for example, extra column volume, particle size, and flow rate (familiarize yourself with the van Deemter curve), ultra-high-pressure LC (UPLC) in combination with either sub 2 μm particle columns or core-shell materials have become standard. When it comes to the available

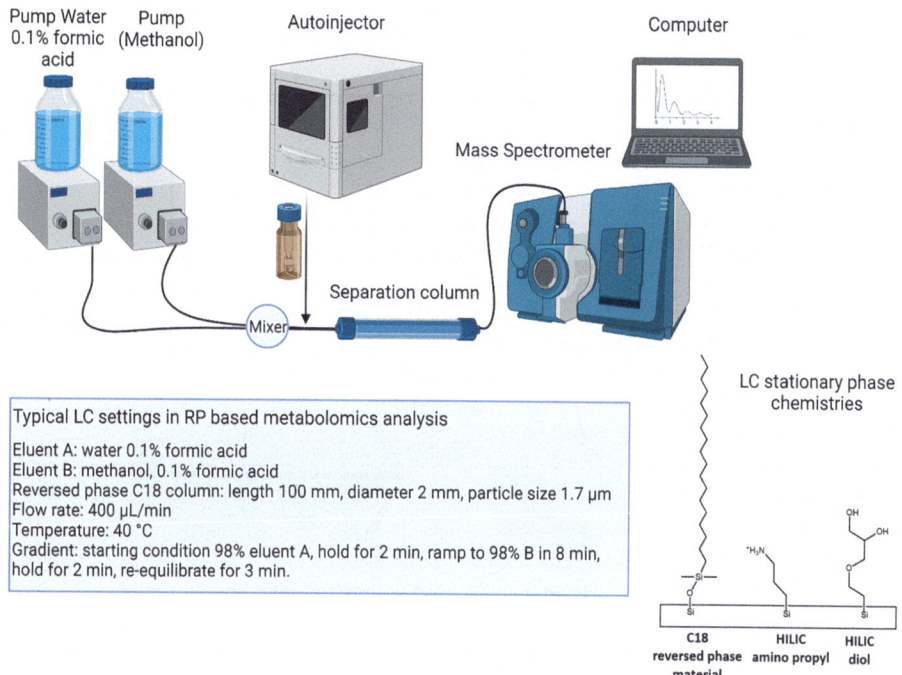

Fig. 1.5 Schematic of an LC-MS system, LC stationary phases and typical settings for RP-based metabolomics analysis (figure partially made with biorender.com)

stationary phases there are two main choices: on one hand, reversed-phase (RP) chromatography mainly employing C18 materials and hydrophilic interaction chromatography (HILIC) using diverse HILIC materials. The difference between the two stationary phases mainly lies in their respective mode of analyte interaction. RP materials are ideal for the retention of lipophilic components, and the stationary phase/analyte interaction is mainly based on nonpolar van der Waals interactions. Compound elution is accomplished by increasing amounts of an organic modifier that competes with the stationary phase and hence will move components along the separation column. HILIC on the other hand is ideal for the retention of very polar, hydrophilic components. HILIC stationary phases, for example, diol or amine-functionalized silica are polar phases that strongly retain polar metabolites. For example, glucose, a very polar metabolite will elute in the void volume of a RP column, while it will have strong retention on a HILIC phase [12].

The hyphenation of LC and MS was mainly accomplished by ESI. Unlike for GC-MS where a gas flow can directly be introduced into an ionization chamber does LC work with large amounts of liquid eluents. Moreover, MS detection only works with ionized components and takes place under high vacuum conditions while elution from the LC column takes place at atmospheric pressure. In turn, the main challenge was to ionize solved

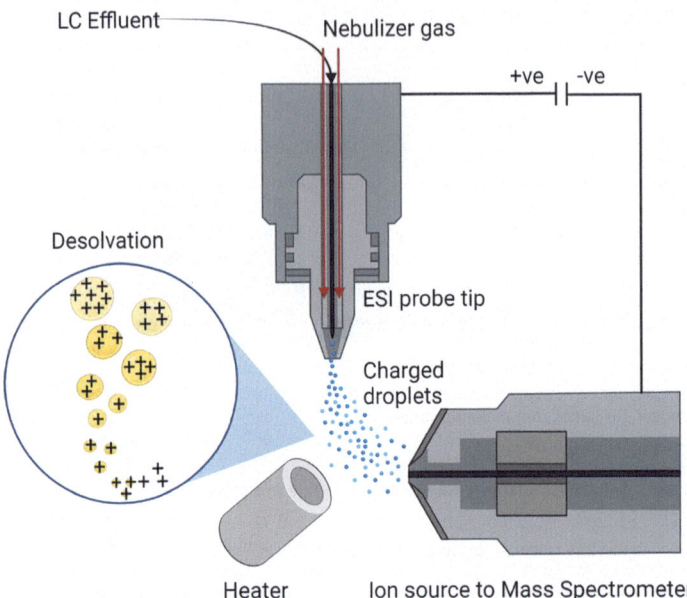

Fig. 1.6 The basic parts of an ESI source and the CRM model for the formation of gas phase ions (figure made with biorender.com)

analyte molecules and transfer them into the MS analyzer without breaking its high vacuum. Notably, aqueous liquids result in very high vapor volumes when being evaporated. Early devices basically build on technology already applied in GC-MS analysis (not discussed here). However, the real breakthrough was the development of the ESI source by Fenn and colleagues. Today, 90% of all LC-MS applications are run with an ESI interface. In ESI, the continuous LC flow is pumped through a charged metal capillary resulting in its dispersion. This process is referred to as electrospray. The resulting fine aerosol subsequently undergoes extensive desolvation, and the charged liquid droplets start to shrink. At the so-called Rayleigh limit, the electrostatic repulsion of like charges, in the steadily decreasing droplets, overcomes the surface tension holding the droplet together, and the droplets basically explode forming significantly smaller droplets. There are two major theories how gas phase ions are ultimately formed after this process: the ion evaporation model (IEM) and the charge residue model (CRM). The IEM argues that the field strength of the continuously shrinking droplets assists field desorption at one point. The CRM on the other hand argues that all solvent will eventually be evaporated in shrinking fission cycles ultimately leaving the charged analytes that were contained in the droplets. Figure 1.6 outlines the main parts of an ESI inlet. Other ionization technologies are atmospheric pressure chemical ionization (APCI) and atmospheric pressure photoionization (APPI).

Following the generation of gas phase molecules, these can be directed toward the mass analyzer. Today many types of mass analyzers exist; besides others these include quadrupoles (Q), triple quadrupoles (QqQ), ion traps (IT), Orbitraps, time-of-flight (TOF), and hybrid instruments such as quadrupole-time-of-flight (QTOF) instruments, or linear ITs. Nowadays, almost all metabolomics research involves the generation of tandem mass spectra (MS/MS). In order to generate MS/MS data, the ions generated during ESI are sent to a collision cell for their fragmentation. Unlike EI, ESI is a soft ionization technique resulting mainly in the formation of molecular ions. However, to obtain characteristic molecular information, analyte fragmentation is of greatest importance not only for compound identification but also for quantification. A collision cell is being introduced after the first mass filter. For example, in a QqQ system, the second Q will be the collision cell. In a QTOF instrument, a collision cell will be present between the Q mass filter and the TOF mass analyzer. In all cases the purpose of the collision cell is analyte fragmentation. This is typically achieved by the introduction of a collision gas, for example, argon. The charged molecules passing the first mass filter will enter the collision cell and collide with collision gas molecules; depending on the kinetic energy and the collision gas density, this will cause the molecules to fragment, producing so-called fragment ions. The so produced fragment ions can now be passed through another mass filter and made visible by an electron multiplier. The entity of all produced fragment ions is called a tandem mass spectrum or MS/MS spectrum. Importantly and just like in EI fragmentation, the produced fragment ions are compound-specific and can be used for analyte identification and quantification. METLIN is the biggest MS/MS database with almost one million physically analyzed components (https://metlin.scripps.edu/). A dedicated chapter of this textbook will explain in detail how to match and identify metabolites using XCMS and METLIN.

Within the field of metabolomics, QqQ instruments are used for wide targeted analysis, whereas particularly QTOF instruments are being applied for untargeted metabolomics assay. The reasons for this are as follows. QqQ instruments have become the gold standard for targeted quantitative analysis, mainly due to the fact that operating QqQ instruments in multiple reaction monitoring (MRM) mode significantly lowers background noise and increases selectivity and thereby also sensitivity (think of the signal-to-noise ratio as the main driver of sensitivity). The choice for QTOF instruments in untargeted workflows is the fact that QTOF instruments typically have a high mass resolving power > 30,000 and are capable of very fast acquisition of tandem mass spectra (up to 200 Hz). While high mass resolution significantly improves analyte identification, as in limiting the relevant search space in terms of analyte mass, the high scan speed allows for excellent coverage of the LC dimension, generating sufficient data points to define a chromatographic peak and simultaneously allow to acquire a large number of MS/MS spectra. Figure 1.7 shows the basic design of QqQ and QTOF mass analyzers and exemplifies the difference in mass resolution.

Importantly, QqQ instruments can be operated in several scan modes; for simplicity we will here only focus on the most widely used mode of multiple reaction monitoring (MRM) analysis. MRM analysis is illustrated in Fig. 1.8. During MRM scanning, the first quadrupole selects an analyte-specific precursor ion, frequently the molecular ion of the compound under investigation. Subsequently, this ion is being fragmented in Q2 (collision cell)

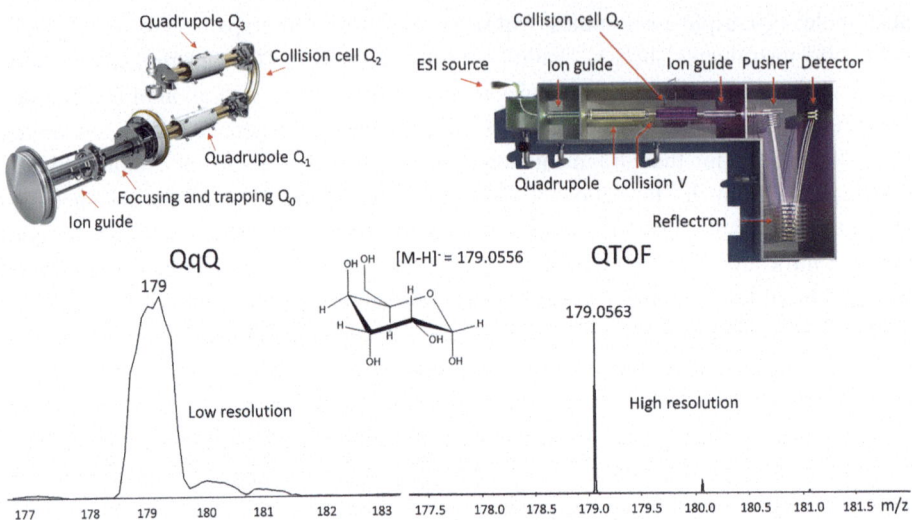

Fig. 1.7 Basic components of a QqQ (left) and QTOF (right) MS system. Below, comparison of mass resolution for the negative molecular ion of glucose, left, QqQ, right TOF MS. QTOF picture reused with permission from [13], QqQ picture courtesy of Sciex, reused with permission

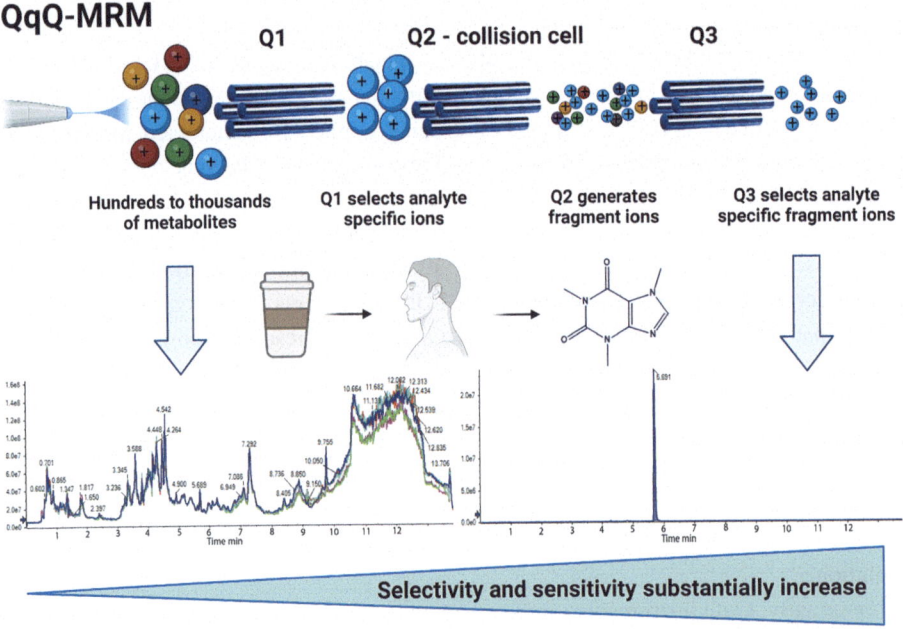

Fig. 1.8 The operation principle of MRM analysis exemplified for the analysis of caffeine in a human saliva sample. Note the substantial increase from a total ion current chromatogram on the left to a single observed peak in MRM mode

producing analyte-specific fragment ions. Of these, one or several analyte-specific ions are selected in Q3; in other words they are allowed to pass Q3 (hence the name mass filter) to reach the detector. During this process selectivity and background noise are constantly improved, which in turn leads to increased signal to noise and hence sensitivity.

Box 1.3 Learning objectives and training GC/MS

LC compare LC- and GC-based separations. What are LC-specific pitfalls and challenges?

LC what are the main parameters influencing LC separations? What does the van Deemter equation describe?

LC-MS how is a MS-compatible buffer characterized?

ESI what are the main differences between ESI and EI?

ESI fragmentation familiarize yourself with ESI fragmentation. What are typical fragmentation processes? How can ESI fragmentation be leveraged for compound identification?

Mass Spectrometry what are the basic operation principles of quadrupoles and time-of-flight mass spectrometers?

QqQ describe all possible scan modes on QqQ MS.

QTOF what are the main differences between QqQ and QTOF MS systems?

1.3.3 Nuclear Magnetic Resonance Spectroscopy (NMR)

A more detailed description of the fundamentals of NMR spectroscopy is given in a dedicated chapter of this book. In short, NMR works by detecting the precession of the nuclear magnetic moments of a sample in a magnetic field. The frequency of the precession is influenced by the chemical environment of the nuclei. Nuclei experiencing a different chemical environment give rise to the NMR spectrum, with each nucleus generating a peak at a specific position in the NMR spectrum, known as the chemical shift. Each distinct chemical compound has its own unique pattern of peaks in the spectrum, the intensity of which is linearly related to the concentration of the compound, which is one of the main strengths of NMR. Of note, only nuclei with an odd number of protons or neutrons present the necessary magnetic characteristics for detection in NMR. In metabolomics, these would typically be ^1H and ^{13}C. As for LC- and GC-based applications, NMR comes with a couple of distinct advantages and disadvantages. One of the biggest advantages of NMR is its nondestructive nature allowing for sample reusage after analysis. Moreover, NMR is the most important cornerstone of de novo substance identification. With the available one- and two-dimensional analysis modes, for example, ^1H, HMBC, CH-COSY, or NOESY, almost any small molecule can unambiguously be identified, given enough and sufficiently pure material is available. Additionally, NMR is a very robust and highly standardized technique allowing to generate quantitative data without the need for isotopically labeled internal standards as well as producing highly reproducible data. Recently, NMR has also become an attractive alternative for the analysis of lipoproteins. The disadvantages of NMR analysis include its lower sensitivity when compared to MS-based approaches (factor 100–1000), the rather high amounts of biological samples needed (e.g., approximately 250 μL blood plasma), and its high investment and maintenance cost. Nevertheless, NMR is an important cornerstone of modern metabolomics analysis and the most important available technology for small metabolite structural elucidation. For further details on NMR-based metabolomics, please refer to the dedicated chapters in this textbook.

Box 4 Summary of Advantages and Disadvantages of GC-MS, LC-MS, and NMR for Metabolomics Analysis

Analytical technology	Advantage	Disadvantage
GC-MS	High separation efficiency Informative MS1 spectra Highly standardized Large libraries Relatively cheap Matrix effects are limited	Analyte breakdown Derivatization necessary Hard ionization
LC-MS(MS)	Large molecular coverage Large libraries Highest sensitivity	Expensive Fragmentation can be instrument-specific Neutral molecules can be hard to ionize/analyze Matrix effects
NMR	Very robust Highly standardized Absolute quantitative De novo substance identification Large coverage	Very expensive Limited sensitivity Complex spectra

1.4   Metabolomics Subfields

During the last decade, several metabolomics subfields have arisen, focusing on either specific organisms, e.g. *plant metabolomics*, or analyte classes, for example, *lipidomics*. Other disciplines gaining significance are *computational metabolomics*, *spatial metabolomics* (e.g., *mass spectrometry imaging*) or *single-cell metabolomics*. Moreover, there is a general trend to integrate and match several omics layers in so-called multi-omics approaches. The most important field might be lipidomics that will be discussed in more detail below.

1.4.1 Lipidomics

Lipidomics, the comprehensive study of lipids is one of the most recent and also significant metabolomics subfields. The importance of lipid analysis can be exemplified by the fact that more than 70% of human plasma metabolites are lipids. Moreover, lipids play very important roles as signaling molecules. A prime example is the eicosanoids, metabolites of the fatty acid arachidonic acid. The eicosanoids play very important roles in inflammation, bronchoconstriction, as well as blood clotting and form the target of nonsteroidal anti-inflammatory drugs (NSAID). Due to their outstanding biological importance, lipids have taken center stage in many disease areas, for example, neurodegenerative diseases, inflammatory diseases, as well as cancer.

The term lipid is rather loosely defined mainly based on a substance solubility in organic solvents. According to the IUPAC gold book, a lipid is defined as "...A loosely defined term for substances of biological origin that are soluble in nonpolar solvents...." Nevertheless, this rather practical definition is reflected in sample preparation techniques for lipidomics analysis making use of organic solvents such as methyl-tert-butyl ether or isopropanol. The main analytical technologies as described above for metabolomics are also being used for lipidomics analysis. However, there are some peculiarities specific to lipidomics. Unlike most metabolites, lipid classes are characterized by a building block structure. In other words, a specific head group (e.g., glycerol or phosphatidylcholine) is combined with fatty acid side chains to form a complex lipid species. This fact has important implications for lipid analysis: (1) to a certain extend, de novo lipid identification can to the most part principally be accomplished by MS/MS analysis without the need for extensive libraries; (2) lipids present an extremely diverse class of molecules with numerous possibilities in terms of head group and side chain combinations; (3) isomers pose a significant challenge in lipid analysis—consider, for example, geometric double-bond isomers, positional isomers, or enantiomers and diastereomers. This latter fact has led to significant efforts in the separation of lipids based on LC but also ion mobility approaches. As described above, lipids can be separated using either RP or HILIC-based LC separations. Importantly, RP separates lipids mainly on their side chain composition, whereas HILIC separates lipids based on their head groups. A very innovative and rapid separation technology for lipids is ion mobility, allowing to separate lipid classes in milliseconds within the gas phase, right before entering the MS. An overview of several lipid classes and their preferred mode of analysis is given below (Fig. 1.9). For further details on lipidomics sample preparation and analysis, please refer to the dedicated chapters of this textbook.

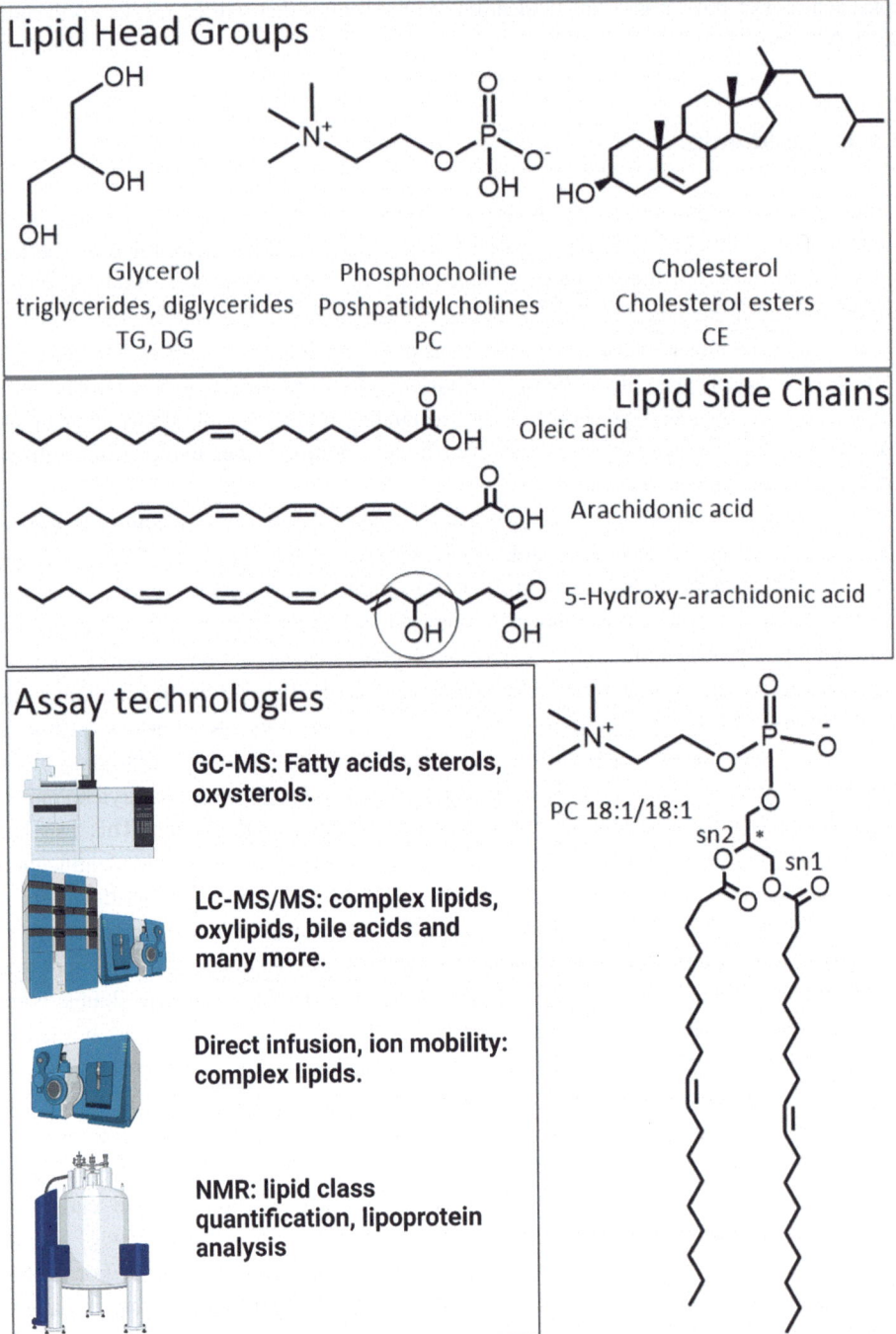

Fig. 1.9 The lipid building block concept (examples of head groups and side chains), lipid analysis technologies, and a phosphatidylcholine lipid with two oleic acid side chains; the circle highlights a stereocenter in 5-hydroxy-arachidonic acid

1.5 Practical Considerations: From Samples to Data

1.5.1 Towards omics scale metabolite analysis

For the past decades, during the era of biochemical genetics, the focus of metabolism research was on genes and their function. However, nowdays, in the post-genomics era of biochemistry, we are focusing back on metabolites and their function in diverse metabolic processes [14]. In other words, we are "rethinking" or revisiting metabolism using metabolomics tools. This "comeback" is due to recent technological developments which allow us to measure metabolites with more specificity and sensitivity than ever before. For this reason, metabolomics is often qualified as a *next-generation metabolic profiling*. Although we are still far from omics scale metabolite analysis, we have made an important step-forward, toward the analysis of a broad range of metabolites, from highly abundant nutrients involved in energy production and storage to low abundant signaling molecules. Importantly, metabolomics approaches have allowed us to discover new biological and potent signaling roles of well-known energy metabolites and compounds we previously thought to be plain by-products of energy metabolism [15, 16]. As David Wishart nicely resumed "While the metabolites may be small...their influence in human physiology and disease is truly profound—they are perhaps the body's most important signaling molecules" [17]. These functionally diverse metabolites are also chemically highly diverse. As Gary Siuzdak noted, "there are no limits on how metabolites can be assembled." In addition, metabolites span an extremely wide concentration range in biological matrices, at least 12 orders of magnitude [18]. This is why we often claim that this high chemical diversity and wide concentration ranges in which metabolites can be present in a biological sample represent the challenge and the aim of metabolomics. This exceptional diversity (compared to other levels of biochemical organization, e.g., genome and proteome) also represents a technological challenge, as there is no technique or combination of techniques which can be used to englobe metabolite diversity present in a cell, tissue, or biofluid. A prominent example is lipid diversity in human blood. More than a thousand lipid species have been reported in human blood plasma, spanning more than nine orders of magnitude, from oxylipins (such as leukotrienes and prostaglandins) present in low picomolar concentrations to highly abundant cholesterol measured at millimolar levels [19]. Beyond the highly conserved *primary metabolites* required for survival, a variety of "exotic" compounds, mainly products of specialized plant and/or microbial metabolism, can also be found in human blood, depending on exposures due to our lifestyle, and the measurement sensitivity, numerous food constituents and additives, phytochemicals, microbial products, toxins, pollutants, cosmetic products, drugs, and drug metabolites [20].

As underlined in the first part of the introduction, during the past decade, mass spectrometry-based approaches have made the most significant imprint in metabolomics, allowing for the measurement of a wide panel of polar and lipid metabolites with high specificity, sensitivity, accuracy, and precision [21, 22]. Mass spectrometers coupled to different (and usually complementary) separation techniques (e.g., LC, GC, capillary electrophoresis, ion mobility) serve today as working horses for the measurement of many hundreds and up to thousand(s) of metabolites (including lipids!) from only minimal amounts of biological samples (e.g., 5 µL of blood plasma).

1.5.2 Pre-analytics or "an Elephant in the Room"

Pre-analytics, encompassing *sample collection, handling, storage, and preparation*, represents a critical step for reliable metabolite measurement and determines the success of metabolomics, along with lipidomics assays [23]. However, compared to analytical protocols which are being constantly improved, to yield ever-better specificity, sensitivity, and productivity of metabolite measurements, pre-analytical considerations remain a true challenge. Analytical technologies and methodologies have significantly evolved, and thereby, instrumental precision and accuracy have been maximized to the extent that there is no need for the analysis of technical replicates. The remaining bias or most of the bias in metabolomics assays is still introduced during the pre-analytical phase which determines the accuracy and precision of probing the metabolome. The key concern represents sample handling following collection, including *quenching to arrest metabolism*[24]. Appropriate sample handling and quenching steps are of utmost importance to produce a polar metabolite or lipid extract which accurately reflects the true metabolic content and status of a model system (e.g., cell, organoid, tissue) without the bias introduced through chemical transformations (due to quencing delay or multiple freeze-thaw cycles) and contaminations, for example. For sample collection and handling, it's strongly recommended to collect samples in a randomized manner and follow a standardized operation protocol (e.g., SOP; Fig. 1.10) in order to treat all samples equally: identical collection time (with respect to organismal physiology, nutritional state, and circadian rhythm), identical collection material (e.g., tubes coated with same anticoagulant, same additives if necessary), identical quenching, and storage procedure [25, 26]. Fast (or, ideally, instantaneous) and appropriate quenching of metabolism is essential due to the inherently transient nature of multiple cellular metabolites which are impacted by the residual enzymatic activity and spontaneous chemical transformations (e.g., oxidation, hydrolysis, etc.) [24]. This challenge does not only apply to cell and tissue samples; biofluid samples can also be significantly affected by lysis of remaining cells, i.e., hemolysis in plasma [23]. For further details on sample handling, please refer to the dedicated chapter of this book.

Finally, subsequent *sample preparation* and applied *extraction protocol(s)*, along with ionization technique (e.g., ESI, EI, CI, matrix-assisted laser desorption, etc.) will further determine "what we see" in a metabolomics and lipidomics assays [27]. Accordingly, the sample preparation protocol serves to extract the metabolites of interest with the highest efficiency possible while simultaneously precipitating proteins and arresting enzymatic activity. Samples should be prepared in a randomized "order" to avoid a systematic bias. Please note that tissue samples require homogenization prior to protein precipitation, often

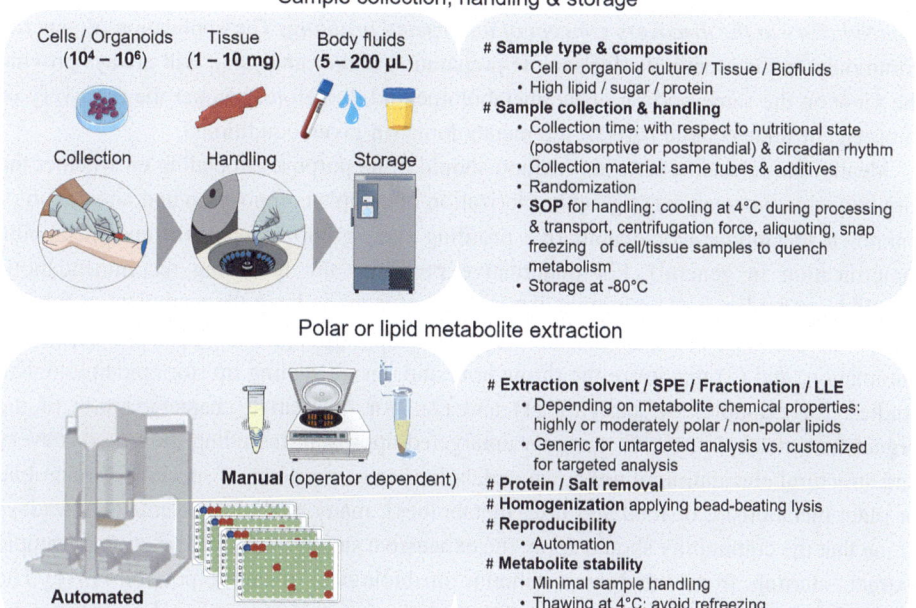

Fig. 1.10 Main considerations in pre-analytics: from sample collection and handling to polar and lipid metabolite extraction for analysis. In addition to the main considerations and recommendations listed in the panels on the right side, the process is illustrated on the left with indications on sample amount with respect to sample type

using a bead beater (to "crush" the heterogeneous tissue into a homogeneous powder or slurry).

Generic extraction protocols are usually applied for untargeted profiling, whereas, in targeted analysis, the extraction is tailored and optimized for the specific class of metabolites of interest. Generic extraction protocols for untargeted analysis are applied to avoid favoring the extraction of one or several classes of metabolites over others, with the aim to achieve as broad metabolite coverage as possible. While the concept of extraction "equality" of as many representative chemical classes as possible is ideal in theory, in reality, it inevitably favors the extraction of the most abundant metabolites, which solubilize the best in the solvents of choice. The selected solvents will further bias the extraction toward more polar (i.e., commonly applied 80% of methanol) or more lipophilic (i.e., BuMe or iPA) metabolites. Therefore, the strategy to prepare sample extracts for polar and nonpolar metabolites separately, such as biphasic extraction or two-step sequential extraction, has been commonly applied [28]. Although this strategy represents a compelling solution to maximize the extraction of a broad range of polar and lipid metabolites simultaneously, reproducibility issues have been reported, in addition to partitioning of amphiphilic metabolites between two phases or extraction steps, respectively [29]. Finally, it should be noted that *the* "standardization" *or application of generic protocols is contradictory to the discovery concept of untargeted profiling*. The application of generic, commonly applied methods (for sample preparation and/or analysis), will always provide the view on the same portion of the metabolome and, therefore, hamper the discovery of "unseen" or undetectable parts of the metabolome (in given conditions).

Ideally, the selected extraction method should fit its purpose depending on whether the aim is qualitative analysis (e.g., characterization of sample composition and annotation of unknown metabolites) or quantitative profiling (i.e., group comparison and metabolite quantification in general). For quantitative profiling, the following recommendations should be considered: (1) minimize the number of steps to keep the analytical variability as low as possible, (2) maximize the reproducibility through the operator-independent automation, and (3) maximize the throughput and ensure scaling up (for epidemiological studies, for example) by applying (1) and (2). For qualitative characterization of the organismal metabolome [30] using an untargeted approach, including the true discovery and structural elucidation of unknown metabolites (e.g., products of specialized microbial or plant metabolism, or food and drug metabolites), many analytical scientists nowadays argue that the community should apply the exhaustive step-by-step fractionation of sample extract, starting from the largest amount of biological material possible [31]. The fractionated extract (enriched in specific classes of interest) can then be analyzed by comprehensive profiling, using several technological platforms, methodologies, and analysis modes to facilitate the detection and annotation of low abundant (unknown) metabolites.

Lastly, as a part of pre-analytical considerations and experimental design, one should carefully think about a *normalization strategy to correct for sample-to-sample amount variation* which will, otherwise, introduce an important bias to metabolite measurement and subsequent data interpretation [32]. Multiple pre- and post-acquisition methods have been reported, and it is crucial to select the most appropriate method depending mainly on the sample type. For homeostatically regulated biofluids, such as blood plasma, normalization to volume is oftentimes sufficient, but for excreted biofluids (like urine, sweat, seminal fluid, bronchoalveolar lavage fluid (BALF) or feces), whose content is heterogeneous and subject to dilution, normalization can prove challenging [33]. There is no consensus on "the best" normalization method to be applied; in turn one should carefully evaluate the applicability of different pre-acquisition strategies (osmolality, protein content, dry weight, etc.) as well as computational (probabilistic quotient normalization, normalization by sum, etc.) ones depending on the sample type and phenotype under investigation (make sure that the normalization strategy is biologically relevant and does not mask the biological variability) [34]. Importantly, pre-acquisition normalization is strongly recommended (when reliable) because it will account for matrix effects and correct it, which cannot be done à posteriori, using post-acquisition statistical methods.

1.5.3 To Remain Untargeted or Use the Power of Targeted Analysis?

As highlighted in the previous section, to resolve the infinite polar and lipid metabolite diversity, we still must combine different sample preparation techniques (e.g., liquid-liquid extraction, solid phase extraction), separation techniques (e.g., LC, GC, capillary electrophoresis, ion mobility), ionization sources (e.g., electrospray, chemical ionization, electron impact, laser desorption), and mass analyzers (triple quadrupole or QqQ, quadrupole time-of-flight or Q-TOF, quadrupole orbital ion trap or Orbitrap, asymmetric track lossless or Astral). In addition to these different techniques and technologies, we also combine complementary modes of analysis (e.g., positive and negative ionization mode, full scan, selected ion monitoring, data-dependent, and data-independent acquisition) and approaches (untargeted vs. targeted) to maximize metabolome coverage. Beyond expanding the metabolome coverage, additional approaches, such as isotopic profiling (or stable isotope-assisted metabolite tracing), are employed to gain insights into pathway activity [35] and MS-based imaging to acquire data on spatial distribution or localization of metabolites of interest within a tissue, an organoid or even down to a single-cell layer [36, 37].

In the first instance, most researchers wish to measure metabolite levels (quenched at a specific point of time) to obtain a snapshot of the metabolic status of the studied model system. The aim of these analyses is usually the relative comparison between two or more sample groups to capture the information about differences in metabolic profiles and potential accumulation and/or depletion of metabolites implicated in a specific pathway and associated with the studied phenotype. The identified pathway can then be further targeted, to validate the results of global profiling; or stable isotope tracing analysis can be performed to conclude on pathway activity under different conditions (e.g., genotype, treatment, etc.) [38]. For comparative analyses, biologists and clinicians will frequently ask for untargeted profiling as many are under the impression that untargeted metabolomics analysis complements modern genomics, transcriptomics, and proteomics analysis in terms of coverage and data output. This is mainly due to the common misconception or the received idea about the "omics scale" and "unbiased" character of the untargeted approach. Unfortunately, this is still far from being real. Although untargeted assays are unbiased in their approach to detect as many metabolites as possible, without a priori hypothesis, their coverage remains limited to the most abundant known and potentially unknown metabolites present in biological samples in rather high, usually μM concentrations (e.g., buildings blocks of cell membranes, "fuel" metabolites involved in energy production and storage, unknown products of xenobiotic metabolism). This largely known metabolome is represented by the "well-trodden foothills" in Fig. 1.1. Accordingly, the size of the unknown metabolome increases as we are getting closer to the "mountaintop"—depicting very low abundant metabolites, sometimes present only in pM concentrations. Importantly, the coverage of untargeted assays is limited due to the lack of sensitivity at MS1 (in full scan mode) and MS2 level which inevitably affects metabolite detection and the acquisition of high-quality MS/MS data, necessary for metabolite annotation. On the other hand, the coverage of targeted assays, while being limited to known metabolites, spans the entire range of concentrations from low abundant signaling molecules (present in fM to nM concentrations) to highly abundant energy metabolites. Highly sensitive and specific methods with a large dynamic range (in analogy to "an intrepid climber with special equipment") must be developed to quantify these specific, low abundant metabolites of interest (Fig. 1.11).

Importantly, the gap between targeted (which traditionally focused on one pathway or one class of metabolites of interest) and untargeted approaches is becoming narrower and methods for targeted quantification of metabolites implicated in multiple pathways are becoming common practice. Using these high-coverage multiple-pathway or *deep targeted* approaches, several hundred known polar metabolites or thousand(s) lipid species can be readily measured in a straightforward manner. There is no need to invest time on re-annotation of known metabolites, as usually done in an untargeted workflow. Importantly, the comprehensive coverage of this multiple-pathway targeted approach allows for an in-depth investigation of central carbon along with lipid metabolism, as an alternative for untargeted analysis. This deep targeted approach fits the purpose of global profiling which can be applied as a data-driven discovery approach to reveal the unanticipated changes (within known pathways) and generate a hypothesis. Moreover, the advantage of targeted analysis lies in significantly higher sensitivity and specificity (compared to untargeted analysis) and the potential for estimated or absolute quantification, using single-point calibration (or internal standard (IS) spike at known concentration) and calibration curves (for commercially available metabolites) matrix. Importantly, the IS spike allows for the correction of matrix effects. This is highly relevant, even when analyzing the same type of sample (e.g., plasma, urine), in human population studies where expected differences or changes are small (also due to inter-individual variability) but statistically significant and physiologically relevant. Otherwise, the sample-to-sample difference in matrix effect might mask the biological variability or result in false positive findings (due to signal suppression or enhancement because of the presence of specific drug, food metabolite, or similar).

The real potential of untargeted approaches as a discovery tool should however not be denied but rather applied in parsimony, to explore the yet unknown routes of metabolism and identify new and yet unknown metabolites and pathways such as products of specialized metabolism, or drug or food metabolism, for example. Its application to study the exposome in response to diverse environmental exposure is emerging; however, one should remain aware about the high sensitivity that is necessary for the detection of usually very low abundant drug or food metabolites, toxicants, pollutants, etc.

Finally, it is also important to consider that metabolite abundances do not reveal pathway activities because increased metabolite levels, for example, can be due to either faster production or slower consumption, and it is often crucial to differentiate between these alternatives. To this end, one can carefully design and apply stable isotope-assisted analysis to track the fate of labeled metabolites by means of label enrichment [35] across different intermediates involved in specific pathways. The isotopic profiling can be done in a targeted or untargeted manner. Ideally, different approaches can be applied in a

Targeted assays
- *Targeted scan of specific precursor: product ions using single pathway or multiple pathway assays (up to 500-1000 polar and/or lipid metabolites)*
- *Coverage limited to "knowns": from low abundant signalling molecules to abundant energy metabolites*

UnTargeted assays
- *Full scan of the entire small molecule range: m/z 50 – 1700 Da at high resolution (i.e., accurate mass measurement)*
- *Coverage limited to abundant "knowns" & potential "unknowns" (e.g., building blocks, fuels, xenobiotics)*

Size of unknown metabolome

fmol

pmol

nmol

µmol

mmol

Concentration range

Fig. 1.11 Untargeted or targeted—that is the question? Breath of coverage of untargeted and targeted assays is illustrated by the "mountain" analogy

systematic way, from global measurement of metabolite levels to isotopic profiling and supplementation assays with the aim to enhance mechanistic insight and unravel the biological role of candidate metabolites [5].

1.5.4 Focus on Applications: From Model Systems to Human Studies (Fig. 1.12)

The great advantage of metabolomics assays (as compared to proteomics, for example) is their high throughput and applicability to a wide range of biological matrices and model systems, from fractions enriched in specific organelles (e.g., lipid droplets, mitochondria) to cell and organoid cultures and tissue and biofluid analysis in model organisms, as well as at the population level. The assays can be scaled up to thousands of samples, specifically in the context of epidemiological studies.

Overall, it's strongly recommended to discuss the experiment in advance with statisticians [40], to determine the *best-suited design* (completely randomized, crossover design, factorial design, block design, repeated measures, etc.) and necessary *sample size* (to detect an expected effect size), and to perform *power* calculations [41, 42].

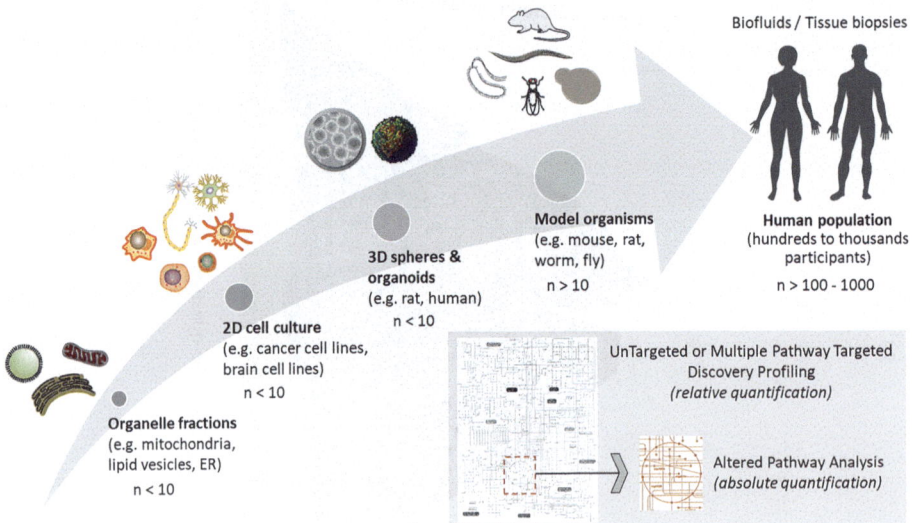

Fig. 1.12 From model systems to human population studies. Due to its high-throughput and phenotyping capacity, different metabolomics and lipidomics approaches, from untargeted screening to targeted quantification, can be applied from different model systems, demanding low numbers of independent biological replicates, to clinical research studies, demanding high numbers of participants due to high human inter-individual variability. An estimated number of replicates is indicated below each studied system. It is mandatory to validate the results of untargeted profiling by targeted quantification, particularly in human population studies, and also to allow for data cross-comparability. The figure was reproduced from [39]

1.6   Summary

Metabolomics has matured into a widely applied technology forming the cornerstone of modern biochemical research. Fields of application include drug discovery and development, biomarker discovery, precision medicine, exposure assessment, and many more. From an analytical perspective, particularly GC-MS, LC-MS/MS, and NMR form the foundation of metabolomics analysis. All these techniques come with specific advantages and disadvantages as discussed in this chapter. Several subfields of metabolomics have emerged in recent years, such as nutritional metabolomics, clinical metabolomics, pharmacometabolomics and others. Nowdays metabolomics has become a major cornerstone of modern physiological and biochemical investigations. Nowadays metabolomics analysis is applied throughout all stages of translational research, from fundamental to preclinical to clinical investigations. Moreover, as for many other scientific disciplines, computational approaches such as machine learning and artificial intelligence (AI) form new highly innovative additions that are rapidly entering the field. Likely, computational advances will also revolutionize metabolomics and together with the other omics technologies lead to digital biology and the generation of digital biochemical and physiological models truly reflecting human physiology. However, for this to happen, the continuous acquisition of high quality experimental data is essential. While fastidious, this experimental work is the groundworks but essential for the generation of data and knowledge, as a prerequisite to any AI approach's accuracy and precision.

References

1. Giera M, Yanes O, Siuzdak G. Metabolite discovery: Biochemistry's scientific driver. Cell Metab. 2022;34(1):21–34.
2. Müller C, Junker J, Bracher F, Giera M. A gas chromatography-mass spectrometry-based whole-cell screening assay for target identification in distal cholesterol biosynthesis. Nat Protoc. 2019;14 (8):2546–70.
3. Johnson CH, Ivanisevic J, Siuzdak G. Metabolomics: beyond biomarkers and towards mechanisms. Nat Rev Mol Cell Biol. 2016;17(7):451–9.
4. Salek RM, Steinbeck C, Viant MR, Goodacre R, Dunn WB. The role of reporting standards for metabolite annotation and identification in metabolomic studies. Gigascience. 2013;2(1):13.
5. Rinschen MM, Ivanisevic J, Giera M, Siuzdak G. Identification of bioactive metabolites using activity metabolomics. Nat Rev Mol Cell Biol. 2019;20(6):353–67.
6. Yore MM, Syed I, Moraes-Vieira PM, Zhang T, Herman MA, Homan EA, Patel RT, Lee J, Chen S, Peroni OD, Dhaneshwar AS, Hammarstedt A, Smith U, McGraw TE, Saghatelian A, Kahn BB. Discovery of a class of endogenous mammalian lipids with anti-diabetic and anti-inflammatory effects. Cell. 2014;159(2):318–32.
7. Brejchova K, Balas L, Paluchova V, Brezinova M, Durand T, Kuda O. Understanding FAHFAs: from structure to metabolic regulation. Prog Lipid Res. 2020;79:101053.
8. Criscuolo A, Nepachalovich P, Garcia-del Rio DF, Lange M, Ni Z, Baroni M, Cruciani G, Goracci L, Blüher M, Fedorova M. Analytical and computational workflow for in-depth analysis of oxidized complex lipids in blood plasma. Nat Commun. 2022;13(1):6547.

9. Halket JM, Zaikin VG. Derivatization in mass spectrometry--1. Silylation Eur J Mass Spectrom (Chichester). 2003;9(1):1–21.

10. Fiehn O. Metabolomics by gas chromatography-mass spectrometry: combined targeted and untargeted profiling. Curr Protoc Mol Biol. 2016;114:30.4.1–30.4.32.

11. Hoving LR, Heijink M, van Harmelen V, van Dijk KW, Giera M. GC-MS analysis of short-chain fatty acids in feces, cecum content, and blood samples. Methods Mol Biol. 2018;1730:247–56.

12. Kohler I, Hankemeier T, van der Graaf PH, Knibbe CAJ, van Hasselt JGC. Integrating clinical metabolomics-based biomarker discovery and clinical pharmacology to enable precision medicine. Eur J Pharm Sci. 2017;109:S15–21.

13. Laganowsky A, Reading E, Hopper JTS, Robinson CV. Mass spectrometry of intact membrane protein complexes. Nat Protoc. 2013;8(4):639–51.

14. van der Greef J, van Wietmarschen H, van Ommen B, Verheij E. Looking back into the future: 30 years of metabolomics at TNO. Mass Spectrom Rev. 2013;32(5):399–415.

15. Guijas C, Montenegro-Burke JR, Warth B, Spilker ME, Siuzdak G. Metabolomics activity screening for identifying metabolites that modulate phenotype. Nat Biotechnol. 2018;36(4):316–20.

16. Mills EL, Pierce KA, Jedrychowski MP, Garrity R, Winther S, Vidoni S, Yoneshiro T, Spinelli JB, Lu GZ, Kazak L, Banks AS, Haigis MC, Kajimura S, Murphy MP, Gygi SP, Clish CB, Chouchani ET. Accumulation of succinate controls activation of adipose tissue thermogenesis. Nature. 2018;560(7716):102–6.

17. Wishart D. Metabolomics for investigating physiological and pathophysiological processes. Physiol Rev. 2019;99:1819–75.

18. Wishart DS, Guo A, Oler E, Wang F, Anjum A, Peters H, Dizon R, Sayeeda Z, Tian S, Lee BL, Berjanskii M, Mah R, Yamamoto M, Jovel J, Torres-Calzada C, Hiebert-Giesbrecht M, Lui VW, Varshavi D, Varshavi D, Allen D, Arndt D, Khetarpal N, Sivakumaran A, Harford K, Sanford S, Yee K, Cao X, Budinski Z, Liigand J, Zhang L, Zheng J, Mandal R, Karu N, Dambrova M, Schiöth HB, Greiner R, Gautam V. HMDB 5.0: the human metabolome database for 2022. Nucleic Acids Res. 2022;50(D1):D622–d631.

19. Bowden JA, Heckert A, Ulmer CZ, Jones CM, Koelmel JP, Abdullah L, Ahonen L, Alnouti Y, Armando AM, Asara JM, Bamba T, Barr JR, Bergquist J, Borchers CH, Brandsma J, Breitkopf SB, Cajka T, Cazenave-Gassiot A, Checa A, Cinel MA, Colas RA, Cremers S, Dennis EA, Evans JE, Fauland A, Fiehn O, Gardner MS, Garrett TJ, Gotlinger KH, Han J, Huang Y, Neo AH, Hyötyläinen T, Izumi Y, Jiang H, Jiang H, Jiang J, Kachman M, Kiyonami R, Klavins K, Klose C, Köfeler HC, Kolmert J, Koal T, Koster G, Kuklenyik Z, Kurland IJ, Leadley M, Lin K, Maddipati KR, McDougall D, Meikle PJ, Mellett NA, Monnin C, Moseley MA, Nandakumar R, Oresic M, Patterson R, Peake D, Pierce JS, Post M, Postle AD, Pugh R, Qiu Y, Quehenberger O, Ramrup P, Rees J, Rembiesa B, Reynaud D, Roth MR, Sales S, Schuhmann K, Schwartzman ML, Serhan CN, Shevchenko A, Somerville SE, St John-Williams L, Surma MA, Takeda H, Thakare R, Thompson JW, Torta F, Triebl A, Trötzmüller M, Ubhayasekera SJK, Vuckovic D, Weir JM, Welti R, Wenk MR, Wheelock CE, Yao L, Yuan M, Zhao XH, Zhou S. Harmonizing lipidomics: NIST interlaboratory comparison exercise for lipidomics using SRM 1950-metabolites in frozen human plasma. J Lipid Res. 2017;58(12):2275–88.

20. Wishart DS, Oler E, Peters H, Guo A, Girod S, Han S, Saha S, Lui VW, LeVatte M, Gautam V, Kaddurah-Daouk R, Karu N. MiMeDB: the human microbial metabolome database. Nucleic Acids Res. 2023;51(D1):D611–d620.

21. Patti GJ, Yanes O, Siuzdak G. Innovation: metabolomics: the apogee of the omics trilogy. Nat Rev Mol Cell Biol. 2012;13(4):263–9.

22. Want E, Nordström A, Morita H, Siuzdak G. From exogenous to endogenous: the inevitable imprint of mass spectrometry in metabolomics. J Proteome Res. 2007;6(2):459–68.

23. Lehmann R. From bedside to bench-practical considerations to avoid pre-analytical pitfalls and assess sample quality for high-resolution metabolomics and lipidomics analyses of body fluids. Anal Bioanal Chem. 2021;413(22):5567–85.

24. Lu W, Su X, Klein MS, Lewis IA, Fiehn O, Rabinowitz JD. Metabolite measurement: pitfalls to avoid and practices to follow. Annu Rev Biochem. 2017;86:277–304.

25. Jonasdottir HS, Brouwers H, Toes REM, Ioan-Facsinay A, Giera M. Effects of anticoagulants and storage conditions on clinical oxylipid levels in human plasma. Biochim Biophys Acta Mol Cell Biol Lipids. 2018;1863(12):1511–22.

26. Ghorasaini M, Mohammed Y, Adamski J, Bettcher L, Bowden JA, Cabruja M, Contrepois K, Ellenberger M, Gajera B, Haid M, Hornburg D, Hunter C, Jones CM, Klein T, Mayboroda O, Mirzaian M, Moaddel R, Ferrucci L, Lovett J, Nazir K, Pearson M, Ubhi BK, Raftery D, Riols F, Sayers R, Sijbrands EJG, Snyder MP, Su B, Velagapudi V, Williams KJ, de Rijke YB, Giera M. Cross-laboratory standardization of preclinical lipidomics using differential mobility spectrometry and multiple reaction monitoring. Anal Chem. 2021;93(49):16369–78.

27. Ivanisevic J, Want E. From samples to insights into metabolism: uncovering biologically relevant information in LC-HRMS metabolomics data. Meta. 2019;9(308)

28. Höring M, Stieglmeier C, Schnabel K, Hallmark T, Ekroos K, Burkhardt R, Liebisch G. Benchmarking one-phase lipid extractions for plasma lipidomics. Anal Chem. 2022;94(36): 12292–6.

29. Medina J, van der Velpen V, Teav T, Guitton Y, Gallart-Ayala H, Ivanisevic J. Single-step extraction coupled with targeted HILIC-MS/MS approach for comprehensive analysis of human plasma lipidome and polar metabolome. 2020;10(12):495.

30. Edison AS, Hall RD, Junot C, Karp PD, Kurland IJ, Mistrik R, Reed LK, Saito K, Salek RM, Steinbeck C, Sumner LW, Viant MR. The time is right to focus on model organism metabolomes. Metabolites. 2016;6(1)

31. Viant MR, Kurland IJ, Jones MR, Dunn WB. How close are we to complete annotation of metabolomes? Curr Opin Chem Biol. 2017;36:64–9.

32. Misra BB. Data normalization strategies in metabolomics: current challenges, approaches, and tools. Eur J Mass Spectrometry. 2020;26(3):165–74.

33. Karu N, Deng L, Slae M, Guo AC, Sajed T, Huynh H, Wine E, Wishart DS. A review on human fecal metabolomics: methods, applications and the human fecal metabolome database. Anal Chim Acta. 2018;1030:1–24.

34. Gagnebin Y, Tonoli D, Lescuyer P, Ponte B, de Seigneux S, Martin PY, Schappler J, Boccard J, Rudaz S. Metabolomic analysis of urine samples by UHPLC-QTOF-MS: impact of normalization strategies. Anal Chim Acta. 2017;955:27–35.

35. Jang C, Chen L, Rabinowitz JD. Metabolomics and isotope tracing. Cell. 2018;173(4):822–37.

36. Chen J, Xie P, Dai Q, Wu P, He Y, Lin Z, Cai Z. Spatial lipidomics and metabolomics of multicellular tumor spheroids using MALDI-2 and trapped ion mobility imaging. Talanta. 2023;265:124795.

37. Alexandrov T. Spatial metabolomics and imaging mass spectrometry in the age of artificial intelligence. Annu Rev Biomed Data Sci. 2020;3(1):61–87.

38. Grankvist N, Watrous JD, Lagerborg KA, Lyutvinskiy Y, Jain M, Nilsson R. Profiling the metabolism of human cells by deep (13)C labeling. Cell Chem Biol. 2018;25(11):1419–1427.e4.

39. Gallart-Ayala H, Teav T, Ivanisevic J. Metabolomics meets lipidomics: assessing the small molecule component of metabolism. BioEssays. 2020;42(12):2000052.

40. Blaise BJ, Correia GDS, Haggart GA, Surowiec I, Sands C, Lewis MR, Pearce JTM, Trygg J, Nicholson JK, Holmes E, Ebbels TMD. Statistical analysis in metabolic phenotyping. Nat Protoc. 2021;16(9):4299–326.
41. Billoir E, Navratil V, Blaise BJ. Sample size calculation in metabolic phenotyping studies. Brief Bioinform. 2015;16(5):813–9.
42. Blaise BJ, Correia G, Tin A, Young JH, Vergnaud A-C, Lewis M, Pearce JTM, Elliott P, Nicholson JK, Holmes E, Ebbels TMD. Power analysis and sample size determination in metabolic phenotyping. Anal Chem. 2016;88(10):5179–88.

Part II

From Sample Collection to Data Acquisition

Pre-analytical Challenges in Clinical Metabolomics: From Bedside to Bench

Isabelle Kohler

> **What You Will Learn in This Chapter**
> - The importance of pre-analytical conditions in the metabolomics and lipidomics workflow
> - The different pre-analytical steps that should be considered when optimizing a method, including sample collection, handling, and storage
> - The effects of inadequate pre-analytical conditions on the metabolome and lipidome
> - General practical recommendations for adequate collection, handling, and storage of biological samples, focusing on blood and urine

I. Kohler (✉)
Division of Bioanalytical Chemistry, Amsterdam Institute of Molecular and Life Sciences (AIMMS), Vrije Universiteit Amsterdam, Amsterdam, The Netherlands

Center for Analytical Sciences Amsterdam (CASA), Amsterdam, The Netherlands

Co van Ledden Hulsebosch Center (CLHC), Amsterdam Center for Forensic Science and Medicine, Amsterdam, The Netherlands
e-mail: i.kohler@vu.nl

2.1   Introduction

Metabolomics, i.e., the analysis of all intermediates and end-products of metabolism, has raised a lot of interest over the last decade in multiple fields, notably clinical research. Tremendous improvements have been achieved with not only the emergence of highly powerful analytical instruments for targeted and untargeted analyses but also the development of sophisticated data analysis pipelines, allowing to process a high amount of complex data. These improvements have led to an increased confidence in the quality of the metabolomics data acquired, resulting in a more widespread use of this approach in clinical research. However, there is one aspect of the entire metabolomics workflow that remains often neglected, namely, the pre-analytical phase. Indeed, even with the most sophisticated instruments or powerful bioinformatic tools, poor data quality will be obtained if the pre-analytical steps are not carefully planned and monitored.

The pre-analytical step englobes all steps performed between the moment the sample is collected from a subject until the sample preparation and actual analysis. The pre-analytical variables notably encompass the type of sample collected, time of collection, sample collection tubes, presence of additives, transport conditions, transport temperature, time before storage, storage temperature, and number of freeze-thaw cycles. Even slight variations in the pre-analytical step can significantly influence the sample integrity and stability of analytes, thereby influencing data quality resulting in data misinterpretation. The clinical consequences of inadequate pre-analytical conditions are therefore high. Most critical steps take place outside of the laboratory where samples are analyzed, which highlights the importance of discussing the pre-analytical considerations with all collaborators involved and implementing standard operating procedures (SOPs) prior to the start of a metabolomics study. Moreover, many large-scale metabolomics studies rely on samples stored in biobanks, which shows the importance for biobanks to guarantee and monitor the quality of biological samples and, in turn, the scientific outcome of subsequent studies.

This chapter focuses on pre-analytical considerations in the analysis of blood-based matrices (i.e., whole blood, plasma, and serum) and urine, discussing the effects of inadequate pre-analytical conditions on metabolome and lipidome. Each pre-analytical step, from bedside to bench, is discussed. For blood-based matrices, special attention is given to the selection of anticoagulants (plasma) or addition of clotting agents (serum). The metabolic differences between plasma and serum, the impact of collection tubes, the temperature at which each step should be performed, and its influence on the metabolome and lipidome are described. Moreover, the effect hemolysis can have on the quality of the measurements performed with nuclear magnetic resonance (NMR) spectroscopy or mass spectrometry (MS) will be discussed. For urine, the focus will be on sampling time, use of additives, and stability at different temperatures during sample collection, handling, transport, and storage. This chapter concludes with a list of general recommendations for pre-analytical considerations to ensure the highest data quality possible.

2.2 Whole Blood, Plasma, and Serum

Blood-based matrices, i.e., whole blood, plasma, and serum, are the most frequently used matrices in clinical metabolomics. Blood shows the advantages of reflecting in vivo physiological states influenced by genetics, epigenetics, lifestyle, environmental factors, and drugs, while being relatively easy to collect [1, 2]. It is therefore not surprising that blood-based matrices are often biobanked for future studies, with analysis taking place up to years or decades after sample collection. This highlights the importance for adequate collection, handling, and long-term storage procedures.

Blood is usually collected in venipuncture tubes, and three types of matrices can be collected, i.e., whole blood, serum, and plasma.

2.2.1 Whole Blood

Whole blood is the most information-rich biofluid but represents a very complex matrix due to its high content in cellular components, i.e., red blood cells (RBCs), white blood cells (WBCs), and platelets. Whole blood therefore contains both intracellular and extracellular pools of metabolites, which comes with additional challenges. For example, the activity of blood cells remains present after whole blood has been collected and exposed to room temperature, which affects sample quality and integrity. Moreover, the analysis and identification of potential biomarker candidates can be complicated by membrane components that originate from the process of cell lysis and isolation [3, 4]. Finally, standardized protocols for preparing whole blood are missing. For all these reasons, plasma and serum are typically preferred over whole blood for metabolomics studies.

2.2.2 Serum and Plasma

2.2.2.1 Serum Collection
Serum is obtained when blood is collected in collection tubes without the addition of anticoagulants. Upon collection, the blood coagulation cascade is activated, with the rapid transformation of fibrinogen into fibrin, which harnesses cells in blood, resulting in the formation of a fibrin- and platelet-based blood clot [5]. After centrifugation, the supernatant (i.e., serum) is collected, while blood cells are eliminated. All the metabolites produced in the body remain in serum, except for fibrinogen and coagulation factors. Moreover, serum also contains metabolites produced during the clotting process.

2.2.2.2 Plasma Collection
In order to obtain plasma, blood is collected in a tube containing an anticoagulant. With the presence of an anticoagulant, the coagulation cascade is inhibited, and no fibrin clot is formed. After centrifugation, the supernatant, i.e., plasma, can be collected, while the blood

cells are pelleted. The composition of plasma is nearly identical to the circulating one, except for the presence of the anticoagulant [5]. Compared with serum, plasma contains a higher number of platelets, which may interfere with the analysis.

2.2.3 Sample Handling

2.2.3.1 Collection Tubes

The types of tubes used to collect blood samples can have a significant impact on the metabolome and should thus be carefully assessed before starting any metabolomics studies.

The venipuncture collection tubes are usually made of plastic (rather than glass). In addition to clot activators (see Sect. 2.2.3.3) or anticoagulants (see Sect. 2.2.3.4), these tubes can also contain a polymer-based gel that provides an easier separation of the supernatant (plasma or serum) from cellular components after centrifugation. The release of plasticizers from these tubes into samples can have a deleterious effect on the quality of the analysis, as they affect the ionization process in MS. A relevant example is polyethylene glycol detected in samples collected with lithium heparinate tubes and serum blood collection tubes, which led to the presence of a typical ion cluster in the MS spectrum [6]. This chemical noise was probably explained by the presence of plastic beads in both tubes.

Moreover, the additives present in these tubes may contain interfering compounds, which can also affect the quality of the data [6, 7]. These interferents are anticipated to be manufacturer-dependent, which highlights the need to select one manufacturer and type of tube within a study. The type of collection tube used within a study (and potential additives) should be carefully selected and tested prior to the study to assess whether they are suitable for the expected application. Besides tubes, the influence of pipette tips and other plastic products, as well as solvents used, should also be evaluated during the planning phase.

Finally, adequate labeling of each tube is obviously necessary. It is worth mentioning that not all labels or markings resist the very low temperatures ($-80\ °C$ or lower) during long-term storage. It is therefore important to take adequate measures to avoid detached labels or unreadable markings.

2.2.3.2 Clotting Time and Temperature During Serum Collection

After blood sampling, billions of cells are still highly active in the tube, releasing, uptaking, and metabolizing metabolites. In serum analysis, the time between collection and separation of cells with serum can therefore have a strong impact on metabolome and lipidome. The temperature at which this process takes place has also an influence on the metabolome composition. The ideal clotting time should be between 30 and 60 min. If the clotting time is too short, the coagulation may be incomplete, and the serum may still contain cellular elements. On the other hand, a longer coagulation time may result in the lysis of the cells in the clot, which will release cellular components into the serum. The coagulation process needs to take place at room temperature, but lower temperatures (i.e., 4 °C) are recommended for the subsequent steps once the clot has been formed, as the activity of cellular metabolism is reduced at those temperatures [6, 8]. It is therefore recommended to keep samples cooled at 4 °C after the formation of the clot and perform transport, handling, and centrifugation at this temperature. Overall, the clotting time and temperature should be strictly controlled and identical for all samples within a metabolomics study. This is particularly important in multicenter studies, where samples are drawn at different places.

Compared with serum, plasma does not require any clotting step at room temperature. Plasma samples can thus be placed onto ice quickly after collection, which is a major advantage compared to serum collection.

2.2.3.3 Clotting Agents and Separator Gels

Clot-activating agents (e.g., silicate-based agents, thrombin) can also be present in the tube, which allows for a shorter clotting time (<30 min) but at higher costs. However, separator gels and clot activators may interfere with the analysis, as demonstrated with NMR [2] and liquid chromatography (LC) hyphenated with MS [1].

2.2.3.4 Anticoagulants in Plasma Collection

Choosing the adequate anticoagulant for plasma collection and analysis represents a crucial step in the metabolomics workflow, as it can significantly influence the quality of the data. Common anticoagulants used in clinical settings include heparin, ethylenediaminetetraacetic acid (EDTA), citrate, and EDTA fluoride. These anticoagulants all prevent the formation of the coagulation clot but with a different mechanism. As a chelating agent, EDTA inhibits several enzymes, which is advantageous to prevent *ex vivo* enzymatic reactions [6]. On the other hand, EDTA and citrate, both chelators of calcium ions, have shown to lead to the formation of sodium and potassium formate clusters that can lead to significant matrix effects in LC-MS or capillary electrophoresis (CE)-MS analysis [8]. Moreover, citrate is an endogenous metabolite involved in multiple metabolic pathways (notably the TCA cycle) and should therefore be avoided if citrate or related metabolites belong to the targeted analytes. Both EDTA and citrate cause strong interference in the NMR spectrum and are therefore less suitable [2, 8]. In addition, EDTA and citrate elute at early times in reversed-phase liquid chromatography (RPLC), which can lead to significant ion suppression or enhancement for poorly retained analytes. Heparin seems to cause less issues related to matrix effects under standard chromatographic conditions [8], but this also depends on the counterion used for this anticoagulant (i.e., Na+ or Li+). Mei *et al.* for instance observed different matrix effects between Na-heparin and Li-heparin plasma, where Li-heparin led to higher matrix effects [7]. Yin *et al.* showed that the presence of Li^+ can increase the signal of plastic polymers and lead to significant matrix effects [6]. Overall, since matrix effects are very dependent on the experimental conditions, it remains difficult to establish general recommendations, as the optimal anticoagulant also strongly depends on the metabolomics applications and target list.

2.2.3.5 Centrifugation Step

The centrifugation force applied to separate erythrocytes, leucocytes, and platelets from the serum/plasma is also relevant, as excessive forces (>4500 g) may lead to hemolysis and should thus be avoided [8]. On the other hand, the centrifugation force and time should be high and long enough to ensure a proper separation of cells from the supernatant. Typically, whole blood samples are centrifuged at 1500–4000 g for 5–10 min. With these conditions, red and white blood cells will be removed, but platelets may remain in the supernatant. These conditions lead to the so-called "platelet-poor" plasma, which is conventionally obtained in clinical labs. The generation of platelet-free plasma, i.e., with less than 10,000 platelets per microliter of plasma, requires a more complex procedure and is usually not used in clinical routine or by biobanks [8].

2.2.3.6 Hemolysis

Hemolysis is defined as the rupture of RBC and release of their contents into the surrounding fluid, which can happen *in vivo* or *ex vivo*. *Ex vivo* hemolysis is one of the most common pre-analytical variables and can happen due to many reasons, e.g., inadequate venipuncture and blood sampling (i.e., too strong aspiration or inadequate diameter of the needle), inadequate cooling during transport (i.e., whole blood transported at a temperature lower than 4 °C), or harsh shaking of whole blood. Hemolysis can lead to a massive release of compounds (e.g., proteins/enzymes, metabolites, and electrolytes) into plasma or serum, which can have a pronounced effect on the blood metabolome and lipidome [1, 8, 9]. The metabolome in hemolyzed samples does not only differ from that of non-hemolyzed samples, but hemolyzed samples also show a greater variability than non-hemolyzed samples [6, 8].

Hemolysis is not always visible, as only concentrations of free hemoglobin above 0.3 g/L (18.8 mM) can be detected by the naked eye (pink to red color) [1]. To date, there is no test or approach that can be used to reliably identify hemolytic samples [8]. This shows the need for adequate procedures that limit the chances for hemolysis, since the metabolome will be already affected even with minimal (non-visible) hemolysis.

2.2.3.7 Storage and Pre-analytical Temperatures

Since sample measurement is most of the time not done on the same day as sample collection, samples need to be adequately stored until further analysis, i.e., for weeks, months, or even years. Similar to hemolysis, storage conditions (short- and long-term), including temperature and freeze-thaw cycles, can have a huge impact on the metabolic composition. The temperature at which the samples are handled and prepared on the day of their analysis is also extremely important.

The ideal long-term storage conditions are to quickly freeze and store several sample aliquots at -80 °C in specific laboratory freezers or biobanks [6, 8]. At -80 °C, most of the metabolites have shown acceptable stability (i.e., no alterations or $\leq 15\%$) for months or even years (5–10 years) [10–14]. Lipids are also usually stable at this temperature if butylated hydroxytoluene (BHT), which acts as a free radical scavenger, is added to the samples directly after plasma/serum separation [11]. In large-scale metabolomics studies, analyzing samples that have been stored for different periods of time at −80 °C is frequent. This may affect the results because of differences in storage duration. However, the expected variability in storage time (i.e., ≤15%) is typically much lower than the biological and inter-individual variability and thus usually acceptable [8].

Typically, −80 °C freezers are not available in study wards. Samples are then often stored temporarily (for hours or days) in −20 °C freezers prior to their transfer to −80 °C freezers. However, a residual enzymatic activity at −20 °C cannot be excluded, which leads to (significant) alteration of the metabolome [1]. Notably, short-term storage at −20 °C has shown to be critical for lipids—and this even for short periods of time; it should therefore be avoided [11, 15, 16]. An interesting and straightforward strategy to increase the stability of bioactive lipids at −20 °C proposed by Giera and co-workers involves the addition of methanol to the samples prior to storage (i.e., protein precipitation), which can preserve sample integrity by preventing (non-)enzymatic degradation [11]. In any case, if a storage step at −20 °C is inevitable, stability studies are recommended prior to the actual study to evaluate the effects of such temperature on the metabolome and lipidome levels.

The time between the collection and freezing procedure is also important, as it is often not possible to immediately process the samples in the minutes after a sample has been drawn. Nevertheless, the metabolome and lipidome are expected to remain mostly stable for up to 2 h after blood collection when tubes are stored at 4 °C [8, 17]. However, samples should be as quickly as possible centrifuged after collection to collect the plasma or serum and store it at −80 °C.

Aliquoting of samples into smaller aliquot directly after collection is recommended. However, it is not uncommon that some aliquots need to be thawed and frozen a number of times dependent on the number of experiments that need to be performed—especially for the most interesting samples. The composition of the metabolome and lipidome may be affected by the number of freeze-thaw cycles, as the samples will be exposed at a higher temperature during the thawing process. For instance, significant changes of the plasma and serum metabolome have been observed when freeze-thawing samples four times at room temperature for 30 min followed by refreezing at −80 °C [12, 18], but metabolites were found to be more stable if the thawing procedure was performed at 4 °C instead of room temperature [6, 19]. The metabolite classes mostly sensitive to repeated freeze-thaw cycles are metabolites from the lipid and central carbon metabolism, antioxidants, nucleotides, and volatile metabolites [8]. Blood samples should therefore be thawed at a temperature ≤10 °C to prevent significant changes in the metabolome and lipidome composition [6, 20]. Overall, these results show the importance of adequately assessing the number and volume of the aliquots prior to a study, depending on the planned experiments, and avoid mixing aliquots within a specific study.

Once samples have been thawed and are ready for sample preparation and analysis, the temperature at which all steps are performed can also significantly influence the composition of the metabolome. This has been investigated by Nishiumi et al., who evaluated the evolution of the concentrations of a panel of metabolites (i.e., cations, anions, and lipids) at room temperature (30 min) and cold temperature (i.e., 4 °C for 8 h) in plasma samples [21]. The results are shown in Fig. 2.1. Some metabolites, such as hypoxanthine, showed dramatic changes in their concentrations at both room and cold temperatures. Notably, plasma concentrations of hypoxanthine were two times higher after storage at room temperature for 15 min only. Another interesting example is pyruvic acid and the trimethylsilylated derivative of glycerol, whose plasma concentrations increased at room temperature over time but decreased at cold temperature, respectively [21]. The instability of some metabolites at room temperature and/or 4 °C has been also reported in other studies [6, 8]. Once more, this highlights the importance of investigating the effect of the temperature on the metabolites or metabolite classes of interest in preliminary studies—not only for long-term storage but also during actual analysis at the bench side, after samples have been thawed.

2.2.4 Comparison Between Plasma and Serum at the Metabolome Level

Recent studies have investigated the differences at the metabolome and lipidome levels between serum and plasma, as well as with different coagulants. For instance, using NMR, Vignoli *et al.* found that 75% of the metabolites quantified showed significantly different concentrations between citrate plasma, EDTA plasma, and serum [2]. Notably, amino acids showed higher levels in serum samples, which may be explained by the inhibitory effect of anticoagulants on plasma proteolytic activities and the release of amino acids by activated platelets during the coagulation procedure. Moreover, lactate and pyruvate concentrations were higher and lower in serum, respectively, compared with plasma samples. A hypothesis for this difference is an ongoing glycolysis during the process of blood clotting. Finally, higher levels of acetone, acetic acid, and formic acid were reported in plasma, likely due to their presence in the collecting tubes or the anticoagulant solution [2].

Another study from Yu *et al.* compared the concentration of 122 metabolites in serum and EDTA plasma samples analyzed using the AbsoluteIDQ TM kit p150 from Biocrates Life Sciences, based on flow injection analysis-MS [22]. Eighty-five percent of those metabolites showed higher concentrations in serum (more than 20% difference for arginine, serine, phenylalanine, glycine, and some lysophosphatidylcholines), while the composition of plasma samples was found to be more stable and repeatable than that of serum. The higher concentrations observed in serum samples may be particularly interesting to detect and quantify low-abundant metabolites.

Sotelo-Orozco investigated the difference in the metabolome between serum and plasma with different anticoagulants using NMR spectroscopy [23]. Heparin and EDTA plasma led to a very similar metabolome than serum, with less than 10% of the targeted 50 metabolites showing a difference. Fluoride plasma showed a higher difference, with 11 metabolites presenting different levels compared with serum. On the other hand, citrate and acid citrate dextrose plasma samples showed major differences compared with serum, largely due to interfering peaks in the NMR spectra caused by the anticoagulants

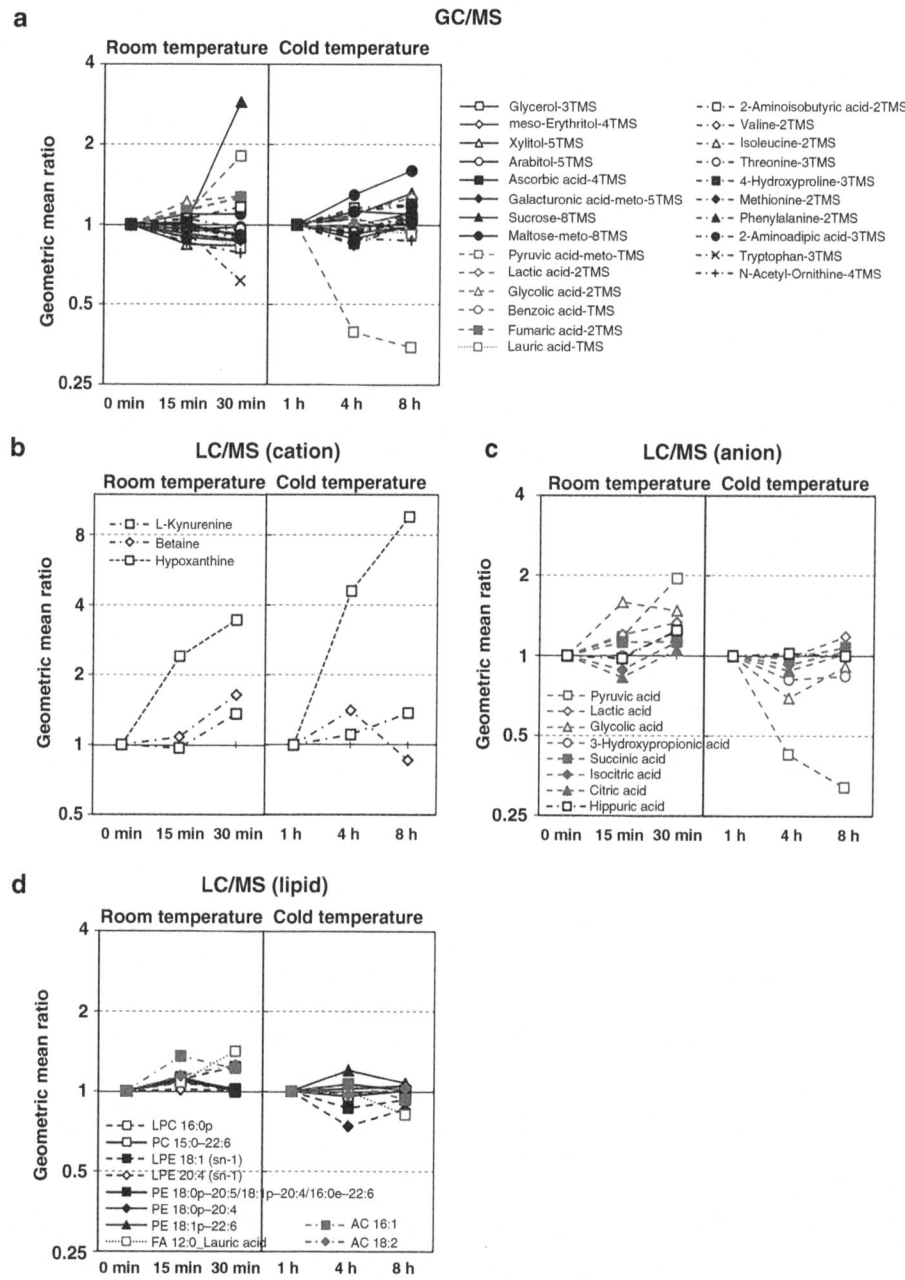

Fig. 2.1 Longitudinal changes in the plasma levels of different metabolites depending on the storage conditions, i.e., room temperature (25 °C) and cold temperature (4 °C). The plots show the geometric mean ration of the levels of each metabolite in plasma to the level seen at baseline, i.e., 0 min at room temperature or 1 h at cold temperature. (**a**) Analysis with gas chromatography—mass spectrometry (GC-MS); (**b**) Analysis of cationic metabolites with liquid chromatography—mass spectrometry (LC-MS); (**c**) Analysis of anionic metabolites with LC-MS; and (**d**) Analysis of lipids with LC-MS. Reprinted from [21] with permissions

themselves (citrate and glucose). Interestingly, most amino acids and derivatives showed higher concentrations in serum compared with plasma—whatever the anticoagulant used, which is also supported by other studies [21–23].

The type of anticoagulant used can also impact the metabolome and lipidome differently depending on the temperature. An interesting example comes from the study of Hahnefeld *et al.*, who showed that the amounts of the endocannabinoids 1-arachydonoyl glycerol, 2-arachydonoyl glycerol, and arachidonoyl ethanolamide (anandamide) increased more markedly in K_3EDTA plasma after storage on ice for 20 min (by 60, 95, and 30%, respectively) than with sodium fluoride/citrate plasma [24]. Endocannabinoids are known to be prone to *ex vivo* formation, and extra caution during sample handling and storage should therefore be taken when targeting this class.

Finally, Giera and co-workers evaluated the effects of two anticoagulants, namely, K_2EDTA and sodium heparin on the concentrations of oxylipins and PUFAs in plasma [11]. They observed that certain eicosanoids and leukotrienes, i.e., LTB_4, LTE_4, TXB_2, 5-HETE, and 12-HETE, were much higher in heparin samples compared with EDTA samples. Some of these analytes were even not detected in EDTA plasma (i.e., LTB_4 and LTE_4).

2.3   Urine

Urine is largely used in clinical metabolomics studies due to the ease of collection and the relatively large quantities that can be collected. As the major route of excretion, urine is a very rich matrix but almost cell-free and with very low protein content. Therefore, it usually requires less sophisticated sample preparation prior to analysis when compared to other protein-rich matrices, such as plasma or serum. However, similar to blood-based matrices, pre-analytical conditions can have a significant impact on the quality of the metabolic data. This is particularly relevant in diseased patients, as the composition of urine can then significantly change and will be more impacted by pre-analytical conditions, due to the possible higher concentrations of proteins/enzymes and presence of metaboli-cally active cells, such as RBC, WBC, bacteria, yeasts, and oxalate crystals [8, 25]. For this population, urine test strips can be considered to assess the presence of these variables, representing an easy and quick option that can be performed directly after collection [8].

2.3.1   Sample Handling

2.3.1.1 Sample Collection
The collection of urine samples is relatively easy and, thus, does not require highly trained staff. Samples are typically collected in plastic containers. Similar to blood collection, it is important to ensure that the container does not release interferents during the analysis and that urinary metabolites do not adsorb to the container surface [8]. Compared with blood,

urinary metabolites are more prone to nonspecific adsorption to the collection container, as less proteins are available to bind to metabolites and keep them in solution. In immobilized patients, a catheter can be used to collect samples, while adsorbent pads can be put in diapers for the collection of urine samples in newborn or pediatric patients. One important recommendation during the usual procedure is to collect urine midstream, which is expected to lead to lower contamination from epithelial cells and bacteria compared with the collection of the first stream [9, 26]. Indeed, a significant amount of metabolic active epithelial cells present on the genital surface are flushed with the first stream urine, resulting in contaminated samples even for healthy donors.

Since urine can be easily collected, it is not uncommon that the sample collection is done by patients or study participants, which may lead to additional errors. In this case, adequate training of the participants based on SOPs is recommended [8, 9].

Compared with blood, the composition of urine varies significantly over the time course of the day, a characteristic that needs to be considered for sample collection. Five approaches can be considered for urine collection, i.e., (1) collection of random samples, (2) collection of spot samples (at specific times), (3) pool of 24-h samples, (4) collection of first morning void (urine obtained directly after getting up), and (5) collection of second morning urine [9, 27]. Each of these approaches shows advantages and disadvantages for metabolomics and requires different considerations. Collection of first morning void shows the advantages that subjects have been typically fasting for several hours before the collection, which lowers the influence of the last meal and/or medication on the sample composition. The second morning urine, collected in a fasted state between 7 and 10 am, shows a different metabolic profile and is even less impacted by the diet of the previous day, which is advantageous [28]. Random samples are collected at any time of the day and will therefore be strongly impacted by the time of collection and effects of meals or other interventions. Spot samples are usually considered in standardized excretion monitoring studies. Finally, 24-h samples (i.e., the collection of all urine voids obtained in 24 h and pooled together) are preferred to compensate for the large intraindividual variability (due to circadian rhythm and diet) in urine composition over the 24-h period. However, they rely on the compliance of the study participants who should not miss any sample and keep the urine collector cool during the whole collection period [9, 27, 29]. Among all these options, the best consensus for metabolomics, as suggested by Lehman and co-workers, seems to be the second morning urine collected after an overnight fast until sample collection [8, 28].

2.3.1.2 Temperature During Collection and Transport

Temperature during collection and transport of urine samples remains a relevant pre-analytical factor for urine analysis, whether urine is collected at the hospital, the study ward, or at home by the patients themselves. During a 24-h collection and/or transport of urine samples, the effects of temperature on the metabolome and lipidome are difficult to predict, as they depend on the targeted analytes. At room temperature, degradation of metabolites will quickly occur for many metabolites, yet not all [9, 30]. Some metabolites did not show relevant changes in their urinary concentrations at temperatures between 4 and 10 °C for up to 72 h [31–34], while others were more significantly impacted [31]. However, it is worth mentioning that most of the studies investigating the effects of temperature on the urinary metabolome focused on healthy individuals and did not include urine collected from diseased patients, which contains enzymes and cellular compounds that can have a profound effect on the metabolome [8].

Overall, to ensure the lowest risk of metabolite degradation, urine samples should be aliquoted and stored immediately at −80 °C (or −20 °C if not available), where the stability of metabolites is mostly ensured [9, 27, 35]. However, this is often not possible for 24-h samples or second morning urine samples collected at the study participant's place. In this case, urine samples should be kept at 4 °C during (repeated) sample collection and sample transport to reduce enzymatic activities and microbial growth and limit metabolite degradation, prior to long-term storage at −80 °C.

The addition of stabilizers, such as borate, thymol, or sodium azide, may be considered to further increase the stability of urine samples and prevent bacterial growth. However, they can lead to interference in both MS- and NMR-based approaches; can complex—in the case of borate—with some metabolites; and can introduce artifacts by changing the pH and ionic strength of urine [8, 27, 35]. They are therefore not commonly used and not recommended, unless needed for a specific metabolite.

2.3.1.3 Storage, Freeze-Thaw Cycles, and Analysis

Urine samples should be stored at −80 °C as soon as possible following collection. Numerous studies have reported that the urinary metabolome is not significantly altered at this temperature, and this is for a long period of time (up to years) [8, 9, 35, 36]. This is particularly the case in untargeted approaches, where the effect of storage typically does not impact the principal component analysis and sample classification [27]. In targeted approaches, it is wise to perform a preliminary study to evaluate the actual stability of the target metabolite classes at this temperature.

Similar to blood-based matrices, the number of freeze-thaw cycles should be reduced to its minimum. The effect of freeze-thaw cycles on metabolite stability depends on the target analyte: for instance, acylcarnitines and hexose, as well as urea urinary levels, were significantly altered after two and three cycles, respectively [31, 37]. Another example includes L-isoleucine, phenylalanine, and 2,3,4-tri-O-acetyl-D-xylono-1,5-lactone, whose urinary levels were significantly altered after three freeze-thaw cycles [38]. Nevertheless, for many metabolites, the effects of the number of freeze-thaw cycles seem to be negligible [39].

The temperature of urine samples is not only important during collection, transport, and storage but also during the actual analysis. It is indeed common to analyze large batches of samples, where (prepared) samples remain for a long period of time in the LC autosampler or the NMR cooling rack. A rule of thumb here is to store samples at 4–10 °C for no longer than 48 h [9].

A relevant strategy to improve the (long-term) stability of urine is to remove cellular debris, bacteria, and other particulate matters prior to storage using centrifugation and/or filtration. The European Consensus Expert Group Report for laboratory management of samples in biobanks recommends a combined use of mild pre-centrifugation (i.e., 1000–4000 g at 4 °C) and filtration as pre-treatment prior to storage at -80 °C [36]. The centrifugation force should be mild to avoid lysis of cells and release of cellular components into the urine but high enough to be effective [40].

2.4   General Recommendations

Pre-analytical conditions can have a significant impact on metabolome and lipidome composition. They should therefore be carefully investigated prior to any metabolomics study to ensure the highest data quality. General recommendations can be established for both blood-based matrices and urine, but many experimental conditions also depend on the targeted analytes, which again highlights the need for preliminary studies.

A crucial aspect is to discuss the pre-analytical conditions between all collaborators involved, i.e., clinical doctors, study nurses, analytical chemists, bioinformaticians, etc. Each step should be carefully documented in SOPs, monitored, and reproduced in different collection locations in the case of multicentric studies. In the latter cases, clear guidelines should be established regarding the material that needs to be used (e.g., collection tubes, pipette tips, etc.), which should ideally be purchased from the same vendor. Finally, the pre-analytical conditions should also be detailed in scientific articles, so that results can be more easily replicated or validated.

Very often, analytical chemists have access to biobanked samples, therefore not controlling study design and pre-analytical considerations. In this context, it is essential that the most detailed information on the pre-analytical conditions is made available and considered during data analysis and interpretation.

Suggestions to minimize pre-analytical issues for blood and urine samples are summarized in Table 2.1, adapted from Lehmann [8] and Yin *et al.* [6, 41].

Take-Home Message
- The pre-analytical phase is of utmost importance in metabolomics for the generation of high-quality data.
- Each pre-analytical step, e.g., type of collection tube, type of anticoagulants, or temperature of storage, can significantly influence the stability of metabolites, thereby influencing the composition of the metabolome and lipidome.
- General recommendations are difficult to draw, as the effects often depend on the targeted metabolites, but common guidelines can be followed to increase the overall stability of metabolites and ensure sample integrity.

(continued)

Table 2.1 Suggestions of experimental procedures to minimize pre-analytical errors in metabolomics. Adapted from [8] and [6, 41]

Parameter	Blood	Urine
Preliminary tests	Test the suitability of material: collection tubes, pipette tips, vials, etc. Use the same brands between collection sites Harmonize sample labelling	
Sample collection	EDTA or heparin plasma (avoid the first tube of the drawing sequence) Cool directly after collection (4 °C, continuous cooling)	Midstream second morning urine Cool directly after collection (4 °C, continuous cooling)
Hemolysis	Exclude results obtained from hemolyzed samples	
Transportation	Continuous cooling at 4 °C (ice water or cold pack)	
Centrifugation	2500 g at 4 °C for 10 min, within 2 h after collection	
After centrifugation	Transfer supernatant into cryotubes as fast as possible while maintaining temperature at 4 °C, followed by snap-freezing	
Storage	At −80 °C	
Thawing	At 4 °C	
Number of freeze-thaw cycles	Limit the number of freeze-thaw cycles to a minimum Record number of freeze-thaw cycle for each sample	
Recording	Monitor every step and document any deviation from the protocol Inspect NMR and MS data for unexpected signal intensities during the entire analysis	

- In any case, preliminary tests should be performed prior to any metabolomics study to investigate the effects of the pre-analytical conditions.
- Pre-analytical conditions should always be carefully discussed between all partners involved in a metabolomics study—from bedside to bench.
- Clear guidelines and SOPs should be established and followed; any deviation from the guidelines should be reported.
- Details on the pre-analytical steps should be included in research articles.

Acknowledgments Marloes van Os is acknowledged for her help in collecting and processing the relevant literature.

References

1. Hernandes VV, Barbas C, Dudzik D. A review of blood sample handling and pre-processing for metabolomics studies. Electrophoresis. 2017;38(18):2232–41.

2. Vignoli A, Tenori L, Morsiani C, Turano P, Capri M, Luchinat C. Serum or plasma (and which plasma), that is the question. J Proteome Res. 2022;21(4):1061–72.
3. Denery JR, Nunes AA, Dickerson TJ. Characterization of differences between blood sample matrices in untargeted metabolomics. Anal Chem. 2011;83(3):1040–7.
4. Kondoh H, Kameda M, Yanagida M. Whole blood metabolomics in aging research. Int J Mol Sci. 2020;22(1)
5. Plebani M, Banfi G, Bernardini S, Bondanini F, Conti L, Dorizzi R, et al. Serum or plasma? An old question looking for new answers. Clin Chem Lab Med. 2020;58(2):178–87.
6. Yin P, Peter A, Franken H, Zhao X, Neukamm SS, Rosenbaum L, et al. Preanalytical aspects and sample quality assessment in metabolomics studies of human blood. Clin Chem. 2013;59(5): 833–45.
7. Mei H, Hsieh Y, Nardo C, Xu X, Wang S, Ng K, et al. Investigation of matrix effects in bioanalytical high-performance liquid chromatography/tandem mass spectrometric assays: application to drug discovery. Rapid Commun Mass Spectrom. 2003;17(1):97–103.
8. Lehmann R. From bedside to bench-practical considerations to avoid pre-analytical pitfalls and assess sample quality for high-resolution metabolomics and lipidomics analyses of body fluids. Anal Bioanal Chem. 2021;413(22):5567–85.
9. Bi H, Guo Z, Jia X, Liu H, Ma L, Xue L. The key points in the pre-analytical procedures of blood and urine samples in metabolomics studies. Metabolomics. 2020;16(6):68.
10. Koch E, Mainka M, Dalle C, Ostermann AI, Rund KM, Kutzner L, et al. Stability of oxylipins during plasma generation and long-term storage. Talanta. 2020;217:121074.
11. Jonasdottir HS, Brouwers H, Toes REM, Ioan-Facsinay A, Giera M. Effects of anticoagulants and storage conditions on clinical oxylipid levels in human plasma. Biochim Biophys Acta Mol Cell Biol Lipids. 2018;1863(12):1511–22.
12. Pinto J, Domingues MR, Galhano E, Pita C, Almeida Mdo C, Carreira IM, et al. Human plasma stability during handling and storage: impact on NMR metabolomics. Analyst. 2014;139(5): 1168–77.
13. Haid M, Muschet C, Wahl S, Römisch-Margl W, Prehn C, Möller G, et al. Long-term stability of human plasma metabolites during storage at −80 °C. J Proteome Res. 2018;17(1):203–11.
14. Wagner-Golbs A, Neuber S, Kamlage B, Christiansen N, Bethan B, Rennefahrt U, et al. Effects of long-term storage at −80 °C on the human plasma metabolome. Meta. 2019;9(5)
15. Pottala JV, Espeland MA, Polreis J, Robinson J, Harris WS. Correcting the effects of −20 °C storage and aliquot size on erythrocyte fatty acid content in the Women's Health Initiative. Lipids. 2012;47(9):835–46.
16. Metherel AH, Aristizabal Henao JJ, Stark KD. EPA and DHA levels in whole blood decrease more rapidly when stored at −20 °C as compared with room temperature, 4 and −75 °C. Lipids. 2013;48(11):1079–91.
17. Reis GB, Rees JC, Ivanova AA, Kuklenyik Z, Drew NM, Pirkle JL, et al. Stability of lipids in plasma and serum: effects of temperature-related storage conditions on the human lipidome. J Mass Spectrom Adv Clin Lab. 2021;22:34–42.
18. Fliniaux O, Gaillard G, Lion A, Cailleu D, Mesnard F, Betsou F. Influence of common preanalytical variations on the metabolic profile of serum samples in biobanks. J Biomol NMR. 2011;51(4):457–65.
19. Goodman K, Mitchell M, Evans AM, Miller LAD, Ford L, Wittmann B, et al. Assessment of the effects of repeated freeze thawing and extended bench top processing of plasma samples using untargeted metabolomics. Metabolomics. 2021;17(3):31.
20. Pizarro C, Arenzana-Rámila I, Pérez-del-Notario N, Pérez-Matute P, González-Sáiz JM. Thawing as a critical pre-analytical step in the lipidomic profiling of plasma samples: new standardized protocol. Anal Chim Acta. 2016;912:1–9.

21. Nishiumi S, Suzuki M, Kobayashi T, Yoshida M. Differences in metabolite profiles caused by pre-analytical blood processing procedures. J Biosci Bioeng. 2018;125(5):613–8.
22. Yu Z, Kastenmüller G, He Y, Belcredi P, Möller G, Prehn C, et al. Differences between human plasma and serum metabolite profiles. PLoS One. 2011;6(7):e21230.
23. Sotelo-Orozco J, Chen SY, Hertz-Picciotto I, Slupsky CM. A comparison of serum and plasma blood collection tubes for the integration of epidemiological and metabolomics data. Front Mol Biosci. 2021;8:682134.
24. Hahnefeld L, Gurke R, Thomas D, Schreiber Y, Schäfer SMG, Trautmann S, et al. Implementation of lipidomics in clinical routine: can fluoride/citrate blood sampling tubes improve preanalytical stability? Talanta. 2020;209:120593.
25. Kirwan JA, Brennan L, Broadhurst D, Fiehn O, Cascante M, Dunn WB, et al. Preanalytical processing and biobanking procedures of biological samples for metabolomics research: a white paper, community perspective (for "precision medicine and pharmacometabolomics task group"--the metabolomics society initiative). Clin Chem. 2018;64(8):1158–82.
26. Emwas AH, Roy R, McKay RT, Ryan D, Brennan L, Tenori L, et al. Recommendations and standardization of biomarker quantification using NMR-based metabolomics with particular focus on urinary analysis. J Proteome Res. 2016;15(2):360–73.
27. Fernández-Peralbo MA, Luque de Castro MD. Preparation of urine samples prior to targeted or untargeted metabolomics mass-spectrometry analysis. TrAC. 2012;41:75–85.
28. Liu X, Yin P, Shao Y, Wang Z, Wang B, Lehmann R, et al. Which is the urine sample material of choice for metabolomics-driven biomarker studies? Anal Chim Acta. 2020;1105:120–7.
29. Kim K, Mall C, Taylor SL, Hitchcock S, Zhang C, Wettersten HI, et al. Mealtime, temporal, and daily variability of the human urinary and plasma metabolomes in a tightly controlled environment. PLoS One. 2014;9(1):e86223.
30. Wang X, Gu H, Palma-Duran SA, Fierro A, Jasbi P, Shi X, et al. Influence of storage conditions and preservatives on metabolite fingerprints in urine. Meta. 2019;9(10)
31. Rotter M, Brandmaier S, Prehn C, Adam J, Rabstein S, Gawrych K, et al. Stability of targeted metabolite profiles of urine samples under different storage conditions. Metabolomics. 2017;13 (1):4.
32. Budde K, Gök ÖN, Pietzner M, Meisinger C, Leitzmann M, Nauck M, et al. Quality assurance in the pre-analytical phase of human urine samples by (1)H NMR spectroscopy. Arch Biochem Biophys. 2016;589:10–7.
33. Lauridsen M, Hansen SH, Jaroszewski JW, Cornett C. Human urine as test material in 1H NMR-based metabonomics: recommendations for sample preparation and storage. Anal Chem. 2007;79(3):1181–6.
34. Roux A, Thévenot EA, Seguin F, Olivier MF, Junot C. Impact of collection conditions on the metabolite content of human urine samples as analyzed by liquid chromatography coupled to mass spectrometry and nuclear magnetic resonance spectroscopy. Metabolomics. 2015;11(5): 1095–105.
35. González-Domínguez R, González-Domínguez Á, Sayago A, Fernández-Recamales Á. Recommendations and best practices for standardizing the pre-analytical processing of blood and urine samples in metabolomics. Meta. 2020;10(6)
36. Yuille M, Illig T, Hveem K, Schmitz G, Hansen J, Neumaier M, et al. Laboratory management of samples in biobanks: European consensus expert group report. Biopreserv Biobank. 2010;8(1): 65–9.
37. Saude EJ, Sykes BD. Urine stability for metabolomic studies: effects of preparation and storage. Metabolomics. 2007;3(1):19–27.

38. Pasikanti KK, Ho PC, Chan EC. Development and validation of a gas chromatography/mass spectrometry metabonomic platform for the global profiling of urinary metabolites. Rapid Commun Mass Spectrom. 2008;22(19):2984–92.
39. Gika HG, Theodoridis GA, Wilson ID. Liquid chromatography and ultra-performance liquid chromatography-mass spectrometry fingerprinting of human urine: sample stability under different handling and storage conditions for metabonomics studies. J Chromatogr A. 2008;1189(1–2): 314–22.
40. Bernini P, Bertini I, Luchinat C, Nincheri P, Staderini S, Turano P. Standard operating procedures for pre-analytical handling of blood and urine for metabolomic studies and biobanks. J Biomol NMR. 2011;49(3–4):231–43.
41. Yin P, Lehmann R, Xu G. Effects of pre-analytical processes on blood samples used in metabolomics studies. Anal Bioanal Chem. 2015;407(17):4879–92.

How to Prepare Your Samples for Polar Metabolite Analysis?

3

Elizabeth Want

What You Will Learn in This Chapter

Biological samples such as urine, blood, faeces, cell lysates and tissue are complex matrices containing thousands of small molecules, lipids and proteins. They are invaluable in metabolomics studies of health and disease, with each matrix providing complementary metabolic information. Certain classes of polar molecules are implicated in different diseases, such as keto acids, fatty acids and other organic acids which have been associated with heart disease, cancer, liver disease, obesity and diabetes [1]. Therefore it is important to recover these compounds from biological matrices in a robust and reproducible manner in order to understand more about human health and disease.

In this chapter you will learn more about why polar metabolites are important and what biological processes they are involved in. You will find out about different biological matrices and how to handle them in terms of extracting polar metabolites. Important considerations will be covered, with step-by-step protocols given for the preparation of samples for polar metabolite analysis. You will discover the importance of the various steps of metabolite quenching and extraction and when they are needed. The challenges of metabolite losses, degradation and the importance of sample diluent for downstream LC-MS analysis will be introduced.

E. Want (✉)
Imperial College London, London, UK
e-mail: e.want@imperial.ac.uk

© The Author(s), under exclusive license to Springer Nature Switzerland AG 2023
J. Ivanisevic, M. Giera (eds.), *A Practical Guide to Metabolomics Applications in Health and Disease*, Learning Materials in Biosciences,
https://doi.org/10.1007/978-3-031-44256-8_3

3.1   Introduction

Metabolomics is the study of small molecules (typically <1.5 kDa) in biological matrices. This area of research has grown extensively over the past 20 years, in part due to advancements in analytical platforms (e.g. nuclear magnetic resonance (NMR) spectroscopy and mass spectrometry), data processing tools and metabolite identification workflows. There are many good reviews on this area to which the reader is directed to [2, 3]. The aim of metabolomics studies can be to elucidate novel biomarkers of diseases such as cancer, liver disease and heart disease, which can be used for prognosis or diagnosis and to understand disease mechanisms and the effects of therapeutic intervention [2–4]. However, sample preparation still poses immense challenges in the field of metabolomics, impacting on the metabolites detected and the subsequent biological interpretation.

It is impossible to understand the complex changes in metabolism brought about by disease, drug intervention etc., without analysing biological matrices, usually several in combination. Common matrices include blood plasma and serum, urine, faeces and tissue. In addition, in vitro studies of disease employ cell models and the subsequent analysis of media and lysates. All of these samples are complex in composition, containing lipids, proteins and small molecules (metabolites). Of these small molecules, many are polar in nature and include amino acids, other organic acids, nucleic acids, carbohydrates, dipeptides and thiols, as well as some vitamins and neurotransmitters (Table 3.1). They are present in almost all biological samples, with roles ranging from regulation of intracellular energy metabolism including glucose, lipid and amino acid oxidation to growth, development and reproduction [5, 6]. Metabolites have diverse physical properties and are present in biological samples spanning a vast dynamic range; thus no single analytical tool can measure them all. Lu et al. reported previously that the Human Metabolome Database (HMDB) had identified >40,000 human metabolites, but at the time of publication, only 680 water-soluble compounds had been mapped to standard metabolic pathways in humans (<2%), leaving a great amount unknown concerning polar metabolites and their roles [5, 7].

In biofluids and tissues, polar molecules are present in concentrations from low levels of neurotransmitters such as dopamine (sub-nM) to high levels of amino acids such as glutamine (high uM in serum). Polar metabolite concentrations are determined by both internal factors, e.g. cellular activity, inflammation, the gut microbiome and external factors such as diet, physical activity, medication and environmental exposures (the exposome). Therefore, monitoring changes in metabolite concentrations (or ratios of metabolites) can serve as a readout of cellular biochemical activity and its response to changes in the environment. In the field of metabolomics, these molecules are measured as part of global metabolite profiles (untargeted assays) or in targeted assays focussing on specific pathways or classes of analytes. Hence, the extraction and measurement of the broadest possible range of polar metabolites in biological specimens are key to better understanding metabolic changes in different physiological conditions.

Table 3.1 Types of polar metabolites and their functions in the human body. Created in Biorender

Polar metabolite class	Description	Function
Amino acids	• Contain both amino and carboxylic functional groups	• Comprise proteins • Serve as backbones for neurotransmitters and hormones
Organic acids	• Contain one or more carboxylic acid groups	• Regulation of basic cellular processes, e.g. modification of cellular pH or redox state
Sugars	• Soluble carbohydrate - mono or disaccharide	• Supply glucose to the brain • Provide energy to cells in body
Nucleic acids	• Biopolymers—main information carrying molecules in the cell • Composed of small, polar nucleobases and sugar phosphates	• Storage and expression of genetic information

Traditionally, polar metabolite extraction can be performed in a single-step using methanol (MeOH) [5, 6]. However, many protocols in the literature offer approaches to extract both polar metabolites and lipids in a single experiment [6, 8], which can be advantageous if there is limited sample available or a large number of samples to extract. However, it may be favourable to extract samples twice to produce separate polar (aqueous) and non-polar (organic) extracts, depending on factors such as the amount of starting material and the biological questions to be asked. Here we will compare both approaches and offer considerations for both.

3.2 Considerations for Polar Metabolite Extraction from Biological Samples

The key steps to consider when extracting metabolites from biological samples are:

1. Sample type
2. Sample handling and storage
3. Is the measurement endpoint a targeted or untargeted assay? Different requirements?
4. Quenching and metabolite extraction
5. Resuspension and choice of diluent

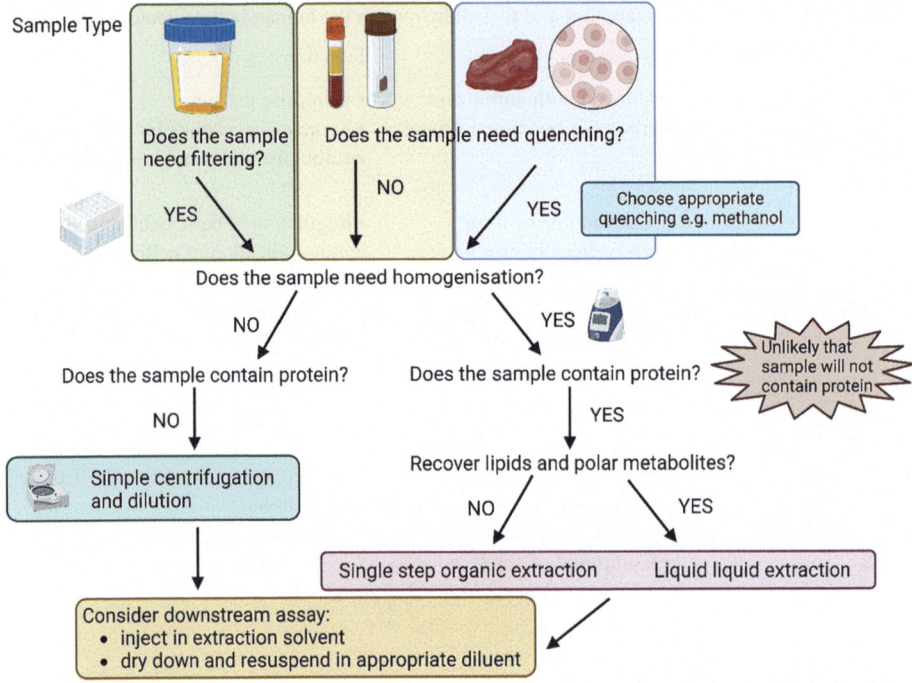

Fig. 3.1 Considerations for sample extraction depending on sample type and downstream analysis. Created in Biorender

Many of these are interlinked (Fig. 3.1). These will be discussed in the following section.

3.2.1 Sample Type: Which Biological Sample Is Being Used?

The best approach for extracting polar metabolites depends on the biological matrix being investigated. For example, a simple protein precipitation step using an organic solvent such as methanol, acetonitrile or isopropanol will work well for serum or plasma [9, 10]. However, this will not be needed for urine samples which, except in cases of specific diseases, such as myeloma, will contain little protein [11]. In the case of urine, dilution and centrifugation can often be used, as discussed later [12]. Faecal and tissue samples present more of a challenge due to their heterogeneity and so will require a more involved sample preparation approach. Additionally, tissue and cell samples require an initial step of quenching to halt metabolism. In this next section, the different types of biological matrices used in metabolomics studies will be described.

Urine

Urine is a commonly used sample type in metabolomics studies and is a useful biological matrix for the study of the gut microbiome [13], drug metabolites and dietary influences [14], autism [15] and bladder cancer [16] amongst others. It has been reported that urinary metabolite changes in disease states are larger than urinary protein changes, providing a clear readout on disease phenotypes [17]. Urine is the most easily collected and stored biofluid, enabling multiple collections within short timeframes and is not usually sample-limited. Whilst being complex, inherently urine contains mostly small, polar molecules (amino and organic acids, sugars, hormones) as larger molecules and lipids are removed by the kidney via filtration. However, urine samples are inherently variable and can differ hugely day to day due to factors such as diet, exercise, diurnal cycle, age and gender [17]. There can also be issues with variations in dilution between samples, which could be dealt with at several stages of the sample preparation, analysis or data handling [17]. One way to address the variability in urine samples is to collect a 24 h urine sample, although often this is not very practical [18, 19]. Alternatively, metabolite levels could be reported as a ratio to creatinine, but this has limitations, as creatinine levels show great variability [20].

Blood Plasma or Serum

Plasma and serum are two of the most frequently analysed biofluids in metabolomics studies, second only to urine, in part due to the minimally invasive collection of blood from humans or animals. Plasma or serum contains many polar metabolites such as amino acids, other organic acids and sugars. However, these constitute in the minority when compared with lipids [21], which account for >70% of the plasma metabolome. Plasma and serum samples have been used extensively in metabolomics studies to study heart disease, cancer and COVID-19 [22–25]. Advantages of using plasma and serum samples are that these biofluids are usually under tight homeostatic control and so the variation seen in urine samples is largely absent in blood products. By analysing plasma or serum, a "snapshot" of the global processes occurring in the organs and tissues of the body can be obtained, although cannot be pinned down to one organ in particular. Perturbations due to disease, therapeutic interventions, ageing, exercise, etc. can be observed [26]. However, it is not feasible to collect as many sequential samples as with urine samples.

Faeces

Faecal samples are not as traditionally well studied as blood or urine; nevertheless valuable information can be gleaned from these samples regarding gut health, colon cancer, inflammatory bowel diseases (IBD) and the gut microbiota-host interactions [4], [27]. Faecal sample collection and preparation are less standardised than for urine or blood plasma and have been reviewed in publications such as [28, 29]. It is known that faecal samples contain potentially thousands of metabolites such as amino and organic acids, sugars and short chain fatty acids (e.g. acetate, butyrate and propionate). Many small molecules in faeces, e.g. amino acids, have been redirected from the diet to important metabolic pathways, such as bile acid synthesis by the gut microbiota [30, 31]. These important polar molecules are valuable in the diagnosis of diseases such as diabetes, metabolic syndrome and obesity [32].

Tissue Samples

Although biofluids can be advantageous for large population studies, the metabolic profiles obtained are not specific to individual organs and so can reflect changes occurring in many parts of the body. Tissue samples can therefore be an important source of metabolic information, revealing changes due to certain diseases, e.g. liver disease, heart disease and cancer [33–35]. They may even provide more sensitive or robust biomarkers of disease than "surrogate" biofluids such as urine and plasma/serum. However, they are much more invasive to collect than blood, urine and faeces, and sample amounts are more likely to be limited. Further, it may be difficult to collect a large number of samples, particularly in human cohorts, and control samples may be limited (many samples are collected via biopsy—less likely to be performed on healthy individuals). The metabolic composition will vary greatly depending on the tissue type; however, most tissues will be rich in lipids and proteins as well as small molecules. As a more biologically active sample type than blood or urine, a quenching step is likely to be needed during preparation. Therefore a more detailed protocol is likely and will vary between tissue types. Saoi provides a very good review on tissue metabolomics [35].

Cell Samples

In vitro studies play an important role in metabolomics as well as other omics studies, providing complementary information to in vivo studies and often reducing the need for animal models [36–38]. Cell models have proved particularly valuable in the study of cancers and the dysregulation of cellular metabolism is a well-recognised hallmark of cancer. Therefore the collection and preparation of cell samples has become more widely documented in recent years. Cell lysates contain a mixture of small molecules, lipids and proteins. They provide their own challenges, with key considerations being efficient sampling, quenching and extraction of metabolites to preserve the metabolome. Different protocols will be needed for the treatment of cell suspensions and adherent cells, as well as cell media.

Less Commonly Used Biological Matrices for Metabolomics

Other, rarer biological matrices used in metabolomics studies include microdialysates [39, 40], breast milk [41] and nipple aspirate [42]. The use of cerebrospinal fluid, saliva and whole blood for metabolomics are described briefly below.

Cerebrospinal Fluid

Cerebrospinal fluid (CSF) is produced by the central nervous system. It is in dynamic exchange with the nerve tissue and circulates metabolites. Due to its proximity to the brain, CSF is most valuable in the study of brain conditions, e.g. Alzheimer's disease and

Parkinson's disease, multiple sclerosis and brain injury [43–45]. The profiling of CSF is key in the search for new biomarkers for these diseases.

Saliva

Saliva is being used more for metabolomics studies as its collection is non-invasive and generally not sample-limited. It is a filtrate of blood, and thus it has been suggested that it could act as an alternative. However, there is little consensus on sampling and extraction approaches, and there may be large variations between samples. Studies have looked at the effect of exercise on plasma, urine and saliva [46], investigating traffic pollution [47], exercise [48] as well as disease severity in COVID patients [49].

Whole Blood

Whole blood is not as commonly used in metabolomics studies. However, whole blood has been profiled in the study of sepsis [50]. This group reported that sample preparation time is shorter than for serum and so the effects of haemolysis are minimised. Further, the analysis of whole blood means that any contributions to the metabolite profile from the erythrocytes are captured. Dried blood spots (DBS) are being employed more in metabolomics studies [51, 52], and Tobin et al. compared DBS and plasma samples for untargeted metabolomics [53]. DBS sampling has historically been used for genetic screening. Sample collection is relatively simple, with small volumes obtained, and the samples can be stored at ambient temperature. Further advantages of DBS are the reduction of the need of expensive laboratory equipment and cold supply chain. However, the stability of metabolites in DBS samples has yet to be fully evaluated. Many microsampling approaches lend themselves to metabolomics, e.g. the hemaPEN, which can collect small volumes of blood for analysis [54].

(1) What is the most commonly used biological sample in metabolomics? Why?

3.2.2   Pre-analytical Considerations: Sample Handling and Storage

Metabolite stability in biological samples ranges from very stable (e.g. lipids) over several days to very unstable, e.g. within minutes after sample collection (energy metabolites such as ATP, NADP). The analytical tools used in metabolomics studies, namely, NMR and MS, offer highly sensitive approaches to metabolite measurement. Therefore it is critical to handle these samples correctly in order to minimise metabolite losses and maximise the biological information obtained. The sample handling (pre-analytical) phase is therefore key to the success of metabolomics studies and varies between sample types [55–59] (Fig. 3.2).

The aim of efficient sample handling is to reduce the time from sample collection to storage in order to minimise metabolite losses and potential interconversions. Inappropriate

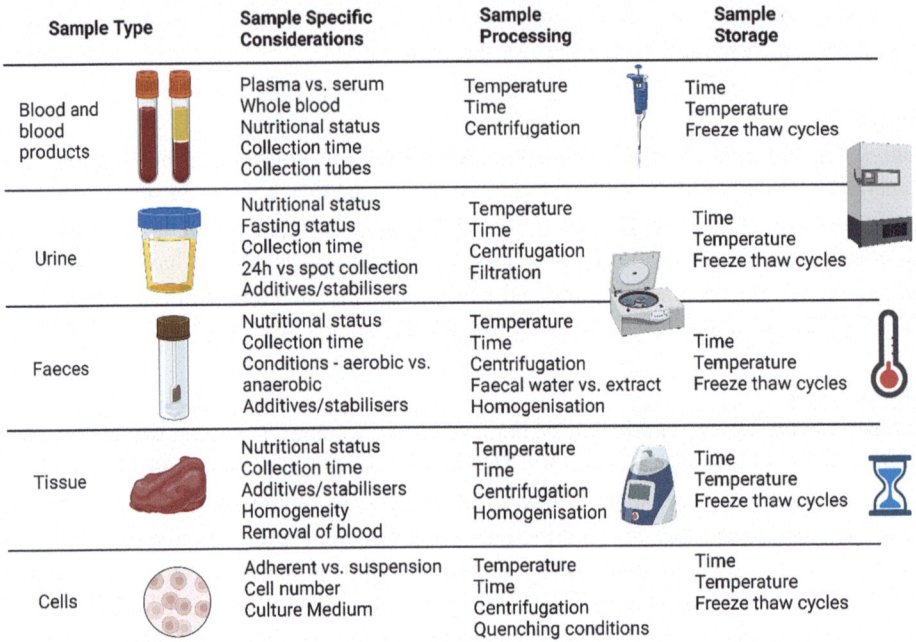

Sample Type	Sample Specific Considerations	Sample Processing	Sample Storage
Blood and blood products	Plasma vs. serum Whole blood Nutritional status Collection time Collection tubes	Temperature Time Centrifugation	Time Temperature Freeze thaw cycles
Urine	Nutritional status Fasting status Collection time 24h vs spot collection Additives/stabilisers	Temperature Time Centrifugation Filtration	Time Temperature Freeze thaw cycles
Faeces	Nutritional status Collection time Conditions - aerobic vs. anaerobic Additives/stabilisers	Temperature Time Centrifugation Faecal water vs. extract Homogenisation	Time Temperature Freeze thaw cycles
Tissue	Nutritional status Collection time Additives/stabilisers Homogeneity Removal of blood	Temperature Time Centrifugation Homogenisation	Time Temperature Freeze thaw cycles
Cells	Adherent vs. suspension Cell number Culture Medium	Temperature Time Centrifugation Quenching conditions	Time Temperature Freeze thaw cycles

Fig. 3.2 Pre-analytical considerations for biological samples for metabolomics studies. Created in Biorender

collection or storage of biological samples will impact the resulting metabolite composition and could reduce sample quality and subsequent analysis and data interpretation. This could result in large intersample variability, impacting downstream on biomarker discovery and elucidation.

Factors such as time of collection, whilst important, are beyond the scope of this chapter, and so the reader is referred to the literature, e.g. [28]. Samples should be collected at the lowest temperature possible and maintained on ice. Ideally samples should be frozen immediately after collection (or post-essential processing), again at the lowest possible temperature.

Urine

Urine can be collected for metabolomics studies in both the clinic and at home. Storage temperature is a key factor in sample stability and reproducibility, as some metabolites will be more prone to degradation at room temperature than others. Rotter et al. showed some amino acids such as arginine, methionine, serine, valine, leucine and isoleucine as well as hexose degraded in urine samples left at room temperature for 24 h [31, 60, 61]. In the case of urine, there may be issues with bacterial growth and so a preservative, e.g. sodium azide could be added to the samples [28, 43, 61]. Boric acid has sometimes been used but has been shown to complex with metabolites in the samples and thus may alter their metabolic composition [62]. Neither approach is commonly used in a hospital setting when collecting urine samples. Samples are usually centrifuged before freezing to remove particulates [28]. Filtration has also been reported to aid sample stability [28, 61, 63] as due to the presence of cells, bacteria and fungi, there may be residual enzymatic activity which may alter the metabolite composition. Samples should be stored at $-80\,°C$ if possible, although studies have shown that shorter-term storage at $-20\,°C$ does not appear to have a detrimental effect on the metabolic profiles [64].

Blood Plasma and Serum

Blood is commonly collected in clinical and epidemiology studies. Blood collection is more invasive than for urine but still relatively straightforward and in the case of humans not sample-limited for metabolomics studies. Impacts of drug intake and diet can be monitored through plasma or serum profiles, and perturbations in metabolism due to disease can be studied [20–23]. Microsampling and dried blood spots can also be studied and are becoming increasingly popular, but are not the focus of this chapter (see previous section [54]).

Plasma and serum are almost equal in use for metabolomics studies of blood. Historically, serum was preferred in clinical studies, and many biobanks contain mostly serum samples. However, they differ in their collection mechanism, and the resulting metabolic composition varies also. It has been reported that serum will provide a higher level of metabolites as it does not contain clotting factors (e.g. fibrinogen proteins) or cells, whilst plasma contains both blood cells and proteins [21]. Blood cells can be removed from plasma through centrifugation. It has also been reported that analysis of plasma profiles is more reproducible due to the absence of clotting factors [21]. However, more detailed research has shown that the metabolite differences are more specific than that, with certain classes of molecules being more abundant in either serum or plasma. Yu et al. conducted a comparison of plasma and serum metabolite profiles through the analysis of >100 metabolites [21]. They found that serum and plasma metabolite profiles differed greatly, with the majority of metabolites being higher in serum samples, potentially indicating a better chance of biomarker detection. Interestingly, reproducibility of measurements was better in plasma samples. Sotelo-Orozco and colleagues reported that amino acids were higher in serum samples than plasma samples [65].

Serum

Serum collection involves leaving the sample of whole blood undisturbed at room temperature for up to 30 min to enable clotting. The sample is then centrifuged at 1–2000 g for 10 min in a refrigerated centrifuge to remove the clot. This results in the activation of platelets which can release proteins and metabolites into the serum, impacting and altering the metabolic composition of the sample [66]. Another consideration is the addition of coagulation enhancers—often silicate-based—in the serum collection tubes which may impact the metabolite profile. Sometimes polymers are present in the gel in the tubes which can also negatively impact the metabolome of the samples. The sample is then aliquoted into a clean tube and stored at $-80\,°C$.

Plasma

A key consideration is the collection tube used for plasma samples, commonly containing the anticoagulants ethylene diamine tetra acetic acid (EDTA), citrate or heparin. In the literature there is no clear consensus on which anticoagulant to use, and it is likely dependent on downstream metabolite analysis. It has been reported that EDTA will impact metabolic profiles to a greater extent than heparin or citrate, by chelating with compounds and producing ion suppression or enhancement during mass spectrometric analysis [67]. It is worth noting that citrate, as an endogenous molecule, could also alter the metabolic composition of the sample in (a lesser) but similar manner to EDTA. Therefore collection with tubes containing sodium or lithium heparin would be advised [65]. Sotelo-Orozco et al. carried out a comparison of blood collection tubes [65] and reported that the metabolite profiles of plasma from heparin tubes were most similar to those of serum. For both plasma and serum samples, lower temperatures will reduce the activity of cellular metabolism. Therefore plasma blood collection tubes should be kept on ice where possible, but this is not feasible for serum as clotting needs to proceed at room temperature. An additional factor to consider is that too high a centrifugation force can cause samples to haemolyse, thus releasing metabolites from platelets. However, too low a speed will mean that cells remain in the samples. Storage of serum and plasma samples should ideally be at the lowest temperature possible, $-80\ °C$. It is important to note that freezing of plasma samples results in platelet lysis and release of intracellular molecules, again impacting the metabolic composition of the sample [66].

Faecal Samples

Faecal sample collection is non-invasive, and samples can be collected directly into a suitable bag or tube in the case of humans or in metabolic cages for animals. As with urine and blood samples, the use of additives may impact the resulting metabolic profile and so should be employed with care. Another key consideration is exposure to aerobic conditions or keeping samples at room temperature, which can cause microbial fermentation to occur rapidly, altering metabolite composition. It is crucial to store faecal samples at $-80\ °C$ as quickly as possible, as changes can occur more rapidly than in blood or urine samples. It is advised to homogenise and aliquot faecal samples before freezing to ensure uniform sample handling [68].

Tissue Samples

A major factor in the collection and preparation of tissue samples is the heterogenous nature of tissue samples. One way to circumvent this is to always collect samples from the same region of the tissue to maximise homogeneity, ideally by the same person. There will likely be contamination of the sample due to blood metabolites, and so a washing step may be needed, e.g. with water or PBS after collection. Samples should be snap-frozen in liquid nitrogen as soon as possible to preserve sample integrity. If several studies are being conducted then it may be useful to cut the sample into smaller pieces before freezing to avoid the impact of freeze-thaw on the samples.

Cell Samples

Cells can either be in suspension or adherent, and so the collection method will differ depending on the cell type. It is important to bear in mind factors such as the number of cells seeded to ensure sufficient yield for metabolite studies, the composition of the media in which the cells are growing and how this may affect the metabolite profiles. Culture media inherently contains factors to aid the growth of cells, such as amino acids and foetal bovine serum (FBS), which will affect metabolite composition and may cause ion suppression in the case of FBS if not removed. It has been reported that culture media including HEPES should be avoided as it can impact NMR metabolic profiling [69]. Washing steps should be implemented to remove the media from the cells themselves, such as using water or PBS. Washing with PBS removes all media nutrients and thus affects the environment and likely the resulting cell metabolome. Some metabolomics studies will remove the media using a pipette and then employ a quenching and adherent cell removal with, e.g. methanol. It has been reported that the cell detachment method (e.g. scraping vs. trypsinisation) had a greater effect on the metabolite profile than the lysis method employed [70]. The quenching and extraction methods will of course impact metabolite yield, reproducibility, etc. and will be discussed later.

The following step should be followed: (1) isolate cells from media. This can be achieved in a couple of different ways. Pelleting is rather slow which could result in metabolite losses/alterations. Therefore fast filtration is recommended for suspension cultures [5], and for adherent cultures it is recommended to aspirate the media and add the quenching solvent, e.g. methanol, directly to the plate.

Cerebrospinal Fluid

CSF is not easy to collect compared with blood and urine samples, and the sample may be contaminated with blood. Often CSF samples are collected in parallel with blood samples so that metabolite comparisons can be made between the sample types. Compared to blood or urine, CSF is more simple in composition [43], with a low lipid and protein content (<500 mg/ml). However, sample handling can still be challenging, with sample pH, dilution, storage duration and temperature causing sample variation and potential issues with data quality [71]. The stability of CSF samples is highly dependent on pH and immediately after sampling CSF can undergo changes in pH [72]. This is particularly the case if CSF samples are exposed to non-physiological levels of CO_2. Some metabolites will be sensitive to this pH change, and thus the metabolic composition of the sample will be altered. However, there appears to be no consensus in the literature as to whether samples should be pH adjusted, with those studies employing such adjustment ranging from 2.5 to 10 final pH. Compared with urine or blood samples, CSF volumes can be limited, also affecting sample handling and processing. After collection, CSF samples are typically centrifuged, e.g. at 2000 g, 4 °C, and then frozen at −80 °C. Once the cellular components are removed, CSF is quite stable, even at room temperature for up to 8 h.

Table 3.2 Types of profiling studies in metabolomics and their advantages and disadvantages

	Targeted	Untargeted
Sample preparation	Tailored to metabolites of interest	Unbiased—although likely divided into polar and non-polar
Assay time	Fast-moderate (e.g. 1–20 min)	Moderate-slow (e.g. 10–30 min)
Range of metabolites	Focussed, so may range from a few metabolites to several hundred	Unbiased, large—not limited to specific classes or pathways
Quantitation	Yes if isotopically labelled standards used. When a larger number of analytes are in a single assay, some may be measured semiquantitatively	Semi-quantitative—some standards may be used but are usually representative of analyte classes
Cost	Expensive if labelled standards are used	Usually cheaper than targeted
Sample amount needed	Very low (low uL)	Typically higher (>10 uL)
Assay sensitivity	Typically more sensitive than untargeted methods	Typically less sensitive than targeted assays but cover a broader range of analytes

(2) What factors affect sample stability?

3.2.3 What Type of Assay to Perform on the Metabolite Extract?

(A) *Global or untargeted profiling*

Global or untargeted profiling employs generic metabolite extraction methods and, later on, generic assays to detect and measure metabolites [73]. The extraction method should be unbiased towards any particular class of molecule. This is most likely to involve protein removal. However, in practice extraction methods would be tailored to polar or non-polar molecules as lipids and metabolites benefit from different chromatographic approaches for optimum separation and subsequent detection (Table 3.2).

(B) *Targeted assays*

Targeted assays offer specific and sensitive measurements of specific classes of analytes or those belonging to a specific pathway [74]. Therefore sample preparation methods may be more specific than for global profiling and may include solid phase extraction for example. Solid phase extraction may be more likely to be used in targeted assays as it can help to concentrate the metabolites of interest, but metabolite losses must be taken into consideration (Table 3.2).

3.2.4 Quenching and Metabolite Extraction

Sample preparation is a critical step in metabolomics and so needs to be considered very carefully [73]. The main goal is to remove the protein from the sample, whilst ensuring that the components of the metabolome are solubilised. This will impact metabolite degradation and therefore yield and reproducibility and may also impact upon later instrument performance. There are several good review papers [28] which detail preparation methods.

The important steps are

- Inactivation of metabolism
- Metabolite stabilisation
- Maintenance of sample integrity

Quenching

Quenching is of most importance for metabolically active biological specimens such as cells and tissues [5]. In the case of cell samples, quenching is critical in order to stop cellular metabolism and to preserve sample integrity by preventing further metabolite turnover. Specifically, a key goal of quenching is the halting of enzymatic activity and ideally preservation of labile metabolites, including sugar phosphates (from the pentose phosphate pathway), nucleotides, coenzymes and cofactors [5]. The stability of these metabolites is affected by both temperature and pH, which need to be carefully considered and monitored [73]. Quenching can also reduce sample variability and improve sample preparation, ideally without altering the cell environment. The method used for quenching is very important, as issues can arise with incomplete quenching or enzyme activity not being stopped rapidly enough. This will result in continued metabolite turnover and an inaccurate picture of the cellular metabolome. Historically, approaches such as increased temperature and addition of acid have been employed, but these result in losses of more labile or pH-sensitive analytes [60, 61]. However, the use of acidic solvents, e.g. 0.1 M formic acid, has been shown to rapidly halt enzymatic activity. This needs to be followed immediately by neutralisation with ammonium bicarbonate to avoid more labile metabolites being degraded [5]. Alternatively, cells can be detached by trypsinisation and then quenched using liquid nitrogen [60, 61]; however, trypsin can cause cell stress as well as structural and protein disruption, resulting in an altered metabolite composition due to cell leakage. Rushing and colleagues reported that trypsinisation caused a decrease in metabolite abundance [70] compared with scraped samples, likely due to cell membrane permeabilisation and metabolite leakage. More recent approaches use water or PBS to wash the cells after removing the media and then detach the cells through the addition of methanol, which also quenches metabolism [75]. Problems may arise, for example, (a) perturbation of metabolite levels during harvesting and (b) incomplete/too slow termination of enzyme activity during quenching. This will be discussed further later.

Metabolite Extraction
This step is nearly universal in metabolite measurement, almost irrespective of the down-stream analysis platform, and is essential/critical to measurement accuracy and high-quality data. From this step a stable extract is required, which quantitatively reflects the metabolites present in the original sample. Practically, polar metabolites will need to be extracted from samples using a different approach to insoluble metabolites. It is therefore key to consider solubility of the metabolites of interest.

Goals
An ideal sample preparation method would aim for maximum metabolite recovery in terms of metabolite classes and abundance. Ideally the method should be simple, reproducible and cheap enough to scale up to hundreds or thousands of samples. It should remove protein rapidly to reduce enzymatic activity and thus metabolite conversion. There are also specific considerations, such as protein-bound metabolites, e.g. NADP+. Other considerations are fast cellular turnover, reactivity—this differs from lipids. Matrix effects also need to be taken into consideration. The main types of extraction approaches are introduced below and discussed with respect to the relevant biological sample in subsequent sections.

(a) *Solid phase extraction*
Solid phase extraction has advantages such as the ability to concentrate the sample to allow the detection of low abundance analytes [9]. However, metabolite losses may occur due to binding to the sorbent or elution during the washing steps. Sitnikov et al. compared solvent and SPE-based extraction approaches for global metabolomics through the use of spiked standards [9]. They concluded that SPE provided acceptable performance for global metabolomics but that these approaches should be employed depending on the desired metabolome coverage.

(b) *Solvent extraction*
The addition of a single solvent or a solvent mixture is a straightforward approach to precipitate proteins from biological samples and has been used exten-sively in the preparation of blood plasma and serum, urine as well as cerebrospinal fluid, seminal fluid and breast milk. There is no clear consensus in the literature for the extraction. It appears that polar metabolites can be extracted with a variety of different solvents. However, it does not guarantee that all labile metabolites such as ATP and NADPH will be recovered. Sitnikov and colleagues found that solvent precipitation resulted in wide selectivity and excellent precision, but these approaches were highly susceptible to matrix effects [9].

(c) *Acid extraction*
This is less commonly used on its own for metabolomics studies but can aid in protein precipitation and metabolite recovery. However, acidified organic solvents can be used for metabolite extraction [76].

(d) *Liquid-liquid extraction (LLE)*

Many LLE methods are used and can be useful for the simultaneous extraction of polar metabolites and lipids [77]. Dual-phase extraction methods such as the Bligh-Dyer method [77] have the advantage of extracting both polar metabolites and lipids in two separate phases, increasing metabolome coverage.

(e) *Deproteinisation and centrifugation*

Methanol is the most commonly used solvent, with reported high metabolite yields [9–11, 78] and as such is included in the example protocol. It is reported to provide a good metabolite yield, high level of deproteinisation and good reproducibility [11, 15, 17, 18]. Combinations of organic solvents could be used, e.g. ethanol/methanol 1:1, but this decreases solvent polarity which in turn may reduce the recovery of highly polar metabolites from the sample. Acetonitrile, whilst being quite commonly used, lacks the reported reproducibility of methanol in part due to less efficient deproteinisation being observed. Isopropanol is an effective solvent for lipid recovery. It is important to use organic solvents which are compatible with subsequent assays, commonly LC-MS, if no drying down and resuspension step is included. Other solvents which could be used directly are 1-butanol/methanol mix and ethanol.

(3) Why is quenching important?

(4) What are the ideal attributes of a sample preparation method?

Detailed Protocols for the Extraction of Polar Metabolites from Biological Samples
This section contains step-by-step protocols for the extraction of polar metabolites from key biological samples.

Extraction of Urine
Urine is considered the waste product of the body, containing much useful information concerning metabolism, health and disease. From urine you can glean knowledge concerning the gut microbiome through the measurement of gut microbial co-metabolites. Due to the filtering actions of the kidneys, healthy human subjects will not have much protein in their urine. However, there may be some patients with a condition, e.g. multiple myeloma, which causes them to excrete protein in their urine [11]. Therefore, solvent extraction is not usually needed in order to facilitate metabolite recovery. However, rodent urine is more proteinuric and so may require an additional extraction step. Typical protocols for urine would be (a) dilution with water and centrifugation [12] or (b) addition of organic solvents, e.g. methanol or acetonitrile [79]. An advantage of the latter approach is that the organic solvents can eliminate microorganisms as well as remove trace amounts of protein which may be present in the sample. Otherwise the sample may be prone to bacterial growth and sodium azide or similar may need to be added, altering the environment of the sample somewhat. The dilution factor is also a key

consideration, with human urine being diluted 1:1 to 1:10 depending on the publication [4]. Rodent urine is generally more concentrated and so may need to be diluted more.

Step-by-Step Protocol for Urine Preparation
Reagents needed:

- Water (LC-MS grade)
- Methanol (LC-MS grade)
  Equipment needed

- 2 mL tubes, e.g. Eppendorf
- Pipettes and pipette tips
- Centrifuge—ideally refrigerated

Protocol 1: Non-proteinuric samples

- Remove 1 ml aliquot of urine from −80 °C.
- Thaw at 4 °C for at least 1 h.
- Centrifuge at 10,000 rpm for 10 min, ideally at 4 °C.
- Remove 900 ul into clean tube.
- Aliquot 50 ul into mass spectrometry vial and dilute as follows:
  – with 50 ul water for human samples.
  – with 150 ul water for non-proteinuric animal samples.

Protocol 2: Proteinuric samples

- Remove 1 ml aliquot of urine from −80 °C.
- Thaw at 4 °C for at least 1 h.
- Centrifuge at 10,000 rpm for 10 min, ideally at 4 °C.
- Remove 900 ul into clean tube.
- Add 900 ul organic solvent, e.g. methanol or acetonitrile.
- Centrifuge at 10,000 rpm for 10 min, ideally at 4C.
- Aliquot 50 ul into mass spectrometry vial and dilute with 150 ul water.

Extraction of Plasma and Serum

A straightforward way to prepare plasma or serum for polar metabolite analysis is through solvent extraction which should effectively and rapidly remove the proteins from the sample. There is no clear consensus on which solvent to use, as publications report using methanol for polar metabolites [9, 80]. Ratios of solvent/plasma/serum vary between 1:3 or 1:4. Usually the solvent is pre-chilled to minimise enzymatic conversion of metabolites as well as aid protein precipitation. This is also beneficial for later LC-MS analysis, as protein buildup on the column will degrade the column—reducing column life—and the presence of proteins will cause ion suppression of the small molecules.

Step-by-Step Protocol for Plasma or Serum Preparation (based on [10, 81])
Reagents needed:

- Methanol (pre-chilled)—LC-MS grade

Equipment needed:

- 2 mL tubes, e.g. Eppendorf
- Pipettes and pipette tips
- Centrifuge—ideally refrigerated

Protocol

- Remove 100uL aliquot of plasma or serum from −80 °C.
- Thaw at 4 °C for at least 1 h.
- Add 300 uL ice cold methanol.
- Vortex mix for 30 s.
- Incubate at −20 °C for 20 min.
- Centrifuge at 10,000 rpm for 10 min ideally at 4 °C.
- Remove 350 ul into clean tube.
- Dry down in Speedvac or similar for ~2 h until dry.
- Resuspend in 100 uL water for LC-MS analysis.
- Aliquot into mass spectrometry vial.

Extraction of Faecal Samples

As with tissue samples, faecal samples are often heterogeneous which can introduce technical variation and affect data quality. Factors such as diet will have a huge impact on microbial composition and metabolism and subsequently the faecal metabolome. Therefore standardised protocols are needed for the analysis of faecal samples for small molecules [30, 68]. Critical parameters for the extraction of faecal samples have not been fully agreed upon, such as sample size, extraction solvent and pH. The amount of recommended starting material varies between publications, with amounts ranging from 50 mg [82] to 500 mg recommended [68]. For metabolomics analyses, faecal water can be obtained through ultracentrifugation of the sample. Alternatively, an aqueous extract can be prepared through the addition of water or an aqueous buffer to the sample followed by homogenisation and centrifugation. Some studies propose the use of an organic solvent, such as methanol which has the ability to extract many analytes from faecal samples across a wide polarity range [30, 68, 83]. Hosseinkhani et al. recommend a more complex solvent mixture of MTBE/methanol/water, 3.6/2.8/3.5, v/v/v) as the extraction solvent of choice [68]. They compared ethanol, methanol, methanol/water and water [82, 84].

Reagents needed:

- Methanol (pre-chilled) LC-MS grade
- Water (LC-MS grade)

Equipment needed:

- 2 ml tubes, e.g. Eppendorf
- Bead-beating tubes
- Zirconium beads
- Pipettes and pipette tips
- Centrifuge—ideally refrigerated

Step-by-step protocol [82, 85, 86]

- Fill 2 mL bead-beating tubes with ~100 uL zirconium beads.
- Add 50 mg freeze-dried faecal sample.
- Add 1.0 ml pre-chilled methanol/water 1:1.
- Homogenise 2 × 6500 cycles of 40 s, keeping on ice in between cycles.
- Centrifuge at 10,000 rpm for 10 min ideally at 4 °C.
- Remove 850 ul supernatant into clean tube.
- Retain pellet for extraction of non-polar metabolites.
- Dry down in Speedvac or similar for ~2 h until dry.
- Resuspend in 100 uL water for LC-MS analysis.
- Aliquot into mass spectrometry vial.

Extraction of Cells

A simple approach for the extraction of metabolites from adherent cells is the addition of an organic solvent such as methanol to quench metabolism once the cell media has been removed. The sample can then be frozen at this stage or water added to result in a solution of methanol/water (80:20). The resulting sample can be sonicated to lyse the cells as shown in the protocol. Cell suspensions can be quenched, centrifuged and then treated in the same way as adherent cells. For cell media, there may be proteins present and so the sample can be treated like a plasma or serum sample, by the addition of cold methanol [69].

Reagents needed:

- Methanol (pre-chilled) (LC-MS grade)
- Water (LC-MS grade)

Equipment needed:

- Scraper to remove adherent cells
- 2 ml tubes, e.g. Eppendorf
- Pipettes and pipette tips
- Centrifuge—ideally refrigerated

Step-by-Step Protocol

1. *Adherent cell lysates*
   (i) *Quenching*
       - Remove cells from plate using 1 ml methanol into clean 2 mL tube.
       - As there may be some evaporation of the methanol, top up to 1 ml if needed.
       - Freeze or continue to extraction step below.
   (ii) *Metabolite extraction*
       - Add 250 uL water to make 80:20 methanol/water solution.
       - Sonicate in ice bath for 30 min.
       - Centrifuge 10,000 rpm for 10 min at 4 °C.
       - Remove 1 mL supernatant to clean tube.
       - Dry down, e.g. in Speedvac for ~2 h.
       - Resuspend in 100 uL water for LC-MS analysis.
       - Aliquot into mass spectrometry vial.
2. *Cell media*
   - Remove 100 uL aliquot of media from −80 °C.
   - Thaw at 4C for at least 1 h.
   - Add 300 uL ice cold methanol.
   - Vortex mix for 30 s.
   - Incubate at −20 °C for 20 min.
   - Centrifuge at 10000 rpm for 10 min ideally at 4 °C.
   - Remove 350 µl into clean tube.
   - Dry down in Speedvac or similar for ~2 h until dry.
   - Resuspend in 100 µL water for LC-MS analysis.
   - Aliquot into mass spectrometry vial.

(continued)

3. *Cell suspensions*
    (i) *Quenching*
        • Quench with 60% methanol (may be buffered, e.g. with AMBIC).
        • Centrifuge at 1000 g for 1 min ideally at 4 °C or even −20 °C to gently pellet.
        • Remove supernatant into clean 2 ml tube—maintain for analysis (*as Cell Media above*).
        • Resuspend pellet in 100% methanol to maintain quenching.
        • Freeze or continue to metabolite extraction step below.
    (ii) *Metabolite extraction*
        • Add 250 μL water to make 80:20 methanol/water solution.
        • Sonicate in ice bath for 30 min.
        • Centrifuge 10,000 rpm for 10 min at 4C.
        • Remove 1 mL supernatant to clean tube.
        • Dry down, e.g. in Speedvac for ~2 h.
        • Resuspend in 100 μL water for LC-MS analysis.
        • Aliquot into mass spectrometry vial.

Extraction of Tissues

As introduced, tissue sample heterogeneity can cause issues with reproducible metabolite extraction. Therefore, a dual extraction approach is particularly important to enable the comparison of polar and non-polar metabolites from the sample part of the tissue [8]. Additionally, some tissue samples may be fatty, and some contain calcified deposits, e.g. blood arteries and veins [85, 87], whilst some may be very fibrous, e.g. placenta. There are several different approaches which can be used depending on the tissue type. Quenching, as discussed for cell samples, is also important for tissue samples and can be achieved through the addition of organic solvent. Some possible approaches are discussed here and then two possible protocols are detailed.

Sample lyophilisation/grinding into powder: One school of thought is that the sample could be lyophilised and ground into a powder. This would then enable technical replicates to be produced easily. However, this brings with it its own disadvantages, such as metabolite losses/issues with resuspension.

Bead beating: Many protocols in the literature recommend homogenisation of the tissue sample through bead beating. This can be achieved through the addition of small zirconium beads, or one large, e.g. stainless steel, bead. Solvents such as water and water/methanol are added, and extraction can be performed as a biphasic extraction, e.g. Bligh Dyer or Folch, or in a two-step extraction where the polar metabolites are extracted first followed by the non-polar metabolites [85, 86].

Therefore two alternative protocols are detailed here, summarising the two most commonly employed approaches.

Reagents needed:

- Methanol (pre-chilled) (LC-MS grade)
- Water (LC-MS grade)

Equipment needed:

- 2 ml tubes, e.g. Eppendorf
- Bead-beating tube
- Zirconium beads
- Pipettes and pipette tips
- Centrifuge—ideally refrigerated

Step-by-Step Protocol

1. *Pulverisation in liquid nitrogen (based on* [88])
   - Cool tubes containing tissue in liquid nitrogen for at least 10 min.
2. *Bead beating with organic solvent*
   - (i) *Aqueous extract for polar metabolites*
     - Fill 2 mL bead-beating tubes with ~100 μL zirconium beads.
     - Weigh out 50 mg tissue into tubes.
     - Add 1.0 ml pre-chilled methanol/water 1:1.
     - Homogenise 2 × 6500 cycles of 40 s, keeping on ice in between cycles.
     - Centrifuge at 10,000 rpm for 10 min ideally at 4C.
     - Remove 850 μl supernatant into clean tube.
     - Retain pellet for extraction of non-polar metabolites.
     - Dry down in Speedvac or similar for ~2 h until dry.
     - Resuspend in 100 μL water for LC-MS analysis.
     - Aliquot into mass spectrometry vial.

Extraction of CSF

Again there is no consensus in the literature for the preparation of CSF samples. However, common approaches are similar in nature to serum and plasma, namely, deproteinisation with an organic solvent.

Reagents needed:

- Methanol (pre-chilled) LC-MS grade

Equipment needed:

- Tubes, e.g. Eppendorf
- Pipettes and pipette tips
- Centrifuge—ideally refrigerated

Protocol

- Remove 100 µL aliquot of CSF from −80 °C.
- Thaw at 4C for at least 1 h.
- Add 300 µL ice cold methanol.
- Vortex mix for 30 s.
- Incubate at −20 C for 20 min.
- Centrifuge at 10,000 rpm for 10 min ideally at 4C.
- Remove 350 µl into clean tube.
- Dry down in Speedvac or similar for ~2 h until dry.
- Resuspend in 100 µL water for LC-MS analysis.
- Aliquot into mass spectrometry vial.

(5) What are the main challenges with tissue samples?

3.2.5   Resuspension and Choice of Diluent

For LC-MS analysis it is important that the metabolites are in a diluent very close to the starting mobile phase. In the case of HILIC analysis, this will be acetonitrile or similar, and so there may be the issue of solubility of polar metabolites, again causing incomplete recovery.

Challenges
(1) Solvents may interfere with downstream analysis, e.g. which solvents are not compatible with mass spectrometry? Incompatible solvents may impact on the chromatographic retention time of compounds and may have to dry down and resuspend, leading to metabolite losses. This has been found to be the case with NADPH and reduced glutathione (GSH) using lyophilisation, evaporation with nitrogen or a room temperature Speedvac by

Lu et al. [5]. This step has also been reported to increase reactions between metabolites as the sample concentration increases.

It may be possible to extract the polar metabolites from the sample and inject these directly onto the chromatographic column or into the mass spectrometer. However, there may be issues with stability in the autosampler and evaporation of organic solvents in well plates or vials, leading to alterations in metabolite composition.

The Problem of Metabolite Degradation

Due to the complex nature of biological samples, the resulting extracts contain a large number of metabolites. Even with an extract of polar metabolites, there will be analytes from many different classes and with very different structures. Some of these metabolites are very labile, e.g. ATP and NADPH and are likely to degrade. There could also be the problem of metabolite interconversion, e.g. in the case of glutathione and NADPH [5].

Solutions:

- Minimise time between sample preparation and analysis.
- Extract at the lowest temperature possible.
- Freeze at the lowest temperature as soon as possible.

Extraction for Both Polar Metabolites and Lipids: A Compromise?

It is becoming increasingly common in metabolomics studies to employ a single method to extract both polar metabolites and lipids from the same sample [e.g. metabolites-10-00495, d0an01319f]. Advantages include minimising the amount of sample used and the ability to correlate changes in polar analytes and lipids from the same sample. Biphasic plasma and serum extractions have the additional advantage of removing glycerophospholipids from the polar extracts, as these lipids are known to cause matrix effects.

Take Home Message
- Pre-analytical conditions impact greatly on downstream metabolic profiles and vary between biological samples.
- Similarly, extraction methods are sample type-dependent. Tissue and cell samples will need quenching to halt metabolism. Protein must be removed from the samples prior to metabolite analysis.
- Some extraction methods may reduce the levels of abundant high-energy metabolites and increase the levels of less-abundant lower energy compounds. Therefore the sample will not be a true reflection of the metabolome.
- Sample diluent for resuspension of metabolite extracts is important for downstream LC-MS analysis.

1. Urine is the most commonly collected biofluid. It is non-invasive, not usually sample-limited and allows for sequential collections.
2. Temperature of collection and storage, storage duration and pH affect sample stability.
3. Quenching is important to (a) halt cellular metabolism and (b) to preserve sample integrity by preventing further metabolite turnover.
4. An ideal sample preparation method should be cheap, reproducible and rapid enough to be scaled up to hundreds or thousands of samples.
5. Tissue samples are heterogenous which can cause variability. Quenching needs to be performed to halt cellular metabolism.

References

1. Fujiwara T, Inoue R, Ohtawa T, Tsunoda M. Liquid-chromatographic methods for carboxylic acids in biological samples. Vol. 25, Molecules (Basel). NLM (Medline); 2020.
2. Walker JM. Methods in molecular biology [Internet]. Available from: http://www.springer.com/series/7651
3. Zhou J, Zhong L. Applications of liquid chromatography-mass spectrometry based metabolomics in predictive and personalized medicine. Front Mol Biosci. 2022;9
4. Martias C, Baroukh N, Mavel S, Blasco H, Lefèvre A, Roch L, et al. Optimization of sample preparation for metabolomics exploration of urine, feces, blood and saliva in humans using combined nmr and uhplc-hrms platforms. Molecules. 2021;26(14)
5. Lu W, Su X, Klein MS, Lewis IA, Fiehn O, Rabinowitz JD. Metabolite measurement: Pitfalls to avoid and practices to follow. Vol. 86, Annual Review of Biochemistry. Annual Reviews Inc.; 2017. p. 277–304.
6. Medina J, van der Velpen V, Teav T, Guitton Y, Gallart-Ayala H, Ivanisevic J. Single-step extraction coupled with targeted hilic-ms/ms approach for comprehensive analysis of human plasma lipidome and polar metabolome. Meta. 2020;10(12):1–17.
7. Thiele I, Swainston N, Fleming RMT, Hoppe A, Sahoo S, Aurich MK, et al. A community-driven global reconstruction of human metabolism. Nat Biotechnol. 2013;31(5):419–25.
8. Hasegawa Y, Otoki Y, McClorry S, Coates LC, Lombardi RL, Taha AY, et al. Optimization of a method for the simultaneous extraction of polar and non-polar oxylipin metabolites, dna, rna, small rna, and protein from a single small tissue sample. Methods Protoc. 2020;3(3):1–14.
9. Sitnikov DG, Monnin CS, Vuckovic D. Systematic assessment of seven solvent and solid-phase extraction methods for metabolomics analysis of human plasma by LC-MS. Sci Rep. 2016;6
10. Want EJ, O'Maille G, Smith CA, Brandon TR, Uritboonthai W, Qin C, et al. Solvent-dependent metabolite distribution, clustering, and protein extraction for serum profiling with mass spectrometry. Anal Chem. 2006;78(3):743–52.
11. Southam AD, Haglington LD, Najdekr L, Jankevics A, Weber RJM, Dunn WB. Assessment of human plasma and urine sample preparation for reproducible and high-throughput UHPLC-MS clinical metabolic phenotyping. Analyst. 2020;145(20):6511–23.
12. Want EJ, Wilson ID, Gika H, Theodoridis G, Plumb RS, Shockcor J, et al. Global metabolic profiling procedures for urine using UPLC-MS. Nat Protoc. 2010;5(6):1005–18.

13. Diallo AF, Lockwood MB, Maki KA, Franks AT, Roy A, Jaime-Lara R, et al. Metabolic profiling of blood and urine for exploring the functional role of the microbiota in human health. Biol Res Nurs. 2020;22(4):449–57.

14. Steuer AE, Brockbals L, Kraemer T. Metabolomic strategies in biomarker research-new approach for indirect identification of drug consumption and sample manipulation in clinical and forensic toxicology? Vol. 7, Frontiers in Chemistry. Frontiers Media S.A.; 2019.

15. Likhitweerawong N, Thonusin C, Boonchooduang N, Louthrenoo O, Nookaew I, Chattipakorn N, et al. Profiles of urine and blood metabolomics in autism spectrum disorders. Vol. 36, Metabolic brain disease. Springer; 2021. p. 1641–71.

16. Wang R, Kang H, Zhang X, Nie Q, Wang H, Wang C, et al. Urinary metabolomics for discovering metabolic biomarkers of bladder cancer by UPLC-MS. BMC Cancer. 2022;22(1)

17. Emwas AH, Luchinat C, Turano P, Tenori L, Roy R, Salek RM, et al. Standardizing the experimental conditions for using urine in NMR-based metabolomic studies with a particular focus on diagnostic studies: a review. Vol. 11, Metabolomics. New York: Springer; 2015. p. 872–94.

18. Aylward LL, Hays SM, Zidek A. Variation in urinary spot sample, 24h samples, and longer-term average urinary concentrations of short-lived environmental chemicals: implications for exposure assessment and reverse dosimetry. J Expo Sci Environ Epidemiol. 2017;27(6):582–90.

19. Shihabi ZK, Shihabi ZK, Schwartz RP, Pugia MJ. Decreasing the variability observed in urine analysis. 2001.

20. Sallsten G, Barregard L. Variability of urinary creatinine in healthy individuals. Int J Environ Res Public Health. 2021;18(6):1–12.

21. Yu Z, Kastenmüller G, He Y, Belcredi P, Möller G, Prehn C, et al. Differences between human plasma and serum metabolite profiles. PLoS One. 2011;6(7)

22. Würtz P, Havulinna AS, Soininen P, Tynkkynen T, Prieto-Merino D, Tillin T, et al. Metabolite profiling and cardiovascular event risk: a prospective study of 3 population-based cohorts. Circulation. 2015;131(9):774–85.

23. Schmidt DR, Patel R, Kirsch DG, Lewis CA, van der Heiden MG, Locasale JW. Metabolomics in cancer research and emerging applications in clinical oncology. CA Cancer J Clin. 2021;71(4): 333–58.

24. Sindelar M, Stancliffe E, Schwaiger-Haber M, Anbukumar DS, Adkins-Travis K, Goss CW, et al. Longitudinal metabolomics of human plasma reveals prognostic markers of COVID-19 disease severity. Cell Rep Med. 2021;2(8):100369.

25. Yu B, Zanetti KA, Temprosa M, Albanes D, Appel N, Barrera CB, et al. The consortium of metabolomics studies (COMETS): metabolomics in 47 prospective cohort studies. Am J Epidemiol. 2019;188(6):991–1012.

26. Kiseleva O, Kurbatov I, Ilgisonis E, Poverennaya E. Defining blood plasma and serum metabolome by gc-ms. Vol. 12, Metabolites. MDPI; 2022.

27. Zhao L, Ni Y, Su M, Li H, Dong F, Chen W, et al. High throughput and quantitative measurement of microbial metabolome by gas chromatography/mass spectrometry using automated alkyl chloroformate derivatization. Anal Chem. 2017;89(10):5565–77.

28. Smith L, Villaret-Cazadamont J, Claus SP, Canlet C, Guillou H, Cabaton NJ, et al. Important considerations for sample collection in metabolomics studies with a special focus on applications to liver functions. Vol. 10, Metabolites. MDPI AG; 2020.

29. Deda O, Gika HG, Wilson ID, Theodoridis GA. An overview of fecal sample preparation for global metabolic profiling. J Pharmac Biomed Anal. 2015;113:137–50.

30. de Zawadzki A, Thiele M, Suvitaival T, Wretlind A, Kim M, Ali M, et al. High-throughput UHPLC-MS to screen metabolites in feces for gut metabolic health. Meta. 2022;12(3)

31. Lehmann R. From bedside to bench-practical considerations to avoid pre-analytical pitfalls and assess sample quality for high-resolution metabolomics and lipidomics analyses of body fluids. https://doi.org/10.1007/s00216-021-03450-0

32. Newgard CB, An J, Bain JR, Muehlbauer MJ, Stevens RD, Lien LF, et al. A branched-chain amino acid-related metabolic signature that differentiates obese and lean humans and contributes to insulin resistance. Cell Metab. 2009;9(4):311–26.

33. Hagyousif YA, Sharaf BM, Zenati RA, El-Huneidi W, Bustanji Y, Abu-Gharbieh E, et al. Skin cancer metabolic profile assessed by different analytical platforms. Int J Mol Sci [Internet]. 2023;24(2):1604. Available from: https://www.mdpi.com/1422-0067/24/2/1604

34. Liesenfeld DB, Grapov D, Fahrmann JF, Salou M, Scherer D, Toth R, et al. Metabolomics and transcriptomics identify pathway differences between visceral and subcutaneous adipose tissue in colorectal cancer patients: the ColoCare study. Am J Clin Nutr. 2015;102(2):433–43.

35. Saoi M, Britz-Mckibbin P. New advances in tissue metabolomics: a review. Vol. 11, Metabolites. MDPI; 2021.

36. Rauth S, Karmakar S, Batra SK, Ponnusamy MP. Recent advances in organoid development and applications in disease modeling. Vol. 1875. Biochimica et Biophysica Acta - reviews on cancer. Elsevier B.V.; 2021.

37. Malinowska JM, Palosaari T, Sund J, Carpi D, Lloyd GR, Weber RJM, et al. Automated sample preparation and data collection workflow for high-throughput in vitro metabolomics. Meta. 2022;12(1)

38. Malinowska JM, Palosaari T, Sund J, Carpi D, Bouhifd M, Weber RJM, et al. Integrating in vitro metabolomics with a 96-well high-throughput screening platform. Metabolomics. 2022;18(1):11.

39. Wang L, Pi Z, Liu S, Liu Z, Song F. Targeted metabolome profiling by dual-probe microdialysis sampling and treatment using Gardenia jasminoides for rats with type 2 diabetes. Sci Rep. 2017;7 (1)

40. Friston D, Laycock H, Nagy I, Want EJ. Microdialysis workflow for metabotyping superficial pathologies: application to burn injury. Anal Chem. 2019;91(10):6541–8.

41. Nolan LS, Lewis AN, Gong Q, Sollome JJ, Dewitt ON, Williams RD, et al. Untargeted metabolomic analysis of human milk from mothers of preterm infants. Nutrients. 2021;13(10)

42. Shaheed SU, Tait C, Kyriacou K, Linforth R, Salhab M, Sutton C. Evaluation of nipple aspirate fluid as a diagnostic tool for early detection of breast cancer. Vol. 15, Clinical proteomics. BioMed Central Ltd.; 2018.

43. Albrecht B, Voronina E, Schipke C, Peters O, Parr MK, Díaz-Hernández MD, et al. Pursuing experimental reproducibility: an efficient protocol for the preparation of cerebrospinal fluid samples for nmr-based metabolomics and analysis of sample degradation. Meta. 2020;10(6): 1–16.

44. Blasco H, Corcia P, Moreau C, Veau S, Fournier C, Vourc'h P, et al. 1H-NMR-based metabolomic profiling of CSF in early amyotrophic lateral sclerosis. PLoS One. 2010;5(10)

45. Carlsson H, Abujrais S, Herman S, Khoonsari PE, Åkerfeldt T, Svenningsson A, et al. Targeted metabolomics of CSF in healthy individuals and patients with secondary progressive multiple sclerosis using high-resolution mass spectrometry. Metabolomics. 2020;16(2)

46. Alzharani MA, Alshuwaier GO, Aljaloud KS, Al-Tannak NF, Watson DG. Metabolomics profiling of plasma, urine and saliva after short term training in young professional football players in Saudi Arabia. Sci Rep. 2020;10(1):19759.

47. Li Z, Sarnat JA, Liu KH, Hood RB, Chang CJ, Hu X, et al. Evaluation of the use of saliva metabolome as a surrogate of blood metabolome in assessing internal exposures to traffic-related air pollution. Environ Sci Technol. 2022;56:6525.

48. McBride EM, Lawrence RJ, McGee K, Mach PM, Demond PS, Busch MW, et al. Rapid liquid chromatography tandem mass spectrometry method for targeted quantitation of human performance metabolites in saliva. J Chromatogr A. 2019;1601:205–13.
49. Saheb Sharif-Askari N, Soares NC, Mohamed HA, Saheb Sharif-Askari F, Alsayed HAH, Al-Hroub H, et al. Saliva metabolomic profile of COVID-19 patients associates with disease severity. Metabolomics. 2022;18(11):81.
50. Stringer KA, Younger JG, McHugh C, Yeomans L, Finkel MA, Puskarich MA, et al. Whole blood reveals more metabolic detail of the human metabolome than serum as measured by 1H-NMR spectroscopy: implications for sepsis metabolomics. Shock. 2015;44(3):200–8.
51. Petrick LM, Uppal K, Funk WE. Metabolomics and adductomics of newborn bloodspots to retrospectively assess the early-life exposome, Vol. 32, Current opinion in pediatrics. NLM (Medline); 2020. p. 300–7.
52. Petrick L, Edmands W, Schiffman C, Grigoryan H, Perttula K, Yano Y, et al. An untargeted metabolomics method for archived newborn dried blood spots in epidemiologic studies. Metabolomics. 2017;13(3)
53. Tobin NH, Murphy A, Li F, Brummel SS, Taha TE, Saidi F, et al. Comparison of dried blood spot and plasma sampling for untargeted metabolomics. Metabolomics. 2021;17(7):62.
54. Nix C, Hemmati M, Cobraiville G, Servais AC, Fillet M. Blood microsampling to monitor metabolic profiles during physical exercise. Front Mol Biosci. 2021;8
55. Sens A, Rischke S, Hahnefeld L, Dorochow E, Schäfer SMG, Thomas D, et al. Pre-analytical sample handling standardization for reliable measurement of metabolites and lipids in LC-MS-based clinical research. J Mass Spectrometry Adv Clin Lab. 2023;28:35–46.
56. Gegner HM, Naake T, Dugourd A, Müller T, Czernilofsky F, Kliewer G, et al. Pre-analytical processing of plasma and serum samples for combined proteome and metabolome analysis. Front Mol Biosci. 2022;9
57. Yin P, Lehmann R, Xu G. Effects of pre-analytical processes on blood samples used in metabolomics studies. Vol. 407, Analytical and Bioanalytical Chemistry. Springer Science and Business Media Deutschland GmbH; 2015. p. 4879–92.
58. Revuelta-López E, Barallat J, Cserkoóvá A, Gálvez-Montón C, Jaffe AS, Januzzi JL, et al. Pre-analytical considerations in biomarker research: focus on cardiovascular disease. Vol. 59, Clinical Chemistry and Laboratory Medicine. De Gruyter Open Ltd; 2021. p. 1747–60.
59. McClain KM, Moore SC, Sampson JN, Henderson TR, Gebauer SK, Newman JW, et al. Preanalytical sample handling conditions and their effects on the human serum metabolome in epidemiologic studies. Am J Epidemiol. 2021;190(3):459–67.
60. Rotter M, Brandmaier S, Prehn C, Adam J, Rabstein S, Gawrych K, et al. Stability of targeted metabolite profiles of urine samples under different storage conditions. Metabolomics. 2017;13(1):4.
61. Saude EJ, Sykes BD. Urine stability for metabolomic studies: effects of preparation and storage. Metabolomics. 2007;3(1):19–27.
62. Smith LM, Maher AD, Want EJ, Elliott P, Stamler J, Hawkes GE, et al. Large-scale human metabolic phenotyping and molecular epidemiological studies via 1H NMR spectroscopy of urine: investigation of borate preservation. Anal Chem. 2009;81(12):4847–56.
63. Bernini P, Bertini I, Luchinat C, Nincheri P, Staderini S, Turano P. Standard operating procedures for pre-analytical handling of blood and urine for metabolomic studies and biobanks. J Biomol NMR. 2011;49(3–4):231–43.
64. González-Domínguez R, González-Domínguez Á, Sayago A, Fernández-Recamales Á. Recommendations and best practices for standardizing the pre-analytical processing of blood and urine samples in metabolomics. Vol. 10, Metabolites. MDPI AG; 2020. p. 1–18.

65. Sotelo-Orozco J, Chen SY, Hertz-Picciotto I, Slupsky CM. A comparison of serum and plasma blood collection tubes for the integration of epidemiological and metabolomics data. Front Mol Biosci. 2021:8.

66. Kaluarachchi M, Boulangé CL, Karaman I, Lindon JC, Ebbels TMD, Elliott P, et al. A comparison of human serum and plasma metabolites using untargeted 1 H NMR spectroscopy and UPLC-MS. Metabolomics. 2018;14(3):32.

67. Khadka M, Todor A, Maner-Smith KM, Colucci JK, Tran V, Gau DA, et al. The effect of anticoagulants, temperature, and time on the human plasma metabolome and lipidome from healthy donors as determined by liquid chromatography-mass spectrometry. Biomol Ther. 2019;9 (5)

68. Hosseinkhani F, Dubbelman AC, Karu N, Harms AC, Hankemeier T. Towards standards for human fecal sample preparation in targeted and untargeted lc-hrms studies. Meta. 2021;11(6)

69. Kostidis S, Addie RD, Morreau H, Mayboroda OA, Giera M. Quantitative NMR analysis of intra- and extracellular metabolism of mammalian cells: a tutorial. Anal Chim Acta. 2017;980:1–24.

70. Rushing BR, Schroder M, Sumner SCJ. Comparison of lysis and detachment sample preparation methods for cultured triple-negative breast cancer cells using UHPLC–HRMS-based metabolomics. Meta. 2022;12(2)

71. Bosman P, Pichon V, Acevedo AC, Chardin H, Combes A. Development of analytical methods to study the salivary metabolome: impact of the sampling. Anal Bioanal Chem. 2022;414(23): 6899–909.

72. Song Z, Wang M, Zhu Z, Tang G, Liu Y, Chai Y. Optimization of pretreatment methods for cerebrospinal fluid metabolomics based on ultrahigh performance liquid chromatography/mass spectrometry. J Pharm Biomed Anal. 2021:197.

73. Gertsman I, Barshop BA. Promises and pitfalls of untargeted metabolomics. J Inherited Metabolic Dis. 2018;41:355–66.

74. Rischke S, Hahnefeld L, Burla B, Behrens F, Gurke R, Garrett TJ. Small molecule biomarker discovery: proposed workflow for LC-MS-based clinical research projects. J Mass Spectrometry Adv Clin Lab. 2023;28:47–55.

75. Prasannan CB, Jaiswal D, Davis R, Wangikar PP. An improved method for extraction of polar and charged metabolites from cyanobacteria. PLoS One. 2018;13(10):e0204273.

76. Reichl B, Eichelberg N, Freytag M, Gojo J, Peyrl A, Buchberger W. Evaluation and optimization of common lipid extraction methods in cerebrospinal fluid samples. J Chromatogr B Analyt Technol Biomed Life Sci. 2020:1153.

77. Malik D, Rhoades S, Weljie AM. Extraction and analysis of pan-metabolome polar metabolites by ultra performance liquid chromatography–tandem mass spectrometry (UPLC-MS/MS). Bio Protoc. 2018;8(3)

78. Rico E, González O, Blanco ME, Alonso RM. Evaluation of human plasma sample preparation protocols for untargeted metabolic profiles analyzed by UHPLC-ESI-TOF-MS. Anal Bioanal Chem. 2014;406(29):7641–52.

79. King AM, Mullin LG, Wilson ID, Coen M, Rainville PD, Plumb RS, et al. Development of a rapid profiling method for the analysis of polar analytes in urine using HILIC–MS and ion mobility enabled HILIC–MS. Metabolomics. 2019;15(2):17.

80. Yang Y, Cruickshank C, Armstrong M, Mahaffey S, Reisdorph R, Reisdorph N. New sample preparation approach for mass spectrometry-based profiling of plasma results in improved coverage of metabolome. J Chromatogr A. 2013:1300, 217–26.

81. Nagana Gowda GA, Raftery D. Quantitating metabolites in protein precipitated serum using NMR spectroscopy. Anal Chem. 2014;86(11):5433–40.

82. Kelly PE, Ng HJ, Farrell G, McKirdy S, Russell RK, Hansen R, et al. An optimised monophasic faecal extraction method for LC-MS analysis and its application in gastrointestinal disease. Meta. 2022;12(11):1110.
83. Vuckovic D. Current trends and challenges in sample preparation for global metabolomics using liquid chromatography-mass spectrometry. Anal Bioanal Chem. 2012;403:1523–48.
84. Erben V, Poschet G, Schrotz-King P, Brenner H. Evaluation of different stool extraction methods for metabolomics measurements in human faecal samples. BMJ Nutr Prev Health. 2021;4(2): 374–84.
85. Vorkas PA, Isaac G, Anwar MA, Davies AH, Want EJ, Nicholson JK, et al. Untargeted UPLC-MS profiling pipeline to expand tissue metabolome coverage: application to cardiovascular disease. Anal Chem. 2015;87(8):4184–93.
86. Want EJ, Masson P, Michopoulos F, Wilson ID, Theodoridis G, Plumb RS, et al. Global metabolic profiling of animal and human tissues via UPLC-MS. Nat Protoc. 2013;8(1):17–32.
87. Anwar MA, Adesina-Georgiadis KN, Spagou K, Vorkas PA, Li JV, Shalhoub J, et al. A comprehensive characterisation of the metabolic profile of varicose veins; implications in elaborating plausible cellular pathways for disease pathogenesis. Sci Rep. 2017;7(1):2989.
88. Salem M, Bernach M, Bajdzienko K, Giavalisco P. A simple fractionated extraction method for the comprehensive analysis of metabolites, lipids, and proteins from a single sample. J Vis Exp. 2017;2017(124)

How to Extract Lipid Metabolites and Perform Lipid Analysis?

4

Pauline Le Faouder, Anaelle Durbec, Océane Delos,
and Justine Bertrand-Michel

What Will You Learn in This Chapter?

Lipids are complex metabolites and key components of physiology. They are ubiquitous and therefore of interest in various sample types. However, a lot of considerations should be taken into account before starting lipidomics analysis; in this chapter we will discover how to apprehend such a study from sample preparation to ultimate annotation of lipids.

4.1   Introduction

Lipids are a group of very heterogeneous hydrophobic and/or amphiphilic complex molecules whose main physicochemical property is solubility in organic solvents. Behind this very broad definition lie numerous families of molecules with very broad physico-chemical properties: some families, such as triglycerides, are soluble in highly apolar solvents (e.g. chloroform or hexane), while other molecules, such as polyphosphoinositide phosphates, are nearly soluble in water. The classification of this complex group of molecules is not easy. However, the LipidMaps consortium (https://www.lipidmaps.org/)

P. Le Faouder · A. Durbec · O. Delos · J. Bertrand-Michel (✉)
MetaboHUB-MetaToul, National Infrastructure of Metabolomics and Fluxomics, Toulouse, France

I2MC, Université de Toulouse, Inserm, Université Toulouse III – Paul Sabatier (UPS), Toulouse, France
e-mail: justine.bertrand-michel@inserm.fr

© The Author(s), under exclusive license to Springer Nature Switzerland AG 2023
J. Ivanisevic, M. Giera (eds.), *A Practical Guide to Metabolomics Applications in Health and Disease*, Learning Materials in Biosciences,
https://doi.org/10.1007/978-3-031-44256-8_4

Table 4.1 Examples of lipid classes and their functions

Lipid class	Description	Functions		
Fatty acid (FA)	Classified by their carbon chain and double bond number, i.e. FA(18:1) for a FA with 18 carbons and 1 double bond	Building block of most lipids	Nutrition	PPAR activation
Sterols (ST)	Pregnane derivatives derived from isoprenoid metabolism	Membrane composition	Lipid absorption	LXR or SREBP activators Hormones
Phospholipid (GP)	Amphiphilic lipids, characterised by a glycerophosphate head group		Lipid bilayer formation	
Glycerolipid (GL)	Glycerol-containing lipids with up to three FA side chains		Lipid absorption and energy storage	
Sphingolipid (SL)	Sphingoid base substituted by a fatty acid on an amide linkage	Membrane composition		Signal transduction Cell recognition

has proposed a nomenclature build on a simplified classification containing eight major families: fatty acids and their derivatives (FA), glycerolipids (GL), glycerophospholipids (GP), sphingolipids (SP), sterols and their derivatives (ST), prenols (PR), saccharolipids (SL) and polyketides (PK) [1]. This nomenclature has now widely become accepted (few examples are given in Table 4.1).

Given the number of fatty acids and families described, theoretically there are over 180,000 possible molecular lipid species [2]. On the LipidMaps website (https://www.lipidmaps.org/), there are currently 47,659 theoretical structures described, of which 25,616 have actually been verified. These molecules are present in very diverse concentration ranges. For example, in plasma some molecules appear in the mmol/L range (triacylglycerides, certain phospholipids) and others in the pmol/L range (free oxylipins, minority sphingolipids) [3].

As outlined above, lipids are ubiquitous and play numerous biological roles in basically every living organism (examples are given in Table 4.1).

Lipids have important structural roles: lipids make up about 70% of plasma membranes. Biological membranes are organised around a lipid bilayer which is a thin polar membrane formed by two layers of phospholipids. They are oriented so that their hydrophilic heads are in contact with water and their hydrophobic tails are aligned through van der Waals interactions. These membranes are flat sheets that form a continuous barrier around the cells. A wide variety of lipids can be found in membranes: cholesterol, phospholipids and sphingolipids with different fatty acid compositions jointly shaping organelle-specific membrane compositions. For example, the endoplasmic reticulum and the plasma membrane have different compositions. The endoplasmic reticulum contains little cholesterol and high amounts of polyunsaturated phospholipids [4], whereas the plasma membrane contains higher levels of cholesterol and sphingolipids [5]. Some organelles are even composed of specific lipids such as cardiolipins [6] or lysobisphosphatidic acid that can be found in the late endosomes [7]. These various compositions are very finely regulated [8], and their deregulation can lead to pathologies found in certain genetic diseases. Lipids exert their function by disrupting the physico-chemical properties of membranes which will result in membrane bending, fluidity, movement or bulging. But most of the membrane trafficking is mediated by the ability of lipids to modulate proteins [9].

Lipids have an energy storage role which involves several stages: digestion, absorption, transport and lipid catabolism. Lipids contained in food (triacylglycerides, sterols, phospholipids) are digested via pancreatic enzymes in the intestinal lumen. The enterocyte takes up secreted lipids via transport protein CD36 or FAT4 for fatty acids and monoacylglycerides or NPC1L1 for cholesterol. They are then transported in the lymph as chylomicrons and distributed to the organs after transformation in the liver [10]. The liver is the hub of lipid metabolism at the interface between nutritional intake and systemic circulation. It is essential in maintaining cholesterol homeostasis, in the absorption of fatty acids or in the synthesis of triglycerides in conjunction with adipose tissue and skeletal muscle via lipogenesis and lipolysis. An imbalance between lipids absorbed from diet, lipogenesis and lipolysis can lead to an accumulation of fat in the liver (nonalcoholic fatty liver disease, NAFLD) which will lead to chronic inflammation, fibrosis and eventually liver cancer. Cellular insulin signalling can also be disrupted and lead to Type II diabetes [11]. In this context of energy storage, another important organ is adipose tissue. It is a much more complex organ than was thought a few years ago. Phenomena, such as exposure to cold, lead, for example, to lipolysis of lipid droplets in white adipose tissue, which will increase the content of fatty acids in brown adipose tissue [12]. This type of mechanism may be of particular interest to understand in regulating obesity.

Lipids are "signal" molecules: they participate in the regulation of many physiological phenomena [13]. Over the last 20 years, the progress in the sensitivity of lipids analysis has enabled a major leap forward in the structural analysis and quantification of these molecules present in very small quantities. The mechanisms of inflammation and its resolution are mediated by metabolites of polyunsaturated fatty acids (PUFA): oxylipins. Derivatives of n6-PUFAs tend to be proinflammatory molecules, whereas metabolites of n3-PUFAs are anti-inflammatory [14]. Sphingolipids can also be involved in signalling: i.e. ceramides produced by sphingomyelinase cleavage of sphingomyelin are involved in cell cycle arrest and cell senescence. Ceramide-1-phosphate produced by phosphorylation of ceramide by a ceramide kinase has mitogenic properties, antagonises the pro-apoptotic action of ceramides and causes inflammation [15]. Eicosanoids, fatty acid derivatives and sterols (such as oxysterols) are able to act through nuclear receptors (LXR: liver X receptor, PPAR: peroxisomes proliferator-activated receptor, FXR: farnesoid X receptor). These lipo-sensitive receptors associated in heterodimers with the retinoid X receptor (RXR) control the metabolic cascade that governs the transcription of most of the genes involved in the metabolism of lipids, their storage, transport and elimination [16].

These few examples show that lipid signalling is at the heart of a large number of mechanisms and enzymatic functions. Their disruption is deleterious to health and can cause cardiovascular, autoimmune or inflammatory metabolic diseases. It is essential to analyse lipids qualitatively and quantitatively in a robust and efficient manner in order to understand these phenomena or carry out therapeutic monitoring.

1. *How many species of lipids are there?*
   a. 1000.
   b. 10,000.
   c. More than 100,000.
   d. We do not know.
2. *Why is it interesting to study lipids?*
   a. They are signalling molecules.
   b. They are involved in membrane structures.
   c. I do not know.

4.2 Consideration for Lipids Analysis: From the Extraction to the Quantification

Several questions need to be asked and discussed prior to lipidomic experiments to be sure to choose the right strategy. All these points are summarised in Fig. 4.1: for each step of the workflow, different possible options are proposed.

1. *The biological question:* before the analysis, biologists must specify their needs. The biological question must be clear and well-defined, which implies an adapted experimental design and sufficient additional information.
2. *The sampling:* this is a crucial stage that requires careful thought because if it is poorly conceived, the obtained data will be unusable.
   - Lipidomic analysis can be performed on *various samples* such as cells (or purified cell compartments), micro-organisms, cell supernatants, tissues and biological fluid, once they have been stored in good condition (under −80 °C).

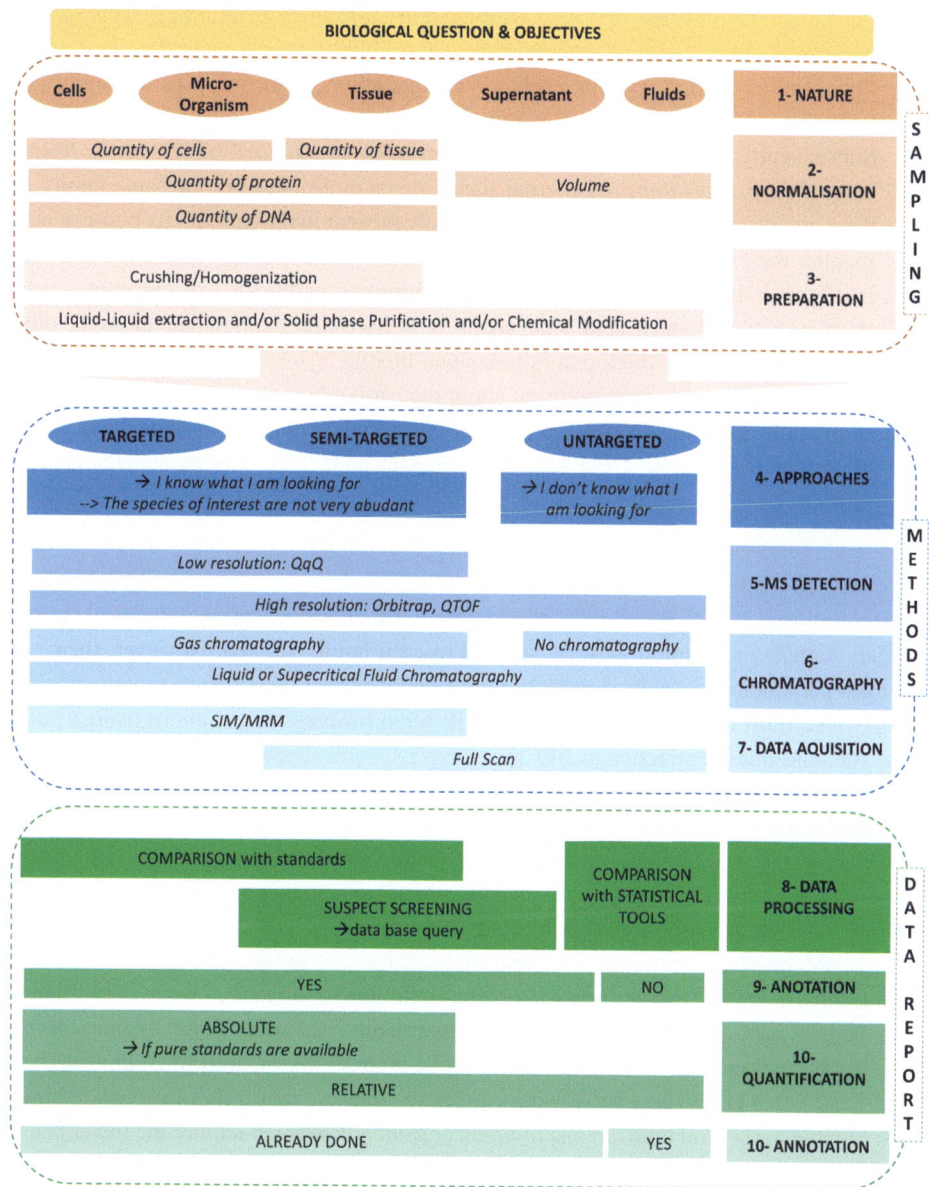

Fig. 4.1 Main steps of a lipidomics workflow (DNA: deoxyribonucleic acid, QqQ: triple quadrupole, QTOF: quadrupole to time of flight, MS: mass spectrometry, SIM: single-ion monitoring, MRM: multiple reaction monitoring)

- For tissue samples, the first step is homogenisation through grinding. This can be performed either in a cold environment using liquid nitrogen or in the presence of liquid (buffer with or without a solvent) using equipment such as a Turax, dounce or ball mill.
- Subsequently, it is essential *to introduce an internal standard (ISTD) at the beginning of the preparation*. An internal standard is a molecule from the same family as the molecules of interest, but it is not naturally present in the sample. Its purpose is to monitor the extraction efficiency. Typically, a molecule labelled with a nucleus that is not naturally occurring (commonly deuterium or carbon-13) or carrying a very short or unusual fatty acid side chain is utilised for mammalian analysis only (excluding bacteria, which naturally contain them).
- It is important at this stage to think about the *way of how to normalise the obtained results*, i.e. relate the results to a quantity of biological material such as the quantity of protein, tissue, cells, DNA (deoxyribonucleic acid). This must be adapted to each project.

To analyse a lipid, which is a "rather soluble" molecule in organic solvents, it needs to be extracted from its biological environment. Due to the enormous physicochemical variety of lipids, sample preparation must be adapted to each family studied. However, there are two main preparation methods (Fig. 4.2): liquid-liquid extraction (LLE) or protein precipitation, which must often be completed by purification (or pre-concentration) using a solid phase (or solid-phase extraction = SPE).

- Liquid–Liquid Extraction

    During a LLE, a mixture of inorganic (generally water) and organic solvents (typically methanol and chloroform) in specific proportions makes it possible to form two phases: an aqueous phase containing polar molecules, salts and some proteins, a semisolid interface containing precipitated proteins and an organic phase containing molecules of interest. Each manipulator adapts this extraction to the family of its interest by adding salts and acid or by modifying the nature and proportions of the solvent. However, the best-known and the most used LLE are the extractions of Bligh and Dyer [17] and Folch [18]. These broad-spectrum extractions are relatively universal and may be suitable for neutral lipids, phospholipids or sphingolipids but require the lower phase to be removed which can be more complicated to perform. Some colleagues have proposed replacing the organic solvents methanol and chloroform (or dichloromethane) with methyl tert-butyl ether [19] or a mixture of butanol and methanol [20] on which the upper phase is extracted making automation significantly easier. For some particular families, other preparations exist, such as hexane extraction allowing selective extraction of neutral lipids. The imagination of chemists is overflowing in this area!
- Solid-Phase Extraction (SPE)

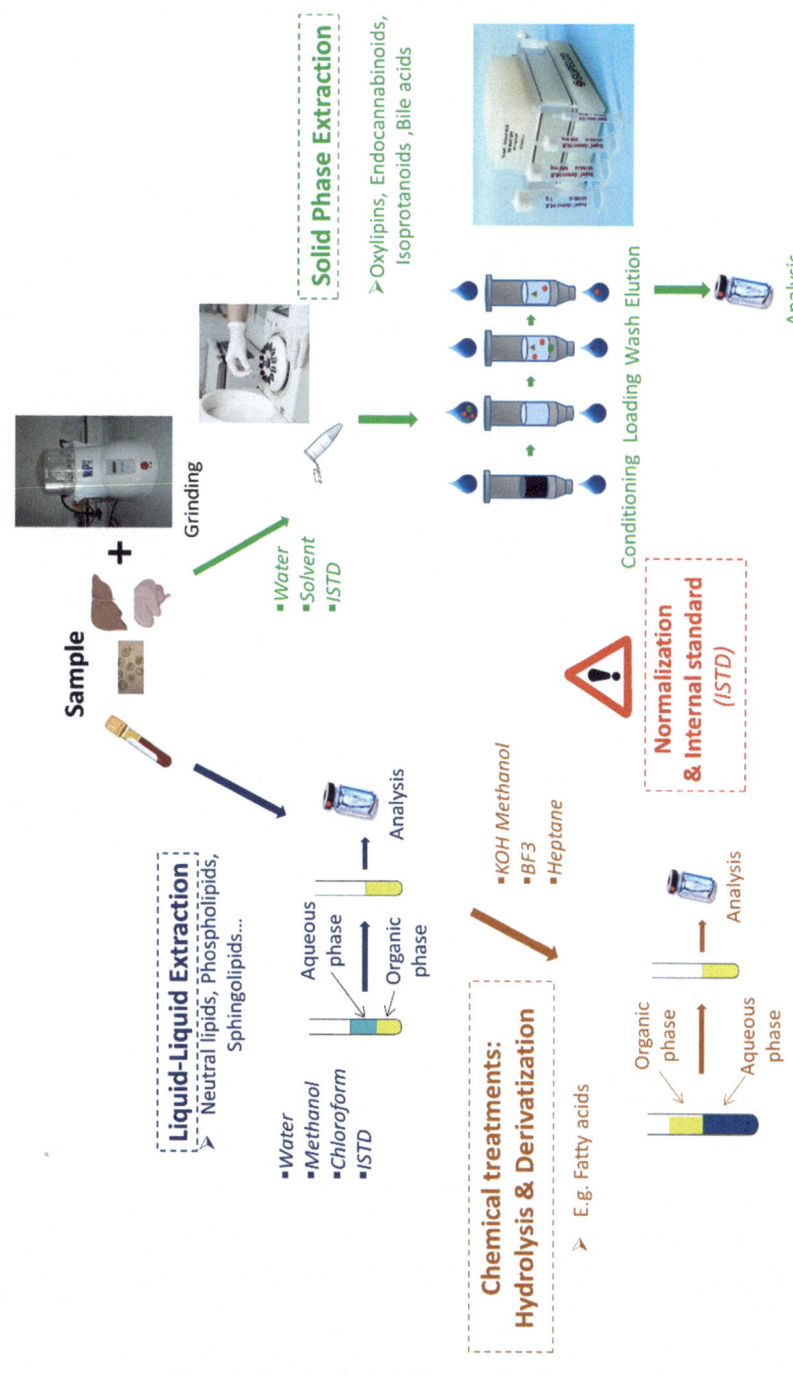

Fig. 4.2 Main protocol for the sample preparation in lipidomic

The alternative method derives from the preparation processes commonly used in metabolomics, where the aim is simply to eliminate proteins by precipitating them in the presence of a mixture of water and solvent: generally, 1 volume of water for 2 to 4 volumes of methanol or acetonitrile. However, the crude extract obtained is usually not suitable for lipidomics as it lacks sufficient purity. Therefore, it needs to undergo pre-purification (or pre-concentration) on a solid-phase extraction (SPE) cartridge. The choice of the cartridge type may vary depending on the specific molecule being purified. For instance, bile acids (unpublished results) and free oxylipins [21] are purified on polymeric SPE columns (HRX or HLB), oxysterols [22] on pure silica columns and isoprostanes [23] on anion exchange SPE columns. These purifications are time-consuming and require excess consumables, but they significantly enhance the sensitivity of the analysis.

• Chemical Modifications

It may also be necessary to modify compounds prior to their analysis. Firstly, some may require release from their backbone to simplify the analysis: for instance, to analyse the total esterified fatty acid, it is necessary to separate them from the glycerol backbone through saponification. This can be achieved by treating the sample with 0.5 N NaOH solution in methanol for 30 min at 50 °C. Secondly, the lipids of interest need to be derivatised which involves modification with a chemical substituent. This is done either to enable its analysis or to enhance the sensitivity of the analysis. This is often essential for gas chromatography (GC) analysis as it reduces the boiling point of compounds: the hydroxyls of sterols or oxysterols are often modified with trimethylsilyl groups [24], the acid functions of fatty acids are usually methylated [25] and some methods need dual derivatisation, for example methylation and trimethylsilylation to analys hydroxy fatty acids [26]. Derivatisation is more rarely used in LC-MS, but it can greatly increase the ionisation efficiency of some polar lipids and therefore improve sensitivity. For example, polyphosphoinositide phosphates or sphingosine-1-phosphate [27] must be methylated [28] to be profiled.

Sample preparation can be complex and time-consuming, but it remains a crucial aspect that needs to be thoroughly discussed and carefully planned before conducting any lipidomic experiment.

3. *How can I normalise brain lipid analysis?*
   a. Per μL.
   b. Per mg of tissue.
   c. Per mg of protein.
   d. I do not need to normalise.
4. *What is the best internal standard to analyse total fatty acid in mice liver?*
   a. Fatty acid C19

(continued)

b. Triglyceride C19
c. Fatty acid 18:1w7
5. *During the sample preparation, I need to prepare more tubes with*:
   a. The internal standard alone
   b. Nothing in the tube
   c. Both
   d. No more tube, no more work!

3. *Method:* the lipidome, like the metabolome, is complex. It covers a variety of structures present in very different quantities. Under these conditions, it is illusory to attempt an analysis in its entirety using a single technique. Three main approaches are therefore possible: a "targeted", "semi-targeted" and "nontargeted" (or "global") approach.

   • *For targeted analysis: the analysis will be focussed on a few particular lipids.* The *sample preparation will be adapted* to the family of interest and can be pretty complex in some cases. The analytical method will then be focused on the lipid family studied. *A high-resolution mass detector* can be used, but *low-resolution detectors* are often preferred, such as triple quadrupoles (QqQ), which are more sensitive and linear. A SIM (single ion monitoring) or MRM (multiple ion monitoring) *scan mode associated with a chromatographic separation* (gas or liquid) is preferred, which improves specificity and sensitivity of the analysis. With the new generation of mass spectrometers, this targeted method can achieve impressive sensitivities to a few picograms of injected compounds.

   • *For semi-targeted methods: the objective is to focus on a "subfamily" of compound such as sphingolipids or phosphatidylinositol (PI). Once again sample preparation will be adapted* but will be less restrictive to be compatible with a larger family. *A high-resolution or low-resolution system* equipped with a *chromatographic system* (more often liquid) can be used with a SIM or MRM *scan mode.* However, a "family" scan mode such as neutral loss or precursor ion scan can be particularly well suited. Indeed, a precursor ion scan on m/z 241 is specific for PI corresponding to an ion loss of m/z 241 and will allow the specific selection of PI subfamily.

   • *For nontargeted methods: the analysis will scan a large lipidome profile because* the researcher has no preconceived idea. In this case, we use a *broad-spectrum sample preparation protocol through LLE which is suitable for most lipid families. But it is far from being "universal"*; some families of compounds will not be extracted under these conditions. The analysis is then carried out *only on high-resolution mass detectors* (Orbitrap or QTOF) with a wide scan of m/z range studied (300–1200 Da) with the best mass resolution by direct introduction (shotgun methods) or after a chromatographic separation (liquid or supercritical fluid). A *full scan mode will be used*, which will allow broad monitoring of lipids of interest.

6. *I would like to run a nontargeted lipidomic analysis, I absolutely need*:
   a. An NMR instrument
   b. A low-resolution mass spectrometer
   c. A high-resolution mass spectrometer
   d. A chromatographic system

4. *Data reporting*: ideally the results need to be qualitative and quantitative which is not so easy depending on the approach used and the lipids studied.
   - *Data processing:* once the sample has been analysed, there are different ways to process the MS data depending on the approach used. To process targeted and semi-targeted data, commercial software is usually well adapted. But for untargeted analyses, due to the large numbers of features (up >1500), two options are possible:
     – *The annotation is done before sample comparison*: the obtained analytical pattern (*m/z* and retention time) is compared with a database. This database is often theoretical because each family of complex lipids has a typical mass spectrum that can be reconstructed in silico. However, it is still better to use a homemade data base which has been manually curated. That is called a *suspect screening approach*.
     – *The annotation is done after the comparison of the sample. The fingerprints of the detected ions* (features) are measured and then *compared using advanced statistical methods*. Discriminating ions are identified and annotated afterwards, which can be long and tedious. This approach has extensively been used in metabolomics, which has led to the development of a range of tools to facilitate this type of workflow.
5. *Annotation—the lipid identification step* is usually done before the quantification for targeted, semi-targeted and untargeted analysis with suspect screening approaches and afterwards for untargeted unbiased methods.

   In lipidomics (and metabolomics), there are four levels of requirements to classify identified and unidentified molecules [29].
   a. Level 1: corresponds to fully identified compounds. It is here necessary to validate the annotation using at least two orthogonal and independent identification methods compared with genuine standard materials.
   b. Level 2: corresponds to putatively annotated compounds. These are generally compounds for which specific annotations are very likely, but the standard is not commercially available, or its synthesis is difficult. It is possible to annotate these

fragments from a database or after analysis of the fragmentation profile of the compound.

c. Level 3: corresponds to compounds where the annotation is incomplete, or where several annotations are proposed for one feature. That is, we know the subfamily of the compound and the molecular formula.

d. Level 4: corresponds to an unknown metabolite with no hypothesis on the structure.

It is very important to indicate the level of identification in any report.

7. *What do I need to annotate a metabolite at level 1*:
   a. *m/z* and a specific retention time
   b. The pure standard, its retention time and MS/MS spectrum
   c. An MS/MS spectrum
   d. The complete structure

6. *Quantification*: ideally the lipidomic data needs to be quantitative; unfortunately mass spectrometers are only quantitative under certain conditions. This is mainly influenced by the ionisation efficiency. The ionisation efficiency can be very different from one instrument to another and from one compound to another even within the same lipid subclass. Indeed a phosphatidylinositol (PI) substituted by two C18:1 fatty acid side chains (PI 36:2) will not ionise in the same way as, for example, a PI with a FA16:0 and FA18:1 side chain (PI 34:1 or PI16:0_18:1). To quantify these compounds in an absolute way, it will be necessary to make calibration ranges with the same molecular species, and only under this condition, we will be able to obtain *absolute quantification* of these compounds. Otherwise we will have to be satisfied with *a relative quantification* by normalising against an internal standard (ISTD) (1 per subclass analysed). Some authors also normalise the quantity of ISTD. This method is called "approximate quantification" or single-point calibration [30]. Therefore, in lipidomics, only simple molecules available commercially, such as sterols, fatty acids and their derivatives, sphingoid bases and few molecular species of complex lipids can be assayed with absolute quantification. All other "complex lipids" such as sphingolipids, phospholipids or atypical molecules will be quantified relative to a control or untreated condition.

For all lipids analysis, it is crucial to consider all these different points. The international lipidomic community proposed a checklist [31] on a dynamic website (https://lipidomicstandards.org/login/) which permit easy summarising and editing of key details of lipidomic analysis using a common language. The idea is to harmonise the field by improving both traceability and reproducibility.

8. *To quantify a lipid, I need*:
   a. Two analytical systems
   b. The pure molecule
   c. A high-resolution system
9. *Why do I need an internal standard?*
   a. For quantification
   b. To obtain the yield of preparation
   c. For quantification and yield of extraction

7. *Which method to use for which lipid family?*

In this part, we will focus on a few families of lipids (the most studied). Briefly we will describe the family of interest and present how to analyse them.

* *Free and total fatty acid:* a fatty acid (FA) is a carboxylic acid with an aliphatic chain, which can be either saturated or unsaturated (mono or poly-unsaturated). Conventional FA are analysed by gas chromatography (GC) coupled with a flame ionisation detector (FID) or mass spectrometry. If you want to analyse free fatty acids, the FA will be methylated for a few minutes at room temperature. If you wish to analyse the total FA pool (free and esterified), a hydrolysis step will be added before methylation. The methylated FA will then be extracted with heptane, concentrated and injected on the GC. A simple ratio between the peak area of measured FA species and an internal standard will be sufficient to obtain a qualitative profile with relative quantification between different conditions. To achieve absolute quantification calibration curves will need to be analysed. The fatty acid profiling by GC is fully described in part "step-by-step" protocols.

The following protocol allows the relative quantitative analysis of conventional fatty acids (FA): C14:0, C15:0, C16:0, C17:0, C18:0, C20:0, C21:0, C22:0, C23:0, C24:0, C14:1w7, C15:1w7, C16:1w7, C16:1w9, C18:1w9 (cis/trans), C18:1w7, C20:1w9, C22:1w9, C24:1w9, C16:2w6, C16:2w4, C16:4w3, C17:1, C22:1w9, C24:1 C18: 2w6 (cis/trans), C18:3w6, C18:3w3, C18:3w4, C20:2w6, C20:3w3, C20:3w6, C20: 4w6, C20:5w3, C22:2w6, C22:3w6, C22:4w6, C22:5w3 and C22:6w3 with internal calibration (internal standard C19:0). C19:0 fatty acid is chosen as an internal standard because it is not naturally present in the matrices studied.
   Sample preparation
   Reagents needed:

* Dichloromethane (CH_2CL_2)
* Methanol (MeOH)

(continued)

- Acetic acid
- Milli-Q water
- Boron trifluoride/methanol (14% BF_3/MeOH)
- Heptane
- Potassium hydroxide (0.5 M KOH in methanol)
- Ethyl acetate EGTA (ethylene glycol-bis(β-aminoethyl ether)-N,N,N',N'--tetraacetic acid) 0.5M in solution in water

  Equipment needed

- Centrifuge
- 2 mL glass tubes
- Pipettes and pipette tips
- Pasteur pipette
- Glass vial, glass insert

1.A Tissue samples	*1.B Fluid samples—plasma example*
• Crush samples in 1 ml of MeOH/ H_2O .5 mM EGTA (2:1)	• Transfer 10 µl of plasma in a glass tube
• Transfer the equivalent of 1 mg of tissue in a glass tube	
• Add 2.5 ml of MeOH (2% acetic acid), 2.5 ml of CH_2Cl_2 and 2 ml of H_2O	
• Add the internal standard 2 µg (C19:0 for free fatty acid, TG19:0 for total fatty acid)	
• Mix vigorously	
• Centrifuge at 2500 rpm for 6 min	
• Collect the lower phase and dry it under nitrogen gaz	

2.A Free fatty acids	*2.B Total fatty acids*
• Add 1 ml of BF_3/MeOH, mix and wait 5 min at room temperature	• Add 1 ml of KOH diluted in MeOH (0.5 M), mix and heat 30 min at 55 °C
• Add 1 ml of H_2O and 2 ml of heptane	• Add 1 ml of BF_3/MeOH and 1 ml of heptane
• Mix and centrifuge at 2500 rpm for 6 min	• Mix and heat 1 h at 80 °C
• Collect the upper phase and dry it under a gentle nitrogen gas stream	• Add 1 ml of H_2O and 2 ml of heptane
• Recover the methylated FA extract with 2×80 µL of ethyl acetate and put it in a glass insert	• Mix and centrifuge at 2500 rpm for 6 min
• Dry it under a nitrogen gas stream	• Collect the upper phase and dry it under nitrogen gas
• Solubilise the dry extract in 10 µl of ethyl acetate	• Recover the methylated FA extract with 2×80 µL of ethyl acetate and put it in a glass insert
	• Dry it under nitrogen gas stream
	• Solubilise the dry extract in 10 µl of ethyl acetate

(continued)

Gas chromatographic analysis: fatty acid separation is conducted on a FAMEWAX column (30 m × 0.32 mm, 0.25 µm). Hydrogen is used as the carrier gas at a flow rate of 1.5 mL/min. The injector and detector temperatures are set at 220 °C and 230 °C, respectively. A volume of 1 µL of standard or sample is injected. The separation has been optimised using the Supelco 37-component FAME mix (Merck), with additional methylated fatty acids included. The oven temperature is initially set at 120 °C for 1 min. It is then increased to 140 °C at a rate of 5 °C/min, followed by a further increase to 180 °C at 3 °C/min for 7 min. Subsequently, the temperature is raised to 195 °C at 2 °C/min and then to 240 °C at 5 °C/min for 3 min. To generate the ionisation flame, hydrogen and air are used at flow rates of 35 mL/min and 350 mL/min, respectively [32].

Data treatment: The manufacturer's software is utilised to integrate each peak of interest and obtain the corresponding area. Fatty acids are quantified using relative quantification, which involves determining the area ratio between the metabolite and the internal standard (ISTD). The obtained values are then normalised using tissue weight, fluid volume, cell number or protein weight.

- *Poly-unsaturated fatty acid (PUFA) metabolites*: these are oxidised compounds derived from the enzymatic oxidation of PUFA or from lipid peroxidation. They are commonly referred to as oxylipins. A wide range of structures are described in the literature, depending on the fatty acid and the type of oxidation. Oxylipins are rather polar lipids, making solid-phase extraction (SPE) the preferred method for extracting these lipids from biological matrices. Nonenzymatically generated oxylipins are typically associated with phospholipids, although an acidic liquid-liquid extraction (LLE) followed by alkaline methanolysis precedes the SPE step. The challenge in quantifying such lipids in diverse biological samples lies in their low quantities and the structural similarities between them. The optimal approach is to utilise liquid chromatography coupled with triple quadrupole mass analysers. Chromatographic separation is usually achieved using a reverse-phase C18 column. To maximise sensitivity, mass spectrometric acquisitions are performed using multiple reaction monitoring (MRM). Finally, commercially available oxylipin standards allow to propose absolute quantification methods using calibration curves [21]. Few authors also employ supercritical fluid chromatography [33] or ion mobility separation improving the separation of isomers [34].
- *Sterols and oxysterols*: The basic structure of sterols is formed by four aromatic rings derived from sterane, with a hydroxyl group substituted at position 3 for cholesterol. This position can be esterified by a fatty acid to form cholesterol esters. In addition to position 3, hydroxyl, ketone or epoxide groups can be present at various positions in oxysterols. Furthermore, an acidic group can be found on the alkyl chain at position 17 to form bile acids. This group of lipids is large and complex, with diverse biophysical properties that require the use of different techniques for analysis. In brief, neutral

sterols such as cholesterol and oxysterols are analysed using gas chromatography (GC), while cholesterol esters and bile acids are primarily analysed using liquid chromatography (LC). Sterols and oxysterols are extracted using liquid-liquid extraction (LLE). To obtain a profile of total sterols, a saponification step is necessary, and a concentration step using solid-phase extraction (SPE) may be required for low-abundance species such as oxysterols. The hydroxyl groups of compounds are then derivatised with trimethylsilyl or tert-butyl dimethylsilyl groups and analysed using gas chromatography-mass spectrometry (GC-MS) on a nonpolar 30 m column, such as 100% dimethyl polysiloxane or 5% phenyl [22]. The extraction of bile acids is relatively simple, involving protein precipitation followed by concentration using SPE on a polymeric phase and analysis on a nonpolar C18 column using liquid chromatography-tandem mass spectrometry (LC-MS/MS) [35]. For these three subclasses of sterols, pure compounds are available enabling absolute quantification through calibration curves. And finally cholesterol ester is extracted using a classical LLE and analysed by LC-MS/MS on polar (HILIC) or reversed phase columns (C8). However, due to absence of pure standards, absolute quantification may pose a challenge.

- *Phospholipids and sphingolipids*: The main phospholipids harbour a common glycerol skeleton with different polar head groups: choline (PC), inositol (PI), ethanolamine (PE) and serine (PS). Sphingolipids have a sphingoid skeleton linked to a fatty acid through an amide bond. The remaining hydroxyl can be unsubstituted, forming ceramides, or linked to choline (sphingomyelin), to one sugar (galactosyl or glycosyl ceramide), two sugars (lactosylceramides) or a several sugars (ganglioside). The sample preparation is quite simple with a conventional LLE might however need adaptation particularly for gangliosides as the addition of sugar molecules starts to shift the physicochemical properties. Different ways can be considered to analyse phospholipids and sphingolipids depending on the specificity required and the level of sensitivity needed. Separation can be performed per family using polar columns (HILIC) or chain length using a reversed-phase column (C8 or C18). Then, the analyses can be carried out by low- or high-resolution mass spectrometry corresponding to semi-targeted methods or global profiling of phospholipids and sphingolipids. Major phospholipids and sphingolipids such as phosphatidylcholine and ceramides have a well-known fragmentation pattern in positive and negative ionisation mode. Thus, it is possible to develop semi-targeted methods using neutral loss scanning (e.g. for PE) or precursor ion scanning (e.g. for ceramides, SM, PC). By scanning the daughter ions or the neutral loss, we can trace the mass of the parent ion and thus characterise the lipids present in the biological samples wich enables MRM transitions to be programmed. This type of analysis combines well HILIC separation [36]. In fact, as lipid families are separated according to their polar head group, all lipids belonging to the same family are gathered under a single chromatographic peak. Thus, structures of the lipids are confirmed by the precise retention time and the fragmentation pattern of the lipid family. Nevertheless, due to the lack of pure standards only relative quantificaion can be proposed.

The HILIC protocol to profile phospholipids (PE, PC, PS, PI) and sphingolipids (ceramides and sphingomyelin) using a targeted approach is fully described in part "step-by-step" protocols.

The following protocol allows the relative quantification of the several species (with the number of carbons and double bonds) of the following subclasses of phospholipids (PL): phosphatidylethanolamine (PE), phosphatidylcholine (PC), phosphatidylserine (PS) and phosphatidylinositol (PI) and sphingolipides (SL): sphingomyeline (SM d18:1) and ceramide (Cer d18:1).

Sample preparation

Reagents needed:

- Dichloromethane (CH$_2$CL$_2$)
- Methanol (MeOH)
- Acetic acid
- Milli-Q water EGTA (ethylene glycol-bis(β-aminoethyl ether)-N,N,N′,N′--tetraacetic acid) 0.5M in solution in water

Equipment needed:

- Centrifuge
- 2 ml glass tubes
- Pipettes and pipette tips
- Pasteur pipette
- Glass vial, glass insert

Tissue samples	Fluid samples—plasma example
• Crush samples in 1 ml of MeOH/H$_2$O.5 mM EGTA (2:1).	• Transfer 10 µl of plasma in a glass tube
• Transfer 1 mg of sample in a glass tube	• Add 2.5 ml of MeOH (2% acetic acid), 2.5 ml of CH$_2$Cl$_2$ and 2 ml of H$_2$O
• Add 2.5 ml of MeOH (2% acetic acid), 2.5 ml of CH$_2$Cl$_2$ and 2 ml of H$_2$O	• Add the internal standard (one per family studied)
• Add the internal standard (one per family studied)	• Mix vigorously
• Mix vigorously	• Centrifuge at 2500 rpm for 6 min
• Centrifuge at 2500 rpm for 6 min	• Collect the lower phase and dry it under nitrogen stream
• Collect the lower phase and dry it under nitrogen stream	• Recover FFA with 2 × 80 µL of MeOH and put it in a glass insert
• Recover lipid extract with 2 × 80 µl of MeOH and put it in a glass insert	• Dry it under nitrogen stream
• Dry it under nitrogen stream	• Solubilise the dry extract in 50 µl of MeOH
• Solubilise the dry extract in 50 µl of MeOH	

(continued)

Liquid chromatography analysis: sample solutions are analysed using a UPLC system coupled to a triple quadrupole mass spectrometer, using the manufacturer's software for data acquisition and analysis. A Kinetex HILIC column (Phenomenex, 50×4.6 mm, 2.6 µm) is used for liquid chromatography. The column temperature is controlled at 40 °C. The mobile phase A is acetonitrile; and B is 10 mM ammonium formate in water at pH 3.2. The gradient is as follows: from 10% to 30% B in 10 min; 10–12 min, 100% B; and then back to 10% B at 13 min for 1 min re-equilibration prior to the next injection. The flow rate is 0.3 ml/min, and the injection volume is 5 µl. An electrospray source is employed in positive (for Cer, PE, PC and SM analysis) and negative ion mode (for PI and PS analysis). Source conditions need to be optimised; for an Agilent 6460, the collision gas is nitrogen, and the needle voltage is set at $+/-$ 4000 V.

Mass spectrometry detection: several scan modes are used. First, we analyse cellular lipid extracts with a precursor ion scan of *m/z* 184, *m/z* 241 and *m/z* 264 for PC/SM, PI and Cer, respectively, and a neutral loss scan of 141 and 87 to monitor PE and PS, respectively. The collision energy optimums for Cer, PE, PC, SM, PI and PS are 25 eV, 20 eV, 30 eV, 25 eV, 45 eV and 22 eV, respectively, on an Agilent 6400 system. Subsequently corresponding MRM transitions are used in order to quantify different PL species for each class. For details please refer to [37].

Data treatment: peaks of interest are integrated using the manufacturer's software to obtain areas. Lipids are quantified using relative quantification (area ratio between the metabolite and the ISTD).

Global profiling of major phospholipids and sphingolipids are usually developed using high-resolution mass spectrometry [38]. MS methods developed are in DDA (data dependant analysis) or DIA (data-independent analysis) mode corresponding to an alternation between MS and MS/MS scans. The characterisation of the lipids is based on the accuracy of mass observed, the fragmentation spectra and the comparison with online or homemade databases. All kinds of separation can be used for global profiling, but separation in reverse-phase mode is more commonly used. Global profiling generally allows the identification of major lipids; thus, to increase the sensitivity of the analysis of minor lipids, a specific sample preparation can be performed. For sphingolipids, for example, two steps are involved in the process, which can include alkaline methanolysis and re-extraction after the initial liquid-liquid extraction. These steps effectively remove glycerolipids and glycerophospholipids, greatly simplifying the lipid extract. Subsequently, a specific elution gradient is developed to separate the desired sphingolipids.

Finally, whatever the method of analysis, semi-targeted, targeted or global profiling, it is important to note that absolute quantification is not possible. In fact, as only a few phospholipids and sphingolipids are commercially available, we cannot use calibration curves. Thus, relative quantification is performed by comparison between the area of the metabolite and the internal standard.

Take-Home Message
- It is crucial to consider the biological question before initiating lipidomics experiments.
- Lipids present a large panel of structures; it is impossible to analyse the lipidome in one shot.
- The sample preparation needs to be adapted to the lipids of interest.
- Absolute quantification is not always feasible.
- The choice between gas or liquid chromatographic systems depends on the lipid family under study.
- The mass detector must be adjusted based on the targeted lipids and their quantity.

1-c/d: If we take the 40 known fatty acids, we can calculate already 180,000 different species [2]; more than 25,000 species have been described on the lipid map website. Lipids are discovered and published every week, so we can easily imagine that more than 100,000 species exist.

2-a/b: Lipids are ubiquitous and have a role in energy storage, membrane structure and signalling.

3-b/c: It is necessary to normalise lipidomic data; for tissue we usually use the quantity of tissue or the quantity of protein.

4-b: The internal standard must not be naturally present but must still represent the family to be studied. So choosing a C19 is the good option (it is not present naturally in mammalian samples), but as most of the fatty acids measured in total fatty acids stem from triacylglyceride (TG) and phospholipids, it is better to take a C19 esterified in a TG.

5-c: It is always much safer to do two controls in a sample series: one tube with the internal standard only and one tube without anything. These two tubes must undergo the whole sample preparation process; this allows checking for contamination during extraction.

6-c: Due to the large quantity of lipid species in global profiling, it is necessary to use a high-resolution mass spectrometer (QTOF, orbitrap or FTICR) to obtain a precise m/z to reduce the chance a "mis-annotation".

7-b: To annotate a metabolite or a lipid at level 1, it is necessary to have the pure metabolite and its retention time and MS/MS spectrum [29].

(continued)

8-b: It is necessary to have the pure compound of interest to perform absolute quantification; indeed a mass spectrometer is not a reliable quantitative tool, and it requires the use of calibration curves for accurate quantification.

9-c: Internal standards are important: first to calculate the yields of extraction of the lipid family of interest and second to quantify metabolites, whether relative or absolute.

4.3　Further Reading

- Reviews:
  - Recommendations for good practice in MS-based lipidomics [39]
  - Contemporary lipidomic analytics: opportunities and pitfalls [40]
  - Lipidomes in health and disease: analytical strategies and considerations [41]
- Interesting web sites:
  - https://lipidmaps.org/
  - https://lipidomicstandards.org/about/
  - https://lipidomicssociety.org/

References

1. Fahy E, Subramaniam S, Murphy RC, Nishijima M, Raetz CRH, Shimizu T, et al. Update of the LIPID MAPS comprehensive classification system for lipids. J Lipid Res. 2009;50:S9–14.
2. Yetukuri L, Ekroos K, Vidal-Puig A, Orešič M. Informatics and computational strategies for the study of lipids. Mol BioSyst. 2008;4(2):121–7.
3. Burla B, Arita M, Arita M, Bendt AK, Cazenave-Gassiot A, Dennis EA, et al. MS-based lipidomics of human blood plasma: a community-initiated position paper to develop accepted guidelines. J Lipid Res. 2018;59(10):2001–17.
4. Antonny B, Vanni S, Shindou H, Ferreira T. From zero to six double bonds: phospholipid unsaturation and organelle function. Trends Cell Biol. 2015;25(7):427–36.
5. Kihara A. Synthesis and degradation pathways, functions, and pathology of ceramides and epidermal acylceramides. Prog Lipid Res. 2016;63:50–69.
6. Gaspard GJ, McMaster CR. Cardiolipin metabolism and its causal role in the etiology of the inherited cardiomyopathy Barth syndrome. Chem Phys Lipids. 2015;193:1–10.
7. Bissig C, Gruenberg J. Lipid sorting and multivesicular endosome biogenesis. Cold Spring Harb Perspect Biol. 2013;5(10):a016816.
8. Harayama T, Riezman H. Understanding the diversity of membrane lipid composition. Nat Rev Mol Cell Biol. 2018;19(5):281–96.
9. De Craene JO, Bertazzi D, Bär S, Friant S. Phosphoinositides, major actors in membrane trafficking and lipid signaling pathways. Int J Mol Sci. 2017;18(3):634.
10. Ko CW, Qu J, Black DD, Tso P. Regulation of intestinal lipid metabolism: current concepts and relevance to disease. Nat Rev Gastroenterol Hepatol. 2020;17(3):169–83.

11. Huang DQ, El-Serag HB, Loomba R. Global epidemiology of NAFLD-related HCC: trends, predictions, risk factors and prevention. Nat Rev Gastroenterol Hepatol [Internet]. 2020 [cité 23 déc 2020]; Disponible sur: http://www.nature.com/articles/s41575-020-00381-6

12. Heeren J, Scheja L. Brown adipose tissue and lipid metabolism. Curr Opin Lipidol. 2018;29(3): 180–5.

13. Wymann MP, Schneiter R. Lipid signalling in disease. Nat Rev Mol Cell Biol. 2008;9(2):162–76.

14. FitzGerald GA. Cardiovascular pharmacology of nonselective nonsteroidal anti-inflammatory drugs and coxibs: clinical considerations. Am J Cardiol. 2002;89(6):26–32.

15. Hannun YA, Obeid LM. Principles of bioactive lipid signalling: lessons from sphingolipids. Nat Rev Mol Cell Biol. 2008;9(2):139–50.

16. Chawla A. Nuclear receptors and lipid physiology: opening the X-files. Science. 2001;294(5548): 1866–70.

17. Bligh EG, Dyer WJ. A rapid method of total lipid extraction and purification. Can J Biochem Physiol. 1959;37(8):911–7.

18. Folch J, Ascoli I, Lees M, Meath JA, LeBARON N. Preparation of lipide extracts from brain tissue. J Biol Chem. 1951;191(2):833–41.

19. Matyash V, Liebisch G, Kurzchalia TV, Shevchenko A, Schwudke D. Lipid extraction by methyl- *tert* -butyl ether for high-throughput lipidomics. J Lipid Res. 2008;49(5):1137–46.

20. Löfgren L, Forsberg GB, Ståhlman M. The BUME method: a new rapid and simple chloroform-free method for total lipid extraction of animal tissue. Sci Rep. 2016;6(1):27688.

21. Le Faouder P, Baillif V, Spreadbury I, Motta JP, Rousset P, Chêne G, et al. LC–MS/MS method for rapid and concomitant quantification of pro-inflammatory and pro-resolving polyunsaturated fatty acid metabolites. J Chromatogr B. 2013;932:123–33.

22. Riols F, Bertrand-Michel J. Analysis of oxysterols. In: Giera M, éditeur. Clinical Metabolomics [Internet]. New York: Springer; 2018 [cité 28 déc 2020]. p. 267–75. (Methods in Molecular Biology; vol. 1730). Disponible sur: https://doi.org/10.1007/978-1-4939-7592-1_19

23. Dupuy A, Le Faouder P, Vigor C, Oger C, Galano JM, Dray C, et al. Simultaneous quantitative profiling of 20 isoprostanoids from omega-3 and omega-6 polyunsaturated fatty acids by LC–MS/MS in various biological samples. Anal Chim Acta. 2016;921:46–58.

24. Griffiths WJ, Abdel-Khalik J, Yutuc E, Morgan AH, Gilmore I, Hearn T, et al. Cholesterolomics: an update. Anal Biochem. 2017;524:56–67.

25. Sobrado LA, Freije-Carrelo L, Moldovan M, Encinar JR, Alonso JIG. Comparison of gas chromatography-combustion-mass spectrometry and gas chromatography-flame ionization detector for the determination of fatty acid methyl esters in biodiesel without specific standards. J Chromatogr A. 2016;1457:134–43.

26. Xia W, Budge SM. GC-MS characterization of hydroxy fatty acids generated from lipid oxidation in vegetable oils. Eur J Lipid Sci Technol. 2018;120(2):1700313.

27. Lee JW, Nishiumi S, Yoshida M, Fukusaki E, Bamba T. Simultaneous profiling of polar lipids by supercritical fluid chromatography/tandem mass spectrometry with methylation. J Chromatogr A Mars. 2013;1279:98–107.

28. Clark J, Anderson KE, Juvin V, Smith TS, Karpe F, Wakelam MJO, et al. Quantification of PtdInsP3 molecular species in cells and tissues by mass spectrometry. Nat Methods Mars. 2011;8 (3):267–72.

29. Sumner LW, Amberg A, Barrett D, Beale MH, Beger R, Daykin CA, et al. Proposed minimum reporting standards for chemical analysis: chemical analysis working group (CAWG) metabolomics standards initiative (MSI). Metabolomics. 2007;3(3):211–21.

30. Khoury S, Canlet C, Lacroix M, Berdeaux O, Jouhet J, Bertrand-Michel J. Quantification of lipids: model, reality, and compromise. Biomol Ther. 2018;8(4):174.

31. McDonald JG, Ejsing CS, Kopczynski D, Holčapek M, Aoki J, Arita M, et al. Introducing the lipidomics minimal reporting checklist. Nat Metab. 2022;4(9):1086–8.
32. Lillington JM, Trafford DJH, Makin HLJ. A rapid and simple method for the esterification of fatty acids and steroid carboxylic acids prior to gas-liquid chromatography. Clin Chim Acta. 1981;111 (1):91–8.
33. Quaranta A, Zöhrer B, Revol-Cavalier J, Benkestock K, Balas L, Oger C, et al. Development of a chiral supercritical fluid chromatography–Tandem mass spectrometry and reversed-phase liquid chromatography–Tandem mass spectrometry platform for the quantitative metabolic profiling of octadecanoid oxylipins. Anal Chem. 2022;94(42):14618–26.
34. Kyle JE, Aly N, Zheng X, Burnum-Johnson KE, Smith RD, Baker ES. Evaluating lipid mediator structural complexity using ion mobility spectrometry combined with mass spectrometry. Bioanalysis. 2018;10(5):279–89.
35. Humbert L, Maubert MA, Wolf C, Duboc H, Mahé M, Farabos D, et al. Bile acid profiling in human biological samples: comparison of extraction procedures and application to normal and cholestatic patients. J Chromatogr B. 2012;899:135–45.
36. Schwalbe-Herrmann M, Willmann J, Leibfritz D. Separation of phospholipid classes by hydrophilic interaction chromatography detected by electrospray ionization mass spectrometry. J Chromatogr A août. 2010;1217(32):5179–83.
37. Chiappini F, Desterke C, Bertrand-Michel J, Guettier C, Le Naour F. Hepatic and serum lipid signatures specific to nonalcoholic steatohepatitis in murine models. Sci Rep. 2016;6(1):31587.
38. Colsch B, Seyer A, Boudah S, Junot C. Lipidomic analysis of cerebrospinal fluid by mass spectrometry–based methods. J Inherit Metab Dis janv. 2015;38(1):53–64.
39. Köfeler HC, Ahrends R, Baker ES, Ekroos K, Han X, Hoffmann N, et al. Recommendations for good practice in MS-based lipidomics. J Lipid Res. 2021;62:100138.
40. Giles C, Takechi R, Lam V, Dhaliwal SS, Mamo JCL. Contemporary lipidomic analytics: opportunities and pitfalls. Prog Lipid Res. 2018;71:86–100.
41. Wei F, Lamichhane S, Orešič M, Hyötyläinen T. Lipidomes in health and disease: analytical strategies and considerations. TrAC Trends Anal Chem. 2019;120:115664.

## Part III

From Data Processing to Polar and Lipid Metabolite Identification

METLIN Tandem Mass Spectrometry and Neutral Loss Databases for the Identification of Microbial Natural Products and Other Chemical Entities

5

Wilasinee Heim, Aries Aisporna, Linh Hoang, H. Paul Benton, and Gary Siuzdak

---

What Will You Learn in This Chapter?
- METLIN database searching
- METLIN liquid chromatography tandem mass spectrometry (LC/MS^2) chemical entity identification
- METLIN neutral loss (NL) chemical entity characterization

5.1 Theoretical Background

In this section the principles of the METLIN tandem mass spectrometry (MS/MS or MS^2) and neutral loss (NL) databases will be outlined.

5.1.1 METLIN-MS^2

The 930 K METLIN tandem mass spectrometry (MS^2) database originated two decades ago [1] to facilitate the identification of known molecules as well as uncharacterized chemical entities. Since 2003 METLIN has expanded into a technology platform that is orders of magnitude larger than other databases (Fig. 5.1). The value of MS^2 data is

---

W. Heim · A. Aisporna · L. Hoang · H. P. Benton · G. Siuzdak (✉)
Scripps Center of Metabolomics and Mass Spectrometry, La Jolla, CA, USA
e-mail: siuzdak@scripps.edu

demonstrated in Fig. 5.1 where the redundancy of hits from "identifications" based solely on precursor mass is high, even with absolute accuracy (no error to four decimal places). Tandem MS^2 data greatly reduces the number of hits (Fig. 5.1). Through this effort to continuously populate METLIN, it has become a comprehensive resource containing over 930,000 molecular standards including lipids, amino acids, carbohydrates, toxins, small peptides, natural products, as well as molecules in over 350 chemical classes. METLIN's high-resolution MS^2 database has been generated solely from reference standards. The METLIN tandem mass spectrometry database was made publicly available in 2005 [1] as a cloud-based technology platform with a few hundred molecules; at that time no such database existed for identifying metabolites or any other chemical entities. In 2012, METLIN grew to more than 10,000 molecules [2], followed by massive growth in the ensuing decade. The total number of molecular standards analyzed is 1.2 million with a success rate of 77%, resulting in over 930,000 compounds, each with experimental MS^2 data collected in positive and negative ionization modes at four different collision energies (0eV, 10eV, 20eV, and 40eV).

METLIN data is broadly useful across multiple tandem mass spectrometry instrument types (e.g., QTOF, QExactive, Orbitrap, Ion Trap, and QqQ). The data is collected in both positive and negative ionization modes at four different collision energies, providing high-resolution spectra systematically acquired, and curated directly from standards; the database also includes a collection of approximately one thousand stable isotope analogs. Tandem mass spectrometry data complements other types of databases, which have been collected for electron impact (EI) or nuclear magnetic resonance (NMR) instrumentation [3–5]. METLIN does not contain any in silico data; while in silico generation of data was originally investigated as a source of data [6–8], all in silico MS^2 spectra were removed due to large inconsistencies between in silico and experimental data.

Since the introduction of METLIN in the early 2000s, numerous other databases have followed with over 20 different databases currently available [5]. The impact of METLIN and these other databases has essentially brought metabolomics from the fringes to what is now a mainstream technology, offering valuable insight into therapeutic drug discovery, clinical diagnostics, mass spectrometry imaging, pharmacology, food safety, sports medicine, toxicology, forensics, environmental analyses, and microbiology [9–15]. For example, these databases serve to identify metabolites as indicators of a microorganism's activity [11], disease onset [11, 16, 17], and disease progression [18, 19] or as responsive elements to therapeutics [20, 21], providing mechanistic insights into biological systems and extending in some cases, to the prioritization and identification of endogenous metabolites for the modulation of phenotype [22–24]. The increasing ability to obtain and process complex data sets has been pivotal to these achievements through the identification of metabolites and other chemicals represented by these dysregulated features. However, the primary obstacle facing the field has now shifted from identifying molecules with known MS^2 spectra to identifying the unknowns that are not present in the databases or are present, yet do not have experimental MS^2 data. METLIN is being designed to meet this challenge.

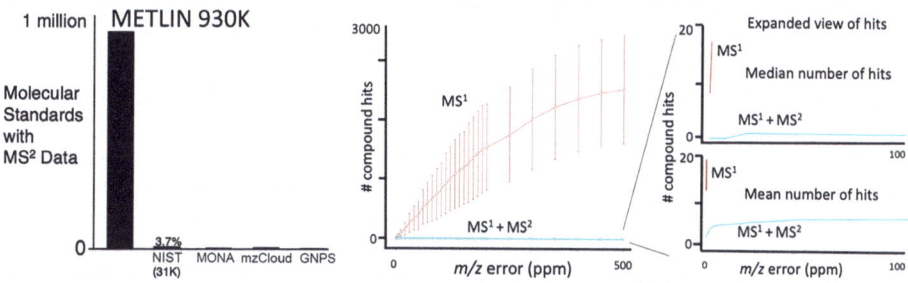

Fig. 5.1 (left) METLIN 930 K molecular standards tandem mass spectrometry database compared to other databases. (middle) The value of combining MS^1 (precursor ion), MS^2, and/or NL data for identification of small molecules. A sampling of METLIN hit rates (middle) as a function of m/z error for MS^1 data alone and combined MS^1 and MS^2 data. The median and mean number of compound hits as a function of mass error created from the METLIN mass spectral library. The plot in red represents the number of hits based on precursor molecular weight as a function of error in part per million (ppm). The plot in blue represents the median and mean number of hits generated as a function of error, precursor molecular weight, and tandem mass spectral fragments. An expanded view (right) of the hits generated from MS^1 precursor ion data and MS^2 fragmentation data. As anticipated, the hit rates are the same using MS^1 precursor ion data combined with NL data

5.1.2   METLIN-NL

METLIN-NL spectral data [25] serves as a mirror of MS^2 data (Fig. 5.2) and represents a valuable yet underutilized resource for molecular discovery and similarity analysis. While MS^2 data is effective in identifying known molecules and proposing the identity of novel, uncharacterized molecules (unknowns), it has limitations in characterizing structurally related compounds. To address this, we have developed an MS^2 to NL converter as an integral part of the METLIN platform.

The primary aim of the converter is to enhance unknown identification and complement the METLIN-MS2 fragment ion database by enabling the characterization of structurally related molecules through NL spectral data. In our efforts, we successfully transformed METLIN's MS^2 data into METLIN-NL for over 930,000 individual molecular standards. The addition of NL spectral data offers a unique dimension for the characterization of chemical and metabolite structures, providing valuable insights into complex molecular relationships. This integration significantly expands the capabilities of METLIN, empowering researchers in their quest for comprehensive and in-depth molecular analyses.

5.1.3   Acquisition of a Tandem Mass Spectrum

Tandem mass spectrometry data acquisition involves the ability of the analyzer to separate different molecular ions, generates fragment ions from a selected ion, and then precisely

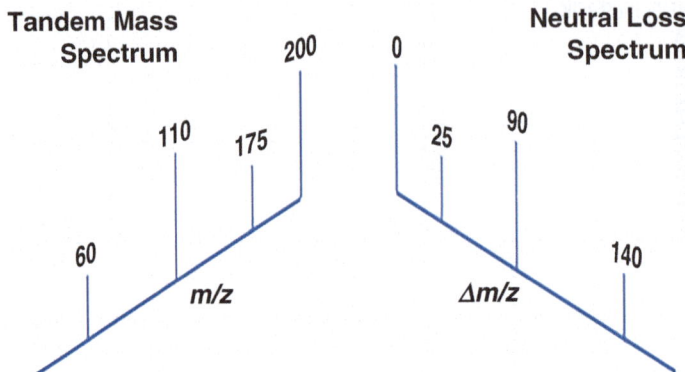

Fig. 5.2 An example of tandem mass spectral data (frag_intensity vs. m/z) and neutral loss spectra (NL_intensity vs. $\Delta m/z$) of the same molecule

measures the masses of the resulting fragmented ions. These fragment ions play a crucial role in determining the structural characteristics of the original molecular ions.

Typically, tandem MS experiments (Fig. 5.3) are conducted by colliding a selected ion with inert gas molecules like argon or helium, leading to the production of fragments that are subsequently analyzed based on their mass. Tandem mass analysis is widely used for characterizing small molecules, sequencing peptides, and determining the structures of carbohydrates, small oligonucleotides, and lipids.

The term "tandem" mass analysis arises from the concept that the events occur either in tandem in space or tandem in time. Tandem mass analysis in space involves consecutive analyzers that sequentially perform fragmentation and mass analysis. On the other hand, tandem mass analysis in time occurs within the same analyzer, where the ion of interest is isolated, fragmented, and then analyzed for the masses of the resulting fragment ions.

5.1.4 Acquisition of Neutral Loss Spectrum

The METLIN NL converter serves as a versatile resource, designed to convert METLIN's extensive MS^2 small molecule molecular standards database (over 930,000 compounds) into a mirrored NL database (METLIN-NL). This conversion enables convenient NL searching. The NL data is derived from a diverse range of standards representing various chemical classes.

The converter takes METLIN's MS^2 data as input and calculates the differences between the precursor molecular ion and the fragment ions in the experimental MS^2 mass spectra. Using this information, it generates METLIN-NL spectra (Fig. 5.4), where NL intensity (NL_intensity) is plotted against the $\Delta m/z$ values of the fragment ions.

The MS^2 to NL converter is readily accessible on METLIN's platform and GitHub, allowing users to view individual MS^2 or NL spectra and conduct comparisons between

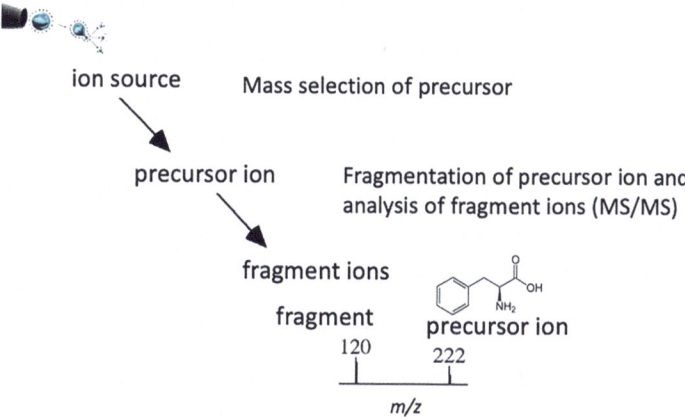

Fig. 5.3 Tandem mass spectrometry analysis where ions are generated in the ion source, precursor ion isolated in the first mass analyzer; precursor ions are then fragmented in the collision cell and the fragments analyzed with the second mass analyzer

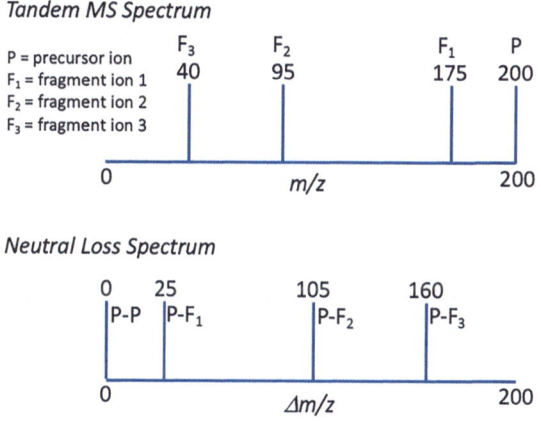

Fig. 5.4 The METLIN-NL mass spectral database was derived from the METLIN-MS^2 data on over 930,000 molecular standards. A representative METLIN-MS^2 spectrum (Fragment$_{intensity}$ vs. m/z) and a METLIN-NL spectrum (NL$_{intensity}$ vs. $\Delta m/z$) generated by calculating the difference between the precursor and fragment ions with NL$_{intensity}$ based on the original fragment ion intensities. "P" refers to precursor ion and "F" refers to fragment ion

two MS^2 or NL data sets. When using METLIN IDs, the MS^2 and NL data are already available, but when employing CSV files (Table 5.1), the MS^2 data is automatically converted to NL.

To facilitate these analyses, METLIN-NL is built on a Linux platform, and its initial graphical user interface (GUI) is created using Highcharts, HTML, JQuery, MySQL, and

Table 5.1 The MS^2 to NL converter operates in the following modes and allows users to create/compare the following data types

Input #1	Input #2	Graph type
METLIN ID	–	Shows MS^2 and NL spectra
METLIN ID	METLIN ID	Compares MS^2 and NL spectra
CSV	–	Shows MS^2 and NL spectra
CSV	CSV	Compares MS^2 and NL spectra
METLIN ID	CSV	Compares MS^2 and NL spectra
CSV	METLIN ID	Compares MS^2 and NL spectra

PHP. The GUI enables comparative analyses between different compounds, presenting NL data (NL_{int} vs. $\Delta m/z$) and MS^2 data ($Frag_{int}$ vs. m/z) in both positive and negative ionization modes. It also allows visualization either at each individual collision energy or as a "composite spectrum" that combines all spectra from multiple collision energies.

Once a spectrum or spectra are generated, users can hover over each peak to access detailed information about m/z, intensity, ionization mode, compound's name, and collisional energy values. To use the website (https://metlin-nl.scripps.edu/), users input a CSV file containing compound names, masses with intensities, collision energy, positive/negative mode, and precursor value. The website provides two downloadable CSV files as examples for proper formatting.

In summary, the METLIN NL converter and METLIN-NL database offer a toolset for researchers to explore and compare MS^2 and NL data, facilitating efficient molecular analysis and supporting a wide range of metabolomics studies.

5.1.5 Identifying Microbial Metabolites

Tandem mass spectrometry data represents a valuable resource for identifying molecules within complex mixtures (Figs. 5.3, 5.4 and 5.5). In a study exploring the impact of the "microbiome" on mammalian blood metabolites, researchers employed MS-based methods to compare plasma extracts from axenic ("germ-free") mice with samples from wild-type animals [27]. The results revealed hundreds of unique features in each sample set, with notable effects on amino acid metabolites.

One intriguing finding was the bacterial-mediated production of bioactive indole-containing metabolites, including the antioxidant 3-indole propionic acid (3-IPA), which exhibited a complete dependence on the presence of gut microflora. Remarkably, this dependency could be established through recolonization with a single species of *Clostridium sporogenes*.

The study showcased the utility of METLIN as a comprehensive database for analyzing tandem mass spectrometry data and identifying key chemical entities, particularly those associated with gut microbial metabolites. This demonstrated the potential of MS-based

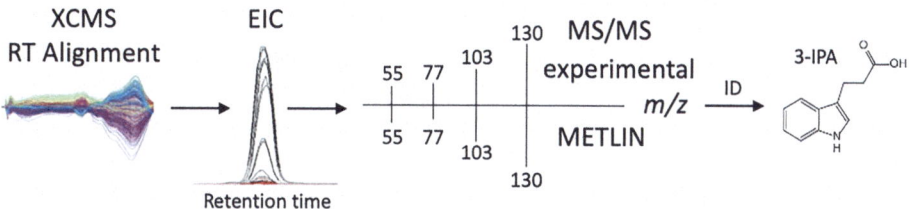

Fig. 5.5 The experimental workflow using LC/MS^2, XCMS [26] (eXtensible Computational Mass Spectrometry), and METLIN in the metabolomics studies, where MS^2 data is used to identify metabolites, here 3-indole propionic acid

methods to shed light on the intricate interactions between the microbiome and mammalian metabolism, offering valuable insights into the complexities of host-microbe relationships.

5.2   Research Question

Use METLIN-MS^2 and METLIN-NL databases to identify microbial natural products, metabolites, and other chemical entities from a complex biological matrix from liquid chromatography tandem mass spectrometry (LC/MS^2) data.

5.3   Hypothesis and Experimental Setup

The adult human intestinal tract hosts an extensive population of commensal bacteria, comprising trillions of microbes collectively known as the "gut microbiome" [28]. This intricate microbial community establishes a mutualistic relationship with its human host, influencing various aspects, including the regulation of the immune system [29, 30], and playing a role in diseases such as inflammatory bowel disease [31, 32]. The diversity of the microbiome can be significantly impacted by factors like changes in diet, antibiotic usage, and probiotics, which involve the consumption of live cultures of beneficial bacteria to modify the existing gut microbial environment [33].

Traditionally, the study of microbial ecology in the gut involved isolating and culturing individual bacterial species associated with this environment [34]. However, more recent advances in metagenomic techniques have enabled researchers to characterize both the composition and potential physiological effects of entire microbial communities without the need to culture each community member individually. These approaches have provided valuable insights into the gut microbiome's functional capabilities.

For instance, metagenomic analyses have revealed the overrepresentation of genes related to specific metabolic pathways, such as amino acid and glycan metabolism, in the distal gut microbiome. This finding reinforces the concept that human metabolism is influenced by a combination of microbial and human attributes [35]. Additionally, studies involving obese mice have demonstrated how alterations in the composition of gut microflora can affect energy extraction from food, leading to increased efficiency compared to their lean counterparts, primarily through a higher complement of genes for polysaccharide metabolism [36]. Another important discovery is the enrichment of bile salt hydrolase encoding genes in the gut microbiome, which results in a wide range of bile acid modifications by gut microbiota, influencing various physiological processes [37].

These metagenomic studies collectively suggest that the metabolites derived from the diverse gut microbial community can play significant roles in human health and disease. Understanding the intricate interactions between the gut microbiome and its human host opens up new avenues for exploring potential therapeutic interventions and dietary strategies to promote better health and well-being.

In this study, we employed XCMS (eXtensible Computational Mass Spectrometry) to statistically identify metabolites specific to the gut microbiome. Subsequently, we utilized METLIN (Fig. 5.5) to analyze the MS^2 data and confidently identify these gut microbiome-derived metabolites.

Our investigation focused on serum samples obtained from two groups of mice: axenic ("germ-free," GF) and conventional ("wild-type," WT) mice. Through mass spectrometry-based profiling, we made a striking discovery—a surprisingly large number of molecules found in the systemic circulation were products of the gut microbiome. Moreover, we found that at least 10% of all detectable endogenous circulating serum metabolites exhibited significant concentration differences of at least 50% between the two mouse lines.

During our analysis, we successfully identified several microbiome-generated molecules present in the serum of WT mice. Among these molecules, some were classified as potentially harmful, such as uremic toxins, while others showed potential beneficial effects, acting as antioxidants that could positively influence the host's health.

Our findings shed light on the extensive impact of the gut microbiome on the systemic circulation and its ability to produce diverse metabolites with both potential advantages and risks for the host. Understanding these microbiome-generated molecules opens up exciting avenues for further research into their roles in health and disease, as well as their therapeutic potential.

5.3.1 Experimental Setup

Experiment 1—Serum from WT and germ-free mice was obtained and analyzed via LC/MS².

Experiment 2—METLIN tandem mass spectrometry database was used to identify metabolites differentially regulated between the two cohorts.

To investigate blood metabolites, a group of ten male Swiss Webster mice aged 8–10 weeks was obtained from Taconic Farms (Germantown, NY). The mice were divided into two sets: wild-type (WT) and germ-free (GF). Both groups were fed autoclaved

NIH-31 chow. Blood was collected by retro-orbital bleed into heparinized tubes, and the serum from each animal was frozen immediately at $-80\,°C$ and shipped on dry ice.

For plasma sample preparation, metabolites were extracted using methanol. Cold methanol was added to plasma samples, followed by incubation at $-20\,°C$ for 1 h. After centrifugation, the supernatant was collected, dried in a SpeedVac, and resuspended in a mixture of water and acetonitrile. The samples were then clarified and prepared for further analysis.

Metabolomics profiling was performed using an HPLC system with specific column dimensions and solvents for positive and negative ion modes. The plasma samples were analyzed alternately between WT and GF mice to reduce systematic error associated with instrumental drift. Data were collected using a TOF (Agilent 6210) operated in full scan mode.

Data analysis and statistics were conducted using XCMS for nonlinear alignment, integration, and peak intensity extraction. Differences between WT and GF plasma metabolites were evaluated using a two-tailed t-test with unequal variance assumption, and statistical plots were generated using Origin software.

For compound identification, LC/MS^2 was conducted using a QTOF (Agilent 6510). Pooled plasma extracts from the same sets of WT and GF mice were used for confirmation. 3-IPA quantitation in mouse serum was performed using an Agilent 6520 Accurate-Mass Q-TOF LC/MS system with specific chromatography conditions.

The study identified numerous metabolites specific to the gut microbiome in the blood, suggesting a significant impact of the microbiome on systemic circulation. Additionally, the concentration of at least 10% of the detectable endogenous circulating serum metabolites differed by at least 50% between WT and GF mice. Several microbiome-generated molecules with potential benefits, such as antioxidants, as well as potentially harmful molecules like uremic toxins, were also identified in the serum of WT mice.

5.3.2   Research Question

1. Can we identify metabolites specifically associated with the gut microbiome?

5.4   Data Analysis and Interpretation

METLIN-MS^2 and METLIN-NL databases were essential in identifying microbial natural products, metabolites, and other chemical entities from a complex biological matrix from liquid chromatography tandem mass spectrometry (LC/MS^2) data and facilitating the biological conclusions (Figs. 5.5, 5.6 and 5.7).

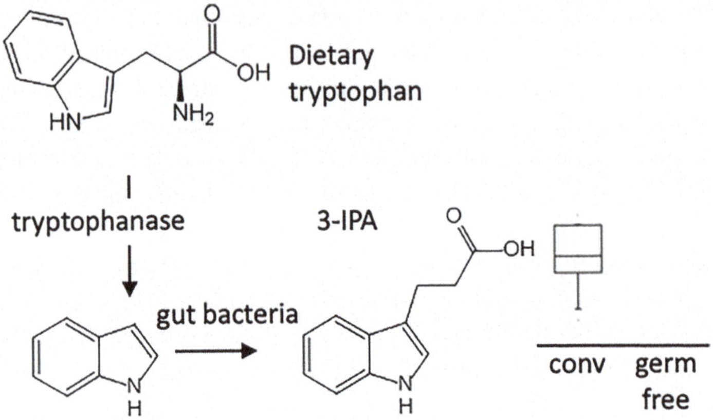

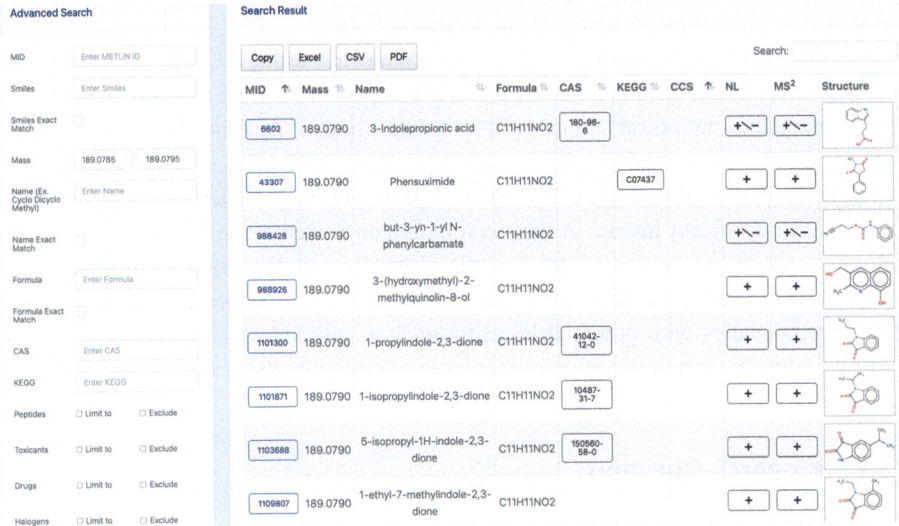

Fig. 5.6 The conversion of dietary tryptophan to indole and then to 3-IPA by the microbiome

Fig. 5.7 An example of an advanced search within METLIN of a molecule within a range of *m/z* 189.0785 to 189.0795 consistent with indole-3-propionic acid; this search primarily produced molecules with an elemental composition of $C_{11}H_{11}NO_2$. Multiple other search options exist as shown on the left panel, including METLIN ID (MID), Smiles, Mass, Name, Formula, CAS number, KEGG number, as well as filters. The right results panel provides the data in multiple formats (Excel, CSV, and PDF) and offers links to the MS^2 and NL data on each molecule. The results panel also provides the MID, Mass, Name, Formula, CAS, KEGG, CCS, NL spectra, MS^2 spectra, and Structure

5.4.1 Mass Spectrometry Reveals the Broad Effect of the Microbiome

The application of mass spectrometry in metabolomics studies has seen significant advancements, primarily due to its broad dynamic range, reliable quantitative capabilities, and its capacity to analyze complex molecular samples. Despite its strengths, the immense chemical diversity within the known metabolome poses numerous challenges for MS-based "global" profiling efforts, as sample preparation, separation, and ionization techniques can introduce analytical biases. In light of these challenges, our objective was to gain an initial understanding of the microbiome's impact on mammalian metabolism by analyzing uniformly processed samples using various ionization methods.

Figure 5.5 depicts the experimental workflow used for the metabolomics studies. Plasma samples from individual WT and germ-free mice were subjected to methanol protein precipitation and subsequent centrifugation, a previously reported effective and reproducible extraction method for such samples. The resulting supernatants underwent analysis by four different methods: reversed-phase liquid chromatography coupled with (1) electrospray ionization (ESI) in the positive mode (+), (2) ESI in the negative mode (−), (3) atmospheric pressure chemical ionization (APCI) in the positive mode, and (4) gas chromatography after derivatization with trimethylsilyl chloride.

To analyze the data, nonlinear data alignment was applied to the raw data from the individual analyses. Multivariate statistics were used to determine group separation and assess global changes, while univariate statistics were employed to evaluate the number and percentage of features that were either unique or exhibited significant variations between the two sample sets.

The principal component analysis (PCA) plot demonstrated excellent separation of the WT and GF sample groups analyzed under + ESI conditions. Similar separations were observed for the - ESI, + APCI, and GC-based analyses. While most individual features detected were present in both WT and GF samples, hundreds of features were exclusive to one sample set, with the majority being unique to the WT samples. Furthermore, summarized in Table 5.2, approximately 10% of all features detected in both sample sets under any analysis conditions exhibited significant changes in their relative signal intensity (defined as a 1.5-fold change with a P-value <0.0001). It's important to note that a given molecule may be represented by several different features, such as naturally occurring isotopic cluster components or nonspecific adduct ions. Nevertheless, these comparative results clearly illustrate the profound impact of the microbiome on mammalian plasma biochemistry.

5.4.2 Metabolite Identification

The most challenging and time-consuming aspect of untargeted MS-based metabolomics profiling studies is identifying features that exhibit significant differences in levels between the two sample sets (Figs. 5.3, 5.4 and 5.5). This process involves integrated analysis, combining accurate mass measurements, tandem MS fragmentation patterns, and extensive literature/database searches to propose candidate structures that still require experimental verification.

Table 5.2 Significant metabolites observed and identified. Compounds designated as WT* were only observed in that group

Metabolite	Fold change	p-value	Compound and metabolism class
Indole derivatives			
Tryptophan	1.7 (GF)	8.42×10^{-12}	Amino acid
N-acetyl tryptophan	2.4 (GF)	3.56×10^{-4}	Acetylated amino acid
Indoxyl sulfate	WT*	1.34×10^{-7}	Hydroxyl sulfate
Serotonin	2.8 (WT)	1.27×10^{-10}	Derived from tryptophan
Indole-3-propionic acid	WT*	7.69×10^{-7}	Bacterial conjugation
Phenyl derivatives			
Tyrosine	1.44 (GF)	1.14×10^{-4}	Amino acid
Hippuric acid	17.4 (GF)	1.98×10^{-9}	Glycine
Phenyl acetyl glycine	3.8 (WT)	4.70×10^{-8}	Glycine
Phenyl sulfate	WT*	9.85×10^{-7}	Sulfate
p-cresol sulfate	WT*	2.00×10^{-3}	Sulfate
Phenylpropionylglycine	WT*	3.07×10^{-7}	Glycine
Cinnamoylglycine	WT*	2.93×10^{-7}	Glycine
Flavones			
Equol sulfate	WT*	1.44×10^{-5}	Sulfate
Methyl equol sulfate	WT*	2.18×10^{-6}	Sulfate
Others			
Urate	1.99 (WT)	1.51×10^{-6}	–
Indole lactic acid – unconfirmed	WT*	3.21×10^{-7}	Bacterial conjugation preliminary ID based on MS^2 and NL
Cholesterol sulfate	WT*		Sulfate
Dihydroxyquinoline glucuronide	WT*	7.64×10^{-6}	Glucuronide
12-hydroxy-5Z, 8Z, 10E, 14Z, 17Z-eicosapentaenoic acid	4.0 (WT)	8.20×10^{-5}	Fatty acid
3-Carboxy-4-methyl-5-pentyl-2-furanpropionic acid glucuronide	3.4 (WT)	1.37×10^{-6}	Glucuronide

In this chapter, we established stringent criteria for feature identification. A feature was considered unequivocally identified only if an authentic sample of the proposed species was independently analyzed in our laboratories, and it exhibited the same chromatographic and mass spectrometric properties as the feature under investigation. For features where an authentic sample was not available, we assigned a highly probable identification status if the experimentally determined properties closely matched those reported in the literature for a particular molecule or were fully consistent with those reported for a closely related analog.

While the identification process remains a work in progress, the results listed in Table 5.2 already showcase several remarkable findings. Despite using a single sample preparation method, we successfully identified a diverse range of chemically distinct compounds, including fatty acids, highly polar charged molecules, and several groups of functionally related molecules. This highlights the complexity and diversity of the metabolites present in the samples and offers valuable insights into the impact of the microbiome on mammalian metabolism.

METLIN can be searched using multiple different parameters and filters as shown in Figs. 5.7 and 5.8. Figure 5.7 shows hits obtained within a specified mass range, and Fig. 5.8 provides the hit when searching the microbial metabolite indole-3-propionic acid. Molecules can be searched based on METLIN ID number (MID), smiles file, mass, name, formula, CAS#, KEGG#, as well as applying filters such as to limit to or exclude peptides, toxicants, drugs, halogenated species, and/or molecules with KEGG IDs.

The primary value of METLIN is in searching MS^2 and NL data as shown in Fig. 5.9. Figure 5.9 represents the identification and similarity analysis part of METLIN where the inputted data (left panel) was searched across all of METLIN's 930 K database. The top hit (creatine as shown in the upper right panel and the lower panel) based on precursor m/z and MS^2 data was significantly higher matching than the other six molecules that had a similar precursor m/z.

5.4.3 The Microbiome Affects the Diversity of Indole-Containing Molecules

Numerous distinct features that exhibited significant differences between WT and germ-free mice were found to correspond to compounds containing indole moieties. For instance, the plasma concentrations of tryptophan and N-acetyl-L-tryptophan in WT mice were 40% and 60% lower, respectively, compared to their respective germ-free counterparts. The decreased levels of tryptophan and N-acetyl-L-tryptophan in WT mouse serum were likely a result of dietary tryptophan metabolism by enteric bacteria that express tryptophanase. Tryptophanase converts tryptophan to indole, pyruvate, and ammonia, and its activity has been shown to increase nearly twofold in WT mice following dosing with tryptophan.

On the other hand, plasma serotonin levels were found to be 2.8-fold higher in WT animals. The increase in serotonin, however, is not directly attributed to a metabolic transformation by gut bacteria, as characterized enteric bacterial species have not been reported to produce serotonin. Serotonin is predominantly produced by Enterochromaffin cells in the gut, making it the primary source of serotonin production in the body. Serotonin

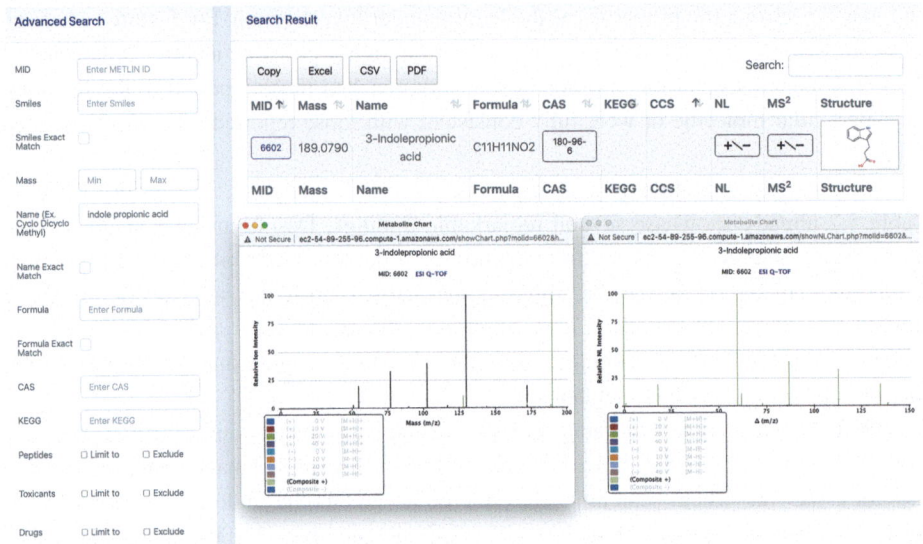

Fig. 5.8 The same Fig. 5.7 example of an advanced search within METLIN of indole-3-propionic acid. This time in the lower right panel showing representative MS^2 and NL composite spectra; however, data at each individual offset voltage representing different collision energies (0 V, 10 V, 20 V, and 40 V) can be shown by clicking on the individual offset voltage

has been associated with gastrointestinal pathologies such as irritable bowel syndrome and Crohn's disease. Therefore, the elevated plasma serotonin levels observed in WT mice may be indirectly linked to an undefined host-microbe interaction.

These findings reveal the complex interplay between the gut microbiome and host metabolism, particularly concerning indole-containing compounds and serotonin levels. The direct influence of enteric bacteria on tryptophan metabolism underscores their role in shaping the host's metabolic profile, while the relationship between gut bacteria and plasma serotonin levels suggests intriguing possibilities for further research on host-microbe interactions and their impact on gastrointestinal health.

The presence of the gut microbiome significantly affected various indole-containing molecules. For instance, indoxyl sulfate (indican), a nephrotoxin that accumulates in the blood of patients with chronic kidney failure, was identified exclusively in the serum of WT mice. This molecule arises from the hepatic transformation of the bacterial metabolite indole. To address this, probiotics containing non-indole-producing bacteria, such as certain Bifidobacterium species, have been tested to reduce plasma levels of indoxyl sulfate in dialysis patients. Conversely, another group of enteric bacteria has been linked to the metabolic conversion of indole to indole-3-propionic acid (IPA). IPA, identified solely in the plasma of WT mice, has been recognized as a potent antioxidant [38].

While previous studies [39, 40] had attributed the presence of IPA in mammals to bacterial metabolic processes, our own investigation was more specific. We found that only *Clostridium sporogenes* produced 3-IPA in culture among representative members of the

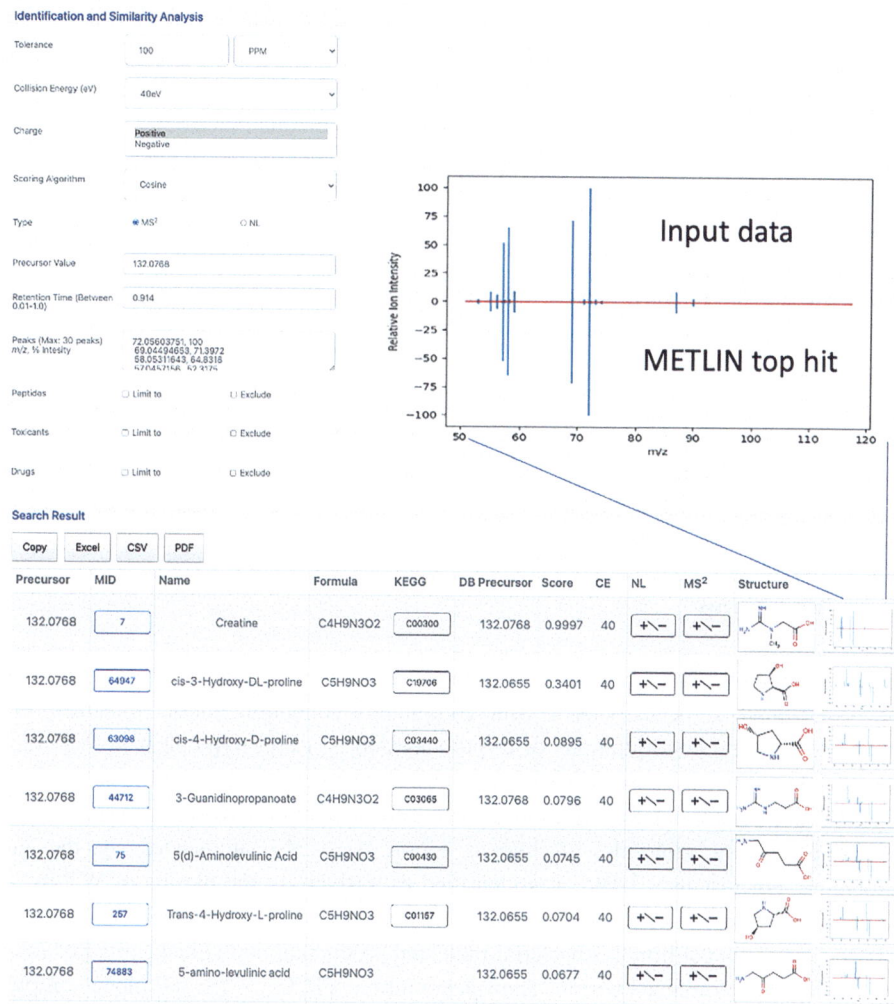

Fig. 5.9 An example of a tandem MS search within METLIN of fragment ions. Inputted data in the top left produced the hit results in the bottom. The top right provides the inputted data in spectral form and the METLIN top hit

intestinal flora. To demonstrate its presence in the bloodstream, we intentionally colonized germ-free mice with *C. sporogenes* strain ATCC 15579. 3-IPA was detectable in the serum 5 days after colonization, and its levels reached those comparable to WT mice by day 10. This experiment confirmed that introducing enteric bacteria capable of IPA production into the gastrointestinal tract was sufficient to introduce IPA into the host's bloodstream.

Furthermore, we injected IPA intraperitoneally into other germ-free animals and monitored its serum concentrations over time. The results showed that IPA was rapidly cleared from the blood, indicating that its presence in the serum of WT animals resulted from continuous production by one or more bacterial species associated with the mammalian gut.

The ability to measure the metabolic profiles of animals selectively colonized with individual bacterial species or complex communities of defined bacterial populations holds great promise as a powerful tool for unraveling the contributions of various species within the microbiome and understanding intricate microbial-mammalian metabolic interactions. This approach opens up new possibilities for research and potential therapeutic applications related to the gut microbiome and its impact on host health.

5.4.4 Indole-3-Propionic Acid Identification

The identification of 3-IPA was facilitated by its MS^2 data; this data has been added as an example in the METLIN Gen2 AWS (Amazon Web Services) cloud technology platform. By clicking on the example button for IPA in the "ID and Similarity Analysis," the "Peaks" including m/z and % intensity will automatically appear. This will demonstrate the top hit and the subsequent similar molecules.

5.5 Discussion

Numerous metabolic features showing significant differences between WT and germ-free mice were found to be associated with compounds containing indole moieties. For instance, the plasma concentrations of tryptophan and N-acetyl tryptophan in WT mice were 40% and 60% lower, respectively, compared to their germ-free counterparts. Interestingly, WT animals exhibited 2.8-fold higher levels of serotonin in their plasma.

Indole-containing molecules in human serum were also influenced by the gut microbiome. For instance, 3-IPA was identified exclusively in the plasma of WT mice. Through culture experiments, it was found that C. sporogenes was responsible for producing IPA. To directly confirm this bacterial-mediated production, individual germ-free mice were intentionally colonized with C. sporogenes strain ATCC 15579, and blood samples were taken at various intervals after colonization. IPA was not detectable shortly after introduction of the microbes but was first observed in the serum after 5 days, reaching levels comparable to those in WT mice by day 10. These recolonization studies provide direct evidence of bacterial-mediated IPA production in vivo.

The ability to analyze the metabolic profiles of animals selectively colonized with specific bacterial species or complex communities of defined populations holds great potential as a powerful tool to untangle the contributions of various species within the microbiome and understand the intricate interactions between microbes and the mammalian host's metabolites. This approach opens up new possibilities for research and potential therapeutic applications related to the gut microbiome and its impact on host health.

METLIN-MS^2 and METLIN-NL [1, 2, 10, 41–45] databases (http://metlin.scripps.edu) were created to address the challenge of identifying metabolites and other chemical entities, with MS^2. To address the limitations of MS^2 data, we have also created a separate NL database within METLIN designed to be used in conjunction with MS^2 data. An example of

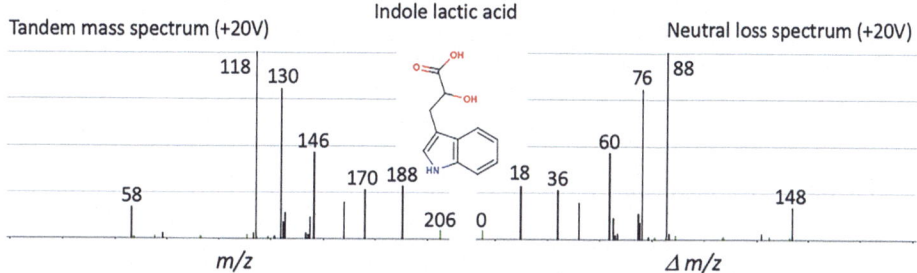

Fig. 5.10 Indole lactic acid, here shown with data acquired at an offset voltage of 20 V, was not identified in the original study, since a standard was not in any database. However, NL data facilitated its ultimate identification via similarity analysis to other indole-containing molecules; this information combined with acquisition of a molecular standard, allowed for the identification of the unknown as indole lactic acid

the utility of METLIN-NL is shown in Fig. 5.10; a molecule that was not conclusively identified in the original study was later identified as indole lactic acid. This was accomplished using NL combined with MS^2 data and final confirmation with a molecular standard.

5.6   Take-Home Messages

- The METLIN 930,000 molecular standards MS^2 and NL databases (METLIN-MS^2 and METLIN-NL) allow for the identification of known chemical entities (including metabolites) and the characterization of unknown molecules.
- Molecular standards and their corresponding experimental MS^2 and NL data at multiple collisional energies and in positive and negative ionization modes are necessary to make accurate characterization of chemical entities.
- Biological questions like the presented microbiota example, can only be answered when the molecules of interest are correctly identified. Speculative identifications, as is performed in in silico characterization of metabolites, are often erroneous and cost biologists years of effort without any fruitful knowledge-based outcome.

References

1. Smith CA, O'Maille G, Want EJ, Qin C, Trauger SA, Brandon TR, Custodio DE, Abagyan R, Siuzdak G., METLIN: A Metabolite Mass Spectral Database. Ther Drug Monit. 2005;27:747–51.
2. Tautenhahn R, Cho K, Uritboonthai W, Zhu Z, Patti GJ, Siuzdak G., An accelerated workflow for untargeted metabolomics using the METLIN database. Nat Biotechnol. 2012;30:826–8.
3. Kind T, Tsugawa H, Cajka T, Ma Y, Lai Z, Mehta SS, Wohlgemuth G, Barupal DK, Showalter MR, Arita M, Fiehn O., Identification of small molecules using accurate mass MS/MS search. Mass Spectrom Rev. 2017; 37(4):513-532.
4. Wishart DS, Tzur D, Knox C, Eisner R, Guo AC, Young N, Cheng D, Jewell K, Arndt D, Sawhney S, Fung C, Nikolai L, Lewis M, Coutouly MA, Forsythe I, Tang P, Shrivastava S,

Jeroncic K, Stothard P, Amegbey G, Block D, Hau DD, Wagner J, Miniaci J, Clements M, Gebremedhin M, Guo N, Zhang Y, Duggan GE, Macinnis GD, Weljie AM, Dowlatabadi R, Bamforth F, Clive D, Greiner R, Li L, Marrie T, Sykes BD, Vogel HJ, Querengesser L., HMDB: the Human Metabolome Database. Nucleic Acids Res. 2007;35:D521–6.

5. Vinaixa M, Schymanski EL, Neumann S, Navarro M, Salek RM, Yanes O., Mass spectral databases for LC/MS- and GC/MS-based metabolomics: state of the field and future prospects. TrAC Trends Anal Chem. 2016;78:23–35.

6. Allen F, Greiner R, Wishart D., Competitive fragmentation modeling of ESI-MS/MS spectra for putative metabolite identification. Metabolomics. 2015;11:98–110.

7. Duhrkop K, Shen H, Meusel M, Rousu J, Bocker S., Searching molecular structure databases with tandem mass spectra using CSI:FingerID. Proc Natl Acad Sci U S A. 2015;112:12580–5.

8. Brouard C, Shen H, Duhrkop K, d'Alche-Buc F, Bocker S, Rousu J., Fast metabolite identification with Input Output Kernel Regression. Bioinformatics. 2016;32:i28–36.

9. Patti GJ, Yanes O, Siuzdak G., Metabolomics: the apogee of the omics trilogy. Nat Rev Mol Cell Biol. 2012;13:263–9.

10. Warth B, Spangler S, Fang M, Johnson CH, Forsberg EM, Granados A, Martin RL, Domingo-Almenara X, Huan T, Rinehart D, Montenegro-Burke JR, Hilmers B, Aisporna A, Hoang LT, Uritboonthai W, Benton HP, Richardson SD, Williams AJ, Siuzdak G., Comparison of Full-Scan, Data-Dependent, and Data-Independent Acquisition Modes in Liquid Chromatography–Mass Spectrometry Based Untargeted Metabolomics. Anal Chem. 2017;89:11505–13.

11. Johnson CH, Dejea CM, Edler D, Hoang LT, Santidrian AF, Felding BH, Ivanisevic J, Cho K, Wick EC, Hechenbleikner EM, Uritboonthai W, Goetz L, Casero RA, Pardoll DM, White JR, Patti GJ, Sears CL, Siuzdak G., Metabolism links bacterial biofilms and colon carcinogenesis. Cell Metab. 2015;21:891–7.

12. Northen TR, Woo H-K, Northen MT, Nordström A, Uritboonthail W, Turner KL, Siuzdak G., High surface area of porous silicon drives desorption of intact molecules. Am Soc Mass Spectrom. 2007;18(11):1945–9.

13. Robinson AR, Yousefzadeh MJ, Rozgaja TA, Wang J, Li X, Tilstra JS, Feldman CH, Gregg SQ, Johnson CH, Skoda EM, Frantz M-C, Bell-Temin H, Pope-Varsalona H, Gurkar AU, Nasto LA, Robinson RAS, Fuhrmann-Stroissnigg H, Czerwinska J, McGowan SJ, Cantu-Medellin N, Harris JB, Maniar S, Ross MA, Trussoni CE, LaRusso NF, Cifuentes-Pagano E, Pagano PJ, Tudek B, Vo NV, Rigatti LH, Opresko PL, Stolz DB, Watkins SC, Burd CE, St. Croix CM, Siuzdak G, Yates NA, Robbins PD, Wang Y, Wipf P, Kelley EE, Niederhofer LJ., Spontaneous DNA damage to the nuclear genome promotes senescence, redox imbalance and aging. Redox Biol. 2018;17:259–73.

14. Marchetti V, Yanes O, Aguilar E, Wang M, Friedlander D, Moreno S, Storm K, Zhan M, Naccache S, Nemerow G, Siuzdak G, Friedlander M., Differential macrophage polarization promotes tissue remodeling and repair in a model of ischemic retinopathy. Sci Rep. 2011;1: 76.

15. Galmozzi A, Kok BP, Kim AS, Montenegro-Burke JR, Lee JY, Spreafico R, Mosure S, Albert V, Cintron-Colon R, Godio C, Webb WR, Conti B, Solt LA, Kojetin D, Parker CG, Peluso JJ, Pru JK, Siuzdak G, Cravatt BF, Saez E., PGRMC2 is an intracellular haem chaperone critical for adipocyte function. Nature 2019;576: 138–42.

16. Priolo C, Pyne S, Rose J, Regan ER, Zadra G, Photopoulos C, Cacciatore S, Schultz D, Scaglia N, McDunn J, De Marzo AM, Loda M., AKT1 and MYC induce distinctive metabolic fingerprints in human prostate cancer. Cancer Res. 2014;74:7198–204.

17. Lim CK, Bilgin A, Lovejoy DB, Tan V, Bustamante S, Taylor BV, Bessede A, Brew BJ, Guillemin GJ., Kynurenine pathway metabolomics predicts and provides mechanistic insight into multiple sclerosis progression. Sci Rep. 2017;7:41473.

18. Hocher B, Adamski J., Metabolomics for clinical use and research in chronic kidney disease. Nat Rev Nephrol. 2017;13:269–84.

19. Roberts LD, Koulman A, Griffin JL., Towards metabolic biomarkers of insulin resistance and type 2 diabetes: progress from the metabolome. Lancet Diabetes Endocrinol. 2014;2:65–75.
20. Armitage EG, Southam AD., Monitoring cancer prognosis, diagnosis and treatment efficacy using metabolomics and lipidomics. Metabolomics. 2016;12:146.
21. Warth B, Raffeiner P, Granados A, Huan T, Fang M, Forsberg EM, Benton HP, Goetz L, Vogt P, Johnson CH, Siuzdak G., Metabolomics Reveals that Dietary Xenoestrogens Alter Cellular Metabolism Induced by Palbociclib/Letrozole Combination Cancer Therapy. Cell Chem Biol. 2018;25(3):291–300.e3..
22. Yanes O, Clark J, Wong DM, Patti GJ, Sanchez-Ruiz A, Benton HP, Trauger SA, Desponts C, Ding S, Siuzdak G. Nat Chem Biol., Metabolic oxidation regulates embryonic stem cell differentiation. 2010;6:411–7.
23. Beyer BA, Fang M, Sadrian B, Montenegro-Burke JR, Plaisted WC, Kok BPC, Saez E, Kondo T, Siuzdak G, Lairson LL., Metabolomics-based discovery of a metabolite that enhances oligodendrocyte maturation. Nat Chem Biol. 2018;14:22–8.
24. Guijas C, Montenegro-Burke JR, Warth B, Spilker ME, Siuzdak G., Metabolomics activity screening for identifying metabolites that modulate phenotype. Nat Biotechnol. 2018;36:316–20.
25. Aisporna A, Benton HP, Chen A, Derks RJE, Galano JM, Giera M, Siuzdak G., Neutral Loss Mass Spectral Data Enhances Molecular Similarity Analysis in METLIN. JASMS. 2022;33:530–4.
26. Smith CA, Want EJ, O'Maille G, Abagyan R, Siuzdak G. XCMS: processing mass spectrometry data for metabolite profiling using nonlinear peak alignment, matching, and identification. Anal Chem. 2006;78:779–87.
27. Wikoff WR, Anfora AT, Liu J, Schultz PG, Lesley SA, Peters EC, Siuzdak G., Metabolomics analysis reveals large effects of gut microflora on mammalian blood metabolites. PNAS. 2009;106:3698–703.
28. Backhed F, Ley RE, Sonnenburg JL, Peterson DA, Gordon JI. Host-bacterial mutualism in the human intestine. Science. 2005;307:1915–20.
29. Turnbaugh PJ, Backhed F, Fulton L, Gordon JI. Diet-induced obesity is linked to marked but reversible alterations in the mouse distal gut microbiome. Cell Host Microbe. 2008;3:213–23.
30. Mazmanian SK, Liu CH, Tzianabos AO, Kasper DL. An immunomodulatory molecule of symbiotic bacteria directs maturation of the host immune system. Cell. 2005;122:107–18.
31. Alverdy JC, Chang EB. The re-emerging role of the intestinal microflora in critical illness and inflammation: why the gut hypothesis of sepsis syndrome will not go away. J Leukoc Biol. 2008;83:461–6.
32. Frank DN, et al. Molecular-phylogenetic characterization of microbial community imbalances in human inflammatory bowel diseases. Proc Natl Acad Sci U S A. 2007;104:13780–5.
33. Hord NG. Eukaryotic-microbiota crosstalk: potential mechanisms for health benefits of prebiotics and probiotics. Annu Rev Nutr. 2008;28:215–31.
34. Zaneveld J, et al. Host-bacterial coevolution and the search for new drug targets. Curr Opin Chem Biol. 2008;12:109–14.
35. Gill SR, et al. Metagenomic analysis of the human distal gut microbiome. Science. 2006;312:1355–9.
36. Turnbaugh PJ, et al. An obesity-associated gut microbiome with increased capacity for energy harvest. Nature. 2006;444:1027–31.
37. Jones BV, Begley M, Hill C, Gahan CG, Marchesi JR. Functional and comparative metagenomic analysis of bile salt hydrolase activity in the human gut microbiome. Proc Natl Acad Sci U S A. 2008;105:13580–5.
38. Karbownik M, Reiter RJ, Garcia JJ, Cabrera J, Burkhardt S, Osuna C, Lewinski A. Indole-3-propionic acid, a melatonin-related molecule, protects hepatic microsomal membranes from iron-induced oxidative damage: relevance to cancer reduction. J Cell Biochem. 2001;81:507–13.

39. Elsden SR, Hilton MG, Waller JM. The end products of the metabolism of aromatic amino acids by clostridia. Arch Microbiol. 1976;107:283–8.

40. Young SN, Anderson GM, Gauthier S, Purdy WC. The origin of indoleacetic acid and indolepropionic acid in rat and human cerebrospinal fluid. J Neurochem. 1980;34:1087–92.

41. Siuzdak G, Ichikawa Y, Caulfield TJ, Munoz B, Wong C-H, Nicolaou KC. Evidence of calcium (2+)-dependent carbohydrate association through ion spray mass spectrometry. J Am Chem Soc. 1993;115:2877–81.

42. Chatman K, Hollenbeck T, Hagey L, Vallee M, Purdy R, Weiss F, Siuzdak G. Nanoelectrospray mass spectrometry and precursor ion monitoring for quantitative steroid analysis and Attomole sensitivity. Anal Chem. 1999;71:2358–63.

43. Go EP, Uritboonthai W, Apon JV, Trauger SA, Nordstrom A, O'Maille G, Brittain SM, Peters EC, Siuzdak G. Selective metabolite and peptide capture/mass detection using fluorous affinity tags. J Proteome Res. 2007;6:1492–14999.

44. Benton HP, Wong DM, Trauger SA, Siuzdak G. XCMS^2: processing tandem mass spectrometry data for metabolite identification and structural characterization. Anal Chem. 2008;80:6382–9.

45. Xue J, Guijas C, Benton HP, Warth B, Siuzdak G. METLIN MS^2 molecular standards database; a broad chemical and biological resource. Nat Methods. 2020;17:953–4.

Untargeted GC-MS Data Processing and Metabolite Identification Using eRah

6

Sara M. de Cripan, Trisha Arora, Adrià Olomí, Jasen P. Finch, and Xavier Domingo-Almenara

---

What You Will Learn in This Chapter

- Process GC-MS-based untargeted metabolomics raw data.
- Detect and deconvolve metabolite's spectra as opposed to the detection of ion peaks stemming from the same or different metabolites.
- Identify deconvolved spectra using spectral matching or retention index calibration.

S. M. de Cripan · T. Arora · X. Domingo-Almenara (✉)
Computational Metabolomics for Systems Biology Lab, Omics Sciences Unit, Eurecat—Technology Centre of Catalonia, Barcelona, Catalonia, Spain

Department of Electrical, Electronic and Control Engineering (DEEEA), Universitat Rovira i Virgili, Tarragona, Catalonia, Spain
e-mail: xavier.domingoa@eurecat.org

A. Olomí
Computational Metabolomics for Systems Biology Lab, Omics Sciences Unit, Eurecat—Technology Centre of Catalonia, Barcelona, Catalonia, Spain

J. P. Finch
Department of Life Sciences, Aberystwyth University, Aberystwyth, UK

6.1 Introduction

Gas chromatography–mass spectrometry (GC-MS) is a ubiquitous platform for measuring volatile and semi-volatile metabolites including those involved in central carbon metabolism like amino acids and amino acid derivatives, carboxylic acids, carbohydrates or lipids including fatty acids, short-chain fatty acids, or sterols. Typically, metabolites in samples are submitted to methoximation (MeOx) and trimethylsilyl (TMS) derivatization to increase their volatility and facilitate their measurement, thus increasing the metabolite coverage when using GC-MS. Although the metabolite coverage of GC-MS is lower than that of liquid chromatography coupled to mass spectrometry (LC-MS), the identification of metabolites using GC-MS is easier compared to LC-MS owing to the highly reproducible electron ionization (EI) source and the robustness of the capillary columns used in GC-MS [1]. The robustness of the capillary columns has facilitated the adoption of calibration methods based on the retention index (RI), i.e., relative retention time. RI data are available in libraries and databases and encode the metabolites' relative retention time difference to a set of reference standards such as n-alkanes or fatty acid methyl esters (FAMEs) [2]. The use of RI values in combination with the "hard" ionization of EI yields rich fragmentation spectra, allowing an easier metabolite annotation or identification compared to LC-MS.

Multiple processing tools for GC-MS use the established peak-picking paradigm widely used in LC-MS [3], where chromatographic peaks or the so-called features—a peak or a group of peaks across samples with a unique m/z and retention time [3]—are detected. However, the fragmentation induced by the "hard" EI source yields a peak ion redundancy where multiple fragment peaks are observed for each metabolite. This led to the emergence of tools that depart from the traditional peak-picking approach and move toward spectra or compound deconvolution. In that sense, these tools do not detect peaks or features, and, instead, these tools detect and deconvolve compounds or spectra—a peak or a group of peaks stemming from the same compound—using either univariate or multivariate statistical methods. For example, the most representative univariate approach—used in the widely known XCMS/CAMERA tandem [4, 5]—groups peaks stemming from the same metabolite on the basis of peak shape similarity [3]. An important distinction between multivariate and univariate approaches is that multivariate approaches are capable of deconvolving mixed signals and, therefore, are very effective at deconvolving peaks that appear co-eluted with peaks from other compounds. Free or open-source tools based on spectra or compound deconvolution that focus on spectra as the analysis entity instead of peaks include MS-DIAL [6] or ADAP-GC 4.0 [7]. However, with the exception of ADAP-GC (from its version 4.0), other tools do not use multivariate algorithms.

eRah [8] is an R package designed to deconvolve mixed peak signals stemming from different metabolites by incorporating multivariate blind source separation (BSS) techniques. eRah provides an accessible framework to tackle the longstanding burden of untargeted metabolomics where thousands of peaks or features are detected and where grouping peaks stemming from each metabolite to match with spectral libraries is a tedious task. In that sense, the analysis and working entity in eRah is not the single peak or m/z feature but the deconvolved spectra (metabolite). Since its initial release in 2016, eRah has been maintained and evolved to incorporate RI determination using either internal or external calibration with reference standards like FAMEs or n-alkanes or using naturally occurring metabolites in samples, therefore bypassing the need for reference standards [9].

In this protocol, we describe eRah's core methods, general workflow and parameters used to analyze GC-MS-based untargeted metabolomics samples. We also describe a practical metabolite identification case by combining spectral matching and RI data using a previously published dataset [2] publicly available in *the MetaboLights repository with accession number MTBLS2841.*

6.2   Overview of eRah's GC-MS Data Processing Workflow

eRah is based on an integrated workflow consisting of four main steps (1) preprocessing and spectral deconvolution using multivariate techniques, (2) alignment of spectra across samples, (3) missing compound recovery, and (4) identification of metabolites by spectral library matching and retention index computation (Fig. 6.1).

6.2.1   Preprocessing and Spectral Deconvolution

An initial preprocessing is applied to the raw data consisting of noise filtering and baseline removal. Next, similar to peak-picking-based algorithms, where peaks are detected, eRah detects spectra or compounds—groups of peaks appearing at the same RT with different m/z ratios that stem from a unique metabolite—using a multivariate matched filter called compound match by local covariance (CMLC) [8]. A subsequent orthogonal signal deconvolution (OSD) method [8, 10] is applied to retrieve the deconvolved spectra of these compounds. As opposed to univariate (i.e., peak-picking) techniques, the use of multivariate methods enables the deconvolution of spectra even when compounds with shared ion fragments appear co-eluted (Fig. 6.2). As seen in the figure, eRah can deconvolve the spectra of different co-eluted metabolites, allowing the use of spectral matching for tentative identification. eRah is capable of finding and deconvolving co-eluted spectra even when the concentrations of metabolites are low. As in the case of compound A in Fig. 6.2, eRah is able to extract a clean spectrum and identify it as a contaminant from siloxane even though the main peak height is below 15,000 counts.

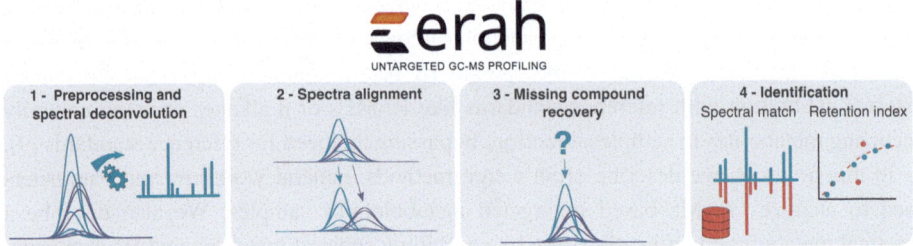

Fig. 6.1 Overview of eRah's processing workflow

6.2.2   Alignment

To account for the retention time (RT) deviation across samples, deconvolved metabolites are aligned by spectral similarity and RT distance. This results in a list of unique spectra appearing across the maximum number of samples and with the least RT deviation and the most spectral similarity. The alignment step enables the quantitative comparison of metabolite's relative concentration across samples by comparing the deconvolved most intense peak of each unique spectra (i.e., unique metabolite).

6.2.3   Missing Compound Recovery

An optional, but recommended, missing compound recovery step can be applied. This process aims at recovering metabolite spectra that are missing in certain samples, i.e., have not been detected or correctly deconvolved in some samples. This step ensures that the final quantitative data table does not have any missing values. Missing compound signals may be the result of an incorrect deconvolution or alignment because of low compound concentration, stronger co-elution in certain samples, or actual absence of compound in the sample.

6.2.4   Identification

Metabolites can be annotated or identified through spectral matching between the mean spectra of each metabolite across samples and a reference library. eRah includes a custom version of MassBank repository, but the user is encouraged to import other libraries (e.g., Golm Metabolome Database). eRah's deconvolved spectra can also be exported for comparison with the NIST library. Additionally, retention index (RI) can be determined using either reference standards (co-injected with samples or external calibration) or using naturally occurring metabolites.

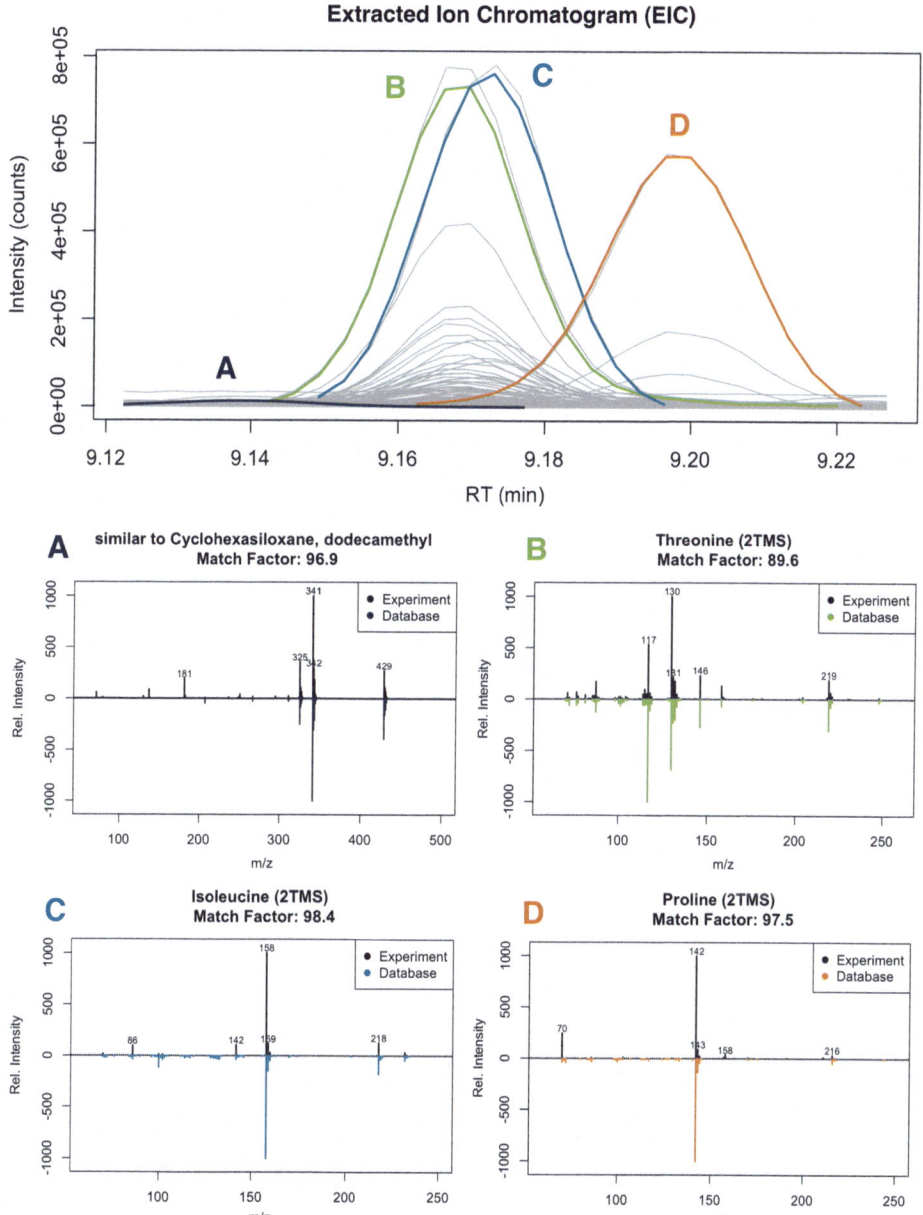

Fig. 6.2 Deconvolved peak profiles for four compounds (**a–d**) (top) and deconvolved spectra (bottom panels). In the top panel, the most intense (deconvolved) peak of the spectra is shown and used for subsequent quantification. In the bottom panels, the deconvolved spectra (black) are compared with the reference spectra from the library (shown inverted and in color)

6.3   GC-MS Data Processing and Metabolite Identification

In this protocol, we will show the eRah's analysis workflow using GC-MS raw data from a comparative analysis of plasma samples from patients with ulcerative colitis (UC) and healthy controls. UC is an inflammatory disease of the large intestine and current therapies for treating UC yield remission rates below 30% [11]. Clinical studies using omics analysis can expedite the identification of druggable targets as well as our understanding about the molecular mechanisms involved in UC. We will use a dataset from an untargeted metabolomics assay consisting of a total of 25 plasma samples from five male and five female UC patients and ten samples from age-, weight-, and sex-matched healthy controls (a total of ten males and ten females), in addition to five technical replicates of a pooled sample used as quality controls. Samples were analyzed using GC-ToF-MS (chromatographic and MS methods are described elsewhere [2]). This dataset was used in a previous publication [2], and GC-MS raw data (mzXML files) are publicly available in the MetaboLights repository with accession number MTBLS2841.[1]

The following is a complete protocol from start (eRah's installation) to finish (statistical analysis and metabolite identification) for the analysis of raw GC-MS samples.

6.3.1   Installation

eRah can be installed from any CRAN repository, by:

```
> install.packages("erah")
```

Of note, there is a development version available in Github, but we strongly recommend to use the CRAN's version as the development version is only for developers and beta testers.

Once installed, the package needs to be loaded by executing:

```
> library(erah)
```

---

[1] https://www.ebi.ac.uk/metabolights/MTBLS2841

Documentation and help can be accessed by executing:

```
> help(package = "erah")
```

6.3.2   Database Import

Users may import their own mass spectral libraries. We strongly recommend using the Golm Metabolome Database (GMD) [12] with eRah. To use the GMD, first, we have to download it from its webpage,[2] by downloading the file "GMD_20111121_VAR5_ALK_MSP.txt"   or   "GMD 20111121_MDN35_ALK_MSP.txt," depending on the type of chromatographic columns (VAR5 or MDN35) being used. If you are not interested in using any retention index information (see Sects. 6.3.7.2 and 6.3.7.3), then either one of the files can be used indistinctly. Then, we can load the library with the function importMSP():

```
  g.info <- "GMD Database: Kopka, J., et al. (2005) Bioinformatics,
21, 1635-1638."
golm.database <-importGMD(filename="GMD_20111121_VAR5_ALK_MSP.txt", DB.
name=
"GMD", DB.version="GMD_20111121", DB.info= g.info, type="VAR5.ALK")
```

The library in R format can now be stored for a faster loading the next time we want to use it:

```
> save(golm.database, file= "golmdatabase.rda")
```

Then, each time that we start a new R session or project, we can replace the default eRah database object mslib, with our custom database:

```
> load("golmdatabase.rda")
> mslib <- golm.database
```

---

[2] http://gmd.mpimp-golm.mpg.de/download/

Other MSP-formatted libraries can also be imported. The procedure is the same as for the GMD database, with the only exception that the function to be used is `importMSP` instead of `importGMD`.

6.3.3 Creating a New Experiment

eRah provides two methods by which a new experiment can be created. The first uses the organization of the raw data files into class subdirectories within an experiment directory. The second requires only the file paths of the raw data files. Either of these methods outlined in Sects. 6.3.3.1 and 6.3.3.2 can be used in this protocol.

6.3.3.1 Using the File Directory Structure

The creation of a new experiment requires organizing the sample raw data files into experimental class subdirectories within a top-level experiment directory. The raw data files for each class should then be stored in the specific class folder. Where there is only a single class, the data files should be placed into a single directory.

For this tutorial, the main directory is called UCExp, and the class subdirectories are named CTR, QC, and UC. The corresponding raw data files must be placed inside these class directories (Fig. 6.3).

In addition to the raw data files, the user must create a ".csv" type file for the instrumental information and optionally another containing the phenotypic information. The instrumental information file must contain at least two columns named "sampleID" and "filename" (case sensitive) and contain the sample identifier and the sample file name, respectively.

The optional phenotypic information file should again contain the "sampleID" sample identifier column but must also have a "class" column that contains the sample class information. Both files can contain additional columns as required.

The instrumental and phenotype information files can both be conveniently created based on the subdirectory structure of experiment directory using the following:

```
> createdt('UCExp')
```

Users have to make sure that the working directory of the R session is set one level above the *UCExp* folder or any other experiment folder. Alternatively, users can replace *UCExp* with the full path of the experiment folder directory. The instrumental and phenotype information should then be parsed from the ".csv" files that were just created:

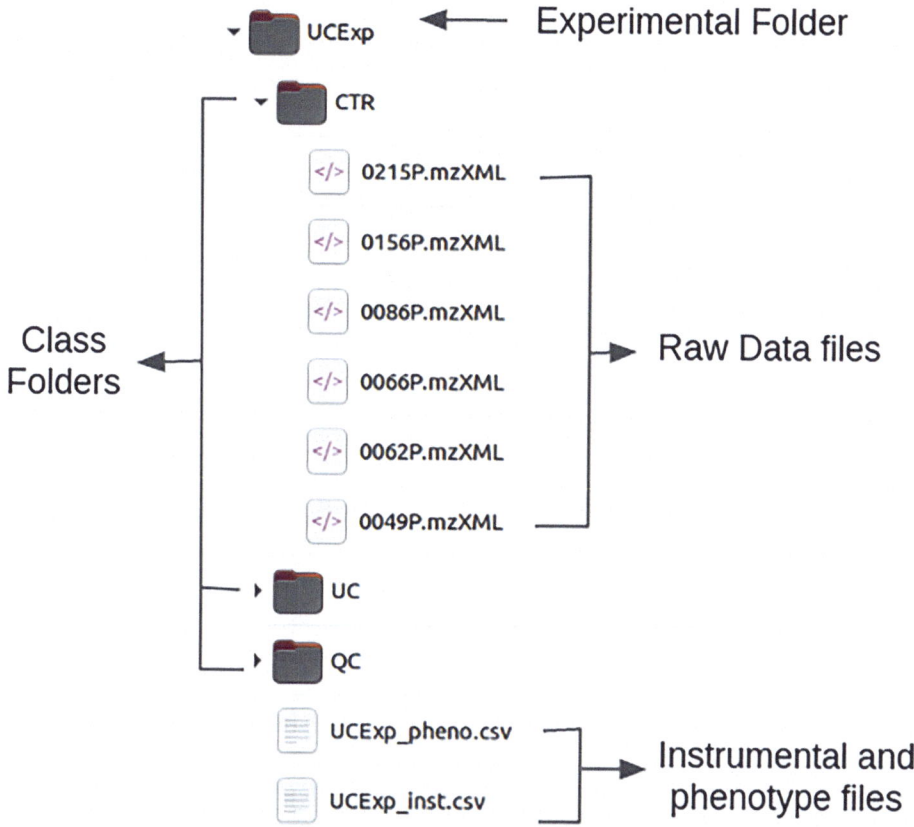

Fig. 6.3 File organization for the UC experiment. First, there is the main experiment directory (UCExp). This contains the class subdirectories (CTR, QC, and UC). Each class subdirectory contains the raw data files (e.g., 0049P.mzXML)

```
> instrumental_info <- read.csv('UCExp/UCExp_inst.csv', sep = ';')
```

```
> phenotype_info <- read.csv('UCExp/UCExp_pheno.csv',
sep = ';')
```

6.3.3.2 Using File Paths Only

Creating an experiment requiring only the file paths to the raw data files could be useful in circumstances such as those where the locations of the data files cannot be easily altered, perhaps where data files are stored in remote data repositories. For the

(continued)

simplicity of this tutorial, it will be assumed that the data files have been organized into the same structure as for Sect. 6.3.3.1; however, this is not an essential requirement for this method.

To begin, first read the file location paths:

```
> file_paths <- list.files(
      path = 'UCExp',
      full.names = TRUE,
      recursive = TRUE,
      pattern = '.mzXML'
      )
```

Using the returned vector of file paths, a data.frame containing the instrumental information for these files can be created using the following:

```
> instrumental_info <- createInstrumentalTable(file_paths)
```

Similarly, a data.frame of the optional phenotype information can be created as shown below. This also requires the input of a vector of the sample class information as the cls argument. Here, this is specified using the classes variable that is created based on the file order that would be returned with the data files organized as for Sect. 6.3.3.1:

```
> classes <- c(rep('CTR',10),
               rep('QC',5),
               rep('UC',10))

> phenotype_info <- createPhenoTable(file_paths,
                                     cls = classes)
```

Additional columns can be added to the instrumental and phenotypic information tables as required. It is also possible for these tables to be parsed from external ". csv"-type files. See Sect. 6.3.3.1 for details on how these files should be formatted.

6.3.3.3 Creating the Experiment

Using the instrumental and phenotypic information objects (`instrumental_info, phenotype_info`) that were prepared using either of the methods in Sects. 6.3.3.1 and 6.3.3.2, the new experiment (`MetaboSet`) object can be created using:

```
> ex <- newExp(instrumental = "UCExp/UCExp_inst.csv", phenotype = "UCExp/
UCExp_pheno.csv", info = "UCExp Experiment
```

6.3.4 Spectral Deconvolution

Deconvolution is performed by the `deconvolveComp` function. This function needs an auxiliary "Deconvolution parameters" object that can be set through `SetDecPar`. This object contains the basic parameters of the deconvolution algorithms such as the minimum peak width or the *m/z* to exclude:

```
> ex.dec.par <- setDecPar(min.peak.width = 1, avoid.processing.mz = c(35:
69,73:75,147:149))
```

The minimum peak width (`min.peak.width`), measured in seconds, is the main deconvolution parameter. A small value will increase the capability to deconvolve co-eluting compounds, but it will increase the detection of false positive peaks or compounds. When using a large value, the algorithm can fail at separating co-eluting compounds, losing the capability of true positive compound detection. Hence, adjusting the parameter will determine eRah's performance. This value should be set to approximately half the mean width of the experimental peaks in samples.

Excluding a specific set of *m/z* values via the *avoid.processing.mz parameter* when processing the samples can improve eRah's efficiency by optimizing the deconvolution process. eRah's deconvolution relies on the existence of selective fragments from each metabolite to distinguish and deconvolve spectra among co-eluted metabolites. The presence of nonselective fragments clutters the chromatogram, masking these selective fragments and hampering eRah's deconvolution performance. When using TMS derivatization, it is recommended to exclude the

(continued)

m/z values 73, 74, 75, 147, 148, 149, corresponding to ubiquitous fragments typically generated from compounds carrying a trimethylsilyl moiety. If these fragments are not removed, a large number of metabolites will share these nonselective *m/z* ions leading to a higher spectral similarity among compounds which hampers the capability of eRah at distinguishing two spectra from two different compounds. In that sense, excluding all *m/z* values from 1 to 69 is also recommended, since fragmentation in this range is highly common for all the metabolites and adds nonselective fragment information. Although fragments in this *m/z* range are generally the most intense, eRah will not consider these *m/z* in any of the subsequent processes, so it will not affect spectral matching and identification.

After that, compounds can be deconvolved as follows:

```
> ex <- deconvolveComp(ex, ex.dec.par)
```

When dealing with a high number of samples, the deconvolution step may lead to a poor run-time performance because of the sequential treatment of each file. The performance can be improved through parallel computing processing. Please see the documentation to use parallel processing.

6.3.5   Spectra Alignment

Spectra (or compound/metabolite) alignment across samples is conducted by the `alignComp` function. As in the deconvolution step, this method needs an auxiliary object containing the specific alignment parameters such as the minimum spectral similarity value (*min.spectra.cor*), a maximum retention time distance (max.time.dist), and the *m/z* range to consider (mz.range). These parameters can be defined through the `setAlPar` function, as follows:

```
> ex.al.par <- setAlPar(min.spectra.cor = 0.90, max.time.dist = 3, mz.
range = 70:600)
```

The algorithm will group together and consider to be the same compound the two or more most similar spectra in different samples sharing the highest spectral similarity

(continued)

(above the minimum spectral correlation value) and with the smallest retention time deviation (within the maximum time distance). The minimum spectral correlation ranges from 0 (no similarity) to 1 (total similarity). The *m/z* comparison range (mz. range) can also be specified, and the default values are 70 to 600 due to the reasons commented before (see Sect. 6.3.4).

Once the parameters have been set, spectra or compound alignment can be executed by:

```
> ex <- alignComp(ex, alParameters = ex.al.par)
```

A strategy to obtain a reliable alignment consists in setting a restrictive (high) minimum spectral similarity value. This will lead to an unalignment of spectra in some samples because, due to an incorrect deconvolution, noise, or their low concentration, will not show a spectral similarity above the minimum spectral similarity value. However, eRah will group the spectra in those samples where the most concentrated and "cleaner" (less affected by noise) spectra of each metabolite appear. This assures the alignment of a representative set of consistent spectra across samples. For example, if we are analyzing, as in this case, 25 samples, and we observe that a unique spectrum has been found across 3 samples, it could indicate that this might be a noisy spectrum from an incorrectly deconvolved compound. However, if we find a group of aligned spectra across 12 samples (almost 50% of the samples), it indicates that there is a high probability that these spectra stem from an existing metabolite, since a group of spectra sharing more than 90% of similarity have been found across these 12 samples. However, due to the use of a high minimum spectral similarity, we are missing the spectra in the other samples and thus the peak area (concentration) of this metabolite in the rest of the samples. This is why the "missing compound recovery" step was designed, to recover missing spectra and retrieve the relative concentration of the metabolite in all samples while retaining the purest group of deconvolved spectra in all samples.

Alignment step could lead to errors or to poor run-time performance if the number of samples is excessively large and especially under Windows OS (due to the virtual memory OS management system). In those cases, alignment can be conducted by block segmentation by using the parameter blocks.size in the function alignComp. For more details access to the alignComp help executing:

```
> ?alignComp
```

6.3.6 Missing Compound Recovery

As mentioned in the previous section, the missing compound recovery assures that all spectra or compounds are found in samples, since the alignment will only find these spectra or compounds across samples that share a spectral similarity above a certain spectral similarity threshold.

This optional step is performed by the "recMissComp" function. This function only needs a minimum value of samples (*min.samples*):

```
> ex <- recMissComp(ex, min.samples = 6)

## recovering
[===============================================]
100% eta: 0s

## Updating alignment table...
## Model fitted!
```

If an aligned spectrum has been found in, at least, the minimum number of samples set via *min.samples*, this spectrum is going to be searched (fitted) in the samples where this spectrum has not been found or correctly aligned.

Users can first inspect the alignment results via the `alignList` function or, alternatively, identify compounds first (see Sect. 6.3.7) and explore the list of identified metabolites through the `idList` function. As we will explain in the next section, the alignment or identification list also includes the number of samples in which a spectrum or metabolite has been found. Based on the results, the user can select an optimal minimum number of samples. We used the rule of 60% of samples from one phenotypic class, i.e., six samples, to set the *min.samples* value. Setting a low *min.samples* value could lead to many false noisy spectra being retrieved. As explained in the previous section, if an aligned spectra group is found only in, e.g., three samples, it could probably stem from an incorrectly deconvolved compound. We do not want to "recover" this noisy compound in the rest of the samples. In other cases, the same metabolite's spectra will be aligned into two different groups by the alignment process. This is because the same metabolite's spectra are more similar within two sets of samples, e.g., one group of spectra from the same metabolite from well-concentrated samples is very similar but less similar to a group of spectra where the metabolite abundance is low. In these cases, the alignment considers that these

(continued)

are two distinct groups of spectra, but, usually, the cleaner aligned spectra group is found in a larger number of samples compared to the aligned noisy spectra group.

In summary, after alignment, and due to the reasons discussed above, there is a large number of spectra that are found in just two or a few more samples (noisy spectra), only a relatively small number of spectra are aligned across all the samples, while in the rest of cases spectra are usually aligned across 50% to 80% of the samples. The missing compound recovery will remove from the alignment or identification list all the aligned spectra groups that were aligned in fewer samples than the *min.samples* value. Using a low *min.samples* number will yield an excessively large list of deconvolved spectra that includes a large portion of noisy spectra and that does not represent the real compound number in the samples.

A recommended strategy to find an optimal *min.samples* value is to first evaluate the average number of samples where compounds are found or aligned, by `alignList` (or `idList` functions after the identification step, see Sect. 6.3.7). This will allow us to find the average number of samples in which the compounds have been found and adjust the *min.samples* value accordingly.

Of note, if the spectra have been already identified via the `identifyComp` function, we have to reidentify the compounds after executing the missing compound recovery step, running `identifyComp` function, as explained in the next section.

6.3.7 Metabolite Identification: Spectral Matching and Retention Index

eRah can provide annotations of the observed spectra through spectral matching against a reference database (see Sect. 6.3.7.1). Also, after the initial annotation via spectral matching, as an additional step to consolidate the annotations, users can compute a retention index (RI) error of the retrieved compounds. eRah's RI calculation is inspired on the Kovats RI calculation method [13, 14] and where an internal (using co-injected standards) or external (using standards analyzed separately from samples) retention time/retention index (RT/RI) calibration is performed using a linear interpolation. To perform an RT/RI calibration curve, we need a set of compounds to be used as reference. Traditionally, this is done by co-injecting a set of FAMEs or n-alkalanes with samples [15]. Alternatively, we introduced a method to use naturally occurring metabolites in samples, where metabolites that are detected in samples can be used as a proxy for co-injected standards, thus avoiding the need of standard co-injection [9]. Ultimately, we need a list of metabolites—either reference standards or correctly identified naturally occurring metabolites in samples—for which we have RI information in databases. In this protocol, we will show how to

(continued)

calibrate the RI/RT curve with these two situations (Sects. 6.3.7.2 and 6.3.7.3). We will also explain how to combine, interpret, and use spectral matching and RI data in a practical metabolite identification case in Sect. 6.4.

6.3.7.1 Identification via Spectral Matching

Through the `identifyComp` method, eRah compares, against a reference database, the mean spectrum of each group of aligned spectra across samples. A version of MassBank MS library [16] is included in this package, but we recommend importing and using the Golm Metabolome DataBase (GMD) [12] as reference database. GMD will be used in this protocol (see Sect. 6.3.2 to import GMD). For each group of aligned spectra, eRah computes the mean spectrum (calculated as the sum of intensities of all the aligned spectra and excluding the spectra recovered by the missing compound recovery function) and matches it against the entire spectral database, returning the top N (by default, n.putative = 3) top most similar identities in the reference database based on spectral similarity:

```
> ex<-identifyComp(ex)

## Constructing matrix database...
## Comparing spectra...
## Done!
```

Of note, the *m/z* excluded in the deconvolution step via the `avoid.processing.mz` parameter are also excluded in both the alignment and identification (these *m/z* are removed from the reference database before matching).

Users can explore the identified metabolites with the `idList` function:

```
> id.list <- idList(ex)
```

Users can inspect the identification list (*id.list*) via the R or Rstudio environment; alternatively, you can export the identification list into a CSV or spreadsheet via the `write.csv` function of R or the `write.xlsx` function of the package *openxlsx*.

The structure of the identification list is as follows:

(continued)

- AlignID: The unique identifier for each group of spectra aligned across samples (i.e., each deconvolved metabolite).
- Tmean: the mean retention time of the spectra.
- FoundIn: the number of samples in which the spectra have been found. If missing compound recovery has been applied, all the spectra (i.e., metabolites) will have a FoundIn value that corresponds to the total number of samples.
- Name.X, where X is the number of hits, being Name.1 the metabolite identity of the first match or hit, Name.2, the second hit, etc.
- MatchFactor.X: the corresponding spectral similarity score for each hit.
- DB.Id.X: the database entry index of each identification hit. Can be used to programmatically retrieve additional information from the spectral database used.
- CAS.X: the CAS number for each hit.
- Formula.X: The chemical formula for each hit.

6.3.7.2 Retention Index Calibration via Reference Standards

The samples used to illustrate this protocol were not co-injected with reference standards; therefore, we will describe the following steps assuming this hypothetical case. Let us then suppose that we have injected or co-injected a mix of FAMEs. In the case of FAME co-injection, we need to identify the FAMEs in the identification list of eRah through the idList function as shown in the previous Sect. 6.3.7.1 and retrieve the AlignID value of each observed FAME. Alternatively, in the case of external FAME injection, we need to retrieve the experimental RT of these FAMEs in our chromatographic method.

For the first case, we will use the AlignID to compute the RT/RI. Please note that executing the following will not work as we are illustrating a hypothetical case:

```
> # This is just an example of the hypothetical case, this will not work with
the use-case analysis shown in this protocol.
> ex <- computeRIerror(ex, mslib, reference.list=list(AlignID = c
(45,67,92,120)))
> ## We need to update the ID list by:
> id.list <- idList(ex)
```

For the second case, we need to find the RI of the corresponding FAMEs using the function findComp():

```
> findComp("FAME")
```

```
DB.Id Compound Name CAS Formula
1 57 Octanoic acid methyl ester (FAME MIX) 111-11-5 C9H18O2
2 133 Nonanoic acid methyl ester (FAME MIX) 1731-84-6 C10H20O2
3 231 Decanoic acid, methyl ester (FAME MIX) 110-42-9 C11H22O2
... [rest of data not shown]
```

Next, we can retrieve the reference RI manually by inspecting each entry in the database, according to each DB.Id, for instance, to retrieve the reference RI of *Octanoic acid methyl ester*, with a database id (DB.Id) number 57, we execute:

```
> mslib@database[[57]]
$Name
[1] "Octanoic acid methyl ester (FAME MIX)"

$Synon
[1] "Octanoic acid methyl ester // Octanoic acid methyl ester (FAME MIX) //
Octanoic acid methyl ester, n- // Octanoic acid, methyl ester"

$RI.VAR5.FAME
[1] 0

$RI.VAR5.ALK
[1] 1120.62

$RI.MDN35.FAME
[1] 0

$RI.MDN35.ALK
[1] 0
... [rest of data not shown]
```

The output displays all the metabolite reference data from the library—in this case, GMD. Among these data, we can find the RI of the VAR5 column calculated using alkanes. Either FAMEs or alkanes RI can be used as long as we keep the same choice for all cases, but it is important that we use the right type of column. Here, MDN35 RI fields are empty because we used the VAR5 file, as explained in Sect. 6.3.2, and, in this case, the field of the RI of the VAR5 using FAME is empty because it is not in the library.

Once we know the RT and RI of the reference compounds, we can execute the following code. Please remember that executing the following will not work as we are illustrating a hypothetical case:

```
> ex <- computeRIerror(ex, mslib, reference.list=list(RT=c(4.4683,
7.4402, 8.8121),
> RI=c(1120.62, 1225.27, 1326.95)))
> ## We need to update the ID list by:
> id.list <- idList(ex)
```

Note that the identification list needs to be retrieved again (updated) as shown in the previous code, via the idList function. The identification list is now updated with the RI errors for all the deconvolved spectra. In Sect. 6.4 we will describe in more detail how to use this information for metabolite annotation or identification.

6.3.7.3 Retention Index Calibration via Naturally Occurring Metabolites

As mentioned before, eRah can use naturally occurring metabolites to compute RI. The use of FAMEs or n-alkanes standard materials is used because they provide unequivocal references for computing RI, and their elution range spans the entire chromatogram. However, we can instead use metabolites that are naturally occurring in the samples as long as there is a list of unequivocally identified metabolites—ideally confirmed with standard materials—wherein the elution range spans the entire chromatogram or the RT range of interest. The process is the same as with the use of internal calibration using reference standards (Sect. 6.3.7.2), using instead the AlignID of the identified naturally occurring metabolites in samples.

6.3.8 Results Visualization and Interpretation

Once the identification process has been performed, we can now proceed with the statistical analysis. While the identification list provides information on the identity of the metabolites found, we need to retrieve their peak area, i.e., relative concentration. This can be obtained through the alignList function, which provides the AlignID, FoundIn, and tmean descriptors together with the peak area (or height) of the deconvolved chromatographic peak corresponding to the deconvolved most intense peak of each metabolite's spectrum. Figure 6.2 shows an example of the eRah's peak deconvolution capability even in compound co-elution cases. Additionally, the dataList function merges both the identification list and the alignment list into one table.

(continued)

The statistical analysis is not covered in detail in this protocol. We will briefly show how to compare UC vs. CTR samples—for the sake of simplicity, without considering the coefficient of variations of the QC samples, an established procedure in untargeted metabolomics data analysis [17]. First, we retrieve the alignment list (peak area values):

```
> al.list <- alignList(ex)
```

We inspect the column names of the list by:

```
> colnames(al.list)
[1] "AlignID" "Factor" "tmean" "FoundIn" "0049P" "0062P" "0066P" ...
```

As shown, the first four columns are the AlignID, the Factor's ID (equivalent to AlignID), the mean RT, and the FoundIn value (the number of samples in which the spectra have been found). Next to these four columns, the sample names are shown. We can use that information to assign each column number to the sample class, i.e., samples 5 to 14 are control, and 20 to 29 are UC (being the samples 15 to 19 the QC). Then, we can compute the P-values:

```
> ctr <- c(5:14)
> uc <- c(20:29)
> pVals <- apply(al.list, 1, function(x) t.test(x[ctr],x[uc])$p.value)
```

Next, we can retrieve the spectral match of the statistically significant dysregulated spectra by:

AlignID	tmean	FoundIn	Name.1	MatchFactor.1
132	6.3216	25	Dodecane	94.01
199	7.0760	25	Alanine, beta- (1TMS)	93.66
249	7.7219	25	Tridecane, n-	91.60
263	7.9218	25	Dodecane	90.71
302	8.6431	25	Serine (2TMS)	93.34
445	10.8666	25	Aspartic acid (2TMS)	91.26
475	11.3419	25	Pentadecane, n-	94.25

```
[rest of data not shown...]
```

```
AlignID tmean FoundIn Name.1 MatchFactor.1
132 6.3216 25 Dodecane 94.01
199 7.0760 25 Alanine, beta- (1TMS) 93.66
249 7.7219 25 Tridecane, n- 91.60
263 7.9218 25 Dodecane 90.71
302 8.6431 25 Serine (2TMS) 93.34
445 10.8666 25 Aspartic acid (2TMS) 91.26
475 11.3419 25 Pentadecane, n- 94.25
[rest of data not shown...]
```

From the list, different spectra seem to be dysregulated, including those that match to β-alanine, serine or aspartic acid spectra in the reference database as the first hit. eRah includes two functions that allow to inspect the profile and spectra of aligned metabolites to corroborate the peak deconvolution and the identity of the candidates. We can visually inspect the deconvolved peaks via the plotProfile function. For instance, for the case of serine (AlignID = 302):

```
> plotProfile(ex, 302)
```

Figure 6.4 shows the result of this function. As observed, all the peaks in samples show a good Gaussian shape which suggests that the peaks have been correctly deconvolved.

Note that the AlignID used in this protocol are just used as an example, and they might not correspond to those retrieved by the user after processing their samples with eRah in another computer and environment. The reader should replace the AlignID in this protocol with those obtained in their experiment.

Additionally, we can use plotSpectra function to plot the metabolite's experimental vs reference spectra comparison. As mentioned before, eRah matches the deconvolved spectra against the entire spectral database and reports the top N most similar matches (by default, $N = 3$, as defined when using the identifyComp function). We can inspect the first, second, or third match via the n.putative parameter of the plotSpectra function, by:

```
> plotSpectra(ex, AlignId=302, n.putative=1)
> plotSpectra(ex, AlignId=302, n.putative=2)
```

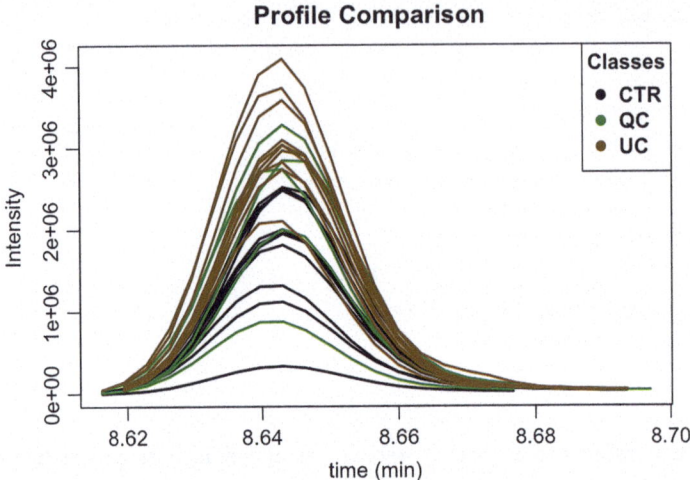

Fig. 6.4 Deconvolved peak profile of serine, where each color represents the peak in a specific sample class: control (CTR), ulcerative colitis (UC), and quality control (QC) samples

Figure 6.5 shows the result of these functions. Visual inspection allows us to prioritize the first hit (serine) as the true identity over alanine (second hit). However, these annotations, based only on spectral matching, should be taken with caution. In the next section, we describe how to combine spectral matching with RI data for metabolite identification.

6.4   A Practical Case: Metabolite IdentificationMetabolite identification in GC-MSGas chromatography-mass spectrometry (GC-MS) with eRah

This section describes a practical case where we will illustrate how to use spectral matching and retention index (RI)–calculated using naturally occurring metabolites–to identify metabolites in GC-MS. Although pure standards should be used to unequivocally confirm metabolite identities, the comparison of two orthogonal properties like RI and spectral matching can yield high confidence metabolite annotations. We will assume that the users have processed all their samples with eRah, performing deconvolution, alignment, missing compound recovery, and identification via spectral matching.

After spectral matching, eRah provides an identification list with the N (by default 3) most plausible identity candidates for each deconvolved spectrum according to the spectral similarity with a reference database. From all identities that match to each deconvolved spectrum, we need to find whether any of the top identities is the true identity of the deconvolved spectra, on the basis of spectral similarity and RI. We will calculate RI using

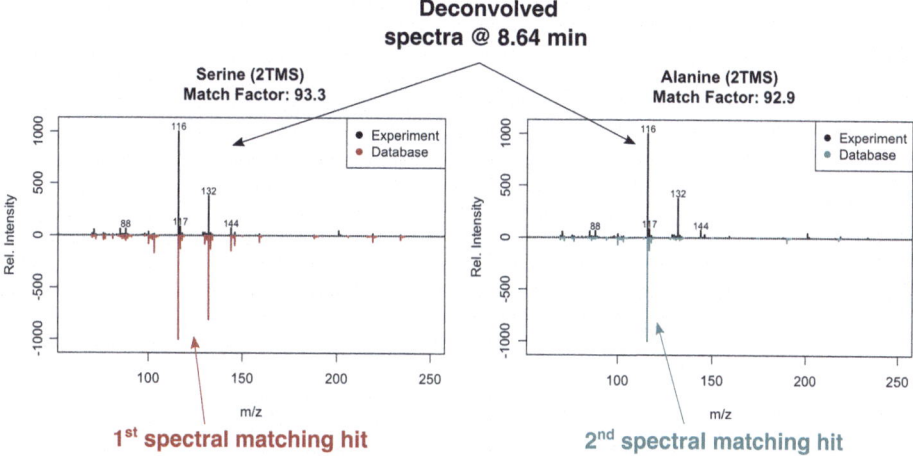

Fig. 6.5 Visual spectra comparison. The experimental deconvolved spectrum (black, top spectra) vs the first most similar match or hit (red, bottom spectra shown inverted, left panel) and the second most similar hit (blue, bottom spectra shown inverted, right panel)

naturally occurring metabolites in our samples, i.e., using a set of correctly identified metabolites. To this end, we will generate an initial list consisting of highly plausible correct annotations of metabolites solely relying on spectral similarity, as it is the only information that we have available to annotate metabolites before computing RI values. We will refer to this list of annotations based on spectral similarity as the initial annotation list. We will then curate this list by filtering out incorrect annotations based on a preliminary calculation of RI data. Finally, we will use the naturally occurring metabolites identified in the curated list to determine the theoretical RI for all the deconvolved spectra in the original identification list and compare them with reference RI values in the GMD database.

The first step to generate the initial annotation list is to review the list of identified metabolites using the identification table (through the `idList` function). We can filter the list using just a subset of identifications with the first hit having a spectral similarity, known in eRah as match factor (MF), above 80% by:

```
id.list[which(id.list$MatchFactor.1>80),]
```

Then, we can review all the entries manually. As explained before, eRah generates an identification list with the most plausible identity candidates for each deconvolved spectrum according to the spectral similarity with a reference database. For the sake of simplicity, only the top 2 ($N = 2$) most similar identities will be used in this example. For instance, a deconvolved spectrum at 7.16 min is matched as first hit to leucine (Leu) with a very high MF (99%, Table 6.1). This is a very high match factor, and one could think that the annotation should be reliable. If we take a look at the second hit, we observe that the spectral match with isoleucine (Ileu) is also very high (99%). It is a well-known fact that

Table 6.1 Initial annotation list

AlignID	RT	Name 1 (1st hit)	MF 1	Name 2 (2nd hit)	MF 2
99	5.85	Lactic acid (2TMS)	88.08	Butanoic acid, 3-hydroxy- (2TMS)	81.50
152	6.44	Alanine (2TMS)	99.72	Tartronic acid, 2-(methylaminomethyl)- (1MeOx) (3TMS)	99.66
171	6.67	Glycine (2TMS)	98.90	Unknown#bth-pae-008	98.58
204	7.16	Leucine (1TMS)	99.30	Isoleucine (1TMS)	99.24
208	7.27	Butanoic acid, 3-hydroxy- (2TMS)	97.22	Lactic acid (2TMS)	87.90
211	7.27	Heptanoic acid, n- (1TMS)	96.63	Alanine, beta- (1TMS)	73.45
224	7.46	Isoleucine (1TMS)	99.62	Leucine (1TMS)	99.53
302	8.64	Serine (2TMS)	93.34	Alanine (2TMS)	92.90
340	9.34	Glycine (3TMS)	98.61	Isobutanoic acid, 3-amino- (3TMS)	97.38
374	9.83	Fumaric acid (2TMS)	96.62	Unknown	92.36
396	10.12	Serine (3TMS)	93.72	Sphingosine (3TMS)	83.59
583	12.55	Cysteine (3TMS)	93.28	Cysteinyl-glycine (3TMS)	73.57
609	12.83	Serine (4TMS)	96.26	Histidinol (4TMS)	90.69
759	14.80	Ornithine (3TMS)	94.08	Ornithine, N2-acetyl- (3TMS)	92.88
773	15.05	Glycerol-3-phosphate (4TMS)	94.18	Unknown	84.21
811	15.56	Ornithine (4TMS)	98.42	Proline (2TMS)	86.53
814	15.64	Citric acid (4TMS)	98.20	Isocitric acid (4TMS)	67.04
911	17.61	Tyrosine (2TMS)	94.57	Lactic acid, 3-(4-hydroxyphenyl)- (3TMS)	93.32
912	17.66	Hexadecanoid acid (1TMS)	87.35	Octadecenoid acid, 9-(E)- (1TMS) similar to inositol (6TMS)	93.03
950	18.41	Inositol, myo- (6TMS)	93.59	Similar to inositol (6TMS)	93.03
952	18.45	Uric acid (4TMS)	95.64	Unknown	66.85
1044	20.11	Cystine (4TMS)	90.62	Tryptophan, 5-hydroxy- (3TMS) MP	80.66

The table shows the AlignID (unique spectrum ID number), the mean RT in minutes, and the names of the metabolite identities of the first and second hit and their corresponding match factor (MF), i.e., spectral similarity value (%). The corresponding number of trimethylsilyl (TMS) and methoxyamine (MeOx) derivative groups for each metabolite is shown in parentheses

Leu and Ileu can be easily mistaken one for the other with mass spectrometry, but this case illustrates why considering the other hits reported by eRah is also important. Since it is difficult to know whether this spectrum is in fact Leu or Ileu, we will not include this metabolite in our initial annotation list. We will however go back to this example later to see how we can know which of the two isomers is the correct identity.

Let's now focus on another deconvolved spectrum at 15.64 min that is matched as first hit to citric acid with an MF of 98%. This is a high match factor which, again, indicates a high likelihood of the identification being correct. The second most plausible hit is isocitric acid, a structural isomer of citric acid, with an MF of 67%. It is very likely that this spectrum corresponds to citric acid, as its isomer fragmentates very distinctively, yielding a lower MF when compared to the reference spectra of citric acid. Evidently, it could be the case that this is not citric acid and that the actual metabolite's spectrum is missing in the database used for spectral matching. However, if that is the case, the actual metabolite would be a (stereo)isomer or a very similar metabolite, since the spectral similarity with citric acid is very high (97%). Given that citric acid's match has a high MF and that we can discard the possibility that the identity of this spectrum is the most structurally similar known isomer of citric acid, we will include citric acid in our initial annotation list.

Let's take a look at another example, where two deconvolved spectra (6.4 and 6.6 min) are matched as first hit to alanine and glycine with an MF of 99 and 98%, respectively (Table 6.1). If we observe their corresponding second hits (tartronic acid, 2-(methylaminomethyl)- (1MeOx) (3TMS) and Unknown#bth-pae-008 in Table 6.1), we observe that their MF is also very high. Tartronic acid-2-(methylaminomethyl) is known to elute at a later retention time; therefore, we can discard the second hit for the first spectra. For the second case, the match to an unknown metabolite is due to the spectral library used, GMD, which contains unknown but observed spectra, i.e., known unknowns. The case for the unknown metabolite is not trivial, but we will assume that this spectrum corresponds to glycine, first, because it is an amino acid that is typically observed in serum with GC-MS and there are no other deconvolved spectra matching to glycine in the identification list and, second, because we will later curate this initial annotation list by computing the RT/RI curve that will allow us to distinguish among metabolites with similar spectra but eluting at different RT (different RI).

In other cases, visual inspection of the spectral matching can contribute to distinguishing the true identity among the different hits. This is the case of lactic acid (2TMS), serine (4TMS), and ornithine (3TMS) (Table 6.1). Using the plotSpectra function as described in Sect. 6.3.8, we can visualize the spectral comparisons among the first, second, or other hits. Figure 6.6 shows the visual spectral comparison of the first and the second hit of lactic acid (2TMS), serine (4TMS), and ornithine (3TMS), where careful inspection of selective fragment peaks can help us determine whether the first or the second hit is the most plausible identity. It is worth noting that the MF differences between the first and the second hit in these cases range from 1 to 6% (Fig. 6.6 and Table 6.1). The case of serine (2TMS) shown in the previous section (Fig. 6.5) serves also as a good example. We will add lactic acid (2TMS), serine (2TMS), serine (4TMS), and ornithine (3TMS) to the initial annotation list.

Finally, and before computing the RT/RI curve, let's go back to the Leu/Ile case, corresponding to a spectrum at 7.16 min. Leu and Ile are essential amino acids that are observed in serum with GC-MS. In fact, there is another spectrum eluting at 7.45 min that matches to Ile. It is then safe to assume that these two spectra are either Leu and Ile or Ile

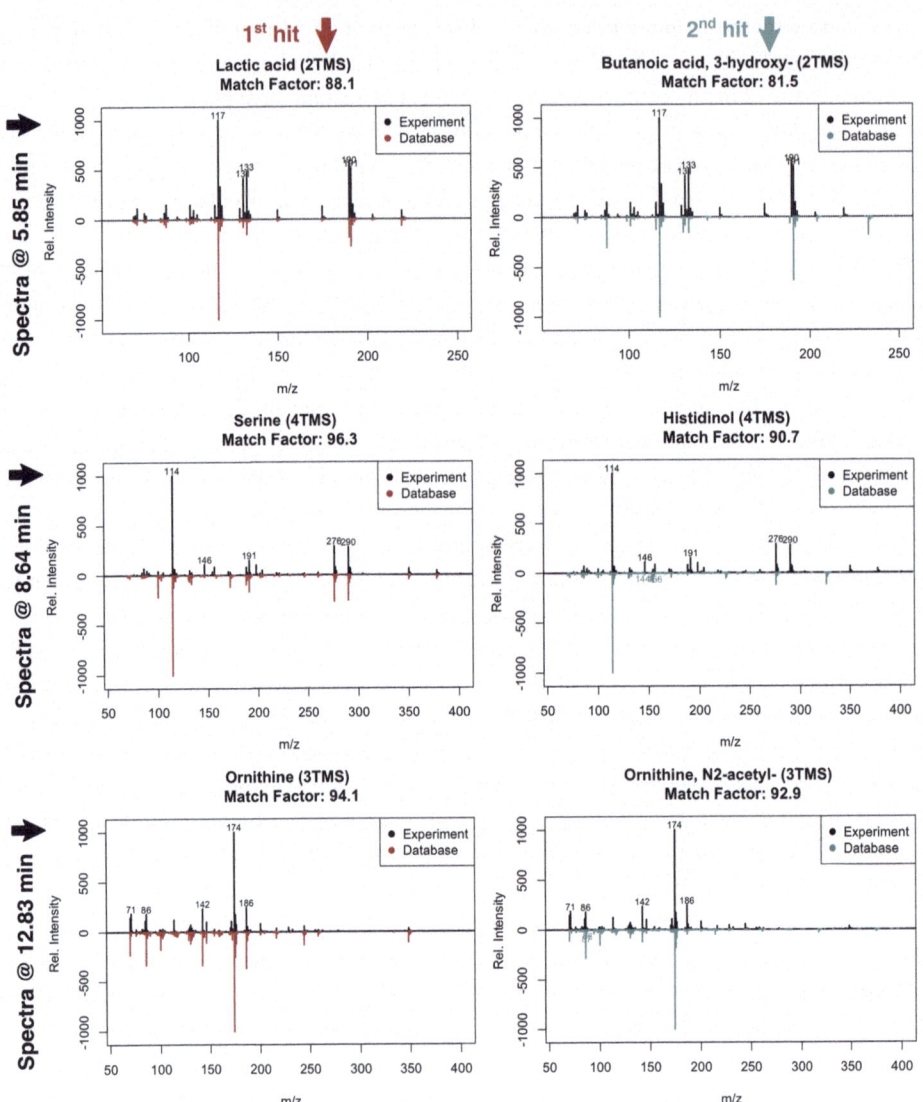

Fig. 6.6 Visual spectra comparisons. Each row shows the same experimental deconvolved (black) vs the first most similar match or hit (red, inverted, left column) and the second most similar hit (blue, inverted, right column). Spectral differences as observed from the presence or absence of some selective ion fragments in either the experimental or database spectra can contribute to associate the most plausible identity to each experimental spectrum

and Leu. By considering the elution order to know which of the two metabolites elutes first, we should be able to assign these two spectra as Leu and Ile or Ile and Leu, respectively. We can find their reference RI value of Leu and Ile by using the DB Index (DB.Ind.1) value in the identification list through the `idList` function):

```
> mslib@database[[100]]$RI.VAR5.ALK
[1] 1151.06
> mslib@database[[109]]$RI.VAR5.ALK
[1] 1174.98
```

Leu has a lower RI which means that it will always elute before Ileu. Knowing that the elution order is Leu/Ileu, we can add these identities to the initial annotation list.

Using all these strategies and considering also the RT of previous experiments in which we have used standards, we can generate our initial annotations list. Table 6.1 shows our suggested annotation list. We will now compute a preliminary RT/RI curve to, first, discard incorrect annotations in our initial annotation list and, second, to compute RI values for all the spectra using the naturally occurring metabolites in the annotation list. By using the AlignID values of our annotation list, we can display the RT/RI curve by:

```
> putAnn <- c(99,302,609,759,204,224,152,171,814,340,374,396,583,773,
811,950,911,912,952,1044,211,208)
> showRTRICurve(ex, reference.list=list(AlignID=putAnn), nAnchors=4, ri.
thrs='1R')
The points outside the RI threshold are: 911, 912, 1044
```

Please note that the AlignID used in this protocol are just used as examples, and they might not correspond to those retrieved by the user after processing their samples with eRah in another computer. Readers should replace the AlignID in this protocol for those obtained with their experiment.

The showRTRICurve function uses two additional parameters. The ri.thrs paramerer defines an RI error threshold (in this case, 1R is used for a 1% relative RI threshold, whereas, e.g., 5A would imply five absolute RI units). This RI threshold is used to help us to identify which annotations fall outside the expected RI range. A 1% relative RI error or five absolute RI units are the established thresholds [18], i.e., the experimental RI values of the correct identities should fall under this threshold. The nAnchors parameter defines the number of values to trace the RT/RI curve: the lower the value, the more precise the curve should be as it will average the RI and RT values of the naturally occurring metabolites annotated—which are also affected by an intrinsic RT drift. However, a lower value will use a lower number of references to trace the RT/RI curve, thus decreasing the importance of the naturally occurring metabolites used to build a curve, e.g., if the RT/RI curve is nonlinear, two anchors will probably fail in correctly modelling the curve.

The function displays the resulting curve (Fig. 6.7) and also prompts the AlignID values that fall outside the defined RI error (red dots). The RI value of each metabolite or analyte encodes the relative RT between the nearest reference compound (n-alkanes, FAMEs or in this case, naturally occurring metabolites) eluting immediately before and after each metabolite or analyte [19]. Visually, this means that the RI of the correct metabolite identity and the two adjacently eluting references should display a linear relation. As observed, a linear relation is observed for most of the annotations. There are three annotations (red circles) that, visually, do not show a linear relation with adjacently eluting metabolites and,

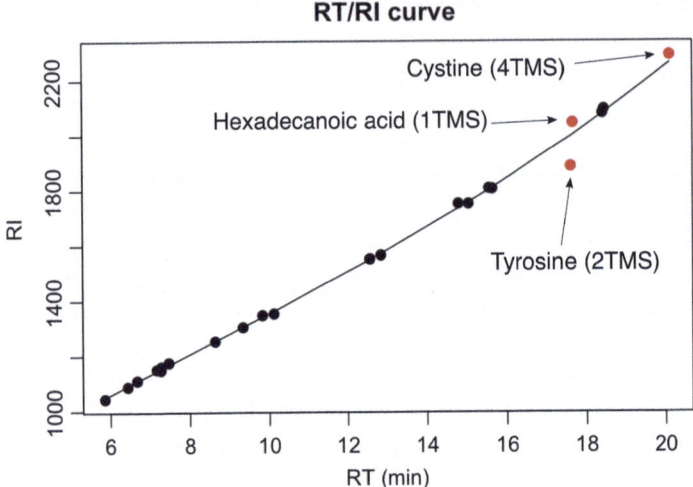

Fig. 6.7 Initial RT/RI curve (black solid line) generated using the initial annotation list. Black dots represent the corresponding RT/RI values of the metabolite's identities used to build the curve that falls below the selected error threshold (1% RI error), whereas red dots are the ones that fall outside the error threshold

quantitatively, show an RI error above 1%. These are the spectra matched to tyrosine, hexadecenoic acid, and cystine. We discard the annotation of hexadecanoic acid. For the case of tyrosine, eluting at 17.61 min, we observe that another spectrum eluting at 16.19 min also matches to tyrosine (MF 98%), and we replace this identity in our annotation list. We keep cystine in the annotation list as the incorrect annotations of hexadecenoic acid and tyrosine could have biased the calculation of the RT/RI curve leading to the annotation of cystine falling outside the RI error. We recompute the RT/RI curve with the updated annotation list:

```
> showRTRICurve(ex, reference.list=list(AlignID=putAnn), nAnchors=4,
ri.thrs='1R')
No points outside the selected RI threshold
```

The new resulting curve (Fig. 6.8) is now consistent. This exercise also shows why identifications based solely on spectral matching have to be taken with caution. We can now use the metabolites identified in our annotation list as reference for building our reference RT/RI curve and compute RI values for all the observed spectra. We can compute RI values, as described in Sects. 6.3.7.2 and 6.3.7.3 using:

```
> ex <- computeRIerror(ex, mslib, reference.list=list(AlignID=putAnn))
> ## We need to update the ID list by:
> id.list <- idList(ex)
```

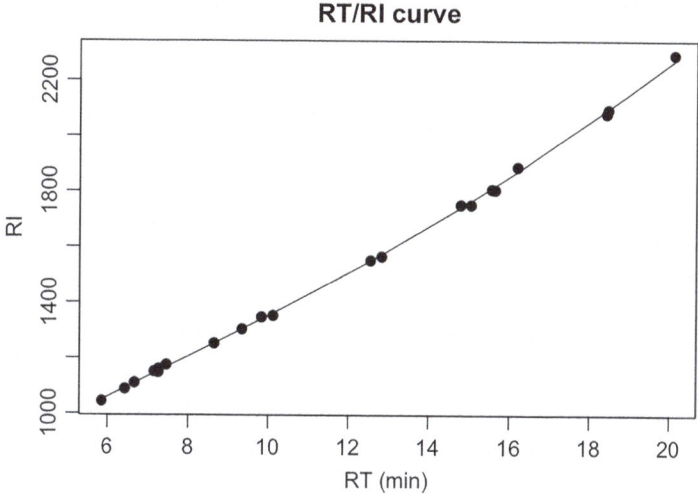

Fig. 6.8 Final RT/RI curve (black solid line) generated using the curated annotation list. Black dots represent the corresponding RT/RI values of the metabolite's identities used to build the curve that fall within the selected threshold (1% RI error)

The id.list now contains the identification list with the MF and the RI errors for all the deconvolved spectra. Users can choose between relative (default) or absolute RI error calculation via the `ri.error.type` parameter of the function `computeRIerror`. The RI errors are included in the columns RI.error.1, RI.error.2, and so on for as many putative hits as retrieved. Table 6.2 shows some examples that we will now discuss.

The spectrum at 10.48 min (Table 6.2) is matched to threonine as first hit (MF = 95%, RI error = 0.78%), but allothreonine is its second hit (MF = 88%, RI error = 1.2%). If we rank the results by MF and RI, both spectral and RI error data suggest that this spectrum is threonine. The spectrum at 12.38 min shows RI errors of 0.99% for the phenylalanine hit and 0.29% for purine hit. Although the low RI error for the purine hit could suggest that this spectrum is purine, by considering both the MF and RI errors, the phenylalanine match is more plausible given that its RI error is below the established 1% RI error threshold and that it shows a higher spectral similarity (MF = 97%, compared to MF = 90% of purine). Visual spectral comparison (using the `plotSpectra` function) also supports the hypothesis that this spectrum corresponds to phenylalanine. On the other hand, the spectra at 13.57 min and 14.80 min match to dodecanoic acid and ornithine, respectively, but both first and second hits have similar MF values, making these spectra hard to identify by MF alone. RI errors clearly indicate that the first hit is the correct annotation. It is not always the case that the first hit corresponds to the true spectra identity. For example, the spectra at 12.14 min and 16.8 min match to 2-piperdinecarboxylic and levodopa as first hit, but to pyroglutamic acid and tyrosine as second hit, with both first and second hits having a similar MF. RI errors clearly indicate that the second hit is the correct identity.

Table 6.2 Identification examples resulting from calculating the RI errors with the RT/RI curve

AlignID	RT	Name 1 (1st hit)	MF 1	RIe 1 (%)	Name 2 (2nd hit)	MF 2	RIe 2 (%)
416	10.48	Threonine (3TMS)	95.56	0.78	Threonine, Allo- (3TMS)	88.16	1.2
546	12.14	2-Piperidinecarboxylic acid (2TMS)	99.21	10.31	Pyroglutamic acid (2TMS)	99.16	0.02
569	12.39	Phenylalanine (1TMS)	97.98	0.99	Purine	90.05	0.29
674	13.57	Dodecanoic acid (1TMS)	94.37	0.56	Lactic acid dimer (2TMS)	90.72	16.02
759	14.80	Ornithine (3TMS)	94.08	0	Ornithine, N2-acetyl- (3TMS)	92.88	14.09
859	16.80	Levodopa (4TMS)	93.76	7.97	Tyrosine (3TMS)	92.56	0.08

The table shows the AlignID (unique spectrum ID number), the mean RT in min, and the names of the metabolite identities of the first and second hit and their corresponding match factors (MF), i.e., spectral similarity value (%) and the relative RI errors (RIe) for the first and second hit. The corresponding number of trimethylsilyl (TMS) derivative groups for each metabolite is shown in parentheses

In summary, by considering both MF and RI errors, in addition to visual inspection of spectral matches to the reference database, users can readily annotate and identify metabolites in samples without the need of specific RI calibrants like FAMEs or alkanes. Of note, the use of n-alkalanes or FAMEs for RI determination is recommended because they elute equidistantly (relative to the temperature gradient) over a wide RT range. The natural occurring metabolites selected should mimic these properties, assuring that there are not underrepresented RT regions where no natural occurring metabolites are used as reference. Also, in this example we used metabolites that covered the RT range from 6 to 20 min (Figs. 6.7 and 6.8). Computed RI values outside the minimum and maximum RT value of the metabolites used to compute the RT/RT curve should be used with caution.

Finally, it is worth assessing the quality of the deconvolved profiles of the identified or statistically significant dysregulated metabolites using the plotProfile function (see Sect. 6.3.8). As shown in Fig. 6.9, in most cases, eRah will correctly deconvolve the original peak shape (in case of co-elution) of the most intense peak of each metabolite spectrum, as in the case of threonine, even when the metabolite's peak intensity is low, as in the case of fumaric acid (Fig. 6.9). In other cases, eRah might fail in correctly deconvolve the original peak shape for low-intensity peaks, peaks affected by noise, or peaks under strong co-elution with other compound peaks, as in the case of cysteine (Fig. 6.9). In other cases, like alanine (Fig. 6.9), eRah is capable of correctly deconvolving the peak profile,

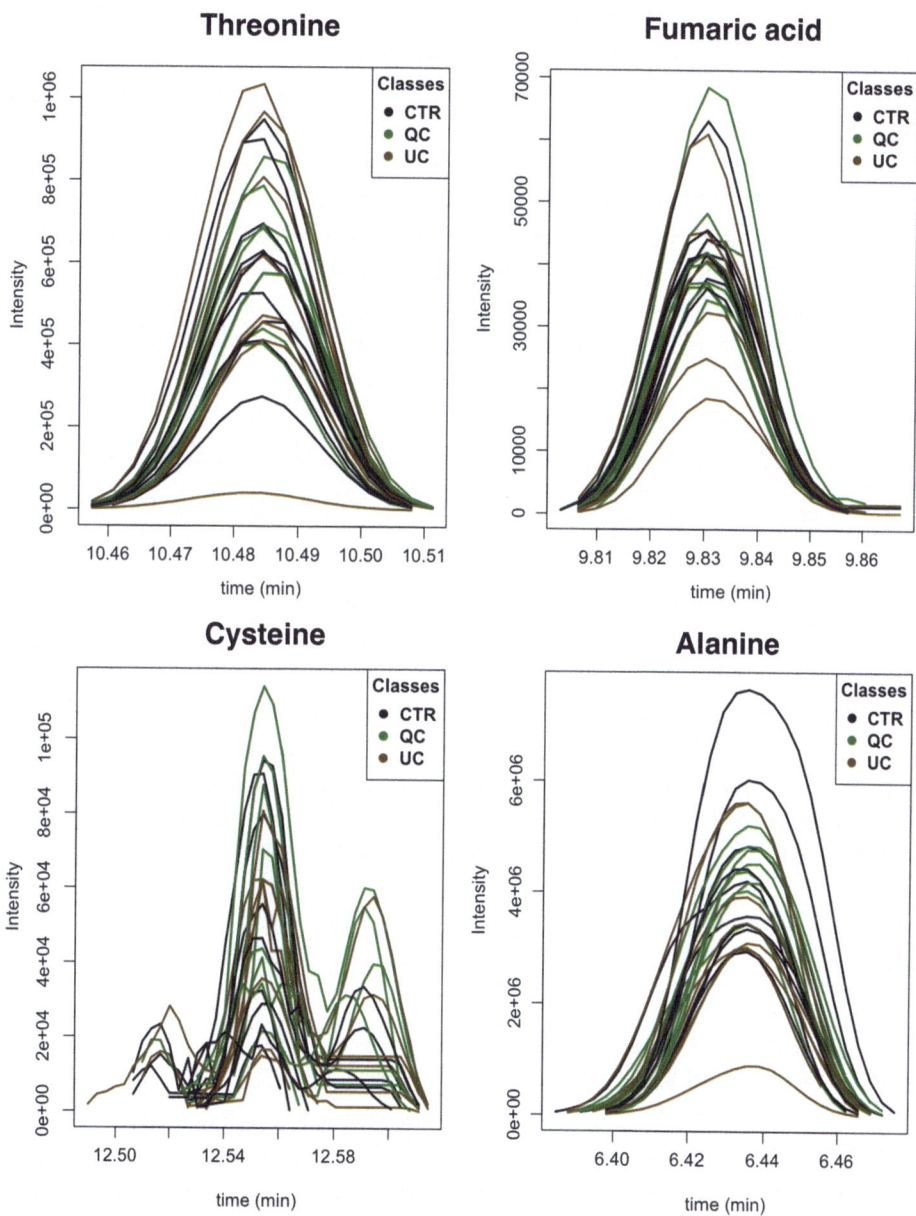

Fig. 6.9 Deconvolved peak profiles of threonine, fumaric acid, cysteine, and alanine, where each color represents the peak in a specific sample class: control (CTR), ulcerative colitis (UC), and quality control (QC) samples. Threonine, fumaric acid and alanine are correctly deconvolved, but, in some samples, alanine peaks are flattened due to the detector saturation. Cysteine peak is incorrectly deconvolved

but, in some samples, the peak flattens due to the saturation of the detector, and this effect should be taken in consideration in the statistical analysis.

Take-Home Message

- eRah processes GC-MS samples for untargeted metabolomics data analysis.
- In eRah, the analysis entity is the deconvolved spectra (a group of peaks stemming from a unique metabolite) as opposed to the single peak or feature as the analysis entity typically used in LC-MS.
- Spectral matching provides a list of metabolite identities for each of the deconvolved spectra in samples.
- eRah can compute retention indices to provide robust metabolite annotations and distinguish these cases where a deconvolved spectrum matches to multiple similar spectra in a database.
- Retention index can be calculated without the need of n-alkanes or FAMEs, using instead naturally occurring metabolites.

Acknowledgments This work was partially funded by the Spanish State Research Agency (AEI/10.13039/ 501100011033) grant PID2019-106277RA-I00 (X.D.-A.), by "la Caixa" Foundation (ID 100010434) via the Junior Leader Fellowship LCF/BQ/PR21/11840001 (X.D.-A.), by the European Commission's Horizon 2020 Research and Innovation Program via the GLOMICAVE project under grant agreement No. 952908 (X.D.-A.) and via the Innovative Training Network Marie Skłodowska-Curie COL_RES project under the grant agreement No 956279 (X.D.-A and T.A.), and by the Centre for the Development of Industrial Technology (CDTI) of the Spanish Ministry of Science and Innovation under grant agreement CER-20191010. S.M.dC. acknowledges the financial support of the Vicente Lopez Fellowship by the Fundació Eurecat. S.M.dC., T.A., and A.O. contributed equally.

References

1. Capellades J, Junza A, Samino S, Brunner JS, Schabbauer G, Vinaixa M, Yanes O. Exploring the use of gas chromatography coupled to chemical ionization mass spectrometry (GC-CI-MS) for stable isotope labeling in metabolomics. Anal Chem. 2021;93(3):1242–8.
2. de Cripan SM, Cereto-Massagué A, Herrero P, Barcaru A, Canela N, Domingo-Almenara X. Machine learning-based retention time prediction of trimethylsilyl derivatives of metabolites. Biomedicine. 2022;10(4):879.
3. Domingo-Almenara X, Montenegro-Burke JR, Benton HP, Siuzdak G. Annotation: a computational solution for streamlining metabolomics analysis. Anal Chem. 2018;90(1):480–9.
4. Smith CA, Want EJ, O'Maille G, Abagyan R, Siuzdak G. XCMS: processing mass spectrometry data for metabolite profiling using nonlinear peak alignment, matching, and identification. Anal Chem. 2006;78(3):779–87.
5. Kuhl C, Tautenhahn R, Böttcher C, Larson TR, Neumann S. CAMERA: an integrated strategy for compound spectra extraction and annotation of liquid chromatography/mass spectrometry data sets. Anal Chem. 2012;84(1):283–9.

6. Tsugawa H, Cajka T, Kind T, Ma Y, Higgins B, Ikeda K, Kanazawa M, VanderGheynst J, Fiehn O, Arita M. MS-DIAL: data-independent MS/MS deconvolution for comprehensive metabolome analysis. Nat Methods. 2015;12(6):523–6.

7. Smirnov A, Qiu Y, Jia W, Walker DI, Jones DP, Du X. ADAP-GC 4.0: application of clustering-assisted multivariate curve resolution to spectral deconvolution of gas chromatography-mass spectrometry metabolomics data. Anal Chem. 2019;91(14):9069–77.

8. Domingo-Almenara X, Brezmes J, Vinaixa M, Samino S, Ramirez N, Ramon-Krauel M, Lerin C, Díaz M, Ibáñez L, Correig X, et al. ERah: a computational tool integrating spectral deconvolution and alignment with quantification and identification of metabolites in GC/MS-based metabolomics. Anal Chem. 2016;88(19):9821–9.

9. Domingo-Almenara X, Brezmes J, Venturini G, Vivó-Truyols G, Perera A, Vinaixa M. Baitmet, a computational approach for GC–MS library-driven metabolite profiling. Metabolomics. 2017;13 (8)

10. Domingo-Almenara X, Perera A, Ramírez N, Cañellas N, Correig X, Brezmes J. Compound identification in gas chromatography/mass spectrometry-based metabolomics by blind source separation. J Chromatogr A. 2015;1409:226–33.

11. Caballol B, Gudiño V, Panes J, Salas A. Ulcerative colitis: shedding light on emerging agents and strategies in preclinical and early clinical development. Expert Opin Investig Drugs. 2021;30(9): 931–46.

12. Hummel J, Selbig J, Walther D, Kopka J. The Golm Metabolome Database: a database for GC-MS based metabolite profiling. In: Topics in current genetics. Berlin, Heidelberg: Springer; 2007. p. 75–95.

13. Vandendool H, Kratz PD. A generalization of the retention index system including linear temperature programmed gas-liquid partition chromatography. J Chromatogr A. 1963;11:463–71.

14. Cuadros-Inostroza A, Caldana C, Redestig H, Kusano M, Lisec J, Peña-Cortés H, Willmitzer L, Hannah MA. TargetSearch – a Bioconductor package for the efficient preprocessing of GC-MS metabolite profiling data. BMC Bioinformatics. 2009;10(1):428.

15. Kind T, Wohlgemuth G, Lee DY, Lu Y, Palazoglu M, Shahbaz S, Fiehn O. FiehnLib: mass spectral and retention index libraries for metabolomics based on quadrupole and time-of-flight gas chromatography/mass spectrometry. Anal Chem. 2009;81(24):10038–48.

16. Horai H, Arita M, Kanaya S, Nihei Y, Ikeda T, Suwa K, Ojima Y, Tanaka K, Tanaka S, Aoshima K, et al. MassBank: a public repository for sharing mass spectral data for life sciences. J Mass Spectrom. 2010;45(7):703–14.

17. Vinaixa M, Samino S, Saez I, Duran J, Guinovart JJ, Yanes O. A guideline to univariate statistical analysis for LC/MS-based untargeted metabolomics-derived data. Meta. 2012;2(4):775–95.

18. Strehmel N, Hummel J, Erban A, Strassburg K, Kopka J. Retention index thresholds for compound matching in GC-MS metabolite profiling. J Chromatogr B. 2008;871(2):182–90.

19. Babushok VI. Chromatographic retention indices in identification of chemical compounds. Trends Anal Chem. 2015;69:98–104.

Part IV

From Statistical Data Analysis to Insights into Biology

Workflow for Knowledge Discovery from Metabolomic Data Using Chemometrics

Miguel de Figueiredo, Serge Rudaz, and Julien Boccard

What You Will Learn in This Chapter

Because of its ability to provide an in-depth characterization of the biochemical events governing living systems, metabolomics constitutes today a potent approach for the investigation of biological phenomena and the discovery of biomarkers related to health and disease. Current analytical instrumentation can simultaneously monitor thousands of metabolites in a single analysis, and grasping meaningful information from these large and complex datasets requires dedicated workflows. For that purpose, chemometrics offers adapted methods accounting for both the nature and the structure of the collected datasets. This chapter aims to present the principles and practical applications of the main chemometric tools that have become essential to extract relevant variable patterns as a prerequisite to knowledge discovery in metabolomics. This includes normalization strategies to make the samples comparable, exploratory approaches to highlight subsets or trends in the data, predictive modeling to extract specific metabolic patterns, as well as the investigation of specific factor effects (e.g., patients, treatment, time) via designed experiments. Particular emphasis is placed on the data structure that can be generated during a longitudinal follow-up and its intrinsic characteristics as well as on methodological guidelines to treat the data in the most efficient way.

M. de Figueiredo · S. Rudaz · J. Boccard (✉)
School of Pharmaceutical Sciences, University of Geneva, Geneva, Switzerland

Institute of Pharmaceutical Sciences of Western Switzerland, University of Geneva, Geneva, Switzerland
e-mail: julien.boccard@unige.ch

7.1   Introduction

As it aims at the concurrent monitoring of a large number of metabolites within a single analysis, metabolomics intrinsically leads to the generation of massive datasets, no matter which analytical platform is used (nuclear magnetic resonance spectroscopy, NMR, or mass spectrometry, MS, with or without prior chromatographic separation) and which data acquisition methodology is implemented (targeted or untargeted). These high-dimensional data matrices necessitate dedicated workflows to extract meaningful patterns and/or trends from the large number of recorded signals. The first preprocessing step is often specific to the nature of the measured data, whether they are one-dimensional spectra (wavelengths or chemical shifts) or more complex structures such as two-dimensional NMR analyses or hyphenations with MS. The annotation of metabolites on the basis of preprocessed signals is also an area that has specificities related to the analytical technique that was employed. Independently of the chosen methods, from the most standard to the most elaborate, these first steps of signal processing give rise to high-dimensional data tables. In that context, chemometrics provides efficient tools to unravel the different sources of variability existing in the data, in order to extract characteristic metabolic patterns to describe clinically relevant conditions [1]. For this process to be effective, it is essential to consider the structure of the data to be analyzed in order to use an appropriate methodology. This knowledge discovery workflow includes processing strategies to make the samples comparable, exploratory approaches to highlight the major trends in the data, and supervised modeling to distinguish specific situations, predict a response or outcome, and investigate the effect of specific factors.

7.2   Case Study Description

An observational monocentric study on chronic kidney disease (CKD) will be used as a working example throughout this chapter. It includes patients suffering from major alterations of renal function characterized by a very low glomerular filtration rate, including end-stage renal disease patients undergoing hemodialysis (HD). Plasma samples were collected from 69 prevalent non-HD CKD patients (CKD group) followed by nephrologists at the Geneva University Hospitals, as well as 35 CKD patients undergoing chronic in-hospital HD. For the latter, samples were drawn before (preHD group) and after (postHD group) hemodialysis. Sample preparation involved protein precipitation using cold methanol and centrifugation. All samples were analyzed by ultrahigh-pressure liquid chromatography in different separation modes coupled to quadrupole time-of-flight MS (Waters H-Class Acquity UPLC and Bruker maXis 3G Q-TOF high-resolution MS). The interested reader is referred to the cited references for a detailed description of the methods [2, 3].

Raw signal processing, including peak picking and alignment, was performed using Progenesis QI 2.3 (Nonlinear Dynamics, Waters, Newcastle upon Tyne, UK). Metabolite identifications with the highest level of confidence (i.e., level 1) were achieved using an in-house library containing experimental data from authentic standards acquired under the same chromatographic conditions [4]. Level 1 annotation was obtained by comparing m/z values, retention times, and isotopic patterns, with confirmatory information from collisional cross-section values and MS/MS spectra, when available. This process led to a dataset of 278 identified metabolites.

7.3   Make the Samples Comparable

As with any biochemical analysis, alterations of metabolic profiles related to a specific condition or due to an experimental factor may be hidden by extrinsic sources of variability. It is therefore essential to guarantee the comparability of the samples considered in a metabolomic study, regardless of the amount of biological material available (e.g., cell cultures), differences in concentration (e.g., urine), or any other artificial source of measurement variability (e.g., potential instrumental drifts). For this purpose, it is possible to use standard reference materials when the analysis is carried out on a limited number of target compounds. The complex and multicomponent nature of metabolic pathways makes this type of approach difficult to implement for obtaining a reliable evaluation of all measured signals in a biological sample. In this context, it is now strongly advised to systematically include quality control samples (QCs) in any analytical sequence to ensure proper comparison of samples by adequately preprocessing the data. Since co-eluting matrix components may modify the measured signal intensities, the composition of the QCs must therefore be biochemically as close as possible to the study samples. Therefore, a mixture of aliquots taken from each of the samples is often recommended to build a representative pool. QCs are then measured at regular intervals over the analytical sequence. These measurements are then specifically monitored to ensure the stability of the signal and to evidence possible issues during data acquisition (i.e., within and/or between batches). The QC mixture can also be diluted (e.g., by a factor 2), forming a diluted QC (dQC), which is analyzed alongside the QCs. Filtering and correction strategies based on chemometric tools can then be applied to ensure the relevance of the measured data [5].

Samples from the clinical study were randomly analyzed in 4 batches, while 15 quality QCs were used for system conditioning at the beginning of each batch. QC and dQC (1:1 dilution in water) samples were injected for data filtering, analytical variability evaluation, and normalization (one injection every five samples).

7.3.1 Feature Filtering

Despite considerable developments in instrumentation and methodology, a non-negligible part of the measured signals may be of insufficient quality, containing noise or artifacts. There are many reasons for this, such as low-concentration compounds at the detection limit of the instrument, overlapping chromatographic peaks, signal saturation, contamination, or degradation. It is therefore often suitable to apply a filtering strategy in order to clean the data from these signals, so as not to compromise the following steps of the workflow. First of all, it is useful to remove certain areas of the spectra or chromatogram of insufficient signal quality. It is also advisable to discard signals for which no intensity could be measured in a large proportion of the samples. More generally, a filter based on the stability of the signal in the QCs is frequently used. This is done by evaluating the relative standard deviation (RSD) in the repeated measurements of the QCs (and potential dilutions) and then applying a threshold (e.g., >30%) above which the signal is considered too unstable to be retained [6]. Furthermore, it is expected that dQCs signal intensity is lower in proportion to the dilution factor. A tolerance range around the theoretical value of the dQC/QC ratio can then be used to keep only the features following this trend.

> A threshold of 50% was applied as the upper limit of the dQC/QC ratio RSD. In addition, a dQC/QC ratio between 0.2 and 0.8 was considered acceptable around the theoretical value of 0.5 (1:1 dilution), and signals outside this range were removed.

7.3.2 Intra-batch and Inter-batch Correction

If the analysis sequence involves a large number of samples, the global signal intensity may be affected by a drift. Clogging, contamination, and many other phenomena can indeed lead to a decrease in the performance of the analytical device, thus compromising the proper characterization of the samples according to their position in the sequence. In this context, chemometric methods taking advantage from regularly injected QCs can be very useful to monitor the signal quality and correct for drifts when necessary. Because systematic variations generate artificial trends in the analytical sequence, they can be easily detected from random fluctuations that are equally distributed around an average value. These typical patterns can then be estimated, modeled, and corrected using different regression approaches, by fitting a mathematical function for each signal in the QCs to characterize its trajectory in the sequence. Standard approaches include ordinary least squares [7] or more sophisticated methods, such as cubic smoothing splines (QC-RSC) [8] and support vector regression (QC-SVRC) [9], as well as locally estimated scatterplot smoothing (LOESS) and its weighted alternative LOWESS [10]. The regression model is then used to correct signal intensity values in the samples. Even if different estimation strategies and functions can be used, they often give rise to comparable results, and simple models are generally preferred. Clustering can also be used to monitor the average intensity of homogeneous subsets of signals over the analytical sequence [11]. In this case, the correction can be done specifically for each cluster, thus allowing different trends in the sequence to be considered.

The occurrence of unwanted systematic differences is particularly crucial when several runs are needed to measure a large number of samples, for example, in a cohort study. Again, these discrepancies can be due to many factors such as variations of buffers or columns, instrument calibration, the degradation of the analytical system, etc. These effects can also be fixed in the same way using correction functions estimated on the basis of QCs or by accounting for the structure of the acquisition scheme during downstream data analysis [12, 13].

> LOESS regression involving a linear fit and an initial smoothing span of 0.75 was used for intra- and inter-batch normalization based on QCs. The span value was then optimized using cross-validation. Figure 7.1 summarizes the post-acquisition data preprocessing workflow.

7.3.3 Sample-Wise Normalization

Another source of unwanted variability is linked with discrepancies in overall signal intensities among samples collected from similar observations. Differences of concentration or amount of biological material, as well as unexpected technical or biological reasons, may necessitate a correction to ensure the comparability of the sample under study. In this case, an appropriate normalization factor needs to be carefully evaluated to correct these so-called size effects. On the one hand, a benchmark parameter can be evaluated, such as creatinine and osmolality in the case of urine analysis [14] or protein amount and cell count for tissue samples and cell cultures. To avoid relying on a single measurement, a group of reference compounds can be used to estimate a correction factor for each sample, but in any case, it should be noted that experimental or pathological conditions may potentially alter any parameter used to perform normalization.

On the other hand, the extensive data collection carried out makes it possible to evaluate normalization factors based on the dataset. The most commonly used approaches include total sum normalization (based on the sum of all signals) and MS total useful signal (based on the sum of "useful" signals common to all samples) [15]. It should be noted that such a global evaluation involves the additional constraint that the signals add to a common constant sum thus creating artificial links, making it more difficult to find biomarkers [16]. To avoid this issue, probabilistic quotient normalization (PQN) was developed as an alternative approach correcting size effects by distinguishing global sample variations related to the majority of measured signals from biological modulations affecting only a subset of biomarkers. To achieve this, ratios of intensities with a reference sample (e.g., the median of the QCs) are computed to estimate a normalization factor, by assuming that experimental factors or conditions should not alter more than half of the measured signals.

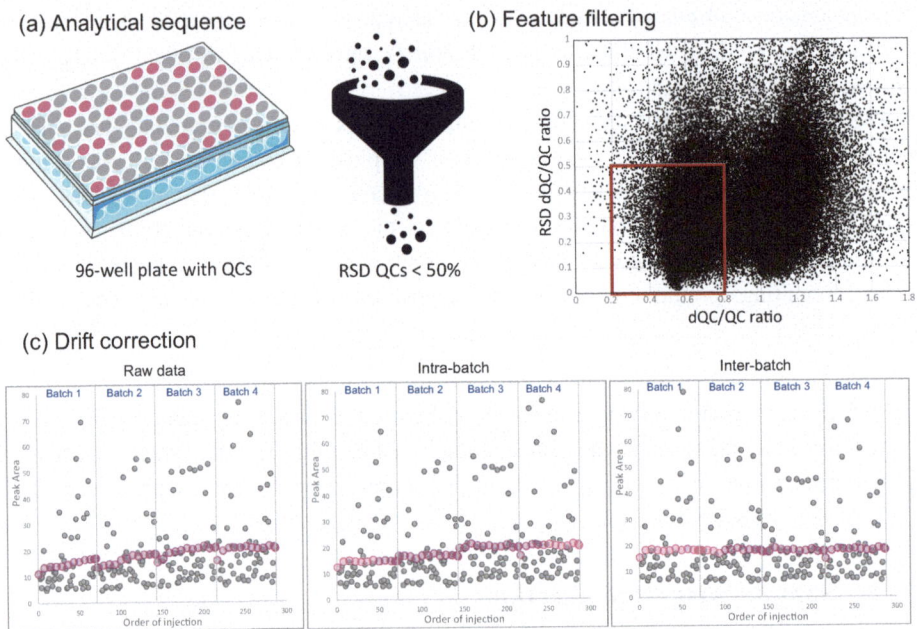

Fig. 7.1 Post-acquisition data preprocessing workflow: (**a**) design of the analytical sequence, (**b**) QC-based feature filtering, and (**c**) intra- and inter-batch drift correction

Alternatively, other methods based on ranks and quantiles constitute nonparametric strategies that can lead to efficient corrections [17].

Because it strongly affects glomerular filtration, CKD may have a major impact on the overall concentration of plasma samples, which is therefore a consequence of the disease. In this context, the choice was made not to apply any normalization strategy because it could have masked the effects of the pathology.

7.4   Assessment and Comparison of Metabolic Profiles

While the first steps of the workflow are crucial to ensure data quality and get rid of irrelevant signals, the discovery of proper metabolic information really starts with the investigation of metabolic patterns among the samples [18]. Starting from the data matrix obtained as output of preprocessing, in which the rows are observations (i.e., samples) and the columns are variables (i.e., signals), both the amplitude and variability of signal intensities related to metabolite abundances can be explored and compared using chemometric tools.

On the one hand, each variable can be considered separately using univariate methods, leading therefore to standard one-variable-at-a-time analysis. This can be of particular interest to have a first overview of potential modulations related to clinically relevant variables. On the other hand, multivariate approaches considering metabolic profiles in their entirety can be very useful when less or no information is available, to generate data-driven biochemical hypotheses from untargeted data acquisitions. In any case, these two approaches should be considered as complementary and can be combined to provide a complete picture of the metabolic events taking place.

7.4.1   One-Metabolite-at-a-Time Investigation

Probably the most standard manner to evaluate changes among the samples under study consists in the investigation of a sole variable, whether it is associated with a peak area, an ion intensity, or the abundance level of an annotated/identified metabolite. While simple descriptive statistics already give information on both the range and the variability of its values, assessing ratios between central tendency values (mean or median) offers a first overview of potential trends among the groups of observations considered (e.g., samples from healthy and diseased people). The obtained fold change is an intuitive metric to compare different conditions against a given baseline, but it lacks the ability to take the variability of the values into account. To improve the objectivity of the comparison, parametric or nonparametric approaches can be used to carry out univariate hypothesis testing and assess whether an observed test statistic can be considered as statistically significant, i.e., due to a real difference between the groups of samples and not to random chance. The subsequent *p-value* of the test is a straightforward and classical parameter to report comparison results, even if its use is regularly criticized. Choosing a suitable test should be done according to the number of observations, as well as the distribution and variability of the values within the different groups. Typical hypothesis tests for continuous variables include Student's and Welch's t-tests, the one-way analysis of variance, the Wilcoxon rank-sum test, and the Kruskal–Wallis test [19].

When many signals are measured in a single experiment, these univariate hypothesis tests can be massively performed in parallel, thus leading to an increasing risk of false positives (type I error). To avoid this multiple-testing problem, different solutions have been developed to correct the threshold used for reporting a difference as significant. While the standard *p-value* threshold to declare an observed difference as significant is 5%, the Bonferroni correction divides this value by the number of tests [20]. However, such an approach can be very restrictive when the number of tests is high, and the false discovery rate (FDR) is often preferred in practice. It is designed to control the error rate among all rejected null hypotheses (i.e., differences considered as significant) to a defined significance level (*q-value*), and different methods can be used to achieve this [21]. The Benjamini-Hochberg approach is probably the most widely used to avoid an increase of the frequency of false negative results (type II error) [22].

When multiple variables are evaluated in parallel, the volcano plot is another efficient tool to get a global overview of the variables and visualize all results at once. It combines statistical significance (usually the negative log10 of the *p-value* or *q-value*) and ratios (usually the log2 of the fold change, FC) between conditions on a scatter plot. As a result, up and down differences of equal magnitudes are equidistant from the center to either the left or the right, while highly significant metabolites with low *p-values* are located at the top, thus enabling an easy detection of the most relevant alterations.

The receiver operating characteristic (ROC) curve is another effective graphical tool to assess the ability of a biomarker to separate two potentially overlapping populations (e.g., case and control) [23]. It is based on the evaluation of two parameters, i.e., the true positive fraction (sensitivity) and the false positive rate (1-specificity), for each value taken by the biomarker in the dataset as a possible decision threshold. While the shape of the curve can be used to evaluate a satisfactory boundary value offering an adequate tradeoff between sensitivity and specificity, the area under the ROC curve (AUROC) is an overall diagnostic statistics of the biomarker predictive ability (an AUROC of one, i.e., 100%, corresponds to perfect discrimination, while random chance leads to 50%).

Univariate comparisons between preHD and postHD were carried out using paired t-tests and *q-values* that were computed following the procedure introduced by Benjamini and Hochberg to estimate the FDR. Figure 7.2a shows a volcano plot where metabolites are deemed significant when their *q-value* and FC absolute value are lower than 0.05 and greater or equal to 2, respectively. Metabolites with a log2 (FC) smaller than -1 or greater than 1 show on average, respectively, values at least twice as low (down-modulated) or twice as high (up-modulated) in the postHD samples with respect to their preHD value. Figure 7.2b shows ROC curves for three metabolites identified in Fig. 7.2a, along with their associated AUROC. The closer to 1 the AUROC is, the greater the discrimination between the two situations, preHD and postHD.

7.4.2 Dimensionality Reduction Strategies

Due to their simplicity of use, one-metabolite-at-a-time approaches overlook metabolic information related to changes in the relationships between variables, i.e., alterations of covariance or correlation patterns. Investigating these synergistic effects is however crucial when complex interconnected metabolic pathways are involved, and this can only be achieved by simultaneously considering the metabolites carrying this metabolic information [24]. To achieve this goal, dimensionality reduction strategies provide efficient tools to extract structures of metabolic alterations related to subsets of variables that may be hidden in the data. The resulting multivariate models can then be used to bring subsets of biomarkers relevant to specific experimental groups or conditions to light and develop new biological hypotheses.

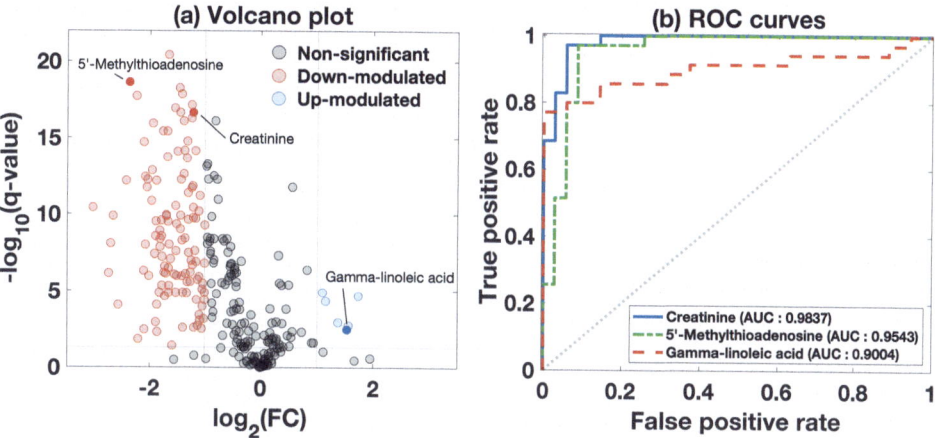

Fig. 7.2 One-variable-at-a-time analysis using (**a**) volcano plot to assert significance of metabolites and whether they are down- or up-modulated; (**b**) ROC curves to evaluate the discriminative power between preHD and postHD samples of three metabolites

7.4.2.1 Make the Variables Comparable

When considering a large number of metabolites simultaneously, the first step in their joint analysis is to make the different signals comparable. A first step of centering is usually performed to make the computation and interpretation of the model more straightforward, by setting the central tendency parameter of each variable, typically its mean, to the value of zero [25].

Moreover, signal intensities may have highly different ranges of variation, including scales and/or units, without direct link to metabolite abundances and their biological relevance. An objective evaluation of the different sources of variability is therefore difficult and a standardization step is useful. To do so, different scaling methods are available to ensure that each variable can contribute to the model, regardless of its magnitude. Comparability of variables is achieved by using a scaling factor that corresponds to the scale or dispersion of each variable. The most commonly used parameters are the mean or median for central tendency and standard deviation or median absolute value for dispersion. Using the latter as standardization factor gives rise to unit variance scaling and leads to build the model on the basis of correlations rather than covariances. When combined with mean centering, it is called autoscaling. In some applications, it may be desirable to retain some of the intensity information, while attempting to bring the variables into a comparable range, and Pareto scaling can be preferred. This strategy is supposed to reduce the dominance of the most intense signals while limiting the influence of noise at the other end of the intensity range. In this case, the standardization factor is the square root of the standard deviation of each variable. Alternative scaling methods based on different criteria to evaluate variable dispersion, such as range scaling and variance stabilization strategies [26], or vast scaling [27], are also available but are of less common use in metabolomics. Note also that it may be preferable not to apply standardization in certain cases, in particular when the information of the intensity of the peaks associated with the concentration of the compound is to be preserved (e.g., with quantitative analysis methods or when subsequent structural elucidation is mandatory) or to maintain the structure of the signals when spectra are measured.

On the other hand, a mathematical function can be used to transform the data and meet distributional assumptions (e.g., normal distribution with uniform variance), consequently improving further data modeling. Most transformations applied in metabolomics are based on log and power functions [28], to correct for the fact that the most intense signals often exhibit the greatest variability. These functions also have a pseudo-scaling effect because they reduce the differences between the most and least intense signals. Moreover, applying a log transform makes additive the multiplicative relations between the variables, which makes it possible to consider these relations in a linear factorial model.

Unit variance scaling was applied to standardize the variables and circumvent any impact of intensity range on the observed differences.

7.4.2.2 Data Exploration

The preliminary approaches used to investigate metabolomic data are very often exploratory. They are usually intended to provide a synthetic overview of the major sources of variability within the data to highlight potential subgroups of samples with comparable metabolic profiles. This can be very useful to ensure the quality of the collected data and the reliability of the preprocessing steps (e.g., by evaluating the distribution of QCs). The methods used are called unsupervised, as they do not consider the presence of classes of observations or the experimental setup, and therefore offer an overview of the samples mainly considering their global variability. This allows the detection of systematic biases, drifts, possible outliers or any other unexpected sources of variability that could impact the dataset.

Principal Component Analysis

Because of its ability to summarize multivariate datasets based on a reduced set of dimensions, principal component analysis (PCA) has become the essential tool to apply as a first step in the exploratory investigation of metabolic data. It provides an easy way to visualize the major variability trends in the dataset, as well as potential groupings or differences between observations without taking explicitly the experimental setup into consideration. PCA maximizes the variance-covariance structure explained by its components as linear combinations of the measured variables. Common patterns of variability shared by several metabolites, i.e., covariances or correlations, are therefore used as a basis to build a new subspace describing the dataset in an efficient way. Principal components (PCs) are related with variability patterns shared among subsets of variables, and they are ranked according to the proportion of the total variance that they capture. The scree plot is a visual diagnostic tool that enables displaying the variance explained by each dimension in a straightforward manner. PCA defines a system of mutually orthogonal axes that describe the largest sources of variance, and a reduced number of PCs is generally used for interpretation. Choosing an adequate number of components depends on the data at hand, and the model size is frequently limited to components with an interpretable distribution of observations, biologically relevant or not. Alternatively, a minimum threshold of explained variance can be chosen to determine the number of PCs.

The coordinates of the observations in this new subspace are called the PCA scores, while the contributions of the variables to build the components are called the loadings. The scores are very useful to detect groupings of observations sharing similar metabolic profiles based on their proximity in the model subspace. The loadings are used to interpret the directions of variation spanned by the PCs and explore the relationships between metabolites, as variables with comparable patterns among the samples are expected to be located close to one another. It is to be noted that the directions of variations are shared between score and loadings, thus allowing their simultaneous interpretation. Because PCA is a linear model, it may have some difficulties to summarize relevant information when the phenomenon of interest is associated with nonlinear relationships. In this case, other methods of dimensionality reduction can be considered.

PCA was carried out as a first exploratory step of multivariate analysis to get an overview of the main sources of variation in the dataset. Model interpretation was based on scores to assess potential sample subgroups as shown in Fig. 7.3a. The three groups of patients were distinguished on the first principal plane (i.e., t_1 vs. t_2). The first principal component t_1 explained 22.9% of the total variability, following a trend separating preHD from the two other groups. In addition, the CKD and postHD groups were differentiated using t_2 (7.3% of the total variability). PCA being unsupervised, it does not aim at separating these groups. However, trends in the data characterizing the main sources of variability could be associated with the groups under scope. Nevertheless, the separation between the groups in this example is not complete as illustrated by the overlap between the convex hulls drawn around each group of samples. Loadings in Fig. 7.3b can be used to evaluate the contribution of the metabolites responsible for the respective separations observed in t_1 and t_2. Metabolites with positive loadings in t_1 have higher values in preHD samples, whereas metabolites with positive loadings in t_2 have higher values in postHD samples.

HCA

Cluster analysis constitutes an alternative unsupervised strategy to detect groups of samples with similar metabolic profiles. It assumes a natural partitioning of the dataset that can be learned from the measured signals. Usually, decision criteria include compactness

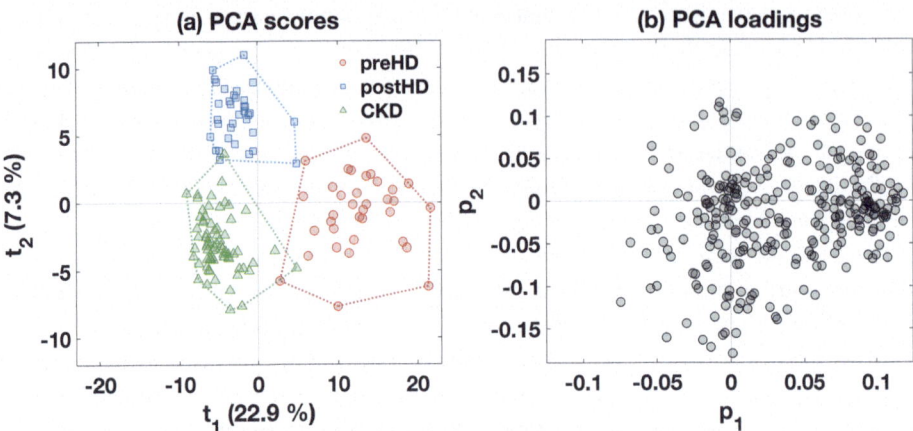

Fig. 7.3 Exploratory multivariate investigation of the data using PCA. First principal plane showing (**a**) the scores of the samples highlighting the separation between the groups; (**b**) the loadings characterizing metabolite contributions in differentiating the groups

(observations in the same cluster as similar as possible) and separability (clusters as different as possible). Hierarchical cluster analysis (HCA) is a widely used method in metabolomics for this purpose, and as implied by its name, it is based on a nested binary partition of the data space grouping observations using a succession of branches organized in a tree structure called dendrogram. The agglomerative approach starting from a series of individual clusters containing a single observation is the most common. Clusters gathering observations with homogeneous profiles are then formed by merging the two most similar groups. By repeating this procedure, a single cluster containing all observations is finally obtained. Interpretation is then done by inspecting the structure of the dendrogram and the length of the different branches it contains.

A key aspect of HCA is the way similarities (or conversely distances) are evaluated using a dedicated metric. Euclidean distance is commonly used, but there are alternatives with different properties (e.g., Mahalanobis, Manhattan, etc.). Similarly, a measure is needed to assess the distance between clusters containing more than one observation and select which groups should be associated. It can range from the simple average distance between observations of the two groups to more complex approaches involving groups barycenters and dispersions, such as Ward's method designed to minimize the increase in intra-class inertia (variability) when gathering two clusters [29]. All these aspects may have a strong impact on the shape of the resulting dendrogram and further interpretation. Resulting classes can be attributed by cutting the tree at an appropriate level to divide the dataset and specific criteria can be applied to objectively estimate the most relevant position (e.g., based on distances distribution or branch length). The dendrogram itself cannot be used directly to find out which variables contribute the most to differences between clusters, but it is possible to calculate average profiles from the different clusters for comparison purposes. Alternatively, a simultaneous analysis of clusters related to both observations and variables can be carried out using co-clustering (or biclustering) methods [30]. A heatmap is often used to link the two resulting dendrograms and visualize specific patterns in the dataset.

Co-clustering heatmaps in the spaces of the samples and metabolites can give visual information on which metabolites are responsible for the clustering of subsets of samples. Rows in Fig. 7.4 show the dendrogram grouping of the samples based on the patient groups (preHD, postHD, and CKD). The columns show the grouping of the metabolites. The upper rectangle in Fig.7.4 indicates that the grouped metabolites tend to have higher values in preHD samples (red) than in the other groups. The middle rectangle shows that the associated metabolites have lower values in postHD samples (blue) than in the other groups. The CKD group (green) is characterized by low values for the metabolites grouped by the upper rectangle and higher values for the metabolites of the bottom rectangle.

7.4.3 Extracting Specific Metabolic Patterns

Unsupervised multivariate methods are efficient investigation tools to handle a dataset and begin to understand which sources are responsible for the observed variability. However, metabolic patterns of specific interest, e.g., biological response linked to a particular condition, may not be directly associated with the major sources of variability in the data. Furthermore, exploratory models may suffer from limited interpretability because of greater, possibly unwanted, trends in the data. In that case, supervised approaches taking explicit advantage of the experimental design may be used to extract a particular source of relevant variation from metabolic profiles composed of a large number of signals using models intended to predict an outcome, e.g., a clinical condition. Modeling starts with a training phase, during which the relationships useful for the prediction of the Y-dependent response are derived from the independent data X, and a subsequent test phase is then achieved to evaluate the prediction accuracy of the model.

PLS and Extensions

Because its graphical outputs, i.e., score and loading plots, are similar to PCA, partial least squares (or projection to latent structures, PLS) regression constitutes a natural next step in multivariate analysis of metabolomic data widely used in practice. It also has the advantage of being able to handle datasets characterized by a limited number of observations but a large number of potentially noisy and correlated variables [31]. The principle of dimensionality reduction is also the basis of PLS regression, as the model will build a projection subspace based on components, called latent variables (LVs) in this case, that are mutually orthogonal [32]. The major difference with PCA is that LVs have the double purpose of explaining the variability of the data and predicting the response. This can be achieved by

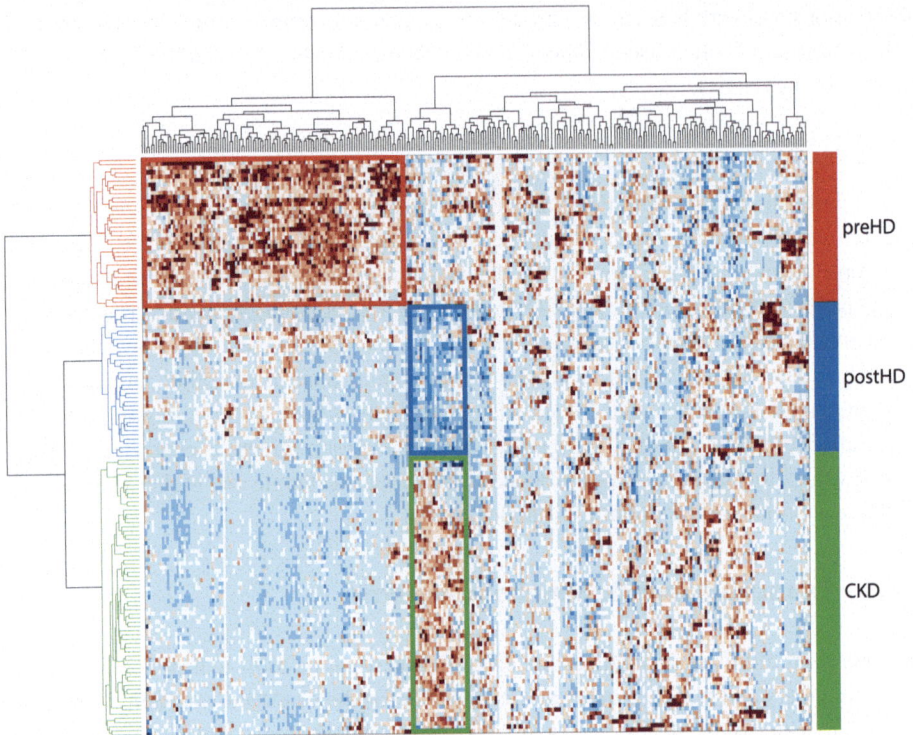

Fig. 7.4 Co-clustering heatmap of the samples (rows) and metabolites (columns) showing which metabolites are responsible for the clustering of sample subsets related to patient groups (preHD in red, postHD in blue, CKD in green). Dark red colors in the heatmap indicate high values of metabolites in given samples, as opposed to light blue, corresponding to low values

maximizing the covariance between X and Y. Successive LVs are computed as linear combinations of the measured variables following a nested scheme to predict the remaining unexplained part of the response(s).

Experimental designs implemented in metabolomics often involve the comparison of different experimental groups. The membership of the observations to the different classes can be coded with zeros and ones to obtain a disjunct matrix that is further used as response to obtain a PLS model aiming to extract specific patterns associated with class separation. In this case, the model maximizes the between-group covariance matrix and is referred to as a PLS discriminant analysis (PLS-DA) [33].

Orthogonal PLS (OPLS) is a PLS-based method that uses specific LVs to distinguish patterns in the data that are suitable to predict the response(s) from other sources of variation uncorrelated to Y, called orthogonal [34]. Corresponding predictive and orthogonal scores and loadings can then be investigated separately to derive biological hypotheses from metabolomic data. It is important to note that an OPLS model does not offer better predictions than a standard PLS model but may facilitate interpretation thanks to projection rotation. As for standard PLS, OPLS can also be used for discriminant analysis (OPLS-DA) using a coded response matrix. In this case, predictive LVs are designed to separate the classes, often allowing a clear distinction of subgroups of observations on the score plot, while biomarkers can be ranked according to their loadings [35]. Other sources of variability, such as inter-individual differences, can be grasped by orthogonal LVs.

Because supervised models such as OPLS could lead to overly optimistic results, it is essential to carefully consider model validation [21]. This is usually done based on diagnostic statistics comparing predicted responses computed by the model and true values. The root mean square error of prediction (RMSEP) is often used in the case of regression, while other parameters, such as classification accuracy, sensitivity, and specificity obtained from the confusion matrix, are usually considered in the case of discriminant analysis. To limit the risk of overfitting and assess its reliability, model validation should involve a new independent set of observations measured in the same conditions. However, this is not always possible in practice, and alternative data-driven strategies including cross-validation and permutation tests are widely used to evaluate the quality of the models in metabolomics. Cross-validation is based on the separation of the dataset into training and test sets. The model is estimated on the basis of the training data, and predictions are made on the test data to evaluate prediction accuracy. Iterative strategies involving multiple cross-validation loops with different training and test sets are usually carried out to obtain robust estimates [36]. Random permutations of the response matrix (i.e., reallocating the class labels in discriminant analysis) constitute another way to assess the global reliability of the true model, by comparing its prediction accuracy to models derived from random response matrices.

A supervised analysis of independent observations collected from CKD and preHD samples was carried out using standard OPLS-DA. Leave-one-out cross-validation was performed to assess the predictive ability of the models, showing high prediction accuracy. As expected based on this performance, the two groups were clearly separated on the score plot as shown in Fig. 7.5a. The loadings bar plot in Fig. 7.5b shows which metabolites are responsible for the discrimination between preHD and CKD patients. Metabolites with negative loadings on the predictive component p_p (green) are characteristic of the CKD group and are responsible for the negative scores of these patients in Fig. 7.5a.

7.4.4 Simultaneous Evaluation of Multiple Factors

Biological systems are affected by numerous sources of variation whose effects are mixed in the acquired data. Extracting relevant information from data in metabolomics is not a trivial task since unsupervised methods (e.g., PCA) lack the ability to disentangle the mixed sources of variation, making interpretation challenging. Supervised methods (e.g.,

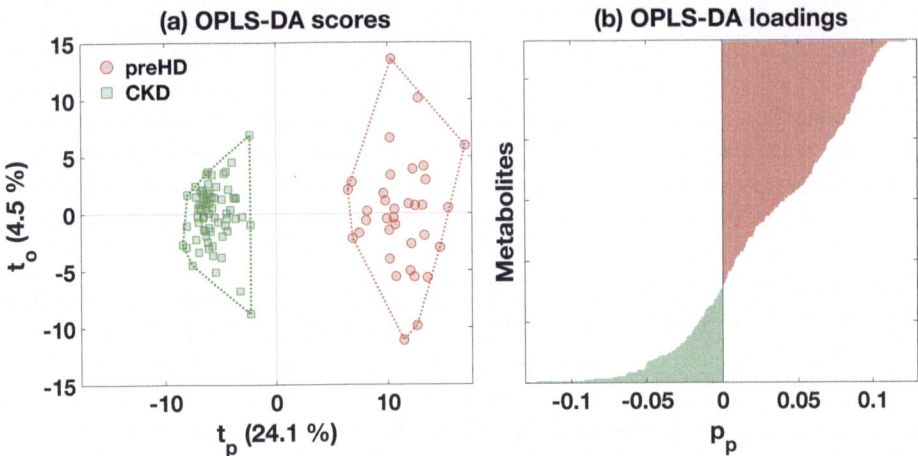

Fig. 7.5 Supervised multivariate investigation of the data using OPLS-DA: (**a**) scores of the samples highlighting the separation between preHD and CKD patients on the predictive component t_p; (**b**) variables predictive loadings characterizing metabolite contributions in differentiating the two groups

PLS) focus on prediction accuracy of the models and do not necessarily describe all the underlying phenomena. For this reason, it has become increasingly common to use design of experiments (DOE) methodologies to investigate factors that are suspected to have an effect on the system under study. Relying on this design, the objective of the data analysis workflow is then to split the data into specific uncorrelated sources of variation associated with the selected design factors and their interaction(s). By these means, it is assured that there is no contribution from other effects included in the design when a specific factor is investigated. However, this property only holds if the decomposed matrix comes from a balanced design. Finally, the variation not explained by the design factors remains after the decomposition as a matrix of residual error.

The primary goal of these methods is to take advantage of the DOE structure using classical analysis of variance (ANOVA) [37]. The latter allows decomposing the data into uncorrelated sources of variation and then analyzing the resulting submatrices using traditional multivariate chemometric tools.

These include ANOVA coupled to PCA [38, 39] and PLS-like [40, 41] approaches, as well as more advanced methods based on multiway [42] or multiblock approaches [43, 44]. Moreover, recent developments were made with respect to the analysis of unbalanced experimental designs [45, 46]. Methods able to deal with more complex experimental designs have also been emerging, with the possibility to analyze designs containing both fixed and random effects [47, 48], or multiple data tables where the same samples were analyzed with the same design using different analytical platforms [49].

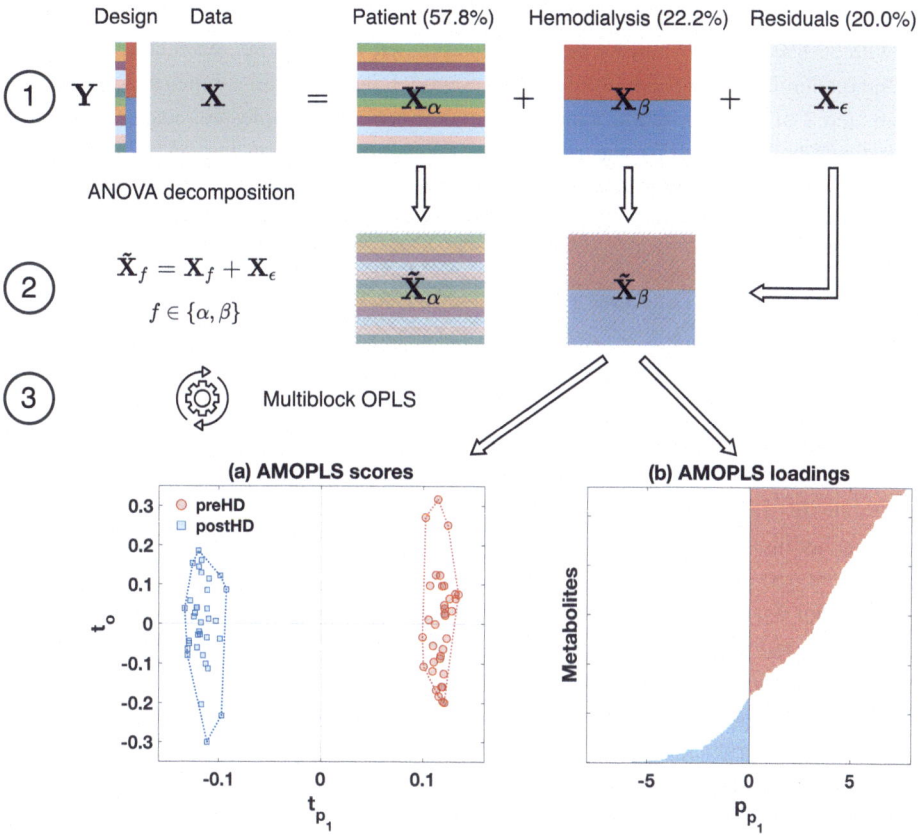

Fig. 7.6 Multivariate investigation of data from designed experiments using AMOPLS: (**a**) scores of the samples highlighting the separation between preHD and postHD patients on the predictive component for the hemodialysis factor; (**b**) variables predictive loadings characterizing metabolite contributions in differentiating the two groups

A multilevel strategy was implemented for the supervised analysis of paired observations involving repeated measurements, i.e., the subset of 70 samples collected from 35 HD patients before (preHD) and after (postHD) hemodialysis. For that purpose, ANOVA multiblock OPLS (AMOPLS) analysis was computed using an experimental design including the *patient* factor (35 levels) and the *hemodialysis* factor (2 levels), thus allowing the *within* to be distinguished from the *between* sources of variability. The results of the ANOVA decomposition in step 1 of Fig. 7.6 indicate that the HD factor represents 22.2% of the total variability in the data, while the *patient* factor summarizing inter-individual differences was related

(continued)

with 57.8% of the overall variability. Unexplained variance, i.e., residuals, involved the remaining 20%. The multiblock OPLS model, including all submatrices obtained in step 2 of Fig. 7.6, led to a clear distinction between samples collected from HD patients before (preHD) and after (postHD) hemodialysis on the AMOPLS score plot. The corresponding loadings were then used for interpretation by highlighting subsets of metabolites associated with this separation.

7.4.5 Biomarker Discovery

Multivariate models aim to extract patterns of variation within large datasets, typically to distinguish subgroups of observations. However, the discovery of biomarkers responsible for the different conditions is very often the objective of the study. Factorial methods based on the construction of components are therefore the starting point for ranking the variables according to their contribution to the different dimensions of the model's subspace. Loading coefficients are directly exploitable outputs for that purpose, but they are linked to a limited part of the variation in the data. Other indices are therefore needed to provide a more global evaluation of the contribution, or predictive merit, of potential biomarkers. Among these, the variable importance in projection (VIP) is a criterion widely used in metabolomics to evaluate the predictive capacity of the variables of a (O)PLS(-DA) model including multiple LVs [50]. It is based on a combination of the loading values associated with all the components of the model for a given variable, using the square root of a weighted sum considering for the predictive ability of each LV. It is a metric that relies on a relative scale specific to each model because it assigns the value of one to the average contribution of the variables in the model. A value greater than one therefore means that the variable considered may be relevant, as more contributive than the average, while smaller values are assigned to variables of little interest. Alternatively, other indices have been developed for ranking variables according to their predictive merit, including the selectivity ratio (SR) and the discriminating variable test (DIVA) [51]. Whatever the parameter chosen, resampling strategies and permutations can be used to evaluate the stability/ robustness of the obtained values [52] and choosing a suitable criterion depends both on the purpose of the investigation and the intrinsic characteristics of the data [53].

As mentioned above, OPLS may help model interpretation by separating predictive from orthogonal sources of variations. This is made possible using different visualization tools allowing straightforward metabolites evaluation and interpretation, including S-plot, Shared and Unique Structure (SUS) plot, among others [54]. SUS plot can be used to compare variable loadings between several OPLS-based models and investigate common or specific differences. To this aim, loading values from both models are scaled as correlation coefficients (pcorr) and displayed on the X-axis and Y-axis, respectively. This graphical tool allows metabolites with similar variations across considered conditions (shared structure on the [(−1;-1), (+1;+1)] diagonal) to be distinguished from specific

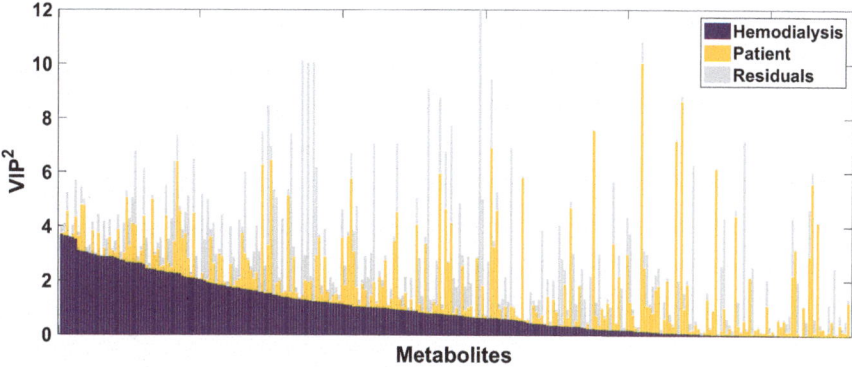

Fig. 7.7 Stacked bar plot showing the squared metabolite VIP for each of the AMOPLS effects, sorted in decreasing order of contribution to the hemodialysis effect

alterations or modulations associated with a given situation (unique structure outside the diagonal).

The squared-VIP (VIP^2) associated with each metabolite for the AMOPLS model was computed and represented in Fig. 7.7 as a stacked bar plot in descending order according to the *hemodialysis* effect contribution. This plot is inspired from the Pareto effect chart computed from squared regression coefficients in DOE. It aims to quantify the importance of each effect in explaining the observed response variability.

A SUS plot was used to compare variable loadings between OPLS and AMOPLS models in Fig. 7.8 and investigate common or specific differences between (1) preHD and CKD, as well as (2) preHD and postHD. This approach highlighted metabolic patterns that could efficiently distinguish different CKD stages and better understand the effect of HD. The biological interpretation of the results is however beyond the scope of this chapter and the interested reader can find all useful information in the corresponding article [3].

Take-Home Message
- Identify and define the biological question to be tackled.
- Implement an adapted experimental design.
      (i.e., sampling strategy, analytical workflow, quality controls, etc.)
- Understand the nature of the acquired dataset and its structure.

(continued)

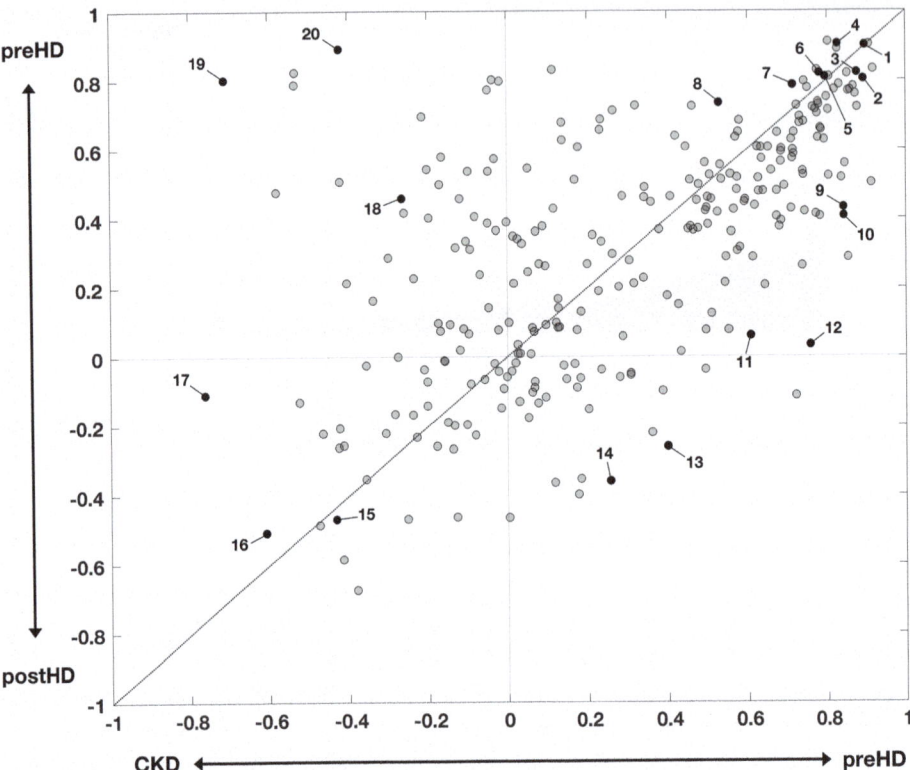

Fig. 7.8 SUS plot between OPLS and AMOPLS variable loadings with the following metabolites highlighted: (1) 5′-methylthioadenosine, (2) creatinine, (3) arabinose, (4) formylmethionine, (5) n-acetylmethionine, (6) myoinositol, (7) N-acetylleucine, (8) 5-hydroxytryptophan, (9) indoxyl sulfate, (10) kynurenic acid, (11) 5a-DHT-17b-glucuronide, (12) cortisol 21-acetate, (13) 1-oleoyl-rac- glycerol, (14) gamma-linolenic acid, (15) biliverdin, (16) keto-isoleucin, (17) tryptophan, (18) guanidoacetic acid, (19) carnitine, (20) uric acid

    (e.g., two-dimensional, paired observations, longitudinal monitoring, multi-platform acquisitions, multiway).

- Identify the information to be obtained and use adapted data analysis workflows (e.g., unsupervised, supervised, or multifactorial strategies).
- Recognize and apply adapted methods for information extraction and interpretation.

References

1. Boccard J, Rudaz S. Harnessing the complexity of metabolomic data with chemometrics. J Chemom. 2014;28:1–9.
2. Gagnebin Y, Boccard J, Ponte B, Rudaz S. Metabolomics in chronic kidney disease: strategies for extended metabolome coverage. J Pharm Biomed Anal. 2018;161:313–25.
3. Gagnebin Y, Jaques DA, Rudaz S, De Seigneux S, Boccard J, Ponte B. Exploring blood alterations in chronic kidney disease and Haemodialysis using metabolomics. Sci Rep. 2020;10:19502.
4. Gagnebin Y, Pezzatti J, Lescuyer P, Boccard J, Ponte B, Rudaz S. Toward a better understanding of chronic kidney disease with complementary chromatographic methods hyphenated with mass spectrometry for improved polar metabolome coverage. J Chromatogr B-Analy Technol Biomed Life Sci. 2019;1116:9–18.
5. Hendriks MMWB, Van Eeuwijk FA, Jellema RH, Westerhuis JA, Reijmers TH, Hoefsloot HCJ, Smilde AK. Data-processing strategies for metabolomics studies. Trac-Trends Analy Chem. 2011;30:1685–98.
6. Naz S, Vallejo M, Garcia A, Barbas C. Method validation strategies involved in non-targeted metabolomics. J Chromatogr A. 2014;1353:99–105.
7. Kamleh MA, Ebbels TMD, Spagou K, Masson P, Want EJ. Optimizing the use of quality control samples for signal drift correction in large-scale urine metabolic profiling studies. Anal Chem. 2012;84:2670–7.
8. Kirwan JA, Broadhurst DI, Davidson RL, Viant MR. Characterising and correcting batch variation in an automated direct infusion mass spectrometry (dims) metabolomics workflow. Anal Bioanal Chem. 2013;405:5147–57.
9. Kuligowski J, Sanchez-Illana A, Sanjuan-Herraez D, Vento M, Quintas G. Intra-batch effect correction in liquid chromatography-mass spectrometry using quality control samples and support vector regression (qc-Svrc). Analyst. 2015;140:7810–7.
10. Dunn WB, Broadhurst D, Begley P, Zelena E, Francis-Mcintyre S, Anderson N, Brown M, Knowles JD, Halsall A, Haselden JN, Nicholls AW, Wilson ID, Kell DB, Goodacre R, C, H. S. M. H. Procedures for large-scale metabolic profiling of serum and plasma using gas chromatography and liquid chromatography coupled to mass spectrometry. Nat Protoc. 2011;6: 1060–83.
11. Brunius C, Shi L, Landberg R. Large-scale untargeted Lc-Ms metabolomics data correction using between-batch feature alignment and cluster-based within-batch signal intensity drift correction. Metabolomics. 2016;12:173.
12. Boccard J, Tonoli D, Strajhar P, Jeanneret F, Odermatt A, Rudaz S. Removal of batch effects using stratified subsampling of metabolomic data for in vitro endocrine disruptors screening. Talanta. 2019;195:77–86.
13. Deng K, Zhang F, Tan QL, Huang Y, Song W, Rong ZW, Zhu ZJ, Li ZZ, Li K. Waveica: a novel algorithm to remove batch effects for large-scale untargeted metabolomics data based on wavelet analysis. Anal Chim Acta. 2019;1061:60–9.
14. Gagnebin Y, Tonoli D, Lescuyer P, Ponte B, De Seigneux S, Martin PY, Schappler J, Boccard J, Rudaz S. Metabolomic analysis of urine samples by Uhplc-Qtof-Ms: impact of normalization strategies. Anal Chim Acta. 2017;955:27–35.
15. Warrack BM, Hnatyshyn S, Ott KH, Reily MD, Sanders M, Zhang HY, Drexler DM. Normalization strategies for metabonomic analysis of urine samples. J Chromatogr B-Anal Technol Biomed Life Sci. 2009;877:547–52.

16. Filzmoser P, Walczak B. What can go wrong at the data normalization step for identification of biomarkers? J Chromatogr A. 2014;1362:194–205.
17. Li B, Tang J, Yang QX, Li S, Cui XJ, Li YH, Chen YZ, Xue WW, Li XF, Zhu F. Noreva: normalization and evaluation of Ms-based metabolomics data. Nucleic Acids Res. 2017;45: W162–70.
18. Boccard J, Veuthey JL, Rudaz S. Knowledge discovery in metabolomics: an overview of Ms data handling. J Sep Sci. 2010;33:290–304.
19. Vinaixa M, Samino S, Saez I, Duran J, Guinovart JJ, Yanes O. A guideline to univariate statistical analysis for Lc/Ms-based untargeted metabolomics-derived data. Meta. 2012;2:775–95.
20. Shaffer JP. Multiple hypothesis-testing. Annu Rev Psychol. 1995;46:561–84.
21. Broadhurst DI, Kell DB. Statistical strategies for avoiding false discoveries in metabolomics and related experiments. Metabolomics. 2006;2:171–96.
22. Benjamini Y, Hochberg Y. Controlling the false discovery rate - a practical and powerful approach to multiple testing. J R Stat Soc Ser B-Methodol. 1995;57:289–300.
23. Fawcett T. An introduction to roc analysis. Pattern Recogn Lett. 2006;27:861–74.
24. Saccenti E, Hoefsloot HCJ, Smilde AK, Westerhuis JA, Hendriks MMWB. Reflections on univariate and multivariate analysis of metabolomics data. Metabolomics. 2014;10:361–74.
25. Bro R, Smilde AK. Centering and scaling in component analysis. J Chemom. 2003;17:16–33.
26. Kohl SM, Klein MS, Hochrein J, Oefner PJ, Spang R, Gronwald W. State-of-the art data normalization methods improve Nmr-based metabolomic analysis. Metabolomics. 2012;8:146–60.
27. Keun HC, Ebbels TMD, Antti H, Bollard ME, Beckonert O, Holmes E, Lindon JC, Nicholson JK. Improved analysis of multivariate data by variable stability scaling: application to Nmr-based metabolic profiling. Anal Chim Acta. 2003;490:265–76.
28. Kvalheim OM, Brakstad F, Liang YZ. Preprocessing of analytical profiles in the presence of homoscedastic or heteroscedastic noise. Anal Chem. 1994;66:43–51.
29. Ward JH. Hierarchical grouping to optimize an objective function. J Am Stat Assoc. 1963;58:236.
30. Bro R, Papalexakis EE, Acar E, Sidiropoulos ND. Coclustering-a useful tool for Chemometrics. J Chemom. 2012;26:256–63.
31. Wold S, Sjostrom M, Eriksson L. Pls-regression: a basic tool of chemometrics. Chemom Intell Lab Syst. 2001;58:109–30.
32. Daszykowski M, Walczak B, Massart DL. Projection methods in chemistry. Chemom Intell Lab Syst. 2003;65:97–112.
33. Barker M, Rayens W. Partial least squares for discrimination. J Chemom. 2003;17:166–73.
34. Trygg J, Wold S. Orthogonal projections to latent structures (O-Pls). J Chemom. 2002;16:119–28.
35. Bylesjö M, Rantalainen M, Cloarec O, Nicholson JK, Holmes E, Trygg J. Opls discriminant analysis: combining the strengths of Pls-Da and Simca classification. J Chemom. 2006;20:341–51.
36. Szymanska E, Saccenti E, Smilde AK, Westerhuis JA. Double-check: validation of diagnostic statistics for Pls-Da models in metabolomics studies. Metabolomics. 2012;8:S3–S16.
37. Searle SR. Linear models. New York: Wiley; 1971.
38. Harrington PD, Vieira NE, Espinoza J, Nien JK, Romero R, Yergey AL. Analysis of variance-principal component analysis: a soft tool for proteomic discovery. Anal Chim Acta. 2005;544: 118–27.
39. Smilde AK, Jansen JJ, Hoefsloot HCJ, Lamers RJAN, Van Der Greef J, Timmerman ME. Anova-simultaneous component analysis (Asca): a new tool for analyzing designed metabolomics data. Bioinformatics. 2005;21:3043–8.

40. Marini F, De Beer D, Joubert E, Walczak B. Analysis of variance of designed chromatographic data sets: the analysis of variance-target projection approach. J Chromatogr A. 2015;1405:94–102.

41. Thissen U, Wopereis S, Van Den Berg SAA, Bobeldijk I, Kleemann R, Kooistra T, Van Dijk KW, Van Ommen B, Smilde AK. Improving the analysis of designed studies by combining statistical modelling with study design information. Bmc Bioinformatics. 2009;10

42. Jansen JJ, Bro R, Hoefsloot HCJ, Van Den Berg FWJ, Westerhuis JA, Smilde AK. Parafasca: Asca combined with Parafac for the analysis of metabolic fingerprinting data. J Chemom. 2008;22:114–21.

43. Boccard J, Rudaz S. Exploring omics data from designed experiments using analysis of variance multiblock orthogonal partial least squares. Anal Chim Acta. 2016;920:18–28.

44. Bouveresse DJR, Pinto RC, Schmidtke LM, Locquet N, Rutledge DN. Identification of significant factors by an extension of Anova-Pca based on multi-block analysis. Chemom Intell Lab Syst. 2011;106:173–82.

45. De Figueiredo M, Giannoukos S, Rudaz S, Zenobi R, Boccard J. Efficiently handling high-dimensional data from multifactorial designs with unequal group sizes using rebalanced Asca (Rasca). J Chemometrics. 2022a;37

46. Thiel M, Feraud B, Govaerts B. Asca plus and Apca plus : extensions of Asca and Apca in the analysis of unbalanced multifactorial designs. J Chemom. 2017;31

47. Madssen TS, Giskeodegard GF, Smilde AK, Westerhuis JA. Repeated measures Asca plus for analysis of longitudinal intervention studies with multivariate outcome data. PLoS Comput Biol. 2021;17:e1009585.

48. Martin M, Govaerts B. Limm-Pca: combining Asca(+) and linear mixed models to analyse high-dimensional designed data. J Chemom. 2020;34

49. De Figueiredo M, Giannoukos S, Wuthrich C, Zenobi R, Rutledge DN. A tutorial on the analysis of multifactorial designs from one or more data sources using Acomdim. J Chemometrics. 2022b;37

50. Kvalheim OM, Arneberg R, Bleie O, Rajalahti T, Smilde AK, Westerhuis JA. Variable Importance In Latent Variable Regression Models. J Chemom. 2014;28:615–22.

51. Rajalahti T, Arneberg R, Kroksveen AC, Berle M, Myhr KM, Kvalheim OM. Discriminating variable test and selectivity ratio plot: quantitative tools for interpretation and variable (bio-marker) selection in complex spectral or chromatographic profiles. Anal Chem. 2009;81:2581–90.

52. Afanador NL, Tran TN, Buydens LMC. Use of the bootstrap and permutation methods for a more robust variable importance in the projection metric for partial least squares regression. Anal Chim Acta. 2013;768:49–56.

53. Farres M, Platikanov S, Tsakovski S, Tauler R. Comparison of the variable importance in projection (Vip) and of the selectivity ratio (Sr) methods for variable selection and interpretation. J Chemom. 2015;29:528–36.

54. Wiklund S, Johansson E, Sjostrom L, Mellerowicz EJ, Edlund U, Shockcor JP, Gottfries J, Moritz T, Trygg J. Visualization of Gc/Tof-Ms-based metabolomics data for identification of biochemically interesting compounds using Opls class models. Anal Chem. 2008;80:115–22.

Using Quantitative Metabolomics and Data Enrichment to Interpret the Biochemistry of a Novel Disease

8

David S. Wishart and Marcia A. Levatte

What You Will Learn in This Chapter

You will learn how to use quantitative metabolomics data and a variety of data enrichment techniques to diagnose a disease, to understand its biochemical features and to use those features to propose treatments. Real, absolutely quantitative plasma metabolomics data, collected from four patients suffering from the same "unknown" disease, will be used as an example. Relevant clinical data will also be provided. From the measured metabolomics data, you will be shown how to use the Human Metabolome Database (HMDB) to compare the measured metabolomic data and identify unusually high or low levels of key metabolites. You will then be shown how to acquire additional information about the most dysregulated metabolites from the HMDB. You will also learn how relevant biochemical and signaling pathways from both the HMDB and a pathway database, called the Small Molecule Pathway Database (SMPDB), can be explored to help identify the disease and uncover its

(continued)

D. S. Wishart (✉)
Department of Biological Sciences, University of Alberta, Edmonton, AB, Canada

Department of Computing Science, University of Alberta, Edmonton, AB, Canada

The Metabolomics Innovation Centre, University of Alberta, Edmonton, AB, Canada
e-mail: dwishart@ualberta.ca

M. A. Levatte
Department of Biological Sciences, University of Alberta, Edmonton, AB, Canada
e-mail: levatte@ualberta.ca

© The Author(s), under exclusive license to Springer Nature Switzerland AG 2023
J. Ivanisevic, M. Giera (eds.), *A Practical Guide to Metabolomics Applications in Health and Disease*, Learning Materials in Biosciences,
https://doi.org/10.1007/978-3-031-44256-8_8

biochemical underpinnings. Additionally, you will be shown how you can learn more about the condition and about known biomarkers associated with the disease, through a database called MarkerDB. Finally, you will learn how potential treatments (drugs, nutritional supplements, intravenous treatments) could be identified and/or used to treat this novel disease based on these biochemical findings.

8.1   Introduction

The application of metabolomics to detect and diagnose disease is far older than most people realize. Indeed, gas chromatography mass spectrometry (GC-MS) and liquid chromatography tandem mass spectrometry (LC-MS/MS) have been used since the early 1990s to perform a clinical test called newborn screening [1]. Newborn screening typically involves taking a blood sample of a baby within hours after its birth and analyzing that blood sample using GC-MS or LC-MS/MS technologies to look for tell-tale biomarkers of certain genetic or metabolic diseases. Many developed countries around the world routinely use newborn screening to detect treatable inborn errors of metabolism (IEMs) [1, 2]. It is estimated that more than 300 million newborns have been screened over the past 25 years and more than 1 million have had their lives saved or substantially improved, thanks to metabolomics [3]. In the United States, nearly 40 IEMs are routinely screened for each newborn child [4]. The reason why metabolomics for newborn screening has been so successful is because the metabolomic assays are targeted and quantitative. That is, they measure the absolute concentrations of a predetermined collection of metabolites known to be associated with specific IEMs. Knowing the concentration of these metabolites (or metabolite biomarkers) and being able to compare them to healthy reference values allow the IEM to be accurately diagnosed. Targeted, quantitative metabolomics also allows those infants diagnosed with an IEM to be routinely monitored over the course of their lifetime, thereby ensuring the treatment (drug or diet) is doing its job [1–4].

Outside of the regular application of quantitative metabolomics to IEM diagnosis and monitoring, it is rare to see any other metabolomics method being used in the clinic. This is because the majority of modern metabolomics research laboratories employ untargeted, nonquantitative metabolomics methods. As a result, most of today's metabolomics discoveries and many of today's metabolomics methods *cannot* be used for medical applications. This is unfortunate. Right now, less than 20% of metabolomics labs around the world use fully quantitative metabolomics methods. These labs know that the only way that metabolomics can be used in the clinic or approved for use in the clinic (by agencies such as the FDA) is through the use of targeted, quantitative metabolomics methods. This is because targeted, quantitative methods provide the rigor, robustness, and reproducibility required for clinical applications and reliable medical diagnoses. Furthermore, these quantitative methods do not require large cohorts of healthy controls to help normalize or scale their data. Instead, the measured values from a single patient can be compared directly to published, healthy reference concentrations.

In this chapter we will show how targeted, quantitative metabolomics can be used to help diagnose and suggest treatment options for a complex, nongenetic disease in adults. To demonstrate how this can be done, we will first describe four patients who came down with the same unknown disorder that left them severely compromised. We will describe their symptoms and conventional efforts to initially identify the disease. Then we will describe how the targeted, quantitative metabolomic data was collected and analyzed from these patients. Using a portion of that metabolomic data, we will show how data enrichment techniques can be used to help identify the disease, to biochemically characterize the disease and to identify possible treatment options for this "unknown" disease. In reality, the unknown disease was identified (using other laboratory methods) shortly after the patients were moved into the intensive care unit (ICU). However, this is a particularly interesting case study that shows how enriched metabolomics data can provide important insights on the biochemistry of complex human diseases and how these insights can guide treatment.

8.2   Patient Background

Four patients exhibiting fever, shortness of breath, low blood oxygen levels, and high heart rates were initially admitted to the emergency room (ER) of a large urban hospital. All four had recently returned from trips to Italy. As all four patients were previously healthy, all four had visited the same country, all four exhibited similar symptoms, and none had experienced any trauma or visible injury, these patients were suspected to have the same unknown infectious lung disease. On admission, age, sex, comorbidities, medications, hematologic labs, arterial-partial-pressure- to-inspired-oxygen ratio, and chest X-ray findings were recorded for all patients. Chest X-rays indicated a lung infection resembling pneumonia. While in the ER, all four patients unexpectedly deteriorated, with severe anoxia becoming most evident. Clinical interventions including use of broad-spectrum antibiotics, vasoactive medications, and high-flow oxygen therapy were tried. These interventions appeared to have little effect. Eventually all four patients were moved to the ICU as their symptoms became worse and their organs (liver, kidneys, lungs) began to fail. Mechanical ventilation was required for all four patients by the time they were admitted to the ICU. As an infectious disease was suspected, the patients were tested for common respiratory pathogens (*Chlamydia pneumoniae*, coronavirus, Entero/rhinovirus, human metapneumovirus, Influenza A, *Mycoplasma pneumoniae*, Parainfluenza 3, respiratory syncytial virus), but no known infectious agent was identified.

To understand why the patients' conditions had worsened and why they were so unresponsive to standard treatments, a quantitative, metabolomic assay was performed on plasma samples collected from all four patients.

8.3   LC-MS/MS Metabolomic Assay

Four plasma samples were collected from each patient on their first day in the ICU, while mechanically ventilated. A targeted, quantitative metabolomics assay (run on an ABSciex 4000 Qtrap tandem MS instrument equipped with an Agilent 1260 series UHPLC system) was used to analyze the samples. A combination of direct injection (DI) MS and reverse-phase LC-MS/MS was employed to measure the plasma concentrations of 141 polar and nonpolar metabolites. The measured metabolites include amino acids and derivatives, biogenic amines, carboxylic acids, acylcarnitines (CAR), lysophosphatidylcholines (LPC), phosphatidylcholines (PC), and sphingomyelins (SM). This absolutely quantitative assay uses chemical labeling with ^{13}C-labeled derivatization agents to help with chromatographic separation and quantification of these metabolite classes. Specifically, ^{13}C-phenylisothiocyanate (PITC) was used for labeling amine-containing compounds, and ^{13}C-3-nitrophenylhydrazinea (3NPH) was used to label compounds with carboxylic acids. Selective mass-spectrometric detection using multiple reaction monitoring pairs and multiple calibration standards was used to identify and quantify all metabolites. The assay was run on a 96-well plate where the first 14 wells were used as blanks or calibrants with one well serving as a solvent blank, three wells as zero-point reference samples, seven wells as calibration standards, and three wells containing quality control samples. Metabolite quantification accuracy was typically within 10–15% from run to run and sample to sample. Additional details regarding the chromatographic, DI, and MS-data acquisition methods are provided elsewhere [5]. The concentrations of the measured plasma metabolites, along with their Human Metabolome Database or HMDB [6] identifiers, from all four patients are provided in Table 8.1. For instructive purposes only, 34 of the 141 measured metabolites are listed.

8.4   Data Enrichment with the HMDB: Part I

Quantitative metabolite measurements are meaningless without being able to compare the measured values to some kind of "healthy" reference values. Standard tables and standard healthy reference values for commonly measured metabolites such as glucose and creatinine are widely available in textbooks or clinical laboratory manuals. However, many of the metabolites measured by modern, targeted metabolomics methods do not have textbook reference values. To determine what the healthy reference values should be for the 34 metabolites listed in Table 8.1, it is necessary to look these up in the Human Metabolome Database (HMDB) [6]. The HMDB is a free, publicly available, searchable database containing information on all known human metabolites. The latest edition of HMDB (Version 5.0) currently contains detailed chemical, molecular biology/biochemical, and clinical data for 253,245 metabolites. First published in 2007, the HMDB was built from extensive literature and online searches of thousands of books, journal articles, and electronic databases. In addition to the extensive literature-derived data, the HMDB also

Table 8.1 Concentrations (in µM) for 34 metabolites measured in plasma for four patients on arrival in the ICU using quantitative metabolomics methods

Compound	HMDB ID	Patient 1 (µM)	Patient 2 (µM)	Patient 3 (µM)	Patient 4 (µM)
Glucose	0000122	9957	9417	6806	5712
Creatinine	0000562	227	567	651	220
Kynurenine	0000684	42.6	40.8	10.7	56.5
Lactate	0000190	637	1270	1300	1150
Arginine	0000517	25.2	82.6	47.2	44.7
Glutamate	0000148	61.9	76.7	66.1	15.6
Glutamine	0000641	305	874	426	273
Glycine	0000123	115	287	180	129
Isoleucine	0000172	38.3	120	62.8	58.2
Leucine	0000687	65	200	90.1	64.7
Lysine	0000182	66.7	224	86.1	110.9
Met-SO	0002005	1.91	2.01	2.10	1.42
Phenylalanine	0000159	89.5	137	102	120
Threonine	0000167	36.7	172	59.6	38.6
Tryptophan	0000929	16.2	39.4	8.2	11.9
Tyrosine	0000158	58.4	114	57	27.6
Valine	0000883	135	295	177	219
LPC 16:0	0010382	27.3	61.2	40.9	16.0
LPC 17:0	0012108	0.47	0.74	0.43	0.22
LPC 18:1	0010385	5.8	9.9	10.1	3.2
LPC 18:2	0010386	5.8	7.1	5.6	2.0
CAR-C0	0000062	15.1	120.7	127.4	63.8
CAR-C2	0000201	1.7	30.1	43.5	19.6
CAR-C3	0000824	0.22	1.58	2.31	0.71
CAR-C3OH	0013125	0.03	0.03	0.04	0.03
CAR-C3:1	0013124	0.06	0.08	0.10	0.08
CAR-C4	0002013	0.16	0.69	1.46	1.35
CAR-C4:1	0013126	0.06	0.08	0.08	0.07
CAR-C4OH	0013127	0.05	0.14	0.19	0.05
CAR-C5	0013128	0.13	0.69	1.05	0.29
CAR-C6:1	0013161	0.01	0.01	0.02	0.02
CAR-C8	0000791	0.10	0.21	0.39	0.16
CAR-C10	0000651	0.17	0.36	0.52	0.24
CAR-C10:1	0013205	0.20	0.38	1.33	0.37

Carnitine (CAR). Lysophosphatidylcholine (LPC). Methionine sulfoxide (Met-SO)

houses experimental MS and NMR data for pure metabolite standards to help with compound identification. Each metabolite in the HMDB has at least 90 data fields including comprehensive *compound descriptions*, chemical names and synonyms, structural information, physicochemical data, reference NMR and MS spectra, *biofluid concentrations*, *disease associations*, *pathway information*, enzyme data, protein sequence data, single-nucleotide polymorphisms (SNP), and mutation data as well as extensive links to images, references, and other public databases including KEGG [7], PubChem [8], MetaCyc [9], ChEBI [10], PDB [11], UniProt [12], and GenBank [13]. For our purposes, the most useful pieces of information in the HMDB are the data on the biofluid concentrations, disease associations, compound descriptions, and pathway diagrams. Here we will show how to search for healthy reference concentrations in the HMDB.

8.4.1 Search the Human Metabolome Database (HMDB) for Normal Reference Concentrations

8.4.1.1. Ensure you have a computer with internet access and an up-to-date web browser.

8.4.1.2. Go to the HMDB home page at www.hmdb.ca (see Fig. 8.1a).

8.4.1.3. At the top right corner in the "Search" box, enter the name of a metabolite (try "Glucose") or the corresponding HMDB identifier (HMDB0000122). Press the blue "Search" button to activate the search. Entering the name will typically return a longer list of compounds with glucose in the name, while entering the HMDB identifier will return only one compound and immediately present the MetaboCard of that compound.

8.4.1.4. If you entered the name ("Glucose"), click on the compound name (D-Glucose) hyperlink and view the MetaboCard for glucose (see Fig. 8.1b).

8.4.1.5. Scroll down the page to find the data category (marked with light brown bars) titled "Normal Concentrations," or click on the blue jump tab (at the top of the page) titled "Concentrations." Clicking on this tab will take you immediately to the data category called "Normal Concentrations."

8.4.1.6. Scroll down the list to find the normal concentration for glucose in blood. As shown in Fig. 8.1c, many compounds in the HMDB have multiple concentrations listed for multiple biofluids. This can be confusing for the uninitiated. The best choice of a given biofluid concentration, or concentration range is the one listed as the TMIC concentration at the top of each table. If that value is not provided, values provided by the Geigy Scientific tables are a good second choice.

8.4.1.7. Enter the healthy normal concentration for glucose (the TMIC value) into a new column in Table 8.1. Repeat this process for all 34 metabolites listed in Table 8.1.

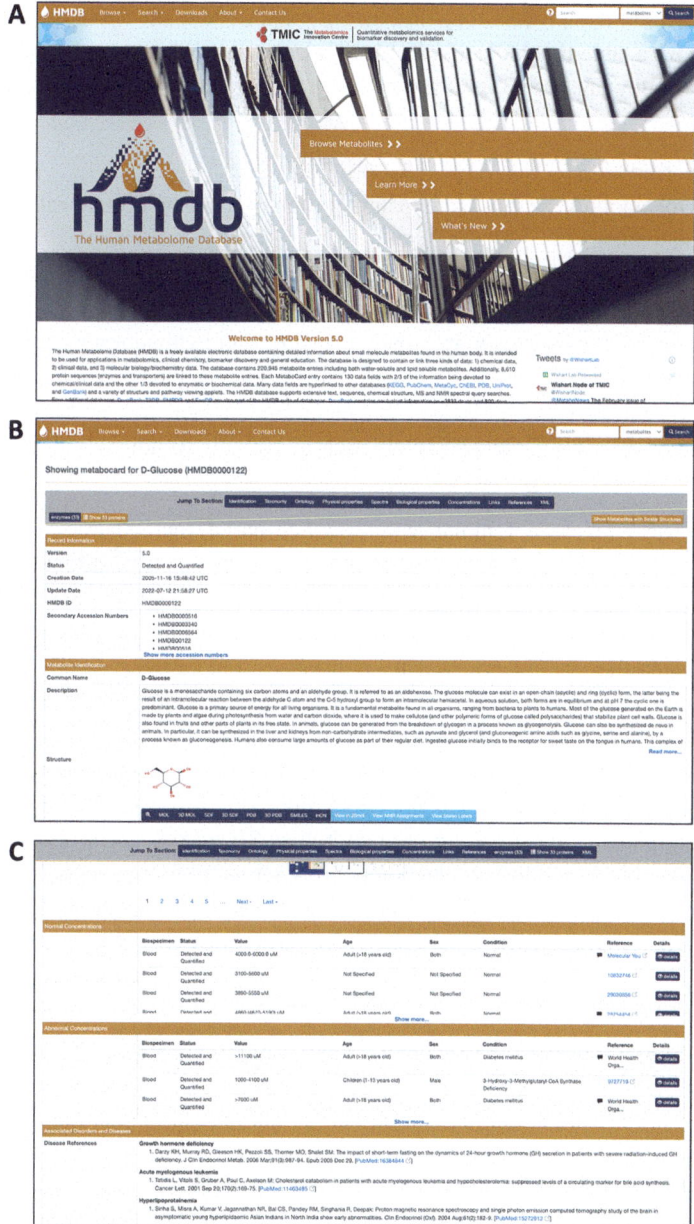

Fig. 8.1 Searching the HMDB for reference values of plasma metabolites. (**a**) The HMDB home page. (**b**) A view of the glucose MetaboCard in HMDB. (**c**) A list of the normal blood concentrations for glucose in the human body

This data enrichment process may take a user about 20 min. At the end of this protocol, you should have a table that looks like Table 8.2. Using these data you should now be able to answer the following questions.

8.4.2 Questions

1. Which metabolites are most significantly different from the normal values in most or all the patients?
2. What criteria are you using or what criteria should you use to identify those metabolites that are significantly different?
3. Which metabolites appear to be largely unchanged or in the normal range for most or all patients?
4. Normally people with shortness of breath or anoxia have very high lactate levels (>3 mM). Why are the lactate levels in these patients so normal?

For those of you more familiar with research-based metabolomics or untargeted metabolomics, you will quickly see that multivariate statistics (i.e., principal component analysis, heat maps, or partial least squares discriminant analysis) are not needed (or even used) to identify the most significantly changed metabolites. This is because in clinical medicine we are often dealing with small sample sizes (one or two, sometimes up to three or four patients), and we normally don't have any controls. Single patient sample sizes are far too small for performing multivariate statistics. Indeed, in clinical metabolomics, the focus is on making diagnoses using one patient at a time employing known or suspected biomarkers. Clinical metabolomics is not about biomarker discovery; it's about diagnosis and improving patient care. Looking through Tables 8.1 or 8.2, you will also notice that there is considerable variability of metabolite levels across all four patients. This is normal, even among very sick patients. This variability is due to a combination of genetics, age, gender, body mass index (BMI), and underlying physiology. It is why metabolite ranges are used rather than single values. You will also see that the normal ranges have maximum values that are often 2–5× the minimum value, which is typical for most metabolites. The exceptions are glucose and phenylalanine, which are among the most tightly regulated metabolites in the human body.

8.5 Data Enrichment with the HMDB: Part II

Now that you have identified which metabolites are dysregulated in these four patients, it is important to identify what disease these individuals may be suffering from. Metabolomics may not be able to identify the infectious organism, but it can identify the secondary diseases arising from that infection as well as the adverse molecular consequences of that infection. By looking more closely at the most dysregulated metabolites from a patient sample, it is often possible to put together a more complete biochemical picture of the

Table 8.2 Concentrations (in µM) for 34 metabolites measured in plasma for four patients on arrival in the ICU using quantitative metabolomics methods

Compound	HMDB ID	Patient 1 (µM)	Patient 2 (µM)	Patient 3 (µM)	Patient 4 (µM)	Normal values (µM)
Glucose	0000122	9957	9417	6806	5712	4000–6000
Creatinine	0000562	227	567	651	220	60–115
Kynurenine	0000684	42.6	40.8	10.7	56.5	1.1–3.2
Lactate	0000190	637	1270	1300	1150	500–2100
Arginine	0000517	25.2	82.6	47.2	44.7	40–130
Glutamate	0000148	61.9	76.7	66.1	15.6	20–110
Glutamine	0000641	305	874	426	273	390–900
Glycine	0000123	115	287	180	129	150–440
Isoleucine	0000172	38.3	120	62.8	58.2	40–100
Leucine	0000687	65	200	90.1	64.7	70–180
Lysine	0000182	66.7	224	86.1	110.9	110–240
Met-SO	0002005	1.91	2.01	2.10	1.42	0.3–1.4
Phenylalanine	0000159	89.5	137	102	120	40–75
Threonine	0000167	36.7	172	59.6	38.6	80–220
Tryptophan	0000929	16.2	39.4	8.2	11.9	40–95
Tyrosine	0000158	58.4	114	57	27.6	35–80
Valine	0000883	135	295	177	219	140–300
LPC 16:0	0010382	27.3	61.2	40.9	16.0	41–150
LPC 17:0	0012108	0.47	0.74	0.43	0.22	0.9–2.6
LPC 18:1	0010385	5.8	9.9	10.1	3.2	6.1–43
LPC 18:2	0010386	5.8	7.1	5.6	2.0	7–59
CAR-C0	0000062	15.1	120.7	127.4	63.8	19–65
CAR-C2	0000201	1.7	30.1	43.5	19.6	3–12.5
CAR-C3	0000824	0.22	1.58	2.31	0.71	0.15–0.7
CAR-C3OH	0013125	0.03	0.03	0.04	0.03	0.02–0.1
CAR-C3:1	0013124	0.06	0.08	0.10	0.08	0.04–0.14
CAR-C4	0002013	0.16	0.69	1.46	1.35	0.10–0.45
CAR-C4:1	0013126	0.06	0.08	0.08	0.07	0.01–0.08
CAR-C4OH	0013127	0.05	0.14	0.19	0.05	0.03–0.09
CAR-C5	0013128	0.13	0.69	1.05	0.29	0.04–0.26
CAR-C6:1	0013161	0.01	0.01	0.02	0.02	0.01–0.035
CAR-C8	0000791	0.10	0.21	0.39	0.16	0.17–0.50
CAR-C10	0000651	0.17	0.36	0.52	0.24	0.16–0.55
CAR-C10:1	0013205	0.20	0.38	1.33	0.37	0.12–0.40

Normal ranges from the HMDB are included. Carnitine (CAR). Lysophosphatidylcholine (LPC). Methionine sulfoxide (Met-SO)

disease. Here we will show how you can use the HMDB to learn more about the dysregulated metabolites and what the underlying disease (and its biochemistry) may be.

8.5.1 Search the Human Metabolome Database (HMDB) for Metabolite Disease Associations

8.5.1.1. Go to the HMDB home page at www.hmdb.ca.

8.5.1.2. in the "Search" box, enter the name of a metabolite (try "Kynurenine") or the corresponding HMDB identifier (HMDB0000684). Press the blue "Search" button to activate the search. Entering the name will typically return a longer list of compounds with kynurenine in the name, while entering the HMDB identifier will return only one compound and immediately present the MetaboCard of that compound (Fig. 8.2a).

8.5.1.3. If you entered the name ("Kynurenine"), click on the L-Kynurenine hyperlink and view the MetaboCard for L-kynurenine.

8.5.1.4. Scroll down the page to find the data category called "Description," and read the information compiled on what is known about kynurenine.

8.5.1.5. Scroll down the kynurenine MetaboCard to find the data category called "Biological Properties," and look at the subcategory called "Pathways." Click on the thumbnail images corresponding to the associated pathways listed for kynurenine. Look at which pathways are associated with any disease or condition. Once you click on the pathway thumbnail image, you can explore the pathways using their expanded views in SMPDB [14]. SMPDB's pathways can be viewed, scrolled, zoomed, and explored more completely using the browsing options displayed on the pathway diagram. An example pathway for kynurenine is shown in Fig. 8.2b. More information about SMPDB is provided below.

8.5.1.6. Scroll down the kynurenine MetaboCard a little further to find the "Abnormal Concentrations" list for kynurenine in blood. Look for conditions or diseases that have concentrations similar to the values measured for these four patients.

8.5.1.7. Repeat this process for other dysregulated metabolites such as glucose, creatinine, tryptophan, arginine, and LPC 18:2. Take notes about your observations or what you have learned

This secondary data enrichment process may take a user about 20–30 min. The intent of this process is to provide you with more biological, biochemical, and disease information about the dysregulated metabolites you identified in part 3 of this exercise. By looking at the abnormal concentrations and assessing how well they match to known diseases, you may be able to identify or "narrow down" what this disease is or what secondary diseases

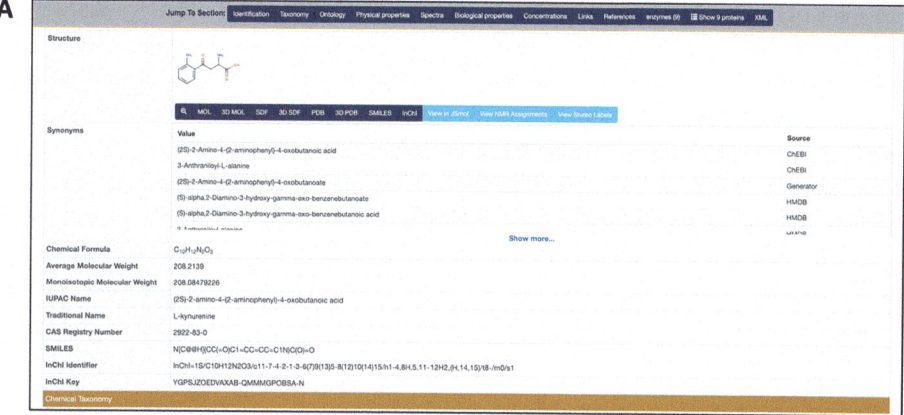

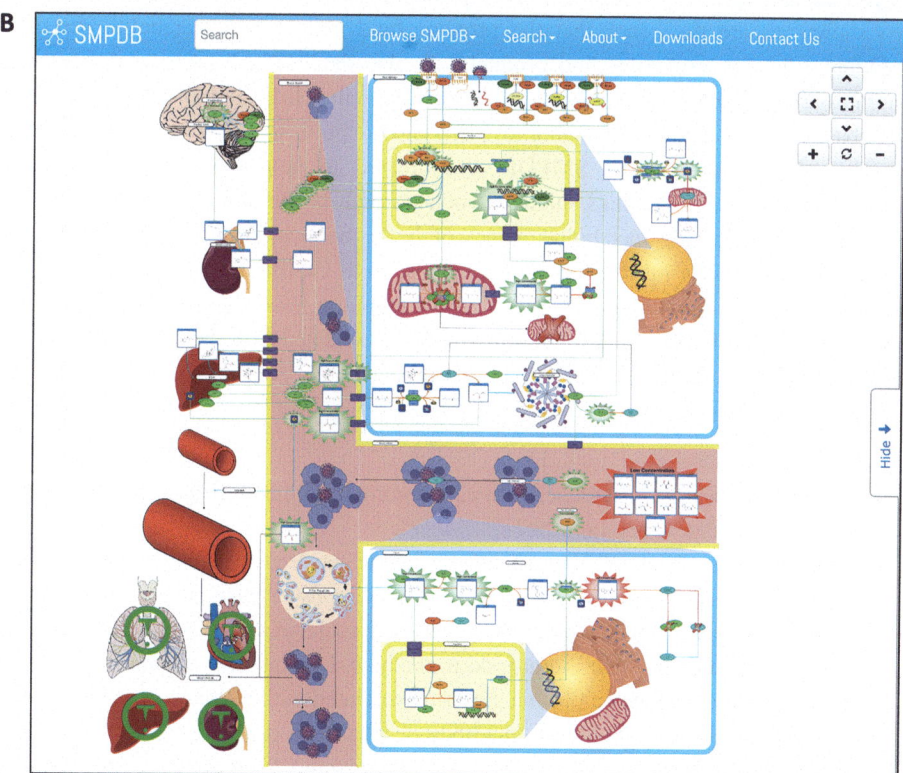

Fig. 8.2 Enriching the annotation of Kynurenine with HMDB and SMPDB. (**a**) A portion of the HMDB MetaboCard for kynurenine. (**b**) An image of one of the many pathways associated with kynurenine as illustrated through SMPDB

the infection is causing. This is an example of using biomarkers to help identify or diagnose a disease. Even if you can't make a detailed diagnosis, you should be able to determine at least some of the conditions or secondary conditions arising from this unknown disease. By exploring the disease pathways associated with the different dysregulated metabolites, you may begin to see how or why these metabolites are increased (or decreased) and how they interact with each other. This fundamentally reveals more about the biochemistry of this disorder. One of the best ways of understanding the biochemistry of a disease (or of a given metabolite) is to use a database called SMPDB. The SMPDB [14] is a database containing >40,000 small molecule pathways found exclusively in humans. Pathways for other model organisms (including humans) such as mice, *Escherichia coli*, and yeast can be found in another pathway database called PathBank [15]. SMPDB contains highly detailed, visually interactive pathways that show the location of metabolites within organs (i.e., the liver, or skeletal muscle) and their cellular and subcellular location (i.e., the cytoplasm, mitochondria, or endoplasmic reticulum) within the cells of that organ. The protein or enzyme complexes, transporters, and other protein regulators that bind the metabolite are also shown, as are the genes that the proteins bind. All known metabolite reactions, including reactants, cofactors, and products are also detailed. All parts of the pathway diagrams are clickable with metabolites hyperlinked to the HMDB or DrugBank (which is a database for drugs and drug metabolites [16]) and the proteins and or enzyme complexes hyperlinked to UniProt. The SMPDB was created to help interpret metabolomics, transcriptomics, and proteomics studies associated with health and disease. SMPDB can be easily searched using metabolite names or lists, chemical structure, protein or gene names or sequences, or other unique identifiers, such as GenBank identifiers. Searches in the SMPDB will return a list of pathways in common with the entered metabolites, proteins, or genes. Using the information you have collected from HMDB and SMPDB, you should now be able to answer the following questions.

8.5.2 Questions

1. Given that antibiotics were ineffective in treating this unknown disease, what kind of infectious disease could it be?
2. What diseases is elevated plasma kynurenine associated with? What causes elevated kynurenine and what might it do to the body or what does it indicate?
3. What diseases is elevated plasma creatinine associated with? What causes elevated creatinine and what might it do to the body or what does it indicate?
4. What diseases is elevated plasma glucose associated with? What causes elevated glucose and what might it do to the body or what does it indicate?
5. What general trends do you see with these other dysregulated metabolites in terms of the diseases or conditions they are associated with?
6. What pathway or pathways do these metabolites seem to converge upon?

The questions you attempted to answer here are the types of questions that any physician or clinical chemist would typically ask, especially for an unknown or uncharacterized disease. While the answers obtained for any single metabolite may not provide a single obvious answer, the weight of evidence gathered covering information from multiple metabolites may provide a clearer picture. This is how disease biomarkers typically work. A single biomarker (such as high glucose) may be indicative of multiple conditions, including diabetes, hyperglycemia, inflammation, a high-sugar meal, or even something as severe as sepsis. However, multiple biomarkers can often provide more specificity and greater sensitivity when diagnosing a disease [17]. This is one of the strengths of using quantitative metabolomics in clinical medicine. It provides dozens, even hundreds of metabolites that allow one to generate a powerful and more accurate biomarker profile for diagnosing and characterizing a disease or condition [18].

8.6 Data Enrichment and Biochemical Interpretation with the SMPDB

One of the best ways of understanding biological or biochemical mechanisms is through the analysis of pathway diagrams. Pathway diagrams are the "Google Maps" of the cell. They illustrate the connectivity of cellular components at different levels of resolution. In terms of resolution, pathway diagrams can be drawn at the "macro" level of organs and tissues, at the "micro" level of cells and organelles, or at the "nano" level of genes, proteins, and metabolites. A good pathway diagram should be able to illustrate events at all three levels. Likewise, a good pathway diagram should help connect the macro to the micro to the nano and the genes to the proteins to the metabolites—and back again. We have already introduced you to the SMPDB. This database was specifically developed to capture this multi-level, multi-omic information and to do it in both a human-readable and a machine-readable fashion. SMPDB is particularly unique in terms of the number of disease and metabolite signaling pathways that it captures. A detailed visual inspection of the SMPDB pathways associated with the individually dysregulated metabolite listed in Tables 8.1 and 8.2 should give you a better idea of the hows and whys arising from this dysregulation. In other words, these pathways help interpret the biochemistry measured via metabolomics. SMPDB also supports multi-metabolite and multigene or multi-protein querying through a search tool called "SMP-MAP," which stands for the Small Molecule Pathway Mapping utility. Through SMP-MAP users can provide a list of dysregulated metabolites (or proteins/genes), and even their concentrations and the SMP-MAP query function will find and sort the pathways that best match the list of metabolites (or genes/proteins). The altered metabolites (or proteins/genes) are then differentially colored on the SMPDB pathway diagram. The top-ranking pathways identified in this manner can be used to identify the likely disorder. In other words, SMP-MAP is essentially a qualitative tool for biomarker discovery. Not only does it allow one to discover potential biomarkers or biomarker-disease associations; it also provides some biochemical rationale for the observed disorder or disease. Here we will show how to use the SMP-MAP function in SMPDB.

8.6.1 Search SMPDB with Dysregulated Metabolites via SMP-MAP

8.6.1.1. Go to the SMPDB home page at www.smpdb.ca (see Fig. 8.3a).

8.6.1.2. Under the "Search" menu at the top of the home page select the option "SMP-MAP Advanced Search."

8.6.1.3. Enter the names of the most dysregulated metabolites, one compound per line, in the search box called "Search for: Compound Names." These would include kynurenine, creatinine, glucose, phenylalanine, tryptophan, methionine sulfoxide, and LPC 17:0 (Fig. 8.3b). You may include others if you wish. If you are unclear how to do this or what format to follow, you can click on the examples given above the text box in the blue buttons.

8.6.1.4. After pressing the "Search" button at the bottom of the page, you should see a table of highly matching pathways along with a list of the number of metabolites matched between the query list and the list of metabolites found in the pathway. As you will see, some of the top scoring hits are pathways for sepsis and immuno-metabolism.

8.6.1.5. Selecting and expanding the pathway view (using the navigation buttons on the top corner of the pathway diagram), reading the accompanying pathway descriptions, or simply visualizing how different metabolites, organelles, cells, and tissues interact in these pathways should provide you with much more biochemical insight into sepsis.

This pathway analysis and biochemical interpretation via SMPDB and SMP-MAP will typically take a user about 20–30 min. The intent of this step is to provide further confirmation that the disease that is afflicting these four patients is virally induced sepsis (accompanied by kidney failure). This SMPDB analysis is also intended to provide you with more biochemical insight into how sepsis evolves at a molecular level and to shed further light on some its physiological consequences. Indeed, these pathway diagrams, along with their detailed explanations and the accompanying HMDB descriptions of each of the dysregulated metabolites, should give you a very detailed picture about what is going on with these four patients.

As shown by the pathway diagrams in SMPDB (see Fig. 8.2b), viral sepsis begins when viral coat proteins activate the Toll-like receptors TLR4 and TLR2 on the membranes of macrophages, T cells, and dendritic cells. In addition to this protein activation, the viral DNA (or RNA if it is an RNA virus) is taken up by macrophage endosomes. Viral DNA fragments (such as CpG DNA) activate the endosomal TLR9, while viral double-stranded DNA fragments activate the endosomal TLR3, and viral single-stranded RNA (if it is an RNA virus) activates endosomal TLR7/8 proteins. Different TRL receptors activate different processes for the innate immune response [19]. The TLR4 activates the production of

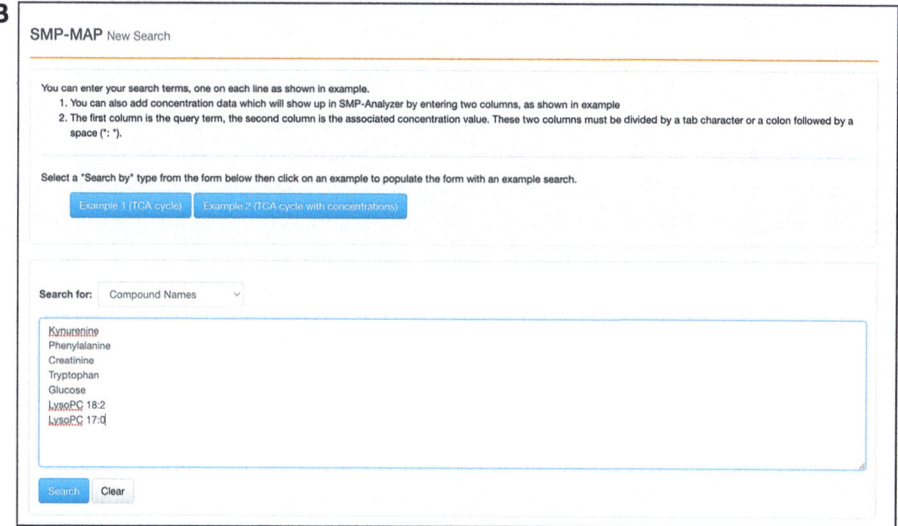

Fig. 8.3 Using the SMPDB query tool known as SMP-MAP Advanced Search to identify potential biomarkers. (**a**) Selecting SMP-MAP from the browser. (**b**) An example of the metabolite entries for SMP-MAP that would create high quality hits to sepsis or immunometabolism pathways similar to those shown in Fig. 8.2b

interferon regulatory factor 3 (IRF3), TIR domain-containing adapter-inducing interferon-β (TRIF), signal transducer and activator of transcription 1 (STAT1), and nuclear factor kappa B (NF-kB) in the cytoplasm, while TLR9, TRL3, and TLR7/8 activate the production of myeloid differentiation primary response 88 (MyD88), TRIF, interferon regulatory factor 7 (IRF7), and NF-kB in the cytoplasm [19]. The NF-kB protein then goes to the nucleus and activates expression of nitric oxide synthase (iNOS) which generates nitric oxide (NO). It also activates aconitate decarboxylase (Irg1), tumor necrosis factor (TNF), interleukin 6 (IL-6), and interleukin 1 beta (IL-1β). These are the pro-inflammatory proteins, while nitric oxide (NO) is also a pro-inflammatory molecule that can lead to the production of oxidized tyrosines (i.e., nitrotyrosine). Similarly, the newly expressed IRF3 and IRF7 proteins go to the nucleus and activate the production of interferon beta (IFN- β), which is another pro-inflammatory cytokine. The other cytokines, TNF, IL-6, IL-1β, and IFN-β, move into the bloodstream and head to the brain and into the hypothalamus, leading to release of the hypothalamic corticotropin-releasing hormone (CRH) [20]. CRH, in turn, activates the release of pituitary adrenocorticotropic hormone (ACTH), which then moves down through the bloodstream towards the adrenal glands (located at the top of the kidneys) to produce cortisol and epinephrine (Fig. 8.2b). Cortisol and epinephrine stimulate the "flight or fight" response, leading to the increased production of glucose from the liver (via glycogen breakdown) and the release of short-chain acylcarnitines (also from the liver) to help support beta-oxidation of fatty acids. These compounds support cell synthesis and growth of the macrophages and neutrophils used in the innate immune response. The liver also produces more IL-6, more TNF, and more NO to further stimulate the innate immune response.

Higher nitric oxide (NO) levels lead to blood vessel dilation and reduced blood pressure, which, in its most extreme form, can be a major problem in sepsis. Higher iNOS expression in macrophages, neutrophils, and dendritic cells consumes the amino acid arginine to produce more NO which disrupts the mitochondrial TCA cycle leading to the accumulation of citrate and the production of fatty acids and acylcarnitines (needed for lipid synthesis). Increased Irg1 (actonitate decarboxylase) expression leads to accumulation of succinate, which results in the succinylation of phosphofructokinase M2 (PKM2) [21]. Succinate also leads to the release of hypoxia-inducible factor 1-alpha (HIF-1α) from its prolyl hydroxylase (PHD)-mediated inhibition. HIF-1α interacts with succinylated PKM2 and induces the expression of glycolytic genes such as Glut1 (the glucose transporter) and the pro-inflammatory cytokine IL-1β [21]. As a result of these metabolic changes and the deactivation of the oxidative phosphorylation pathway in their mitochondria, macrophages, neutrophils, T cells, and dendritic cells shift to aerobic glycolysis [22]. This leads to the production of more reactive oxygen species (ROS) which results in the oxidation of certain amino acids, such as methionine. This leads to the increased production of methionine sulfoxide (Met-SO). As the inflammatory response continues, more glucose and arginine in the bloodstream are consumed by dividing white blood cells to produce more lactate and more NO to further push the aerobic glycolytic pathway [22]. This aerobic glycolysis occurs primarily in white blood cells leading to active cell division and rapid white cell propagation (growing by a factor of three to four in a few hours). Hexokinase (HK) along with increased levels of lactate from aerobic glycolysis activates the inflammasome inside macrophages and dendritic cells, leading to the secretion of IL-1β. This cytokine further drives the aerobic glycolysis pathway for these white blood cells (Fig. 8.2b). All these signals and effects combine to lead to the rapid and sustained production of large numbers of macrophages, neutrophils, dendritic cells, and T cells to fight the viral infection. This often leads to a reduction in essential amino acids (threonine, lysine, tryptophan, leucine, isoleucine, valine, arginine) and a mild reduction in gluconeogenic acids (glycine, serine) in the bloodstream. The reduction in essential amino acids is intended to "starve" the invading viruses (and other pathogens) of the amino acids they need to reproduce [22]. Some of the reduction in amino acid levels is moderated by the proteolysis of myosin in the muscle and the proteolysis of serum albumin in the blood (the most abundant protein in the blood, which is produced by the liver). These proteins act as amino acid reservoirs to help support rapid immune cell production. The loss of serum albumin in the blood to help support amino acid synthesis elsewhere can lead to hypoalbuminemia, a common feature of infections, inflammation, late-stage cancer, and sepsis.

At some point during the innate immune response, the kynurenine pathway becomes dysregulated, potentially through overstimulation by interferon gamma (IFNγ). This hyperstimulation leads to large reductions in tryptophan levels as the indole dioxygenase (IDO) enzyme becomes more active. IDO activation results in the generation (from tryptophan) of large amounts of kynurenine (and its other metabolites) through a self-stimulating autocrine process. Kynurenine binds to the aryl hydrocarbon receptor (AhR) found in most immune cells [23–25]. In addition to increased kynurenine production via IDO-mediated synthesis, hypoalbuminemia can also lead to the release of bound kynurenine (and other immunosuppressive LPCs) into the bloodstream to fuel this kynurenine-mediated process. Regardless of the source of kynurenine, the kynurenine-bound AhR will migrate to the nucleus to bind to NF-kB which leads to more production of the IDO enzyme, which leads to more production of kynurenine and more loss of tryptophan (Fig. 8.2b). High kynurenine levels and low tryptophan levels lead to a shift in T-cell differentiation from a TH1 response (pro-inflammatory) to the production of Treg cells and an anti-inflammatory response [23–25]. High kynurenine levels also lead to the production of more IL10R (the interluekin-10 receptor) via binding of kynurenine to the aryl hydrocarbon receptor (AhR). Activated AhR effectively increases the anti-inflammatory response from interleukin 10 (an anti-inflammatory cytokine). Low trypto-phan levels also lead to the activation of the general control non-depressible 2 kinase (GCN2K) pathway, which inhibits the mammalian target of rapamycin (mTOR) and protein kinase C signaling. This leads to T-cell autophagy and anergy. High levels of kynurenine also lead to the inhibition of T-cell proliferation through induction of T-cell apoptosis [23–25].

In other words, kynurenine leads to a blunted immune response as neither sufficient B cells and macrophages nor T cells (which are needed for B-cell production) are produced, leading to further immune suppression. This allows for uncontrolled viral propagation. As a result, the invading viruses are *not* successfully cleared. This leads to a "vicious" or futile cycle where the growing virus population pushes the body to produce more B cells and T cells, and various organs (muscles, heart, liver) exhaust themselves to produce more metabolites to fuel the pro-inflammatory response, while the kynurenine/tryptophan cycle keeps on killing off T cells and blunting the immune response [23–25]. This "futile" cycle of producing ineffective B and T cells leads to heightened lactate production resulting in lactic acidosis. Likewise, as more NO is produced, this leads to a further loss of blood pressure – both lactic acidosis and hypotension can lead to organ failure. The continuous release of pro-inflammatory cytokines through the failed fight to eliminate the virus can also damage the alveolar-capillary barrier in the lungs. Loss of integrity of this lung barrier leads to influx of pulmonary edema fluid and lung injury or fluid in the lungs (Fig. 8.2b). Excessive, long-term release of glucose, short-chain acylcarnitines, and fatty acids from the liver along with higher amino acid production from the blood and liver via proteolysis of albumin (leading to more extreme hypoalbuminemia) results in reduced uremic toxin clearance and increased levels of uremic solutes in the blood. High levels of uremic toxins lead to liver, heart, brain, and kidney injury [26]. Likewise excessive release of acylcarnitines from the heart and liver leads to heart and liver injury (Fig. 8.2b). Organ failure often develops in end-stage sepsis, leading to death.

8.7   Disease Interpretation with MarkerDB

A number of the dysregulated metabolites we identified in Table 8.2 are known biomarkers of both bacterial and viral sepsis. As highlighted earlier, biomarkers are measurable substances that are indicative of a condition, a disease, a diet, an intervention, or an environmental exposure. Most clinical tests used today are designed to measure specific biomarkers to help diagnose specific diseases. In many cases, measuring the concentration of a given biomarker or combination of biomarkers is key to making a proper diagnosis. Therefore, knowing which biomarker is specific to a disease or which concentration cutoff is specific for a disease is often considered essential knowledge for practicing physicians. Unfortunately, the number of known biomarkers and the number of diagnosable medical conditions now numbers in the thousands. This makes it almost impossible for a single individual to memorize which markers or which concentrations of biomarkers are associated with specific diseases. MarkerDB [27] is a database that was developed to help address this challenge. MarkerDB is a freely available electronic database that consolidates information on all known clinical molecular biomarkers into a single resource. It covers chemical, protein, genetic, and karyotypic biomarkers that can be used for diagnostic, predictive, prognostic, or exposure assessments. MarkerDB includes information on biomarker names, images or structures, associated conditions, detailed disease descriptions, detailed biomarker descriptions, and data on biomarker performance. MarkerDB contains >27,000 biomarkers covering more than 670 human diseases or conditions. Approximately 1100 of these biomarkers are chemical or metabolite biomarkers. Here we will show you how to use MarkerDB to learn more about sepsis and the known biomarkers for sepsis.

8.7.1 Explore MarkerDB to Learn More About Sepsis

8.7.1.1. Go to the MarkerDB home page at www.markerdb.ca (see Fig. 8.4a).

8.7.1.2. Go to the text search box on the top right of the home page and ensure that the selection box indicates "Conditions." Type in the name "sepsis" into the text box and press the "Search" button.

8.7.1.3. You should see several disease entries listed including "Sepsis" and "Septic shock." Click on "Sepsis." The following page should be visible (see Fig. 8.4b).

8.7.1.4. Scroll through the page to learn more about sepsis and some of the known or clinically approved biomarkers for diagnosing sepsis (including protein and metabolite markers) as well as potential genetic susceptibility traits. Check to see if the markers you have identified in the above exercises match to those listed in MarkerDB.

8.7.1.5. Click on the "back" button for your browser, and go to the list of diseases and click on "Septic shock." See what else you can learn about this condition and its known biomarkers.

8.7.2 Questions

1. Given what you now know about the biochemistry of viral sepsis, how might you treat it or prevent it?
2. What is currently done to treat sepsis?

8.8   Conclusion

As many of you may have figured out by now, the unknown viral disease that afflicted these four patients was COVID-19. In particular, the data used here (with some minor modifications to keep things simple) was taken from an early study on COVID-19 done in 2020 [5]. When COVID-19 first appeared, many affected patients unfortunately died due to the effects of "undetected" viral sepsis. It appears that this condition was rarely, if ever, detected. As a result, most patients were only treated for the anoxia or breathing difficulties they experienced. This is why so many patients were put on mechanical ventilation and given nothing more than some palliative care to reduce pain and deal with progressive organ deterioration. In a septic state, mechanical ventilation is simply not enough to change the outcome. Later treatments for COVID-19 included corticosteroids such as dexamethasone. Indeed, corticosteroids are commonly used to treat sepsis and can significantly reduce

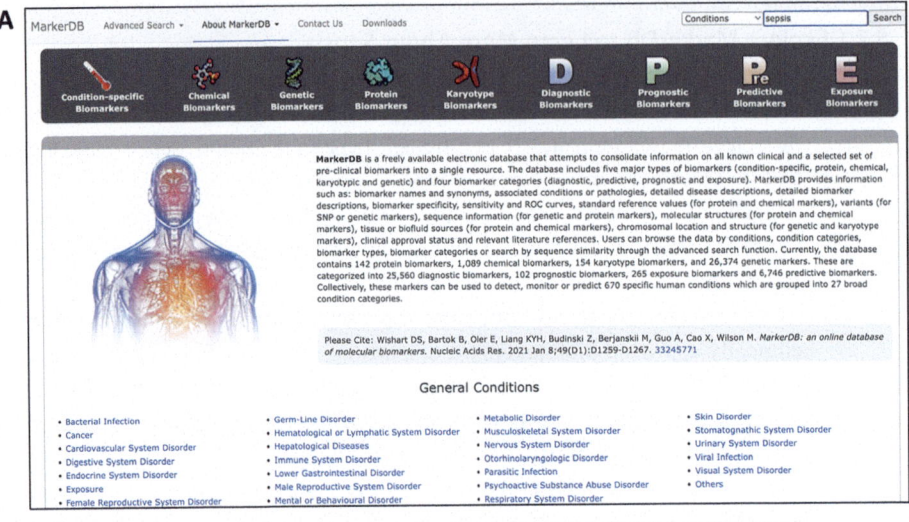

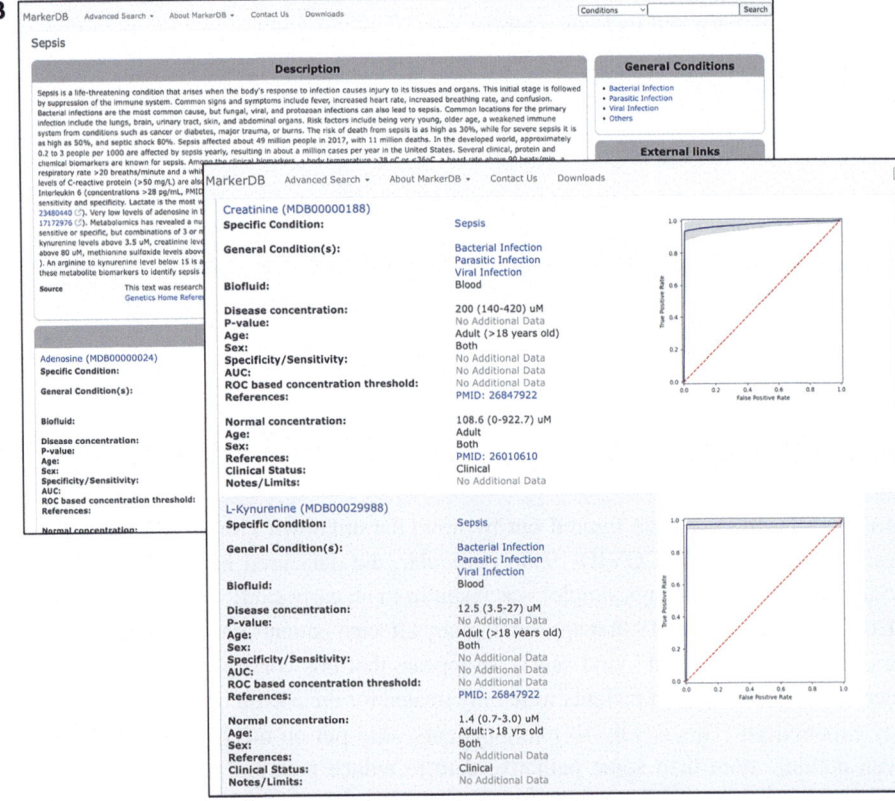

Fig. 8.4 Exploring sepsis and sepsis biomarkers with MarkerDB. (**a**) An example of how MarkerDB can be queried with a suspected disease or condition name. (**b**) Example screenshots of the known biomarkers for sepsis along with estimates of their ROC (sensitivity vs. specificity) curves

mortality from sepsis or septic shock [28]. Interestingly the same efficacy in treating COVID-19 with dexamethasone was seen as previously noted in treating "regular" sepsis [29]. As can be seen from the quantitative metabolomic analyses described here, severe forms of COVID-19 exhibit classic biochemical features and share many common biochemical markers of virally induced sepsis. What's more, the information provided through this data enrichment process suggests some potentially novel routes to treat or possibly prevent COVID-19-induced sepsis. Certainly, had more complete metabolomic analyses been performed early in the COVID-19 pandemic, improved patient care and improved patient outcomes may have been possible.

> **Take-Home Message**
> Quantitative, targeted metabolomics is essential to using metabolomics in the clinic. The ability to accurately measure multiple metabolite concentrations and compare these values to known, healthy values allows metabolomics data to be interpreted using only a single patient at a single time point. This is critical to offering timely medical care. Data enrichment strategies that exploit the HMDB, SMPDB, and MarkerDB allow for very detailed data analysis and interpretation of measured metabolite levels in many human biofluids. The use of these open-access databases can provide important new biochemical insights into specific diseases. Furthermore, this information may be used to help suggest new therapies or new treatments to combat specific diseases.

8.4.2 Answers

1. The most significantly different metabolites are kynurenine (four of four patients), creatinine (four of four), methionine sulfoxide (four of four), phenylalanine (four of four), tryptophan (four of four), and LPC 17:0 (four of four). Other metabolites that differ substantially from normal for three of the four patients are glucose, threonine, leucine, LPC 16:0, LPC 18:2, and most of the short-chain acylcarnitines (CAR C2-C5). There is also a trend for uniformly low arginine levels. The most dysregulated metabolite is kynurenine, which is uniformly 5–20× higher than normal.

2. Identifying abnormal or dysregulated metabolites should be done by looking for metabolites that are substantially (>30%) and uniformly higher or lower than the normal values across all or most patients. Those that are consistently higher/lower across patients are most notable. Those that are high in some patients and low in other patients are less notable, although these metabolites may indicate the

(continued)

existence of divergent metabolic processes. Any metabolite that is 2× higher or 2× lower than normal is very significant. Values that are 10–20% higher or lower than the normal maximum or normal minimum are typically within the instrumental measurement error, although they are worth noting or considering when making a diagnosis or a biochemical interpretation.

3. The metabolite that appears to be most normal is lactate, with all four patients having normal values. Some metabolites such as valine and arginine are normal in three of four patients.

4. The reason why all four patients have normal lactate levels, despite having severe lung disease, is because they are being mechanically ventilated. Mechanical ventilation supplies oxygen to oxygen-poor tissues and quickly normalizes lactate levels. Had these patients not been ventilated, their lactate levels would all likely be >3 mM.

8.5.2 Answers

1. Since broad-spectrum antibiotics were ineffective in treating this disease and tests for conventional infectious organisms turned out negative, it must be assumed that the disease is viral in origin and that the virus is novel or significantly mutated from known variants.

2. According to the HMDB, kynurenine is associated with Alzheimer's disease, kidney failure or kidney dysfunction, inflammation, and sepsis. Elevated kynurenine is normally indicative on immunosuppression or poorly functioning kidneys. Kynurenine arises from the catabolism of tryptophan and is produced to serve as an immunological brake. The degradation of tryptophan is an innate immune response intended to reduce the supply of essential amino acids (like tryptophan) so that dividing pathogens may die or slow their growth. High kynurenine levels are not useful for combatting severe infections.

3. According to the HMDB, elevated plasma or serum creatinine is indicative of poorly functioning kidneys, infected kidneys, or kidney failure. It is also associated with sepsis and other forms of severe infection. Creatinine is a uremic toxin. It is not necessarily toxic, but it does indicate elevated levels of protein breakdown and poorly functioning kidneys.

4. According to the HMDB, elevated plasma or serum glucose is indicative of many conditions including diabetes, hyperglycemia, the consumption of a high carbohydrate meal, inflammation or infection, and sepsis. Glucose serves as a signaling molecule for the body to release insulin and to initiate cell division or growth. Chronically high levels of glucose can lead to the glycation of proteins and their eventual damage, loss of function, or misfolding.

(continued)

5. According to HMDB and SMPDB, most of the dysregulated metabolites seem to be associated with inflammation, infection, or sepsis.
6. Based on the weight of evidence, it appears that these four patients are suffering from a form of viral sepsis.

8.7.2 Answers

1. If the viral (or bacterial load) is high, the simple solution is to reduce kynurenine levels and stop immunosuppression from competing with immune activation. However, if the viral (or bacterial load) is already low or the pathogens have been cleared, then the task is to stop immune activation and let immunosuppression occur. For the first scenario (helping the innate immune response), the best way to reduce kynurenine is to add more serum albumin to "soak" up this uremic toxin. Albumin will also reduce the levels of immunosuppressive LPCs. Alternately, kidney dialysis can also help reduce kynurenine levels (while at the same time helps reduce the heavy load on the kidneys). Increased levels of tryptophan (via intravenous infusion) may also help reduce the autocrine AhR/IDO activity of kynurenine. Adding more tryptophan (along with more arginine) also helps make up the clear deficits of these amino acids in the septic body as both amino acids are pro-inflammatory. Increasing their levels allows immune activation to continue and B-cell/T-cell production to increase so that the pathogens can be cleared. If glucose levels are low or cortisol/epinephrine levels are low, these should be increased. For the second scenario (helping immunosuppression), the best approach is to reduce glucose, reduce lactate, reduce NO levels, reduce levels of cortisol and epinephrine, and provide anti-inflammatory medications. Lowering glucose can be done via insulin, reducing lactate can be done via oxygenation, and reducing NO, cortisol, or epinephrine can be done through various anti-inflammatory or analgesic medications.
2. Sepsis is often characterized by dangerously high levels of lactate (lactic acidosis) and glucose (hyperglycemia), along with low blood pressure (hypotension), and very low levels of albumin (hypoalbuminemia) along with high levels of pathogens (bacteria or viruses) in the blood or certain organs. Controlling pathogen levels with antibacterials or antivirals helps to reduce the pathogen load, thereby mitigating the body's overactive immune response. Providing patients with norepinephrine helps increase blood pressure as does providing solutions of albumin to help restore blood fluid levels and blood serum albumin levels. Supplemental oxygen or mechanical ventilation can help reduce lactate levels while insulin can help control excessively high glucose. Parenteral nutrition can also be used to replace lost or low levels of amino acids.

References

1. Bhattacharya K, Wotton T, Wiley V. The evolution of blood-spot newborn screening. Transl Pediatr. 2014;3(2):63–70.
2. Chace DH, Kalas TA, Naylor EW. Use of tandem mass spectrometry for multianalyte screening of dried blood specimens from newborns. Clin Chem. 2003;49(11):1797–817.
3. Therrell BL, Padilla CD, Loeber JG, et al. Current status of newborn screening worldwide: 2015. Semin Perinatol. 2015;39(3):171–87.
4. Pourfarzam M, Zadhoush F. Newborn screening for inherited metabolic disorders; news and views. J Res Med Sci. 2013;18(9):801–8.
5. Fraser DD, Slessarev M, Martin CM, et al. Metabolomics profiling of critically ill coronavirus disease 2019 patients: identification of diagnostic and prognostic biomarkers. Crit Care Explor. 2020;2(10):e0272.
6. Wishart DS, Guo A, Oler E, et al. HMDB 5.0: the human metabolome database for 2022. Nucleic Acids Res. 2022;50(D1):D622–31.
7. Kanehisa M, Furumichi M, Tanabe M, Sato Y, Morishima K. KEGG: new perspectives on genomes, pathways, diseases and drugs. Nucleic Acids Res. 2017;4 5(D1):D353–61.
8. Kim S, Chen J, Cheng T, et al. PubChem in 2021: new data content and improved web interfaces. Nucleic Acids Res. 2021;49(D1):D1388–95.
9. Caspi R, Billington R, Keseler IM, et al. The MetaCyc database of metabolic pathways and enzymes - a 2019 update. Nucleic Acids Res. 2020;48(D1):D445–53.
10. Hastings J, Owen G, Dekker A, et al. ChEBI in 2016: improved services and an expanding collection of metabolites. Nucleic Acids Res. 2016;44(D1):D1214–9.
11. wwPDB consortium. Protein data Bank: the single global archive for 3D macromolecular structure data. Nucleic Acids Res. 2019;47(D1):D520–8.
12. UniProt Consortium. UniProt: the universal protein knowledgebase in 2021. Nucleic Acids Res. 2021;49(D1):D480–9.
13. Sayers EW, Cavanaugh M, Clark K, et al. GenBank. Nucleic Acids Res. 2022;50(D1):D161–4.
14. Jewison T, Su Y, Disfany FM, et al. SMPDB 2.0: big improvements to the small molecule pathway database. Nucleic Acids Res. 2014;42(Database issue):D478–84.
15. Wishart DS, Li C, Marcu A, et al. PathBank: a comprehensive pathway database for model organisms. Nucleic Acids Res. 2020;48(D1):D470–8.
16. Wishart DS, Feunang YD, Guo AC, et al. DrugBank 5.0: a major update to the DrugBank database for 2018. Nucleic Acids Res. 2018;46(D1):D1074–82.
17. Bayes-Genis A, Ordonez-Llanos J. Multiple biomarker strategies for risk stratification in heart failure. Clin Chim Acta. 2015;443:120–5.
18. Xia J, Broadhurst DI, Wilson M, Wishart DS. Translational biomarker discovery in clinical metabolomics: an introductory tutorial. Metabolomics. 2013;9(2):280–99.
19. Chen F, Zou L, Williams B, Chao W. Targeting toll-like receptors in sepsis: from bench to clinical trials. Antioxid Redox Signal. 2021;35(15):1324–39.
20. Silverman MN, Pearce BD, Biron CA, Miller AH. Immune modulation of the hypothalamic-pituitary-adrenal (HPA) axis during viral infection. Viral Immunol. 2005;18(1):41–78.
21. Zheng S, Liu Q, Liu T, Lu X. Posttranslational modification of pyruvate kinase type M2 (PKM2): novel regulation of its biological roles to be further discovered. J Physiol Biochem. 2021;77(3):355–63.
22. O'Neill LAJ, Kishton RJ, Rathmell J. A guide to immunometabolism for immunologists. Nat Rev Immunol. 2016;16(9):553–65.
23. Bello C, Heinisch PP, Mihalj M, Carrel T, Luedi MM. Indoleamine-2,3-dioxygenase as a perioperative marker of the immune system. Front Physiol. 2021;12:766511.

24. Herrera-Van Oostdam AS, Castañeda-Delgado JE, Oropeza-Valdez JJ, et al. Immunometabolic signatures predict risk of progression to sepsis in COVID-19. PLoS One. 2021;16(8):e0256784.
25. Guarnieri T. Hypothesis: emerging roles for aryl hydrocarbon receptor in orchestrating CoV-2-related inflammation. Cell. 2022;11(4):648.
26. Okusa MD. The changing pattern of acute kidney injury: from one to multiple organ failure. Contrib Nephrol. 2010;65:153–8.
27. Wishart DS, Bartok B, Oler E, et al. MarkerDB: an online database of molecular biomarkers. Nucleic Acids Res. 2021;49(D1):D1259–67.
28. Annane D, Bellissant E, Bollaert PE, Briegel J, Keh D, Kupfer Y. Corticosteroids for treating sepsis. Cochrane Database Syst Rev. 2015;2015(12):CD002243.
29. Kino T, Burd I, Segars JH. Dexamethasone for severe COVID-19: how does it work at cellular and molecular levels? Int J Mol Sci. 2021;22(13):6764.

Part V

Metabolomics to Decipher Physiology in Health and Disease: Clinical Research Studies

Automated Sample Preparation for Blood Plasma Lipidomics

9

Jing Kai Chang, Wai Kin Tham, Peter I. Benke, Markus R. Wenk, and Federico Torta

> **What You Will Learn in This Chapter**
> - Principles of sample preparation in blood plasma lipidomics
> - Experimental design for automated lipid extraction
> - Automated extraction optimization

This chapter describes the experimental steps involved when performing automated lipid extraction for lipidomics-based mass spectrometry. We cover topics such as the importance of including quality control procedures, considerations prior to the use of automation, proposed plate configuration and sample location, and possible issues during the execution of the reported workflows and their solutions. While the procedures described here are optimized for blood plasma lipidomics, they can be easily adapted for lipid extraction from other sample types, such as cells or tissues.

Jing Kai Chang and Wai Kin Tham contributed equally with all other contributors.

J. K. Chang · W. K. Tham · P. I. Benke · M. R. Wenk (✉) · F. Torta (✉)
Singapore Lipidomics Incubator and Precision Medicine Translational Research Program,
Department of Biochemistry, Yong Loo Lin School of Medicine, Life Sciences Institute, National
University of Singapore, Singapore, Singapore
e-mail: bchmrw@nus.edu.sg; bchftt@nus.edu.sg

© The Author(s), under exclusive license to Springer Nature Switzerland AG 2023
J. Ivanisevic, M. Giera (eds.), *A Practical Guide to Metabolomics Applications
in Health and Disease*, Learning Materials in Biosciences,
https://doi.org/10.1007/978-3-031-44256-8_9

9.1   Theoretical Background

Mass spectrometry analysis of lipids is the most used methodology to study the role of lipids in biology and human health. As all the steps in an analytical process are based on accurate sample preparation, the correct sampling is vital for any of the omics approaches. This section outlines the principles of sample preparation for lipidomic analyses, with a particular focus on automated workflows due to the increasing demands for the analysis of large sample cohorts (hundreds to thousands of samples) in life and medical sciences.

9.1.1   Planning for Automated Sample Preparation

Sample preparation for lipidomics is usually executed manually. However, a demand for high-throughput, fast, and more precise methods has risen over the years, challenging existing procedures. Due to the increasing number of samples to be analyzed in preclinical research, the interest in automated sample preparation, one of the main rate-limiting factors for lipidomics, grew exponentially. Automated extraction systems from manufacturers such as Agilent, Camag, Hamilton, and Tecan and have been developed to serve this purpose [1, 2]. Based on mechanical arms that can reproducibly perform liquid transfer, these systems commonly make use of 96- or 384-well plates. These systems can also utilize tubes and vials with different volumes; however, physical constraints may limit the operational number of sample containers, typically down to 24 in the standard rack formats. The use of automation helps to save considerable time and effort, while giving improved or comparable precision compared to a manual extraction procedure. Wang et al. [3] showed how automated and manual extractions are well correlated, with the manual one showing higher standard deviations. This is further iterated in Jung et al. [4] when manual and automated extraction of fresh frozen plasma was compared, with the robotic setup showing considerably smaller CV values after analysis of common endogenous lipids.

Despite these advantages, various factors (described below) can influence the final outcome. Thus, before the implementation of automated sample extraction procedures, the researcher should carry out preliminary comparison of an identical set of samples processed by manual and automated preparation.

Size of the Study
The total number of samples in a study is an important factor to consider before deciding to set up an automated workflow. While for large batches of samples it is obvious to consider this possibility, for smaller experiments the choice might be unclear. Considering that an automated lipid extraction is usually done in a 96-well plate format, the waste of unused positions/wells and of prepared solvents—especially if spiked with standards—that will not be fully used, could be relevant factors that increase the final cost of the experiment. Furthermore, the time required for manual extraction of a small batch might not be

considerably different when compared to the automated extraction for the same set of samples.

As a rule of thumb, cost in terms of consumables and time used for sample preparation should be estimated in both cases before the decision.

Sample Containers and Volume

Since robotic systems, used in automated sample preparation, require specific formats of containers for samples and solvents, these details need to be included in the experimental design. For example, when using one of the systems in our lab, we found that above a certain transfer volume (50 μL), systematic errors decreased significantly. Thus, if feasible, the required sample volume must be above a certain value.

Possible issues deriving from containers that are not compatible or efficient with the robotic system should potentially be sorted out prior to collection or delivery of samples, by requesting them to be aliquoted in a compatible tube or plate format and volume.

For example, while the efficiency of the process is higher when using a 96-well plate format, samples may be delivered in the more commonly used Eppendorf tubes (e.g., 1.5 mL). Adaptors are available for their use, but due to physical constraints (such as the positioning of the tube lid when open inside the rack), they are limited to lower throughput.

Extraction Protocol

Over the years, various lipid extraction protocols have been developed for lipidomic studies. Single-phase extraction methods, such as the butanol/methanol (1:1, v/v) extraction [5], might be preferred for practical reasons, as the lipid extract can be quickly recovered after protein precipitation. However, single-phase methods generate a dirtier extract that is suitable for LC-based methods but not for direct infusion/shotgun ones. Two-phase extraction methods such as the Folch, Bligh, and Dyer (1:2, v/v chloroform/methanol) [6] and the Methyl-*tert*-butyl ether (MTBE) method [7] would return cleaner extracts due to the additional steps and retrieval of lipids from a separated organic layer. However, they require more time and attention to avoid mixing the two phases.

Intuitively, single-phase methods can be implemented more easily for automated preparation procedures because optimizing an automated extraction process for a single-phase method is much faster as the robotic system will only have to transfer enough supernatant. In the case of a two-phase approach, the system will have to be instructed about which phase to recover. This must be based either on sensing a different density of the solvents employed (only possible for selected robotic systems) or by operating at a fixed height, at which the tip will aspirate the extract. Both approaches have been successfully adopted on different systems, but being more complex and tedious processes, the two-phase methods benefit significantly from automation.

Considering a more complex scenario, some lipid classes might require a solid phase extraction (SPE) step, using cartridges or resins immobilized into 96-well plates. While loading the solid phase with liquid samples and recovering the eluate from a collection plate is a simple task for any robotic system, the elution process would require the use of a vacuum manifold that may not be suitable for all the equipment. Furthermore, the assembly of the SPE device requires the use of a robotic arm for full automation. Nevertheless, automation of such procedures is still possible and has been implemented on several commercial platforms.

Questions
1. Beyond the scope of basic research, suggest a practical or translational application that would justify the use of automated extraction.
2. You are conducting a lipidomics cohort study on plasma samples from 2000 subjects. You received 500 μL of plasma in Eppendorf tubes from each subject and intend to extract the samples using a two-phase extraction method. Would you use automation for this study? Explain your rationale.

Answers
1. Batch analysis of lipid biomarkers in a hospital or diagnostic lab setting, where the time dedicated to sample preparation should be minimized to be compatible with the demand for fast and reproducible results.
2. Automation would be preferred for this study as the sample volume is high, the cohort size is large, and it requires a two-phase extraction. The automated procedure should start with aliquoting a smaller volume into the appropriate containers for extraction and proceed with the lipid extraction process, saving significant time for sample preparation.

9.1.2   Quality Control in Lipidomics

As part of quality assurance (QA) and quality control (QC) procedures in lipidomics, specific types of sample should be included in the experimental workflow [8]. This section describes these sample types.

Synthetic Lipid Standards
In liquid chromatography mass spectrometry-based (LC-MS) lipidomics, single molecular species or several lipid classes are measured using untargeted or targeted approaches, with either high-resolution instruments (Q-TOF and Orbitrap) or low-resolution ones (QQQ).

For the purposes of quantification, internal standards are spiked into all samples at known concentrations to correct for extraction recovery and possible MS signal fluctuations. These standards are non-endogenous species that might include short- or odd-chain fatty acid containing lipids or isotopically labelled molecules. Due to the heterogeneity and the high number of lipids present in biological samples, standards are not available for every single lipid. As a compromise accepted in the field, samples are spiked with one standard per lipid class, sometimes as pre-made mixtures at the correct concentrations. Nevertheless, it is highly recommended to use two deuterated standards for each class of lipids when possible, to better reflect the structural heterogeneity of the endogenous molecules that influences their chromatographic elution time and ionization. Indeed, the tendency of lipids to ionize is not only dependent on the head group characteristic of each class but also on the chain length and degree of unsaturation of the side chains [9].

System Suitability Samples

To assess performance and sensitivity of the analytical system, a system suitability test (SST) should be performed prior to injection of any QC or study sample [10]. The SST makes use of samples that are known to give a consistent response with the applied analytical method. SST samples could consist of in-house, pre-extracted reference plasma samples or a pure synthetic standard mix. Conventionally, a few injections of the pooled plasma extracts (TQCs) are first used to equilibrate the MS system, followed by injections of the SST samples. To ensure that the LC system is fully equilibrated, it is important to obtain reproducible retention times from the TQCs and SST. Several analytes (usually 10–20 molecules) with a different range of abundances and retention times are then evaluated in the SST, using the same analytical method that will be used in the main study. The goal is to have reproducible and good (in terms of intensity) signals from the analytes of interest. Various criteria, such as the CV (%) and retention time to test for the reproducibility of the system, the peak area or signal intensity to test for optimal response, and full width at half maximum (FWHM) of the peak of interest to test for optimal resolution may be used to determine whether the SST can be accepted. Common values for acceptance of these parameters are not strictly defined, as highly dependent on the method and the instrument of choice. If the SST fails, it is advisable to avoid proceeding with analysis of the study samples. Rather, troubleshooting of the instrument should be done to resolve the issues reflected by the SST (e.g., Adjustment of EMV (Electron Multiplier Voltage) in cases of low instrument sensitivity, changing the LC column in cases of peak splitting or high FWHM). Figure 9.1 shows a comparison of SST data obtained in the event of an issue with the LC column compared to those obtained with the LC-MS system working properly.

Quality Control Samples

LC and MS instrument performances might deteriorate as the length of analysis increases, contributing to the variability between samples. To measure this variation of the LC-MS signal, technical quality controls (TQCs) should be acquired periodically at regular

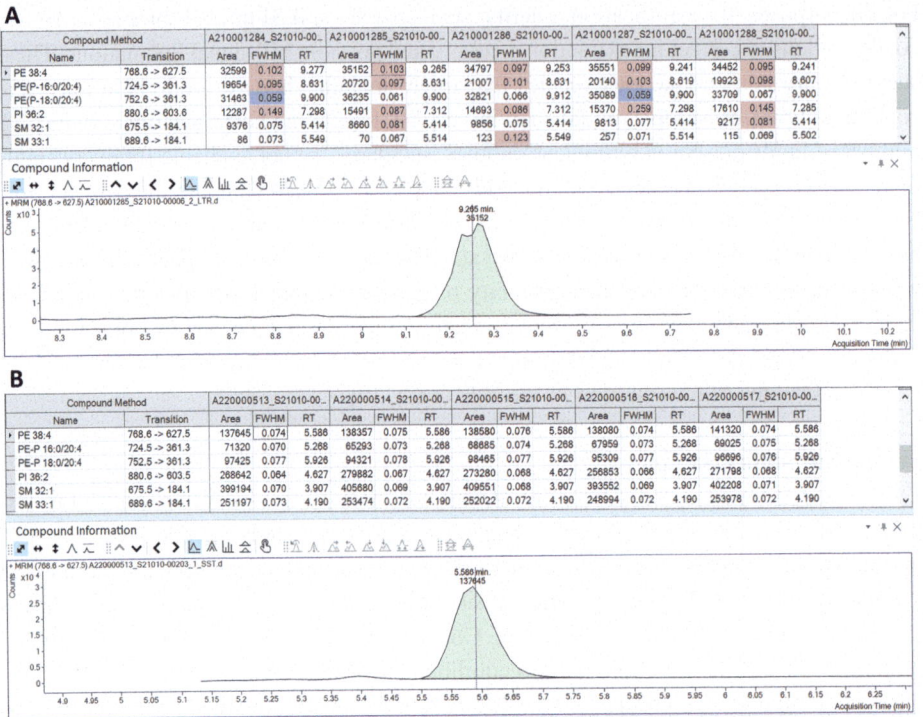

Fig. 9.1 Quantitative results of common lipids in five SST replicates measured by MRM. (**a**) FWHM was found to be >0.08 in most replicates for most lipid species (flagged in red), and peak splitting was observed. In this situation, the column needs to be changed. (**b**) FWHM was found to be <0.08 in all replicates for all lipid species. Narrow chromatographic peaks were observed with no peak splitting. The SST passed and the user could proceed with LC-MS analysis

intervals throughout the analysis, allowing users to monitor the intensity drift of the machine over time or even sudden changes in the signal measured. A TQC sample is usually prepared by pooling equal volumes of lipid extracts from most (or all) samples within the same study, so that the TQC is representative of the cohort under investigation [11, 12]. A possible alternative is to use a mixture of pure synthetic standards as TQC. Since another portion of the analytical variability between samples can originate from the extraction procedure, batch quality controls (BQCs) consisting of pooled aliquots from all the original samples within the study should also be prepared and analyzed at regular intervals.

The number of TQCs and BQCs for each analytical batch may vary depending on the total number of samples. A general rule of thumb is that at least ten TQC and ten BQC should be included in each study whenever possible for better post-acquisition signal correction when needed. This usually translates into a QC injection every 10–20 study samples. Of course, this number is arbitrary and might be smaller in smaller studies. The coefficient of variation (CV (%)) of the TQCs and BQCs after normalization (with corresponding internal standards) may be used as a filtering criterion for any signal (lipid level) measured and can be calculated as follows:

$$CV~(\%) = \frac{\text{Standard Deviation of Normalized Peak Area of Analyte in QCs}}{\text{Mean Normalized Peak Area of Analyte in QCs}} \times 100$$

A maximum acceptable value for the CV (%) of the TQCs and BQCs after normalization should be set by the user following guidelines in the field. For example, LC-MS analysis of eicosanoids generally uses a higher cutoff due to a higher variation caused by biological and chemical stability and low abundance of eicosanoids. As such, there isn't a standardized cutoff adopted by the lipidomics field, though it is usually set below 20–30% to strike a balance between the number of compounds with analyzable data and the quality of data [10].

The same TQC samples may also be injected at different dilutions or injection volumes, to be used as response quality controls (RQCs) at the beginning and the end of each batch, to test possible deviations from linearity of the measurements. This ensures that the MS signal response increases proportionally with injection volume and eliminates potential system overloading and interpretation errors that may result from saturated signals. As a rule of thumb, around six to eight injection volumes ranging from low to high volumes, with each volume injected in triplicate. For example, in an LC-MS method with injection volume of 2 μL, *injection volumes of 0.2 μL (10%), 0.4 μL (20%), 0.8 μL (40%), 1.2 μL (60%), 1.6 μL (80%), 2.0 μL (100%), 4.0 μL (200%) and 8.0 μL (400%) may be used for the RQCs.* Various mathematical or statistical software are available for calculation of the linearity for each transition. For example, Microsoft Excel can be used to estimate an R^2 value as a measure of how well a linear regression model fits the data, with R^2 values >0.8–0.9 considered as acceptable, according to common practice. Some examples of approaches available for linearity assessment include the RSQ function on Microsoft Excel and the Linearity Assessment and Curve Explorer Version 0.1.0, an open-source code accessible via Github [13]. Newer versions with improvements will be released periodically in the future, and users are advised to always check and use the latest code for their analysis.

Blanks

"Blanks" should always be incorporated into routine checks for data quality, as they give an indication on potential sources of background noise present in the solvents which may interfere with data analysis and carryover [10]. Different types of blanks should be analyzed at both the start and end of the study. To investigate interferences that might affect the signals of spiked synthetic standards, one should include a matrix blank (MBLK), a pooled sample that does not contain internal standards. To determine whether the extraction process introduces any background noise or contamination, process blanks (PBLK) and unextracted blanks (UBLK) should also be incorporated in the extraction

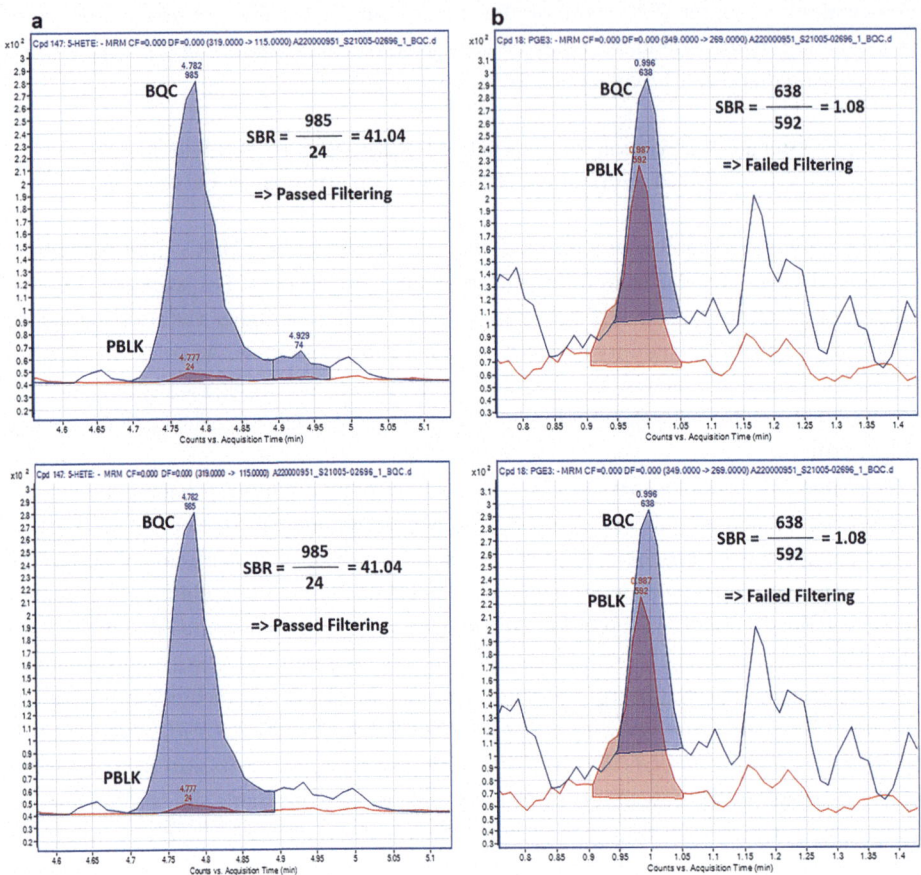

Fig. 9.2 Illustration of SBR calculation for two different lipid species, using BQC and PBLK. For each sample, the top value in the chromatographic peak defines the retention time and the bottom value defines the raw peak area. (**a**) The lipid shows a high SBR with a well-defined chromatographic peak in the BQC, thus passing the SBR-based filtering. (**b**) The compound has a very low SBR, thus failing filtering

batch. These consist of a blank sample (usually water) extracted using solvent with and without internal standards, respectively. An important parameter to filter data is the signal-to-noise (or blank) ratio (SNR or SBR). The SBR value for each compound of interest against the respective signal in blanks may then be calculated using the formula:

$$SBR = \frac{Median\ Peak\ Area\ of\ BQC}{Mean\ Peak\ Area\ of\ Blank}$$

A general criterion would be to accept all signals with SBR > 3–10. Figure 9.2 illustrates examples of the SBR calculation and filtering for real lipid measurements.

Compounds that do not meet the SBR cutoff against the PBLK should be excluded from subsequent data analysis, while internal standards that do not meet the cutoff for SBR against the MBLK should not be used for normalization. In such cases, another standard, as close as possible in terms of chemical structure, retention time, and mass, should instead be chosen.

Long-Term Reference and Reference Material Samples

To ensure that the extraction and LC-MS analysis worked properly, a reference material that has been previously characterized in the lab (long-term reference (LTR) plasma sample) or a sample that has been previously characterized by others, for example, the National Institute of Standards and Technology Standard Reference Material 1950 Plasma (NIST SRM-1950), should also be prepared and analyzed at the start, middle, and end of the sample worklist. These samples should return the same results previously obtained, as a positive control. These reference materials also help the harmonization of lipidomic data between different analytical batches of the same study or between different studies [14].

Table 9.1 summarizes all the quality controls that have been discussed in this chapter.

Questions

1. The following results were obtained from the measurements of QC samples in a targeted LC-MSMS analysis of three lipids. Would you accept the quantitative data relative to these compounds in the study samples for downstream statistical analysis?

   Lipid #1: CV (%) of TQC after normalization = 16.8%, CV (%) of BQC after normalization = 18.0%, SBR 1.5, $R^2 = 0.92$

   Lipid #2: CV (%) of TQC after normalization = 27.4%, CV (%) of BQC after normalization = 43.9%, SBR 15.0, $R^2 = 0.95$.

   Lipid #3: CV (%) of TQC after normalization = 8.9%, CV (%) of BQC after normalization = 9.4%, SBR 89.6, $R^2 = 0.97$

2. You are conducting untargeted analysis for a biological study and intend to use the deuterated phosphatidylserine internal standard (PS 33:1-d7) for normalization of all phosphatidylserine measurements in your study (please see Liebisch et al. [15] if you are not familiar with lipid nomenclature). Following LC-MS acquisition, you find that the MS1 chromatographic peak of this standard produces a well-defined peak in both matrix blanks at the same retention time as the QCs with corresponding SBR of 2.4. Would you proceed to use this internal standard for normalization? If not, suggest possible alternatives.

3. What sample would you use for an eicosanoids targeted panel system suitability test? Explain your reasoning.

Table 9.1 Types of quality controls and blanks included in lipidomics studies

Quality control	Composition	Injection sequence	Rationale
Technical Quality Control (TQC)	Pooled extracts from study samples	Regular intervals between study samples	Test for variability introduced by instrument
Batch Quality Control (BQC)	Extracts from pooled study samples	Regular intervals between study samples	Test for variability introduced by extraction procedure
Response Quality Control (RQC)	Pooled extracts from study samples	Different injection volumes prior to analysis of study samples with triplicates for each volume	Test for saturation effects
Process Blank (PBLK)	Extraction of water using solvent with internal standards	Before and after analysis of study samples	Test for background signals introduced by extraction procedure
Matrix Blank (MBLK)	Extraction of pooled study samples using solvent without internal standards	Before and after analysis of study samples	Test for background signals of internal standards introduced by the sample matrix
Universal Blank (UBLK)	Extraction of water using solvent without internal standards	Before and after analysis of study samples	Test for raw background signals
Long Term Reference (LTR)	Pooled reference plasma from healthy subjects	Before, in the middle and after analysis of study samples	Reference measurements and batch correction
National Institute of Standards and Technology Standard Reference Material 1950 (NIST SRM-1950)	Pooled reference plasma from healthy subjects	Before, in the middle and after analysis of study samples	Reference measurements and batch correction
Equilibration Control (EQC)	Pooled extracts from study samples	Ten injections at the start of LC-MS analysis	Equilibration of column
System Suitability Test (SST)	In-house pre-extracted reference material or standard mix	5 injections after equilibration	Test for instrument performance

Answers

1. Lipid #1 and Lipid #2 should not be accepted for downstream analysis as these compounds show low SBR and high CV (%) in the BQC. Lipid #3 passes all filtering criteria and may be accepted for statistical analysis.
2. The internal standard should not be used for normalization as the low SBR indicates the presence of matrix effect. A different phosphatidylserine standard may be used. Otherwise, standards from other phospholipid species that have similar structure and retention time may be used.
3. An internal standard mix or solution of eicosanoids standards would be preferred as the SST sample. In-house pre-extracted long-term reference material may be a less viable option as eicosanoids may degrade or form oxidative byproducts after prolonged storage, due to their chemical and biological instability. Hence, an internal standard mix or standard solution would be a better option as it is more stable.

9.2 Implementation of Automation

Once the decision to utilize automation for extraction of samples has been confirmed, several steps of planning, preparation, and execution are necessary. This section covers key steps of the automated extraction process, potential challenges, and how to overcome them.

9.2.1 Plate Formatting

When using 96- or 384-well plates, a regular arrangement of study and QC samples is beneficial in reducing potential errors. Plate formatting should consider different aspects, such as adaptability, ease of sample tracking, and storage. A possible 96-well format samples plate (Fig. 9.3a) would include randomized and stratified study samples (SS) distributed in the first ten columns of a plate, while having the last two columns "reserved" for quality controls (BQC and TQC), blanks, and reference materials (LTR and NIST SRM). A separate system plate (Fig. 9.3b) could be used only at the beginning and end of a study and include different QC samples, such as matrix blanks (MBLK), standard blanks (SBLK), as well as dilution series required for post-acquisition processing and QCs for equilibration of the system. This intuitive layout allows for minimal user errors during the aliquoting as well as sample preparation stages. Samples and QCs are allocated at defined positions across plates allowing for ease of worklist generation and minimizing the chances of misplacement of samples and need for visual inspections. Samples are acquired horizontally across from column 1 to column 12 with a periodic acquisition of BQC every 10 samples and a TQC every 20. We also recommend to inject at least two TQC and two

Sample Plate	1	2	3	4	5	6	7	8	9	10	11	12
A	SS	SS	SS	SS	SS	SS	SS	SS	SS	SS	BQC	TQC
B	SS	SS	SS	SS	SS	SS	SS	SS	SS	SS	BQC	TQC
C	SS	SS	SS	SS	SS	SS	SS	SS	SS	SS	BQC	NIST
D	SS	SS	SS	SS	SS	SS	SS	SS	SS	SS	BQC	TQC
E	SS	SS	SS	SS	SS	SS	SS	SS	SS	SS	BQC	LTR
F	SS	SS	SS	SS	SS	SS	SS	SS	SS	SS	BQC	TQC
G	SS	SS	SS	SS	SS	SS	SS	SS	SS	SS	BQC	TQC
H	SS	SS	SS	SS	SS	SS	SS	SS	SS	SS	BQC	TQC

Sample Plate	1	2	3	4	5	6	7	8	9	10	11	12
A	Blank	SBLK	MBLK									
B	Cond.	Cond.	Cond.	Cond.	Cond.	Cond.	Cond.	Cond.	Cond.	Cond.		
C	Cal1	Cal2	Cal3	Cal4	Cal5	Cal6	Cal7					
D												
E												
F												
G												
H												

Fig. 9.3 (**a**) Proposed 96-well plate layout of study samples and required QCs for automated extraction: randomized study samples (SS); Batch quality controls (BQC); pooled extract technical quality Control (TQC); Long-term reference plasma sample (LTR); NIST SRM 1950 plasma (NIST). (**b**) Proposed system plate which contains quality control samples essential for post-acquisition processing (blank, solvent blank; SBLK, standard blank containing solvent and internal standards; MBLK, extracted matrix sample containing no internal standards; Cal1–7, TQC samples that are injected at various volumes to replicate a dilution series used for monitoring of signal response relative to sample dilution for post-acquisition processing)

BQC before the first and after the last sample of the study, to facilitate the use of algorithms for batch correction.

Depending on the type of study, considering groups or individuals' characteristics, the samples should be stratified according to accepted guidelines, both within a plate and between different plates and batches. We suggest the use of deep well plates with a reasonable working volume (500 µL–2000 µL) to ensure sufficient volume during the lipid extraction procedure or to avoid contamination during shaking or liquid transfer. Plates with a smaller working volume (such as PCR plates) with a tapered bottom might be

	1	2	3	4	5	6	7	8	9	10	11	12
A	Blank	Zero	QC3	03	05	08	10	Zero				
B	Cal1	QC1	QC3	03	06	08	QC2	Cal1				
C	Cal2	QC1	01	03	06	08	11	Cal2				
D	Cal3	QC1	01	04	06	09	11	Cal3				
E	Cal4	QC2	01	04	Zero	09	11	Cal4				
F	Cal5	QC2	02	04	07	09	12	Cal5				
G	Cal6	QC2	02	05	07	10	12	Cal6				
H	Cal7	QC3	02	05	07	10	12	Cal7				

Fig. 9.4 Alternative 96-well plate layout for automated extraction, as adopted by Thompson et al. [16]. QCs, blanks, and calibrators specific to the assay can be arranged at the left and right-most columns to mimic the LC-MS acquisition in vertical order, while study samples and reference material are placed in between, in columns 3–7. QCs and calibrators are numbered according to increasing concentrations. Samples are labelled from 01–12. Blank and zero represent matrix and solvent blanks

considered when aliquoting extracts for MS analysis, to reduce the volume of extracts used for injection and create multiple aliquots that could be reanalyzed at a later stage.

Alternatively, it is possible to adopt a layout format reproducing the LC-MS acquisition in vertical order, i.e., blanks, QCs, and calibrators placed before and after the study samples, like the one reported by Thompson et al. [16] and used for an international ring trial (Fig. 9.4). However, it must be noted that this format requires more attention during sample preparation.

As mentioned before, users are not limited to well plates for automated sample preparation and can also utilize individual sample tubes. These should be prepared following the injection order, to simplify operations. Using individual tubes instead of well plates will significantly reduce the overall throughput of prepared samples.

9.2.2 Preparation for Automated Extraction

Note: The automated liquid handling system should be installed inside fume hoods (or similar) equipped with appropriate filters, due to the nature of the materials (biological samples, organic solvents) used during experiments. Ideally, Class II Type B2 fume hoods should be used as these incorporate HEPA filters and prevent contamination from the external environment. Ducting to the chemical exhaust will ensure that the fumes generated from the volatile solvents are safely dealt with.

Sample Aliquoting and Pre-extraction Procedures

While the biological samples ideally should be in a format that is compatible with an automated liquid handler that utilizes a 96-barrel pipetting channel for more efficient sample preparation, samples in Eppendorf tubes (or cryotubes) can still be transferred to a plate either manually or by the robotic system. Shifting from single tubes to plates allows for more samples to be extracted per "pass" by the pipetting arm, thus increasing the efficiency of the procedure.

When working in single tubes and dealing with large number of samples, the labelling of individual tubes is one of the most time-consuming activities. Rather than writing or printing labels, users should consider the use of pre-labelled QR-coded tubes or plates. These labelled containers allow for individual tracking by a barcode reader or a scanner. The coded labels can then be associated to specific sample names by the operator. This procedure also helps to avoid mistakes in writing or reading labels and prevents the deterioration of labels over time.

It is highly recommended to validate the compatibility of the tubes, plates, and plate sealing and plate material with the specific extraction methods and test for potential plastic leeching, which may lead to contaminating peaks and signal suppression, as it was shown by Benke et al. [17] for vial caps in LC-MS lipidomic analysis. Based on most of the manufacturers' data, polypropylene or polyethylene material tubes are resistant to common solvents used in lipidomic workflows, such as chloroform, methanol, butanol, and formic acid at 20 °C. The containers must be compatible with long-term storage at low temperatures. If possible, glass always would be considered as the best option.

After considering all the aforementioned factors, a general pre-extraction procedure outlined below with the finalized system setup shown in Fig. 9.5 can be used.

Step-by-Step Protocol for Pre-extraction System Setup

1. Place filled tip boxes and extraction plates into the automated extraction system at defined positions.
2. Thaw sample plates at 4 °C, centrifuge, and place them into the appropriate positions.
3. Pre-cool the centrifuge to 4 °C for the later sample centrifugation step during the extraction process.
4. Fill reservoirs with the synthetic standard containing extraction solvents. This can be kept at room temperature (covered with silicon mat and plastic lid when not in use) and refrigerated to avoid evaporation. Place them into the automated extraction system.

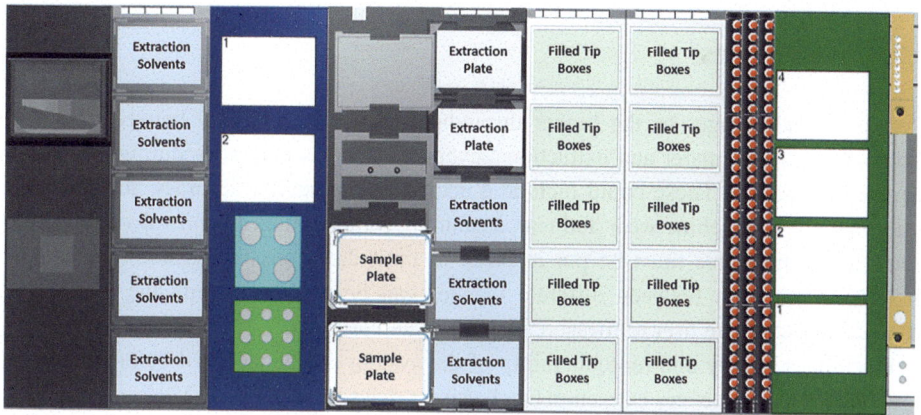

Fig. 9.5 Example of layout and positioning of consumables inside an automated liquid handling system. (Extraction solvents, reservoirs or plates containing solvents used in the extraction process; sample plate, plates that contain study samples, usually in a 96-well format; extraction plate, additional plates used specifically in the extraction process (e.g., 96-well plates for dilution or elution of samples); filled tip boxes, carrier filled with pipette tips compatible with the system, which are used in the extraction process)

9.2.3 Lipid Extraction from Samples

Following the experimental design procedures, lipid extraction from the samples may proceed using the following general steps.

Step-by-Step Protocol for Single-Phase Extraction of Lipids

1. Using the automated system, transfer a sufficient volume of samples from the master plate to a deep well extraction plate. After transferring the samples, seal the master plate and store it at −80 °C. Record the total number of freeze-thaw cycles experienced by the samples.
2. When it's time for lipid extraction, thaw the samples on ice, transfer a defined volume of solvent (butanol/methanol 1:1, v:v) for one-phase extraction, including lipid standards, into the deep well extraction plate that contains the samples.
3. Seal the deep well extraction plate with aluminum or plastic seals.
4. Place the sealed extraction plate on a shaker and shake for 20 min at 1500 rpm.
5. Centrifuge the extraction plate at 4 °C for 30 min at 3000 g.
6. Peel off the seal and transfer a defined volume of organic phase into the MS injection plate. Seal the plate and store at −80 °C till LC-MS analysis.

9.2.4 Challenges

Despite the advantages of the higher throughput and reproducibility between different rounds of sample transfer and extraction, several challenges may arise during automated sample preparation. These could include a decrease of pipetting accuracy and precision with time due to machine wear and tear, sustaining sample integrity and solvent evaporation. Hence, measures must be put in place to limit these possible issues. The sample processing precision between different batches can be monitored with the use of QC samples, allowing the user to measure plate-to-plate variation.

Pipetting accuracy and reproducibility of the system can be assessed before starting a new study or prior to every extraction batch using a dye solution and a UV-Vis plate reader. Users could validate specific pipetting volume accuracies by transferring a dye solution (e.g., tartrazine in 0.05% w/v in DMSO), via the liquid handler system into an empty plate followed by addition of water at three different volumes at the lower, middle, and upper end of the liquid handler's working volume range (e.g., 10 µL, 100 µL, and 200 µL for an operating range of 10–200 µL) in multiple replicates for each volume. The plates should be shaken for at least 1 min at 1000 rpm and subsequently centrifuged at low speeds for at least 2 min and then transferred to a UV-Vis plate reader and the absorbance measured at the optimal maximum (e.g., 259 and 425 nm fir tartrazine dye). The CV (%) of pipetting accuracy and reproducibility across all 96 wells should be <5%. The dye solution tests can also be applied to lower and higher volumes to validate the machine's lower and upper limit of precision and accuracy.

Pipetting accuracy can significantly worsen when using highly volatile solvents in liquid handling systems, as this can introduce dripping during liquid transfer. Having an additional air gap before and after sample aspiration can improve reproducibility and avoid contamination of other samples. Some commercial systems also include specific anti-droplet control features.

If a two-phase extraction is required, one should consider a liquid handling system with liquid density detection. This ensures that the correct layer has been drawn. If the system is not able to perform density detection, a fixed height (usually in millimeters) at which the pipette tip should draw the extract can be easily set. Opting for a solvent mixture where the analytes are recovered in the upper phase of the two-phase extraction, such as in MTBE extraction, can greatly simplify the operations.

As degradation of certain lipid species has been observed at prolonged exposure to room temperature due to enzymatic and chemical processes, the greatest concern during the extraction procedure is maintaining sample integrity [18]. Therefore, users should consider the installation of cooling modules that can keep samples and solvents at low temperatures (4 °C) during the whole process.

The evaporation of the extraction solvents is another major concern when using organic solvents. Should the extraction solvents contain internal standards, the evaporation can influence post-analysis normalization as concentrations of these standards might change over time. Possible solutions would include spiking the diluted internal standards into the samples separately, use of less volatile extraction solvents—such as 1:1 (v/v) butanol/methanol—sealing and chilling the solvent reservoir or using self-contained solvent dispenser bottles.

9.3   Analysis of Samples

To obtain high-quality lipidomic data, there are key considerations for storing and handling of the prepared samples before LC-MS analysis.

9.3.1   Sample Storage

Due to the high number of samples that can be quickly generated by automated systems, the resulting samples are rarely analyzed all at the same time, and thus, they need to be stored for a certain amount of time. It is highly recommended to use thicker aluminum foil seals for the long-term storage of sample plates, while thinner aluminum or plastic seals should be used when samples are placed into the MS autosampler system for analysis. This is to ensure that the seal thickness will not interfere with the puncturing by the injection needle. If multiple injections are used or the analyzed sample needs to be recovered for further analysis, self-sealing plate covers should be used to avoid evaporation (i.e., Rapid Slit Seal). Seals can be adhesive based or sealed using heat. Application of the aluminum foil seals should ideally be done according to the manufacturer's specifications. If samples are stored in QR-coded tubes with plastic caps, users must refer to the vendor's specifications about their suitability for storage at −80 °C or in liquid nitrogen.

Although plates reduce the overall storage space "footprint" compared to the conventional cryoboxes used for sample tubes, they require the purchase of freezer racks that are specific for the plates used (e.g., PCR plates or deep well plates) to fully take advantage of their use.

It is also highly recommended to test the efficacy of the sealing in terms of possible sample evaporation during long-term storage in the freezer. Users could evaluate this by aliquoting a standard mixture into multiple plates that is stable at the desired storage conditions and covering them with the desired plate seals and sealing procedure. All the prepared plates should be stored at the tested condition while one plate can be periodically analyzed and compared with the original signal intensity of standard mix [19].

Questions
1. You obtained 60 μL of lipid extract per sample after using automated extraction. You will be running LC-MS analysis at an injection volume of 20 μL for these samples immediately after extraction and plan to run the same samples on a separate LC-MS method a few weeks later. What approach would you adopt in terms of plate sealing?

Answers
1. Since volume of the obtained extract is low, the injection volume is high, and another future LC-MS analysis is planned, sample recovery is desired. Use self-sealing adhesive seals during LC-MS analysis to prevent evaporation and loss of sample. Following LC-MS analysis, the adhesive seal should be exchanged with a thick aluminum seal and stored at −80 °C.

9.3.2 LC-MS Analysis

As automation can decrease the analytical variability arising from aliquoting and lipid extraction, this approach should also increase the final quality of the lipidomic data, resulting in a higher number of lipids passing the conventional QC filtering criteria. To obtain better results, the samples that need to be analyzed by mass spectrometry should be prepared according to standard procedures for thawing, resuspension, and centrifugation, like the one reported below:

Step-by-Step Protocol for Sample Thawing, Resuspension, and Centrifugation:
1. Unless the samples are analyzed directly after extraction, thaw the frozen (−80 °C) samples at room temperature on a shaker for 30 min.
2. Centrifuge the samples to ensure protein and particulate precipitation, thus preventing clogging of the analytical system. If lipid extracts have been aliquoted into plates, centrifuges with swing bucket rotors and adaptors that can support the use of 96-well or 384-well plates should be used. Due to the significantly lower relative centrifugal force value that can be used with those rotors, the centrifugation of plates requires longer centrifugation time.
3. Transfer an aliquot of the supernatant to a new plate for MS analysis.
4. Before starting the MS analysis, ensure that a sufficient volume of sample is present in all the wells; thus the injection needle will always draw the right amount of sample from all the positions. If necessary, adjust (lower) the injection needle height or use a bottom-sensing setup in the LC method.

Questions

1. You intend to conduct a lipidomic analysis of lipid extracts that were stored at − 80 °C for 7 days in a 96-well plate. Prior to data acquisition, you thawed the samples at room temperature for 10 min, centrifuged the plates for 2 min, and transferred the samples to a new 96-well plate. In the middle of the LC-MS analysis, you find that the pressure of the binary pump increased significantly, and this persisted even after removing the column. Suggest one possible cause for this observation.

Answers

1. Since the high pressure persisted even after removing the column, it is likely that the LC analytical system, valves, or capillary was clogged. This could have been caused by insufficient precipitation of proteins, as centrifugation of samples was only performed for 2 min. During LC-MS analysis, these proteins could have accumulated in the system as more samples were injected, resulting in a higher pressure.

- To ensure a high quality of the data, various quality control steps have to be implemented.
- Automation saves considerable time and effort with improved or comparable precision when compared to a manual sample preparation approach.
- When working with plates or tubes, the layout of the samples should be carefully designed before starting the automated preparation process. Every batch should contain study samples in a stratified, randomized order, and all the different types of QCs and blanks.
- Considering the numbers involved in big lipidomic studies, all consumables should be prepared well in advance to avoid running out of the required items during sample preparation. This can affect the outcome of an entire study.
- Procedures should be implemented to circumvent challenges in automated lipid extraction, which include system wear and tear, evaporation of solvents, and maintenance of sample integrity.
- The containers (plates or tubes) of the lipid extracts should be appropriately sealed and stored after preparation and LC-MS analysis to maintain sample integrity and reduce analytical variability.

References

1. Fingerhut R, Silva Polanco ML, Silva Arevalo Gde J, Swiderska MA. First experience with a fully automated extraction system for simultaneous on-line direct tandem mass spectrometric analysis of amino acids and (acyl-)carnitines in a newborn screening setting. Rapid Commun Mass Spectrom. 2014;28:965–73.
2. Surma MA, Herzog R, Vasilj A, Klose C, Christinat N, Morin-Rivron D, Simons K, Masoodi M, Sampaio JL. An automated shotgun lipidomics platform for high throughput, comprehensive, and quantitative analysis of blood plasma intact lipids. Eur J Lipid Sci Technol. 2015;117:1540–9.
3. Wang LYSK, Rodriguez-Canas C, Mather I, Patel P, Eiden M, Young S, Forouhi NG, Koulman A. Development and validation of a robust automated analysis of plasma phospholipid fatty acids for metabolic phenotyping of large epidemiological studies. Genome Med. 2013;5:39.
4. Jung HR, Sylvanne T, Koistinen KM, Tarasov K, Kauhanen D, Ekroos K. High throughput quantitative molecular lipidomics. Biochim Biophys Acta. 2011;1811:925–34.
5. Alshehry ZH, Barlow CK, Weir JM, Zhou Y, Mcconville MJ, Meikle PJ. An efficient single phase method for the extraction of plasma lipids. Meta. 2015;5:389–403.
6. Bligh EG, Dyer WJ. A rapid method of total lipid extraction and purification. Can J Biochem Physiol. 1959;37:711–7.
7. Matyash V, Liebisch G, Kurzchalia TV, Shevchenko A, Schwudke D. Lipid extraction by methyl-Tert-butyl ether for high-throughput lipidomics. J Lipid Res. 2008;49:1137–46.
8. Evans AM, O'donovan C, Playdon M, Beecher C, Beger RD, Bowden JA, Broadhurst D, Clish CB, Dasari S, Dunn WB, Griffin JL, Hartung T, Hsu PC, Huan T, Jans J, Jones CM, Kachman M, Kleensang A, Lewis MR, Monge ME, Mosley JD, Taylor E, Tayyari F, Theodoridis G, Torta F, Ubhi BK, Vuckovic D, Metabolomics Quality Assurance, Q. C. C. Dissemination and analysis of the quality assurance (Qa) and quality control (qc) practices of Lc-Ms based untargeted metabolomics practitioners. Metabolomics. 2020;16:113.
9. Koivusalo M, Haimi P, Heikinheimo L, Kostiainen R, Somerharju P. Quantitative determination of phospholipid compositions by Esi-Ms: effects of acyl chain length, unsaturation, and lipid concentration on instrument response. J Lipid Res. 2001;42(4):663–72.
10. Broadhurst D, Goodacre R, Reinke SN, Kuligowski J, Wilson ID, Lewis MR, Dunn WB. Guidelines and considerations for the use of system suitability and quality control samples in mass spectrometry assays applied in untargeted clinical metabolomic studies. Metabolomics. 2018;14:72.
11. Burla B, Arita M, Arita M, Bendt AK, Cazenave-Gassiot A, Dennis EA, Ekroos K, Han X, Ikeda K, Liebisch G, Lin MK, Loh TP, Meikle PJ, Oresic M, Quehenberger O, Shevchenko A, Torta F, Wakelam MJO, Wheelock CE, Wenk MR. Ms-based lipidomics of human blood plasma: a community-initiated position paper to develop accepted guidelines. J Lipid Res. 2018;59:2001–17.
12. Olshansky G, Giles C, Salim A, Meikle PJ. Challenges and opportunities for prevention and removal of unwanted variation in lipidomic studies. Prog Lipid Res. 2022;87:101177.
13. Selva JJ, Torta F, Burla B, Wenk MR. Lancer: Linearity Assessment And Curve Explorer [Online]. 2022. https://Github.Com/Slinghub/Lancer. Accessed 27 Jan 2023.
14. Triebl A, Burla B, Selvalatchmanan J, Oh J, Tan SH, Chan MY, Mellet NA, Meikle PJ, Torta F, Wenk MR. Shared reference materials harmonize lipidomics across Ms-based detection platforms and laboratories. J Lipid Res. 2020;61:105–15.
15. Liebisch G, Fahy E, Aoki J, Dennis EA, Durand T, Ejsing CS, Fedorova M, Feussner I, Griffiths WJ, Köfeler H, Merrill AH Jr, Murphy RC, O'donnell VB, Oskolkova O, Subramaniam S, Wakelam MJO, Spener F. Update on lipid maps classification, nomenclature, and shorthand notation for Ms-derived lipid structures. J Lipid Res. 2020;61(12):1539–55.

16. Thompson JW, Adams KJ, Adamski J, Asad Y, Borts D, Bowden JA, Byram G, Dang V, Dunn WB, Fernandez F, Fiehn O, Gaul DA, Huhmer AF, Kalli A, Koal T, Koeniger S, Mandal R, Meier F, Naser FJ, O'neil D, Pal A, Patti GJ, Pham-Tuan H, Prehn C, Raynaud FI, Shen T, Southam AD, St John-Williams L, Sulek K, Vasilopoulou CG, Viant M, Winder CL, Wishart D, Zhang L, Zheng J, Moseley MA. International ring trial of a high resolution targeted metabolomics and lipidomics platform for serum and plasma analysis. Anal Chem. 2019;91: 14407–16.
17. Benke PI, Burla B, Ekroos K, Wenk MR, Torta F. Impact of ion suppression by sample cap liners in lipidomics. Anal Chim Acta. 2020;1137:136–42.
18. Ulmer CZ, Koelmel JP, Jones CM, Garrett TJ, Aristizabal-Henao JJ, Vesper HW, Bowden JA. A review of efforts to improve lipid stability during sample preparation and standardization efforts to ensure accuracy in the reporting of lipid measurements. Lipids. 2021;56:3–16.
19. Matson SL, Chatterjee M, Stock DA, Leet JE, Dumas EA, Ferrante CD, Monahan WE, Cook LS, Watson J, Cloutier NJ, Ferrante MA, Houston JG, Banks MN. Best practices in compound management for preserving compound integrity and accurately providing samples for assays. J Biomol Screen. 2009;14:476–84.

NMR-Based Metabolomics: Monitoring Metabolic Response to Physical Exercise

10

Aswin Verhoeven, Rico J. Derks, and Martin Giera

What You Will Learn from This Chapter
- Exercise physiology
- NMR-based metabolomics
- Multivariate statistical analysis

10.1 Theoretical Background

In this section, the principles of nuclear magnetic resonance (NMR) based metabolomics as well as the metabolic changes during and after exercise will be outlined. Using a straightforward, non-invasive experiment with an amateur road cyclist as a test subject, we will illustrate how NMR-based metabolomics, related data analysis, and biochemical pathways can be integrated and interpreted.

10.1.1 NMR-Based Metabolomics

NMR spectroscopy is based on detecting the precession of the nuclear magnetic moments of a sample (often a liquid solution) in a magnetic field [1]. The frequency of the precession

A. Verhoeven (✉) · R. J. Derks · M. Giera (✉)
Center for Proteomics and Metabolomics, Leiden University Medical Center, Leiden, The Netherlands
e-mail: A.Verhoeven@LUMC.NL; M.A.Giera@LUMC.NL

is determined by the strength of the magnetic field, the nuclear isotope (in metabolomics, this is typically ^{1}H), and the chemical environment of the monitored nuclei. Nuclei experiencing different chemical environments make up the NMR spectrum, with each nucleus generating a peak at a specific position in the spectrum. This position is known as the chemical shift. Each distinct chemical compound has its own unique pattern of peaks in the spectrum, the intensity of which is linearly related to the concentration of the compound, which is one of the main strengths of NMR. In case a mixture of various compounds is measured instead of a pure solution of a single compound, the NMR spectrum will be the sum of the spectra of the individual molecules. For NMR-based metabolomics, this mixture is frequently a biofluid such as plasma or urine that typically consists of hundreds of metabolites with various concentrations, with most of the compounds generating multiple peaks. Thus, biofluid NMR spectra are very complex and extracting useful information is challenging and requires both experimental optimization as well as specific data analysis workflows. Figure 10.1 shows an example of an NMR spectrum of a urine sample.

10.1.2 Acquisition of an NMR Spectrum

An NMR signal is recorded by detecting the voltage over an electrical coil induced by the processing nuclear magnetic dipoles from the sample. To obtain a spectrum from a raw NMR signal, a Fourier transform needs to be applied, which converts the raw "time domain" signal (commonly called free induction decay, FID) to a "frequency domain" spectrum that is easier for humans to interpret. In order to obtain high-quality NMR spectra that are suitable for metabolomics analysis the following points of the experimental procedure require optimization:

- The magnetic field homogeneity (shimming) needs to be properly optimized for each sample prior to data acquisition, as bad shimming results in broader or distorted peaks and less resolving power.
- As water is both the largest and the least interesting component of most biofluids, most NMR experiments include techniques that suppress the water signal to prevent it from completely dominating the NMR spectrum. Yet, some residual water signal always remains. This signal is typically positioned in the center of the ^{1}H spectrum, needs to be sufficiently small and properly in phase with the rest of the spectrum to allow nearby metabolite peaks to be analyzed. To achieve this, the spectrometer frequency needs to be optimized for each sample type.
- The spectrum needs to be properly phase-corrected, a procedure that removes asymmetric components (dispersion) from the peaks. Inaccurate phase correction leads to spectra with twisted peaks, affecting the peak areas and hence metabolite quantification.
- Spectra with high-intensity broad peaks sometimes have baseline distortions that need to be corrected.

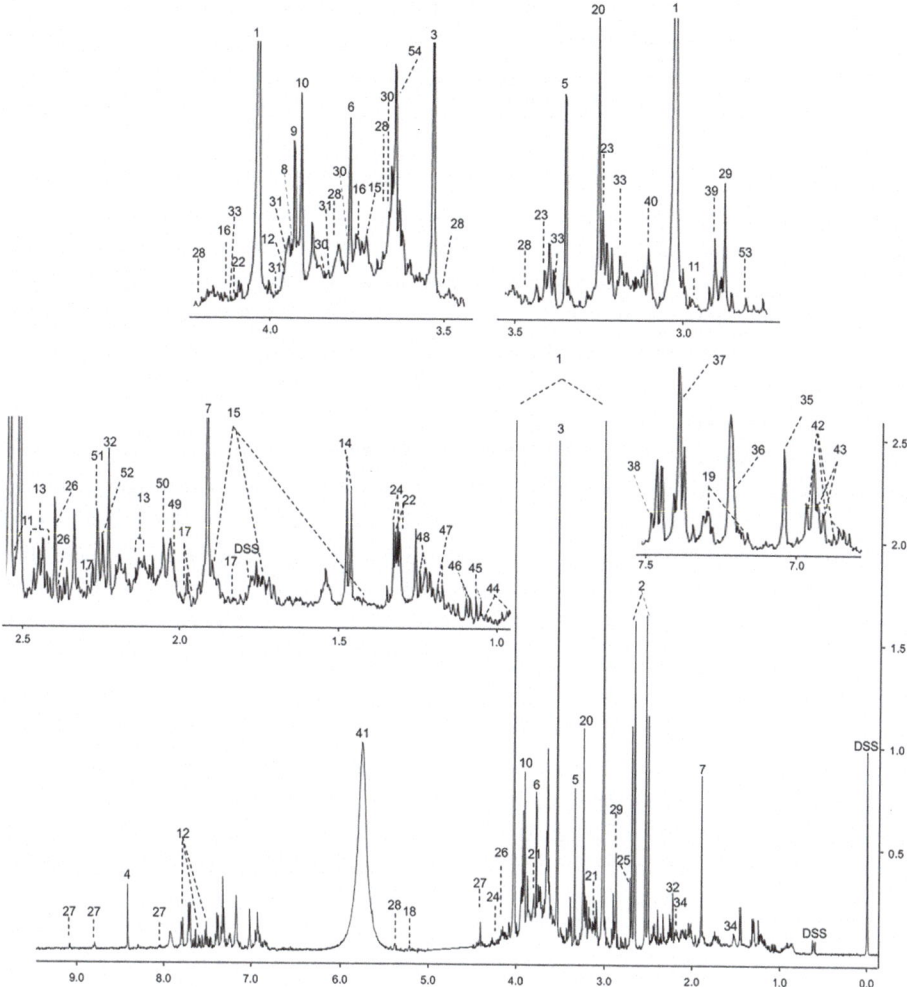

Fig. 10.1 Example of a ^{1}H NMR spectrum of a urine sample, taken with permission from [2]. The numbers indicate various identified metabolites, with some peaks remaining unidentified, illustrating the complexity of a biofluid NMR spectrum

10.1.3 Data Pre-processing

After acquisition, the NMR spectrum must be processed in order to generate a data table, with the samples as rows and the variables as columns. The variables of this data table either reflect "targeted" or "untargeted" metabolite data. To obtain untargeted variables for analysis, the spectrum is usually divided into small segments called bins (see Fig. 10.2). These segments can either all be of equal (equidistant/uniform bins) or of variable width (adaptive bins) [3], with the edges determined by the contents of the spectrum. The signal area within each segment is then integrated. These integrals are the untargeted variables that will be used for further analysis. In the case of targeted analysis, sets of peaks are assigned to specific metabolites using dedicated libraries, and these peaks are deconvoluted and integrated in order to obtain a table consisting of individual metabolite concentrations (areas). Targeted and untargeted analysis both have advantages and disadvantages (see also introductory chapter of this book). For targeted analysis, it is easy to directly draw conclusions about a subject's biology as the identity of metabolites responsible for a specific phenotype in response to a physiological condition or intervention is known. However, in a complex biofluid spectrum identification and deconvolution of all detectable peaks can be a formidable task that requires specialized software (e.g., Chenomx) and perseverance. For untargeted analysis, any significant variation anywhere in the spectrum can be revealed, but as bins can combine the signals of multiple metabolites, important variation of a low-concentration metabolite can be masked. Other times the metabolite responsible for the variation can be hard to identify, sometimes necessitating either two-dimensional NMR experiments (not discussed here) or orthogonal analytical techniques as, for example, fractionation, liquid chromatography, and mass spectrometry.

Another step often applied during data pre-processing is normalization. Normalization is a key procedure to remove dilution effects from the data. For example, in the case of cellular metabolomics analysis [4] metabolite concentrations will depend on the actual cell numbers in culture. In turn, normalization to cell numbers as well as total protein (determined after extraction) has become widely accepted. For biofluids such as plasma and serum which are homeostatic, generally no normalization is required and concentration differences between groups can directly be compared if accurately measured volumes have been used for analysis. In the case of urine, however, normalization is essential due to large intra- and inter-individual differences in sample dilution. Consider the following key questions: (i) how much urine does an individual produce over a given amount of time (renal function)? (ii) Is fluid intake controlled throughout the study (dilution)? and (iii) How much fluid is lost by physical activity (concentration)? It is clear that these and other factors influence urine dilution and thus urinary metabolite concentrations. Hence, normalization should be performed to correct for the dilution differences between urine samples. Several normalization approaches are available, e.g., urine volume, creatinine concentration, specific gravity, or total signal [5]. Depending on the sample and/or data these methods can have drawbacks. Alternative methods are probabilistic quotient normalization (PQN) and support vector regression. However, PQN is a good starting point and a widely accepted and applied approach [6]. In PQN, the dilution problem is addressed by calculating the most probable dilution factor by investigating the distribution of the quotients of the amplitudes of the spectrum to be normalized with those of a reference spectrum. Normalization can be applied to the original data points, the bins, or the peak areas obtained after deconvolution. Each option has its advantages and disadvantages. Using the original points is sensitive to peak shifts and differences in shimming quality. Bins are less sensitive to these effects, but in this case normalization may be affected by the choice of bins. Peak areas are the least sensitive to peak shift and shimming problems but the post hoc addition

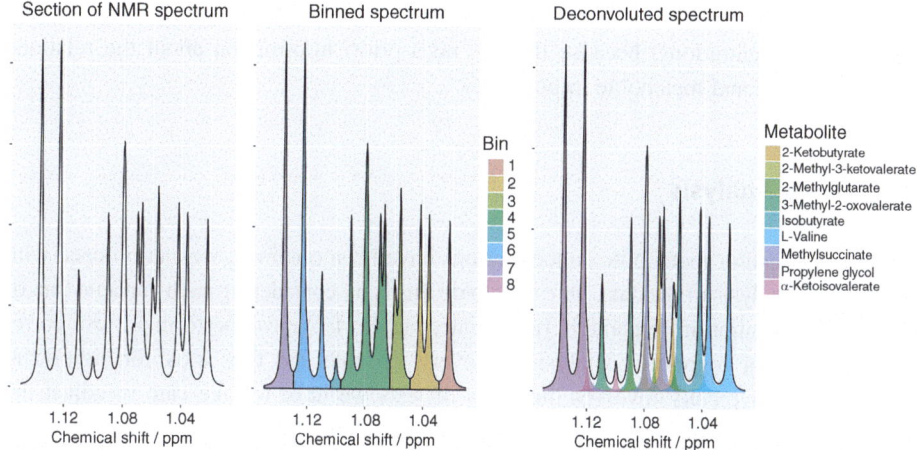

Fig. 10.2 A visualization of extracting untargeted and targeted data from an NMR spectrum. To obtain untargeted data, the spectrum is divided into small regions (bins) and the signal area within those bins is determined and used for further analysis. In this example, adaptive bins [3] instead of uniform bins are used. Note how some bins contain the signal from multiple metabolites. Obtaining targeted data involves the deconvolution of sometimes overlapping peaks and multiplets yielding relative metabolite concentrations and a more directly connected with biology/physiology

of additional metabolites will consequently change the normalization. Next to normalization, statistical analysis and in particular multivariate statistics (see below) require data pre-treatment that includes scaling, centering, and transformation steps [7]. The goal of the pre-treatment steps is to focus on the relevant information in the data, decrease the influence of noise and limit skewing by large concentration differences. In other words, data transformation is done to obtain a more symmetric distribution and to correct for heteroscedasticity. Log or power transformations are the most commonly used transformation methods. Centering and scaling are always combined. With centering, the mean values of each variable (e.g., bin or metabolite) are subtracted from all individual values of that variable so that the focus is on the variation only. Scaling refers to the handling of variance differences between the variables. Unit variance (UV) scaling and pareto scaling are the most commonly used scaling methods. Both methods use a measure of data dispersion as the scaling factor. When UV scaling is used all variables become equally important and as such, UV scaling will increase the influence of metabolites and bins with small variances. A drawback of UV scaling is that the measurement error is relatively large for signals and bins with low intensity, and these will influence the result just as much as a signal or bin with a high intensity and low measurement error. In pareto scaling this problem is reduced by using the square root of the standard deviation as scaling factor, instead of the standard deviation that is used in UV. Pareto scaling is an intermediate scaling between no scaling and UV scaling and is a good option when working with untargeted data (bins) as bins with large variances are more likely to hold important metabolite peaks instead of noise, while

UV scaling is often the best choice when performing multivariate analysis on targeted data (metabolite concentrations) because there is no a priori information about the relations between variance and metabolite importance.

10.1.4 Data Analysis

After the bins and/or metabolite concentrations have been obtained, we can proceed with the statistical analysis of the data. One way to do this is by considering each variable (bin or metabolite concentration) separately (univariate statistics). We can perform a t-test, correlation analysis, or regression analysis for every variable and thus look for significant signals. As this typically involves hundreds of tests we have to take care to adjust the significance requirement for the p-values accordingly by multiple testing corrections (e.g., Bonferroni correction). Such an approach is suitable in situations where we expect only a small part of metabolism to be affected by a certain condition, for example, when investigating the excretion of a drug metabolite. In case we subject the variables to a t-test, we can visualize the result by plotting the p-value against the fold change for each variable in a log-log scatter plot, resulting in what is commonly called a volcano plot (see Fig. 10.3). The most interesting variables will appear in the top right and left corners of the plot.

The other category of statistical analysis methods is multivariate statistical analysis, which involves simultaneously analyzing multiple outcome variables. It is often the case in metabolomics studies that the number of observations (i.e., the samples) is very low compared to the number of variables (i.e., the metabolites or their bins). In these situations, univariate analysis often cannot achieve significance for any of the variables after multiple testing corrections and multivariate analysis can be more effective in revealing significant features of the dataset. Multivariate analysis (for example, principal component analysis, PCA) attempts to construct a small set of new variables that summarize the large number of variables from the original dataset. This tends to work well in situations where the condition under investigation affects a large part of metabolism, for example, genetic defects in an organism's amino acid metabolism (e.g., maple syrup urine disease). These so-called latent variables (often called scores) are commonly ranked according to the proportion of the total variance in the original dataset they explain. Plotting the most important scores in a scatter plot gives a good sense of the behavior of the dataset as a whole. A dedicated chapter of this book explains multivariate statistics in great detail, please refer to this chapter for further reading. An example is given in Fig. 10.4.

10.1.5 Exercise Metabolism

According to Claessen et al., exercise is defined as "...a subset of physical activity that is planned, structured, and repetitive and has as a final or an intermediate objective the

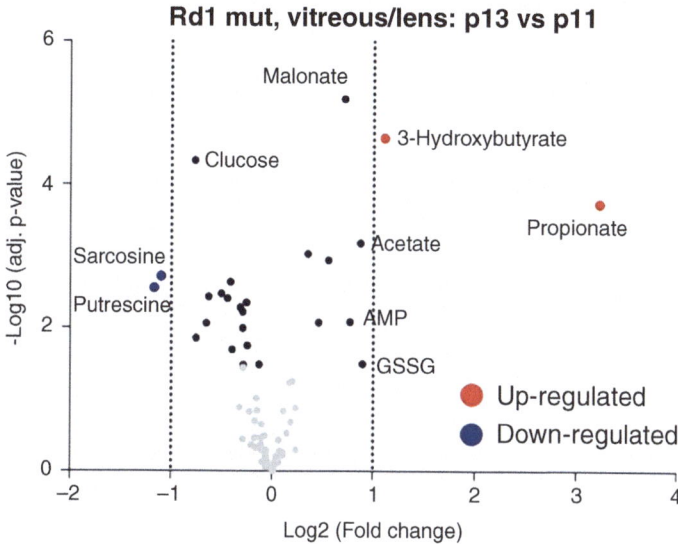

Fig. 10.3 Example of a volcano plot taken with permission from Murenu et al. [8] Metabolites that are not significant are indicated by the gray dots in the lower middle, while significant variables characterized by an adjusted *p*-value <0.05 (*y*-axis) and a fold change >2 (*x*-axis) appear in the upper left and right corners

improvement or maintenance of physical fitness. Physical fitness is a set of attributes that are either health- or skill-related..." [10]. Exercise involves the voluntary activation of skeletal muscles and affects numerous physiological/biochemical pathways as, for example, adenosine triphosphate (ATP) production, glycogenolysis, lipolysis and many more [11]. Figure 10.5 illustrates some of the main physiological and metabolic adaptations to voluntary, dynamic exercise.

Many factors such as duration, intensity, and type of activity influence the exact physiological response to exercise. In this chapter, we will compare biochemical adaptations to repetitive bouts of exercise under fasted and non-fasted conditions and explain how NMR-based metabolomics of urine samples can be used to understand underlying biochemical processes.

10.1.6 Energy Production

The most important molecule for the contraction of skeletal muscles is ATP. The break-down of ATP into its metabolite adenosine diphosphate (ADP) releases energy necessary for muscle contraction [12]. However, intra-muscular ATP storages are limited and hence ATP must be sustained by metabolic activity. The most important biochemical pathways for the sustained production of ATP are, the donation of a phosphate group

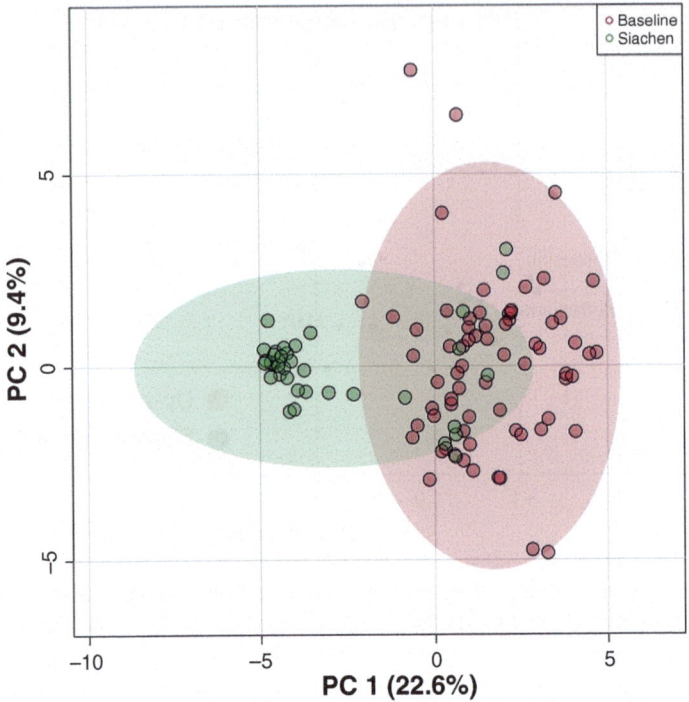

Fig. 10.4 An example of a PCA scores plot taken with permission from Gandhi et al. [9], showing two latent variables that summarize the variation in 36 urine metabolites in samples from people living at sea level (baseline) and participants living at high altitude (Siachen camp, 3700 m above sea level). The percentages on the axes represent the proportion of the total variance in the data table that the latent variable represents

from phosphocreatine to ADP, breakdown of muscle glycogen, and oxidative phosphorylation [12]. The main difference between these pathways lies in the involvement of oxygen (O_2). Phosphocreatine and glycogen breakdown are anaerobic pathways not involving O_2. Oxidative phosphorylation on the other hand is an aerobic process, involving O_2. Short-term exercise mainly involves anaerobic phosphocreatine breakdown and glycolysis (Embden-Meyerhof-Parnas pathway) with the latter producing pyruvate that can either be shuttled into the tricarboxylic acid (TCA) cycle for aerobic consumption or become anaerobically consumed to produce lactate. During prolonged exercise extra-muscular substrates, mainly glucose and free fatty acids are released by the liver into the bloodstream [11]. Both substrates fuel aerobic muscular ATP production mediated by the TCA cycle in conjunction with oxidative phosphorylation [12]. Specifically, both glucose and free fatty acids lead to the production of acetyl-coenzyme A (AcCoA), which in turn, enters the TCA cycle and enables oxidative phosphorylation. For the production of AcCoA, pyruvate resulting from glycolysis is converted to AcCoA by pyruvate dehydrogenase. Fatty acids undergo a process called fatty acid β-oxidation, likewise resulting in the production of AcCoA. During extended exercise or during fasting and carbohydrate-limited diets when

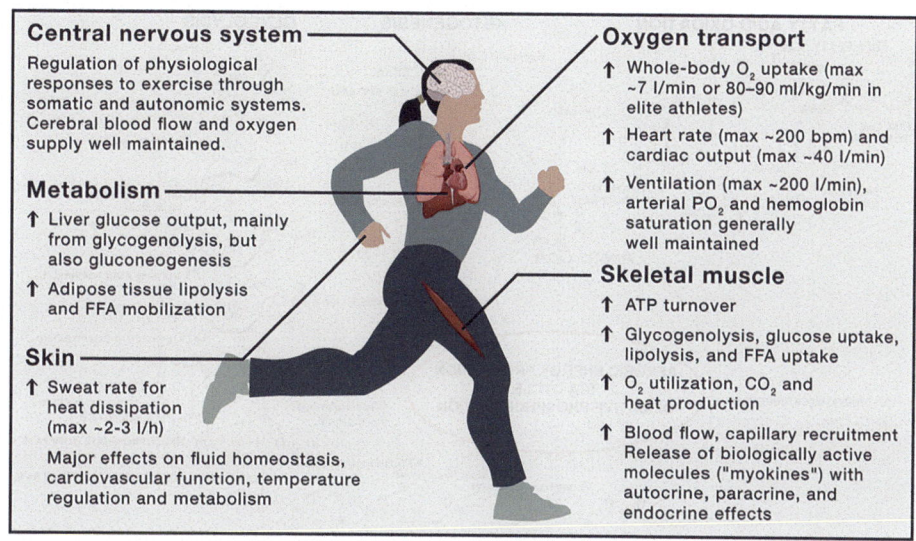

Fig. 10.5 The physiological responses to voluntary, dynamic exercise, taken with permission from [11]

glucose (glycogen) pools become exhausted, liver mitochondria extensively metabolize fatty acids by β-oxidation to produce AcCoA. In parallel, blood glucose levels have to remain within certain limits, triggering the endogenous biosynthesis of glucose (gluconeogenesis) causing pyruvate and oxaloacetate to be diverted away from the TCA cycle. This creates a metabolic challenge to recycle coenzyme A from AcCoA necessary for continuous β-oxidation. To resolve this, a metabolic process called ketogenesis is initiated. During ketogenesis, two units of AcCoA are being condensed to form acetoacyl-CoA, which is subsequently conjugated to another molecule of AcCoA to form 3-hydroxy-3-methylglutaryl-CoA (HMGCoA). HMGCoA is then cleaved to acetoacetate and AcCoA. Eventually, acetoacetate is reduced to 3-hydroxybutyrate (3-HB) or non-enzymatically decarboxylated to produce acetone. Acetoacetate as well as 3-HB are readily released into the bloodstream and absorbed by extra-hepatic organs. In target organs 3-HB can be oxidized back to acetoacetate, which after conversion to its CoA ester can be cleaved to release two units of AcCoA, readily available for the TCA cycle and oxidative phosphorylation within the target tissue [13]. In summary, the end products of ketogenesis are, acetone, acetoacetate, and 3-HB [14]. These three compounds are often collectively called ketone bodies, although the latter is chemically speaking not a ketone. Importantly, ketone bodies enter the circulation and are also secreted via the urine. Taken together, the presence of ketone bodies is a direct indicator of fatty acid β-oxidation which becomes apparent after glycogen pools during exercise have been depleted. The urinary increase of ketone bodies after exercise has been termed post-exercise ketosis resulting from latently increased free fatty acid levels in the bloodstream [15]. Moreover, ketogenesis is also the main process underlying diabetic ketoacidosis—a clinically relevant health condition mainly found in

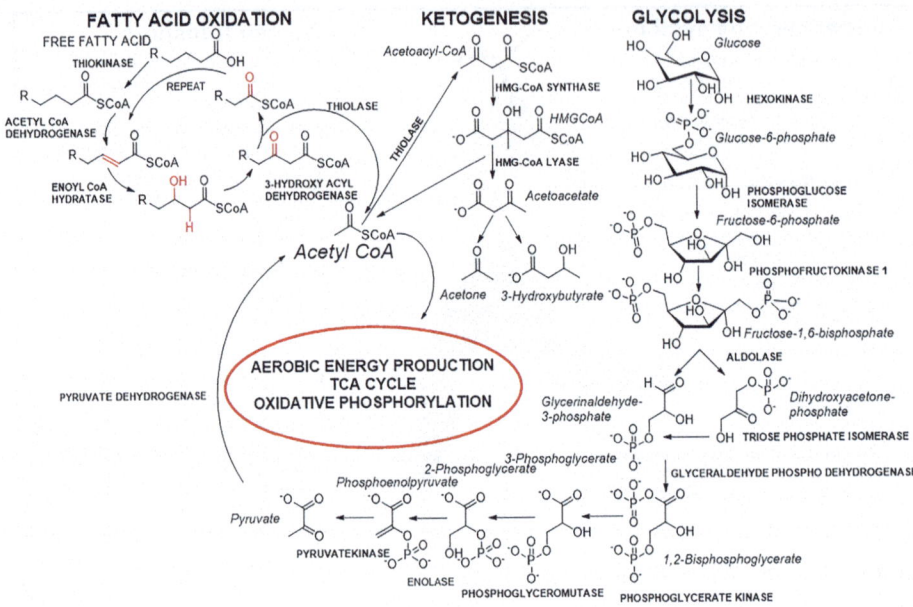

Fig. 10.6 Main pathways involved in aerobic energy production. Enzymes are labelled with capital letters and metabolites with italic letters

type 1 diabetic patients [16]. In diabetes, insulin deficiency together with counterregulatory hormones can lead to increased circulating free fatty acids, undergoing fatty acid β-oxidation in the liver, ultimately producing ketone bodies through ketogenesis. Clinically this fact can be used as a urine test for ketone bodies mainly in type 1 diabetes patients indicating possible ketoacidosis. Figure 10.6 summarizes the two major metabolic pathways fuelling aerobic ATP production as well as ketogenesis.

10.2   Research Question

Establish a metabolomics experiment for monitoring exercise physiology and in particular ketogenesis. Application of non-invasive NMR-based metabolomics analysis investigating urinary ketone body excretion as a marker for exercise-induced ketogenesis.

10.2.1 Hypothesis and Experimental Setup

Ketone bodies reflect physiological activity and fatty acid β-oxidation.
The main source of energy for short (seconds) and intermediate (half-hour) exercise is glucose (glycogen) metabolism. Fat consumption mediated by the liver is initiated after

free glucose and muscular glycogen pools have been exhausted. Gluconeogenesis in combination with β-oxidation of fatty acids triggers ketogenesis, producing acetone, acetoacetate, and 3-HB. The latter metabolites are water-soluble and diagnostic for active β-oxidation and ketogenesis.

10.2.2 Experimental Setup

To investigate the production and urinary excretion of ketone bodies as markers of β-oxidation and ketogenesis we design and execute the following experiments:

Experiment 1—limited carbohydrate intake, physical activity, energy initially produced by glycolysis whereafter β-oxidation and ketogenesis take over.

Experiment 2—limited carbohydrate intake, desk work, limited physical activity, and carbohydrate pools suffice for energy production.

To investigate a switch from glycolytic energy production to β-oxidation and ketogenesis we will compare two metabolic situations, limited carbohydrate intake combined with bouts of cycling (day 0) versus desk work and limited physical activity (7 days after). For both experiments, we will use fresh morning urine as timepoint 0 (t_0). When sampling urine, use sterile containers and immediately refrigerate or freeze the samples in order to prevent bacterial growth. Subsequently, we will take the t_1 sample roughly one hour after t_0 and the intake of water and electrolytes (standardize this if possible). For our cycling experiment, we will now cycle 15 km (30–40 min), consume plenty of water and electrolytes and collect urine after the exercise bout. We will repeat this for a total of 6 times ($t_2, t_3, \ldots, t_7$) and collect urine samples after each round of cycling. As a control experiment, we will go through exactly the same routine however with desk work instead of cycling. The experimental setup is outlined in Fig. 10.7. We will also create one pooled sample from a mixture of all single-point urine samples. Thus we will analyze a total of 17 samples.

Next, we will analyze the collected urine samples using a 600 MHz NMR (or equivalent) spectrometer according to the following protocol:

We prepare a potassium phosphate buffer by adding a 1.5 M KH_2PO_4 solution in D_2O to a 1.5 M K_2HPO_4 solution in D_2O until we reach pH 7.4. As a quantification standard and chemical shift reference, we add 4 mM deuterated trimethylsilylpropionate (TSP-d4). We also add 2 mM of sodium azide (NaN_3) in order to prevent microbial activity. We thaw the urine samples at room temperature on the lab bench. Immediately after thawing we transfer 1.0 mL of each sample into 1.5 mL Eppendorf tubes and centrifuge at 1.550 × g. We create a pooled sample by combining 100 μL of each sample, this mixture then undergoes the same treatment as the other samples. Next, we mix 630 μL of urine with 70 μL of buffer by inverting the container 10 times and transfer 565 μL of the urine–buffer mixture to 5 mm Bruker SampleJet NMR tubes and seal the tubes by inserting POM balls into the holes. Finally, we place the tube rack containing all NMR tubes in the SampleJet autosampler where the samples are cooled to 6 °C while queued for acquisition.

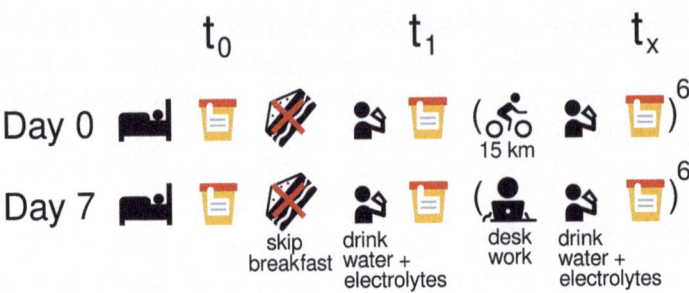

Fig. 10.7 Experimental outline. Urine sample collection is carried out on a single male middle-aged participant in two different sessions, 7 days apart. The first session (day 0) involves multiple cycling sessions, the second session (day 7) only desk work

We will perform the measurements in a 600 MHz NMR spectrometer equipped with a 5 mm TCI cryogenic probe head and a z-gradient system. In order to prepare the spectrometer, we need to optimize the magnetic field homogeneity by shimming and calibrating the sample temperature to 300 K (27 °C). We apply the 3D shimming algorithm of the Bruker Topshim tool to the pooled sample and try to achieve a linewidth of less than 1 Hz at the half maximum height (FWHM) of the TSP peak. We save the shim values for later use. Then, we insert a 5-mm tube filled with 565 µL 99.8% methanol-d5 and perform a simple one-pulse one-scan ^{1}H NMR measurement. After confirming that the shim values are optimized properly by confirming that the J-coupling of the CHD_2 multiplet at 3.35 ppm is clearly resolved, the sample temperature can be measured by applying the Bruker CALCTEMP tool in the apodized spectrum. If the measured temperature is further than 0.1 K off the set temperature, the temperature correction in the Topspin needs to be adapted. The spectrometer is now properly set up for recording the spectra of the urine samples. One-dimensional ^{1}H-NMR spectra of the samples are recorded using the first increment of a NOESY pulse sequence with presaturation (γB1 = 50 Hz) during a relaxation delay of 4 s and a mixing time of 10 ms for efficient water suppression. Sixteen scans of 65,536 points covering 12,335 Hz are accumulated. Exponential line broadening of 1.0 Hz is applied to the raw time-domain data prior to Fourier transformation. The spectra are subsequently phase corrected and referenced to the peak of TSP (0.0 ppm).

10.2.3 Necessary Software and Exemplary Data Set

Several options are available to perform the data processing and analysis, both in the form of software with user-friendly GUIs (AMIX, Chenomx, SIMCA, SPSS), workflow systems (Workflow4Metabolomics [17] for Galaxy [18], KIMBLE [19] for KNIME [20]), or web services (MetaboAnalyst [21]). Bruker even offers targeted metabolomics as a service for specific biofluids (Bruker IVDr, In Vitro Diagnostics for research). For the purposes of this book, we will perform all processing and analysis in a scientific notebook. The programming languages R, Python, and MATLAB are all in use by the metabolomics community, but as R has superior statistical tools, we will use that as the kernel for our notebook. The notebook can be downloaded from either GitHub (https://github.com/CPM-Metabolomics-Lipidomics/NMRcycling) in the form of a Jupyter/RMarkdown notebook, which requires a suitable locally installed Jupyter/RStudio environment. Alternatively, the notebook can be run straight in the browser using Google Colaboratory, which requires only a Google account (link provided on GitHub). The GitHub account also contains the NMR spectra of the cycling dataset. This dataset is identical to the one used in one of our previous publications [19]. Urine used in this study was donated by a healthy adult volunteer and did not involve recruitment or enrollment of human subjects. A KIMBLE workflow version of the notebook is also available on GitHub.

10.2.4 Research Questions

With an exemplary data set, one can train him/herself in data treatment and multivariate statistics as outlined below. We will explain data processing for targeted as well as untargeted analysis of NMR-based metabolomics data. We aim to answer the following central research questions with the data at hand:

1. Can we measure physiological response to exercise by urinary metabolomics analysis? Specifically, can ketone bodies be observed under the described experimental conditions and ketogenesis monitored in urine specimens?
2. What other observations can be made under carbohydrate restriction combined with exercise? To do so, we will apply binning and multivariate analysis to broadly investigate our NMR metabolomics data set.
3. Does multivariate analysis identify ketone bodies as the main drivers?
4. As an example of targeted metabolomics analysis, we will plot the ketone bodies' longitudinal changes for both experiments (conditions).

10.3 Data Analysis and Interpretation

Our notebook will broadly follow the workflow outlined in Fig. 10.8. Additional processing and analysis steps are possible and may yield more information. However, already without these extra steps we will be able to draw interesting conclusions from the dataset, as we will soon discover.

10.3.1 R Code and Explanations

We will now turn to the actual code that will be used to process and analyze the data. The notebook consists of a series of cells with snippets of R code. We will go through the code

Fig. 10.8 Data processing and analysis workflow

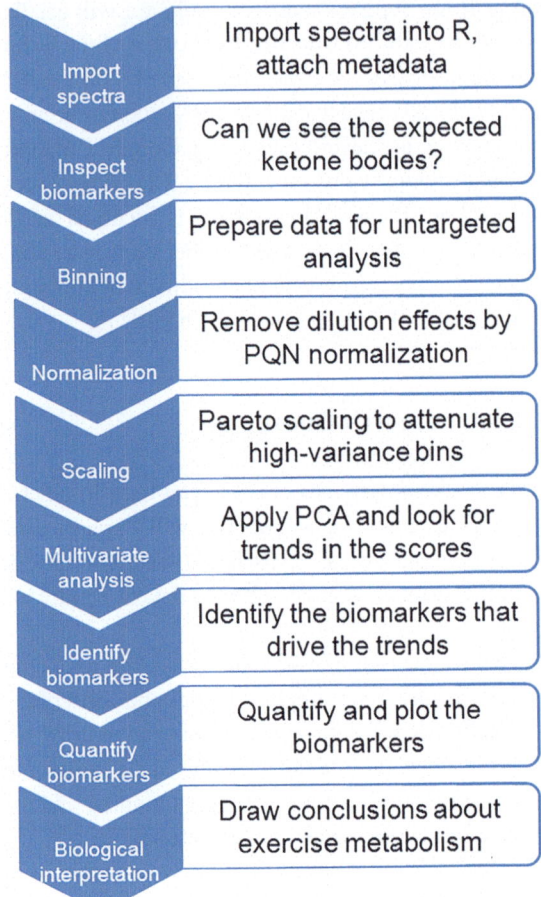

Import spectra → Import spectra into R, attach metadata

Inspect biomarkers → Can we see the expected ketone bodies?

Binning → Prepare data for untargeted analysis

Normalization → Remove dilution effects by PQN normalization

Scaling → Pareto scaling to attenuate high-variance bins

Multivariate analysis → Apply PCA and look for trends in the scores

Identify biomarkers → Identify the biomarkers that drive the trends

Quantify biomarkers → Quantify and plot the biomarkers

Biological interpretation → Draw conclusions about exercise metabolism

cell by cell, by explaining our objectives and then showing the code and its output. Code snippets and output are marked gray in the text. The first cells import various libraries that provide the R environment with the features necessary for our purposes. Most of these are very general libraries for multivariate analysis and plotting, while the AlpsNMR [22] library provides important metabolomics functions. The notebook also makes a connection to the associated GitHub account for importing the NMR data and metadata. We refer the reader to the notebook itself for these initial steps, and start our detailed description at the point where we read the NMR data into the memory.

First, we import all spectra into the R environment.

```
[27] datasets <- nmr_read_samples(sample_names = datasetnames)
```

A digital spectrum consists of a number of points with associated chemical shifts. Because we cannot assume that points in different spectra line up exactly, we perform an interpolation to make sure that they do. At the same time, we dispose of points outside of the range where we expect metabolite peaks, which is downfield from 10 ppm and upfield from −0.5 ppm with a resolution of 0.00023 ppm.

```
[28]  datasets <-nmr_interpolate_1D(datasets,
   axis = c(min = -0.5, max = 10, by = 2.3E-4))
```

Next, we import the previously downloaded metadata. AlpsNMR allows us to automatically join the NMR spectra with the associated metadata.

```
[29]  meta <- read_excel("metadata.xlsx")
  datasets$metadata$external$path <-
datasets$metadata$info$info_sample_path
   datasets <- nmr_meta_add(datasets, meta, by = c("path" = "Path"))
```

Due to the fact that urine is excreted at various levels of dilution, we need to normalize the spectral data to compensate for this before we can continue with analysis. The most well-known method in clinical chemistry is to normalize with respect to creatinine, however, we will apply total area normalization which takes into account all metabolites. Excluding the water peak region between 4.5 and 5.0 ppm prevents the water peak from influencing the normalization and allows us to have a thorough look at the urine NMR spectrum itself. We also exclude the reference peak (TSP) as it is not a metabolite and therefore should not be used for the normalization.

```
[30]  regions <- list(water = c(4.5, 5.0),
   upfield = c(-10, 0.5), downfield = c(10, 20))
  datasets <- nmr_exclude_region(datasets, exclude = regions)
  datasets <- nmr_normalize(datasets, method = c('area'))
  plot(datasets, NMRExperiment = c('Sample9999'))
```

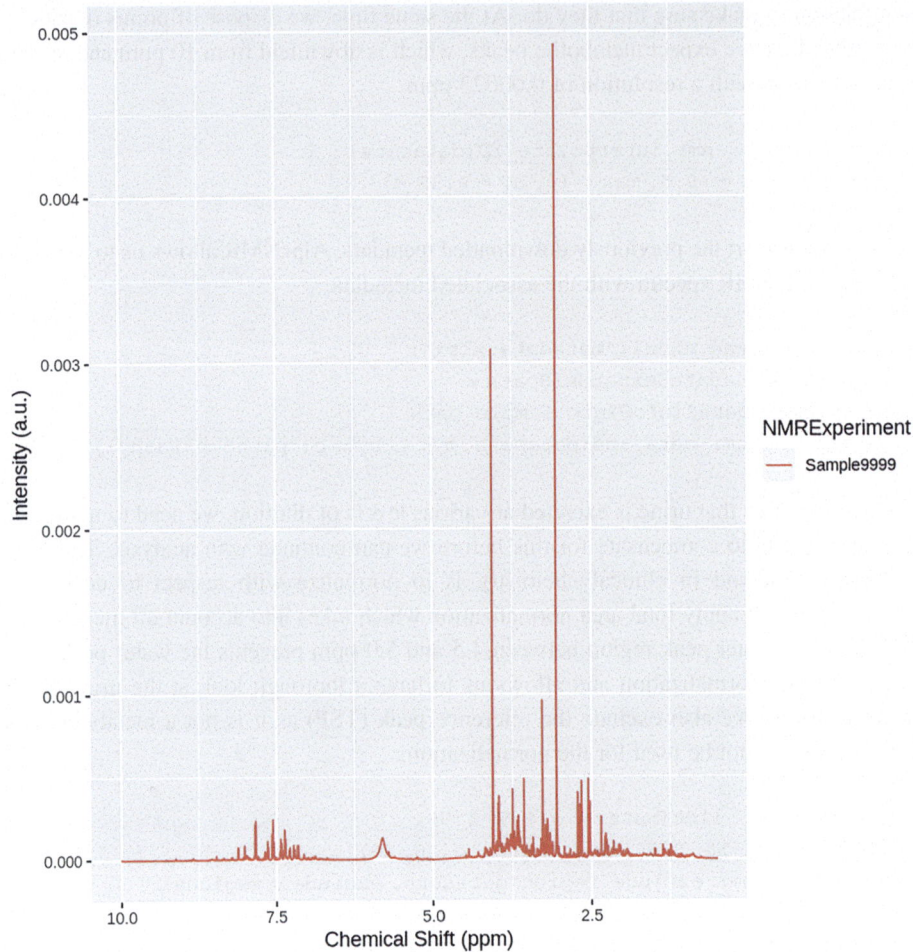

Next we will target the ketone bodies. The AlpsNMR package conveniently contains the chemical shift database from the Human Metabolome Database (HMDB). We correct the chemical shifts in this database using the alanine methyl group position in our spectra (1.486 ppm) and the one in the HMDB (1.47 ppm). On the website of the HMDB, we can find the indices (accession numbers) of the ketone bodies: 11, 60, and 1659 for 3-HB, acetoacetate, and acetone, respectively. Using these accession numbers, we can extract the chemical shifts of the ketone bodies.

```
[31]  fn <- system.file("data", "hmdb.rda", package = "AlpsNMR")
      load(file = fn)
      hmdb$pos_in_ppm <- hmdb$pos_in_ppm - 1.47 + 1.486
      hmdbket <-
        hmdb[grepl('HMDB0000011|HMDB0000060|HMDB0001659',hmdb$Accession),]
      hmdbket[,c('Metabolite', 'pos_in_ppm', 'Accession', 'Urine')]
```

Metabolite	pos_in_ppm	Accession	Urine
(R)-3-Hydroxybutyric acid	1.220	HMDB0000011	TRUE
(R)-3-Hydroxybutyric acid	2.342	HMDB0000011	TRUE
(R)-3-Hydroxybutyric acid	2.418	HMDB0000011	TRUE
(R)-3-Hydroxybutyric acid	4.176	HMDB0000011	TRUE
Acetoacetic acid	2.286	HMDB0000060	TRUE
Acetoacetic acid	3.446	HMDB0000060	TRUE
Acetone	2.236	HMDB0001659	TRUE

Using the chemical shifts extracted from the HMDB, we can plot the regions of the ketone body peaks. By overlaying the spectra from different timepoints, we can see the production of ketone bodies increasing over time.

```
[32]  dsexcday <- datasets[datasets$meta$external$Day == "Exercise day"]
     hmdbsel <- hmdbket[c(1,5,7),]
     rn1 <- hmdbsel[[1,"pos_in_ppm"]]+c(-1,1)*0.03
     rn2 <- hmdbsel[[2,"pos_in_ppm"]]+c(-1,1)*0.03
     rn3 <- hmdbsel[[3,"pos_in_ppm"]]+c(-1,1)*0.03
     pl1 <- plot(dsexcday, chemshift_range = rn1, color = "Timepoint") +
      ggtitle(hmdbsel[[1,"Metabolite"]])
     pl2 <- plot(dsexcday, chemshift_range = rn2, color = "Timepoint") +
      ggtitle(hmdbsel[[2,"Metabolite"]])
     pl3 <- plot(dsexcday, chemshift_range = rn3, color = "Timepoint") +
      ggtitle(hmdbsel[[3,"Metabolite"]])
     pl1 <- pl1 + guides(color = FALSE) +
      theme(axis.title.y=element_blank(),axis.text.y=element_blank(),
      axis.ticks.y=element_blank())
     pl2 <- pl2 + guides(color = FALSE) +
      theme(axis.title.y=element_blank(),axis.text.y=element_blank(),
      axis.ticks.y=element_blank())
     pl3 <- pl3 +
      theme(axis.title.y=element_blank(),axis.text.y=element_blank(),
      axis.ticks.y=element_blank())
     grid.arrange(pl1, pl2, pl3, nrow=1)
```

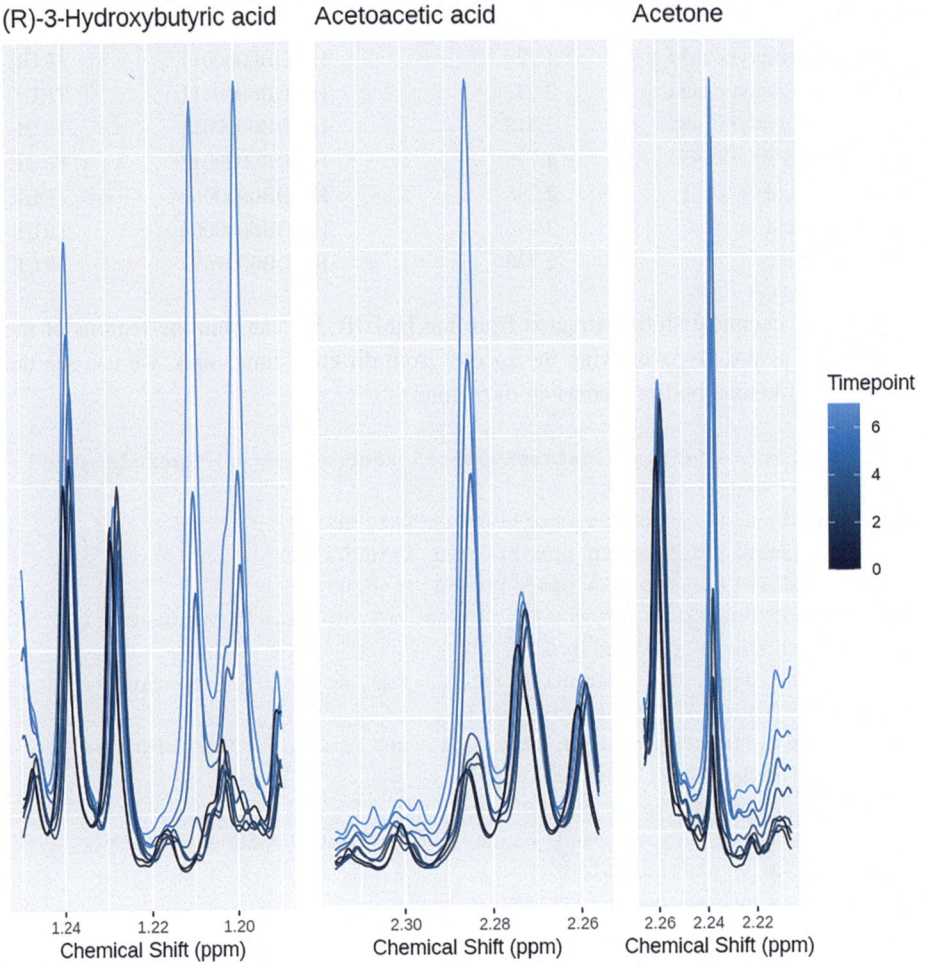

Our hypothesis is that the body activates ketosis during prolonged exercise while fasting. We want to know if this induces any other interesting changes to metabolism. For that purpose, it is convenient to limit the number of variables, thus we integrate the signals over small regions of the spectra in order to obtain 200 bins. Now that we have a table of bins we apply PQN to improve the normalization of the data.

```
[33]   ds <- datasets
       ds$data_1r <- binning(ds$data_1r, bin.size=200)
       ds$axis <- as.vector(binning(ds$axis, bin.size=200))
       df <- ds$data_1r
       ref <- colMeans(df)
       dfscaling <- df/ref[col(df)]
       dfscaling <- rowMedians(dfscaling)
```

```
df = df / dfscaling
ds$data_1r <- df
rownames(df) <- ds$metadata$external$RowID
colnames(df) <- round(ds$axis,2)
```

With the binned spectral data and associated metadata we can perform untargeted analysis. The simplest way to do this is by subjecting all bins to a univariate test, for example, a paired t-test that tests for differences between the exercise day and the rest day after the start of exercise (t_2 and later). A volcano plot (a plot of the base-2 logarithm of the fold change versus the negative of the base-10 logarithm of the p-value) is a useful way of visualizing the result. As we are performing a great number of tests, the usual requirement of $p < 0.05$ is not strict enough; we have to apply multiple testing correction to the significance level requirement. In this example, the Bonferroni correction is used. As it turns out, only very few bins show statistically significant differences between the rest and the exercise day. Those bins also have low fold changes, reducing the likelihood that these are physiologically interesting effects.

```
[35]    day <- ds$metadata$external$Day
    tim <- ds$metadata$external$Timepoint
    nopool <- (day != "Pool")
    noearly <- (tim > 1)
    cn <- colnames(df)
    dfv <- NULL
    for(t in cn) {
    col <- df[,t]
    dfstat <- data.frame(col=col, time=tim, day=day)
    dfstat <- dfstat[nopool & noearly,]
    dfstat <- pivot_wider(dfstat, names_from=day, values_from=col)
    res <- t.test(dfstat$`Exercise day`, dfstat$`Rest day`, paired=TRUE)
    logp <- -log10(res$p.value)
    logfc <- log2(mean(dfstat$`Exercise day`) / mean(dfstat$`Rest day`))
    dfr <- data.frame(cn=t, logp=logp, logfc=logfc)
    dfv <- rbind(dfv, dfr)
    }
    ggplot(data=dfv, aes(x=logfc, y=logp, label=cn)) + geom_point() +
    geom_text_repel() + theme_minimal() +
    geom_vline(xintercept=c(-1, 1), linetype = "dashed") +
    geom_hline(yintercept=-log10(0.05/nrow(dfv)), linetype = "dashed")
```

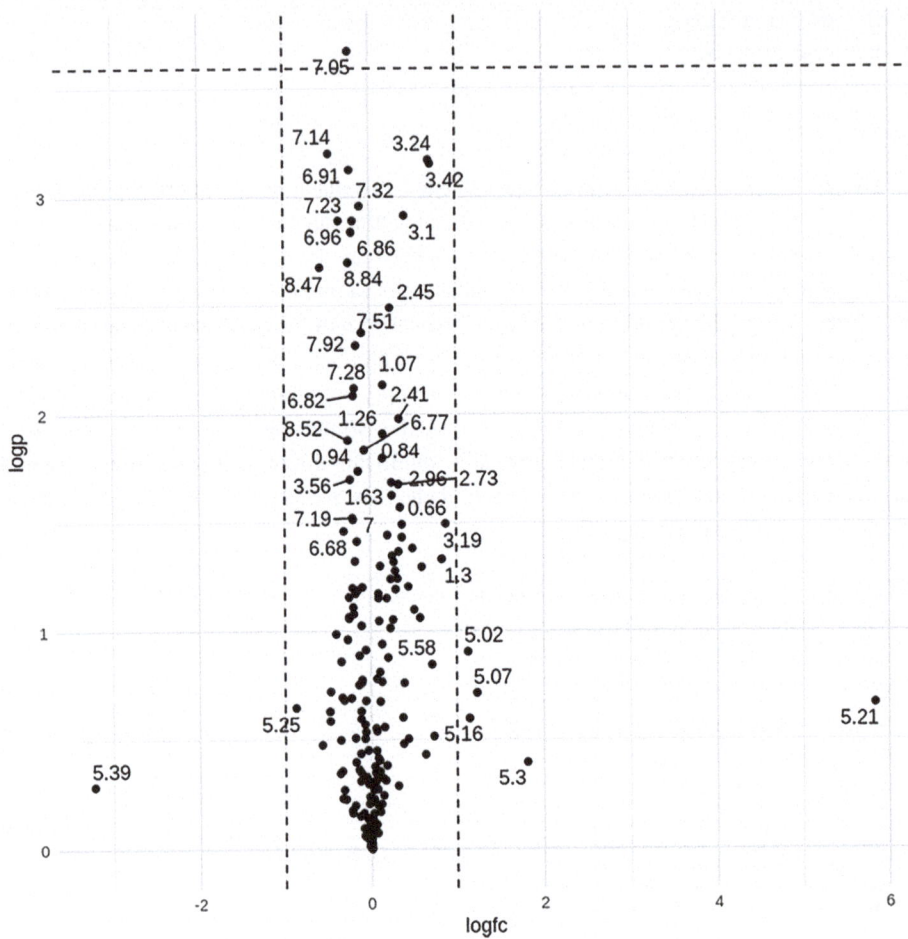

This outcome can be somewhat improved by using bin-wise ANOVA or linear regression (results not shown). However, as our experiment only uses a very limited set of observations, no matter what univariate test we perform, variables are unlikely to appear in the upper left and right rectangles of the volcano plot where we expect the most interesting and significant biomarkers. Much more appropriate for this type of data is multivariate analysis. Our multivariate method of choice is PCA. First, we reduce the importance of the metabolites with a higher concentration with respect to the lower ones by dividing each bin with the square root of its standard deviation (pareto scaling). Subsequently, we perform PCA (this also implicitly mean-centers the data) and plot the scores of the first and second dimensions. This shows a clear grouping of the later timepoints on the exercise day in the first dimension.

```
[38]  groups <- ds$metadata$external$Phase
      df <- ds$data_1r
```

```
rownames(df) <- ds$metadata$external$RowID
colnames(df) <- round(ds$axis,2)
df2 <- df[rownames(df) != 'Pool',]
groups2 <- groups[groups != 'Pool']
df2 <- pareto_scale(df2)
df2 <- df2 * 0.01
res <- (prcomp(df2))
fviz_pca_ind(res, axes = c(1,2), col.ind = groups2, addEllipses = TRUE,
ellipse.type = 'confidence', legend.title = 'Groups', repel = TRUE)
```

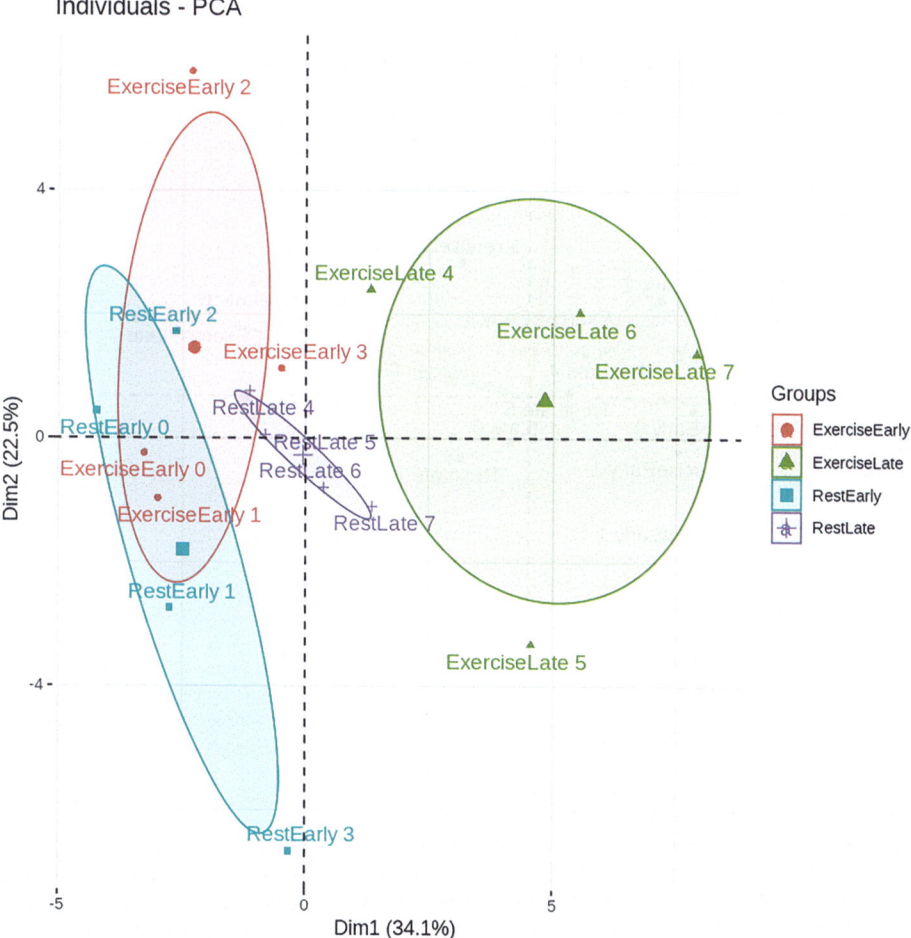

In order to reveal which bin(s) contribute most to the observed grouping we plot the component loadings. We can see that the bin that contributes most to the grouping is the bin centered around a chemical shift of 3.19 ppm. Interestingly, this does not correspond to one of the ketone body peaks.

```
[39]  fviz_pca_biplot(res, repel = TRUE, axes = c(1,2), col.var = 'contrib',
      gradient.cols = c('#00AFBB', '#E7B800', '#FC4E07'),
      select.var = list(contrib = 15))
```

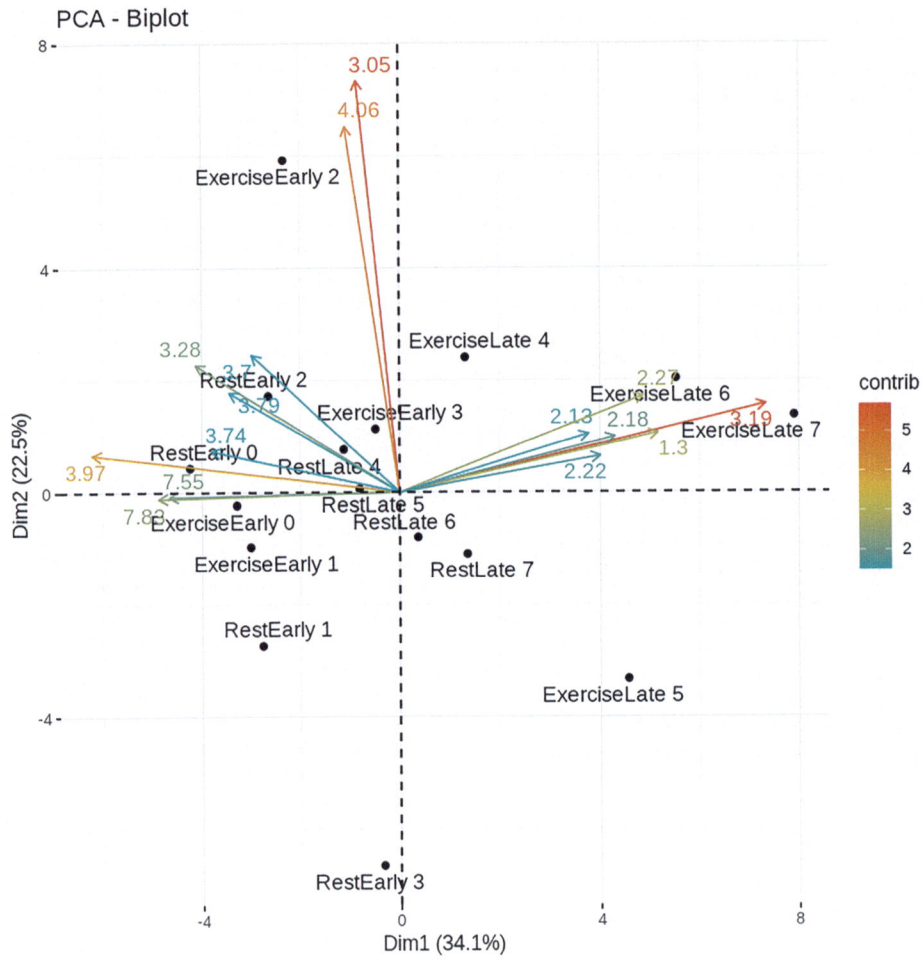

We can plot the bin and color it by timepoint to inspect its behavior on the exercise day.

```
[40]  rn <- ds$axis[which(colnames(df) == '3.19') + c(-1, 0, 1)]
      rn <- 0.5 * c(rn[1]+rn[2], rn[2]+rn[3])
      plot(datasets[datasets$meta$external$Day == "Exercise day"],
```

```
chemshift_range = rn, color = "Timepoint")
```

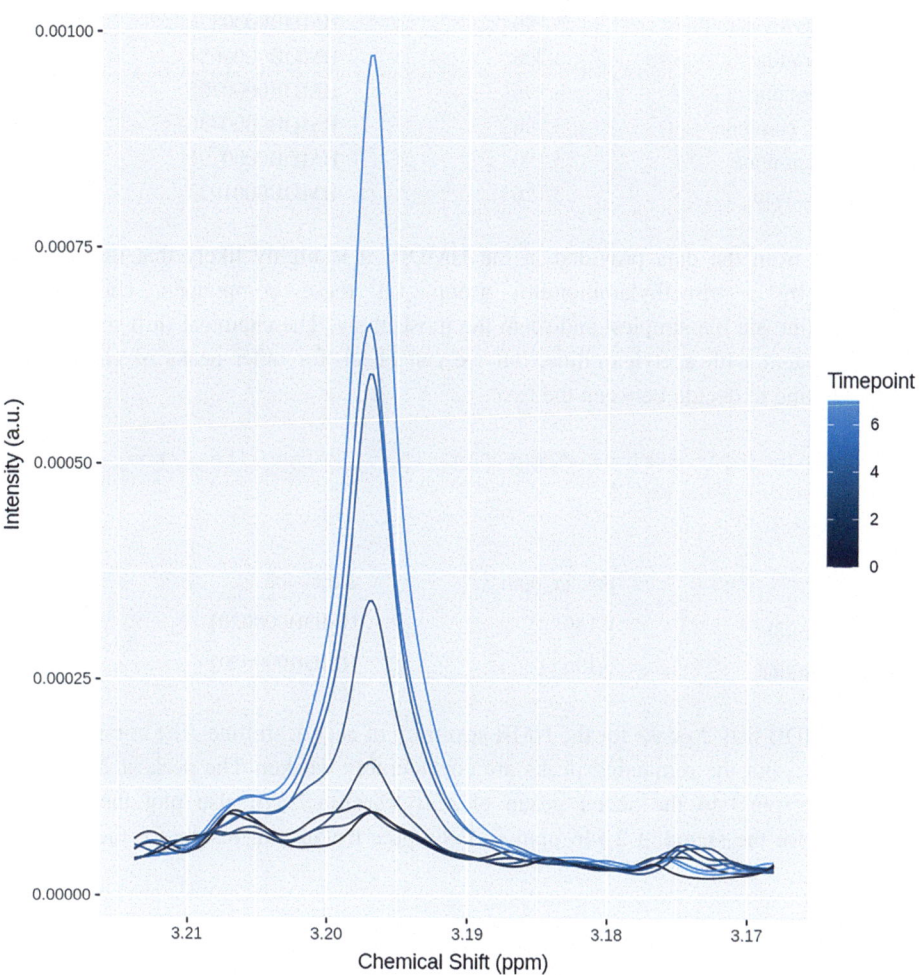

As can be seen above, there is a huge increase in the concentration of this metabolite on the later timepoints, but which metabolite is it? We can filter the chemical shifts of the HMDB to get a few suggestions.

```
[41]  hmdbsel <- hmdb[hmdb$pos_in_ppm>min(rn),]
      hmdbsel <- hmdbsel[hmdbsel$pos_in_ppm<max(rn),]
      hmdbsel <- hmdbsel[is.na(hmdbsel$J_constant),]
      hmdbsel <- hmdbsel[grepl('HMDB000....',hmdbsel$Accession),]
      hmdbsel[,c('Metabolite', 'pos_in_ppm', 'Accession', 'Urine')]
```

Metabolite	pos_in_ppm	Accession	Urine
Choline	3.205	HMDB0000097	TRUE
L-Acetylcarnitine	3.196	HMDB0000201	TRUE
2-Methylbutyroylcarnitine	3.196	HMDB0000378	TRUE
Decanoylcarnitine	3.206	HMDB0000651	TRUE
Hexanoylcarnitine	3.196	HMDB0000705	FALSE
Isobutyryl-L-carnitine	3.196	HMDB0000736	TRUE
L-Octanoylcarnitine	3.196	HMDB0000791	TRUE
9-Methyluric acid	3.176	HMDB0001973	TRUE

Judging from the data provided in the HMDB, it is highly likely that this peak is generated by a trimethylammonium group. Of these compounds, choline and acetylcarnitine are the simplest and seem the most likely. The chemical shift seems to be most consistent with acetylcarnitine, but we can check for other peaks of choline and acetylcarnitine to decide between the two.

```
[42]  hmdb[hmdb$Accession=="HMDB0000201", c('Metabolite', 'pos_in_ppm',
      'Accession', 'Urine')]
```

Metabolite	pos_in_ppm	Accession	Urine
L-Acetylcarnitine	2.146	HMDB0000201	TRUE
L-Acetylcarnitine	3.196	HMDB0000201	TRUE

The HMDB lists 2 peaks for the NMR spectrum of acetylcarnitine. The compound has in fact more, but the remaining peaks are considerably weaker. The peak at 2.146 ppm would correspond to the acetyl group of acetylcarnitine. We also plot the different timepoints for the signal at 2.146 ppm and compare it with our results for the signal at 3.196 ppm.

```
[43]  hmdbsel<-hmdb[hmdb$Accession=="HMDB0000201",]
      hmdbsel<-hmdbsel[round(hmdbsel$pos_in_ppm*1000)!=3196,]
      pos <- hmdbsel[[1,"pos_in_ppm"]]
      rn <- 0.02 * c(-1,1) + pos
      plot(datasets[datasets$meta$external$Day == "Exercise day"],
       chemshift_range = rn, color = "Timepoint")
```

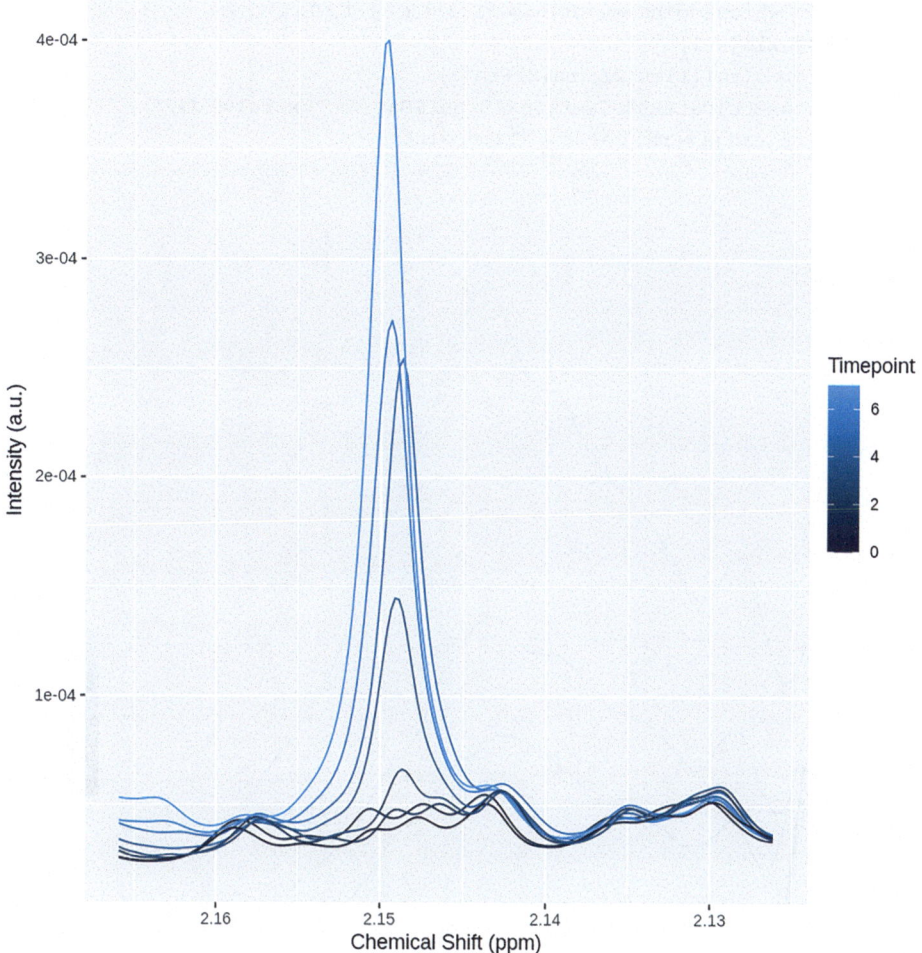

Both peaks show similar behavior, consistent with the hypothesis that the peak at 3.196 ppm is generated by the trimethylammonium group of acetylcarnitine. The choline peaks do not show this behavior (not shown).

A second bin that contributes to the PCA grouping is the bin at 1.3 ppm, which we plot with the following code.

```
[44]  sel <- which(colnames(df) == '1.3') + c(-1, 0, 1)
      rn <- ds$axis[sel]
      rn <- 0.5 * c(rn[1]+rn[2], rn[2]+rn[3])
      plot(datasets[datasets$meta$external$Day == "Exercise day"],
      chemshift_range = rn, color = "Timepoint")
```

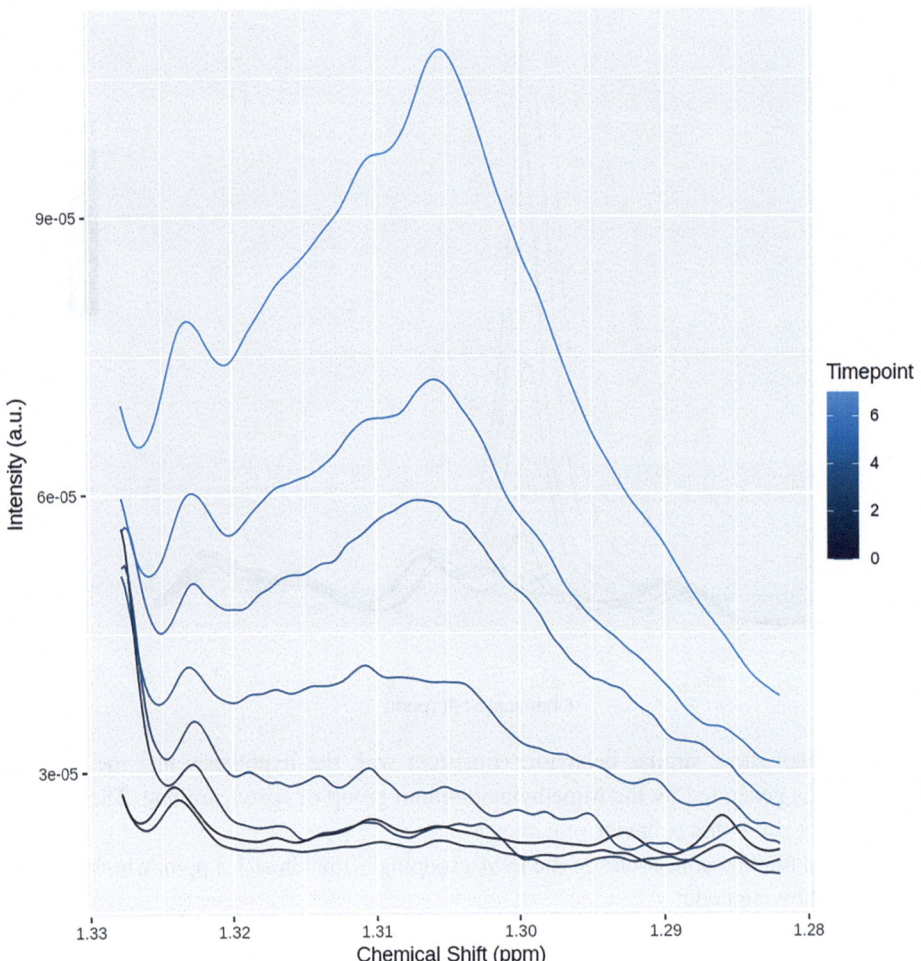

In contrast with the well-resolved sharp singlets and doublets that we have seen previously, here we are confronted with a broad hump, indicating a mixture of similar compounds with broad J-coupling patterns. Again we query the HMDB for candidate molecules.

```
[45]   hmdbsel <- hmdb[hmdb$pos_in_ppm>min(rn),]
   hmdbsel <- hmdbsel[hmdbsel$pos_in_ppm<max(rn),]
   hmdbsel <- hmdbsel[hmdbsel$Urine==TRUE,]
   hmdbsel <- hmdbsel[grepl('HMDB000....',hmdbsel$Accession),]
   hmdbsel[,c('Metabolite', 'pos_in_ppm', 'Accession', 'Urine')]
```

Metabolite	pos_in_ppm	Accession	Urine
3-Hydroxymethylglutaric acid	1.326	HMDB0000355	TRUE
Caprylic acid	1.286	HMDB0000482	TRUE
Capric acid	1.286	HMDB0000511	TRUE
Sebacic acid	1.316	HMDB0000792	TRUE
Traumatic acid	1.326	HMDB0000933	TRUE
Naringin	1.296	HMDB0002927	TRUE

The most likely candidates in this list are caprylate, caprate, regular fatty acids, and sebacate, a dicarboxylic acid. These are likely to be components of a mixture of various fatty acids. We can distinguish between mono- and dicarboxylic acids by looking for the ω-methyl peak of the monocarboxylic acids, which happens to be the most upfield peak in the caprate spectrum.

```
[46]   hmdbsel<-hmdb[hmdb$Accession=="HMDB0000511",]
   pos <- min(hmdbsel[,"pos_in_ppm"])
   rn <- 0.02 * c(-1,1) + pos
   plot(datasets[datasets$meta$external$Day == "Exercise day"],
   chemshift_range = rn, color = "Timepoint")
```

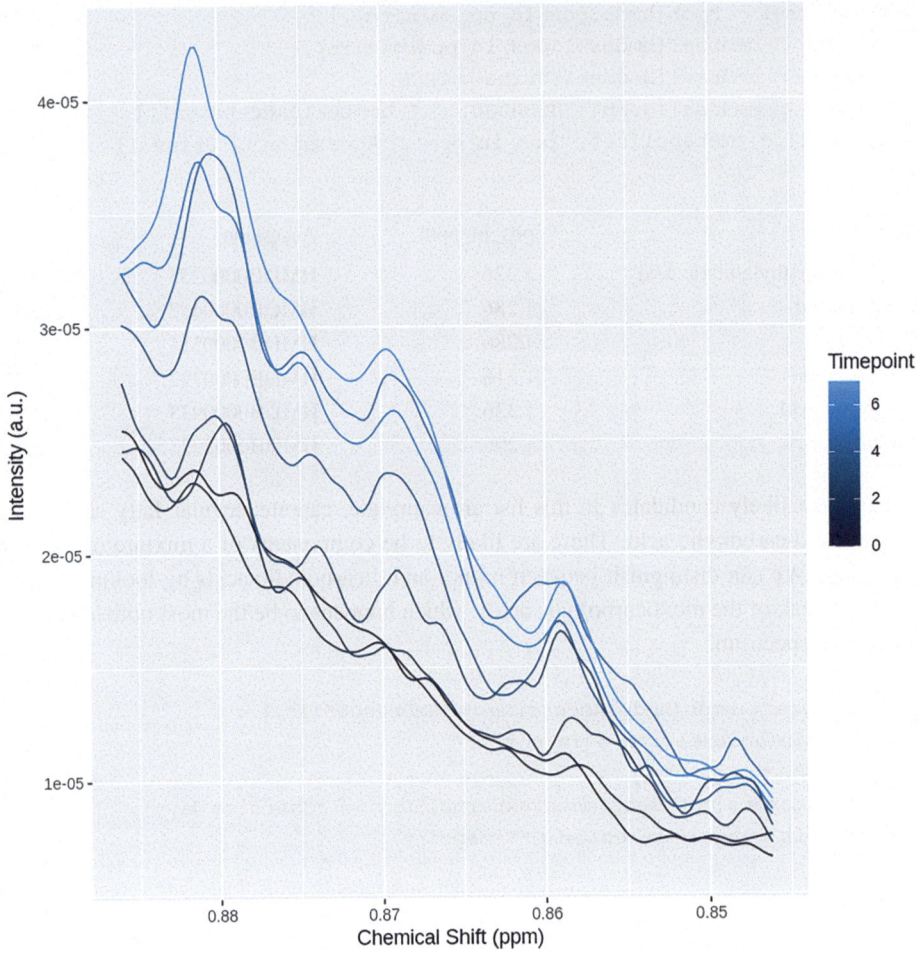

We see only small changes in this region of the spectrum, suggesting that the broad peak at 1.3 ppm is generated mainly by dicarboxylic acids. This makes sense, as dicarboxylic acids are much more water soluble than monocarboxylic acids.

Now that we have identified several interesting metabolites, we will perform targeted analysis to study exactly in which quantities these metabolites are produced as exercise progresses. For that, we will perform a non-linear least-squares fit of the NMR peaks of the metabolites (peak fitting code not shown). We start by fitting one of the two creatinine NMR peaks (at 3.05 ppm) in the spectrum of the pooled sample, as metabolite concentrations in targeted urine analysis are commonly given relative to creatinine.

```
[51]  sp <- datasets[datasets$metadata$external$RowID == "Pool"]$data_1r
      ax <- datasets[datasets$metadata$external$RowID == "Pool"]$axis
      sel <- (ax > 3.03) & (ax < 3.07)
```

```
pl <- peakfit(ax[sel], 3.051, 1/600, 0, 1, 0, sp[sel])
spfit <- do.call("peak", pl)
dffit <- data.frame(ax=ax[sel], sp=sp[sel], spfit)
ggplot(dffit, aes(x=ax, y=sp)) + geom_point() + geom_line(aes(y=spfit))
```

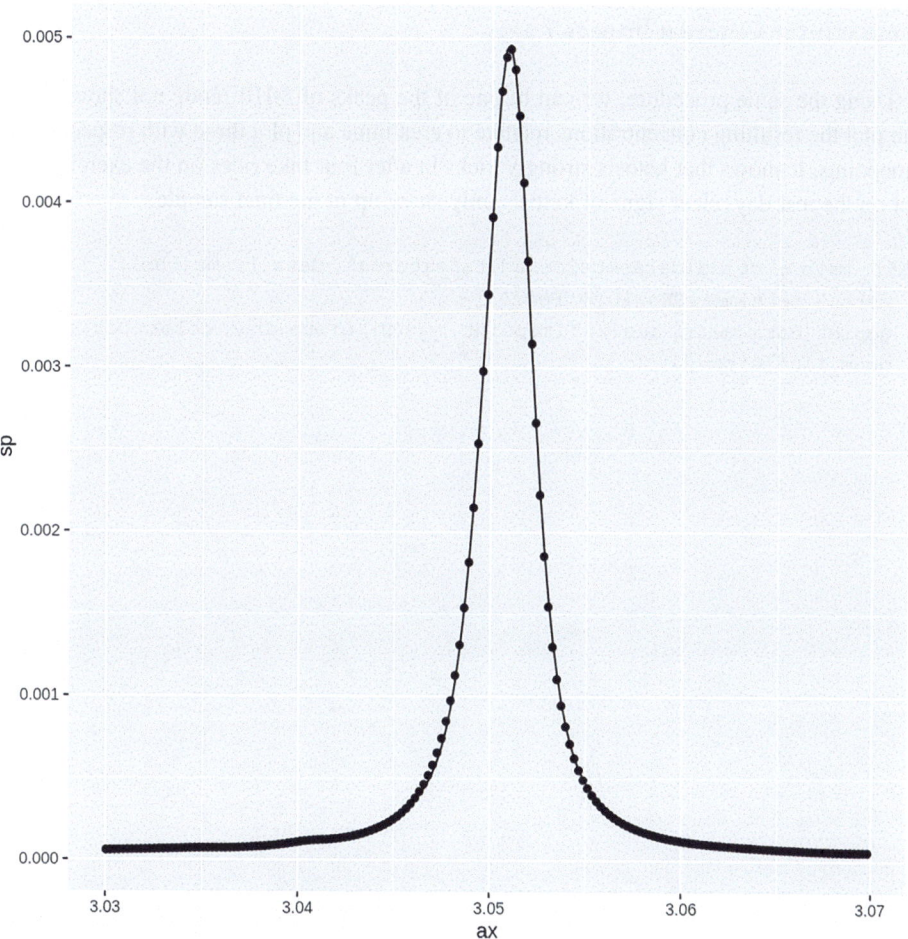

The fit parameters of the creatinine peak in the pooled sample can then be used as a starting point for the fit of the creatinine peaks in the other samples.

```
[52]  creatinine <- NULL
      id <- datasets$metadata$external$RowID
      for(t in id) {
      pos <- which(datasets$metadata$external$RowID == t)
      dffit <- data.frame(sp=datasets[pos]$data_1r[sel],
```

```
ax=datasets[pos]$axis[sel])
   result <- peakfitsimple(dffit$ax, pl$v0, pl$fwhh, pl$p,
 pl$a, pl$b, dffit$sp)
result$v <- NULL
result$id <- t
creatinine <- rbind(creatinine, data.frame(result))
}
creatinine <- (creatinine$a / 3)
```

Using the same procedure, we can fit one of the peaks of 3-HB (code not shown). We can plot the resulting concentrations relative to creatinine and plot these with respect to the timepoints. It shows that ketosis strongly kicks in after four bike rides on the exercise day, but on the rest day, clear signs of ketosis only show up in the final sample.

```
[55]  mets <- cbind(datasets$metadata$external, data.frame(bhb))
   mets <- mets[mets$RowID!="Pool",]
   ggplot(data=mets, aes(x=Timepoint, y=bhb, group=Day, color=Day)) +
   geom_line(size=2)
```

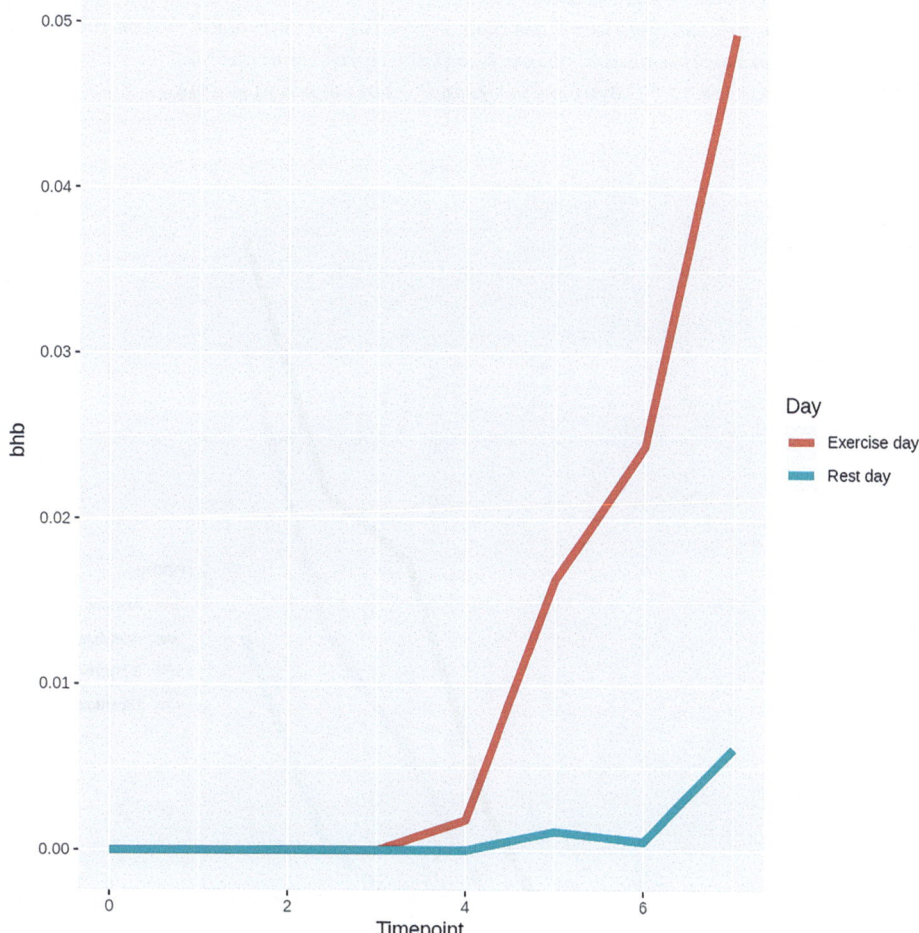

For now, we can skip acetoacetate because the strong overlap with other peaks would require more complex code. We can fit the other metabolites we identified as being regulated as well as the other ketone bodies in a similar way. The dicarboxylic acid peak has an unusual shape as a result of being generated by a mixture of dicarboxylic acids of different lengths and complex J couplings. For the sake of simplicity, we fit it as if it were a regular-shaped single NMR peak. Plotting the concentration of these metabolites on the exercise day as a function of time, we see that the excretion of acetylcarnitine and dicarboxylic acids start a bit earlier than acetone and 3-HB.

```
[64]  mets <- cbind(datasets$metadata$external, data.frame(acetone, bhb,
      acetylcarnitine, dicarboxylate))
      metsxc <- mets[mets$Day == "Exercise day",]
      metsxc <- pivot_longer(data=metsxc, cols=c("acetone", "bhb",
```

```
"acetylcarnitine", "dicarboxylate"))
ggplot(data=metsxc, aes(x=Timepoint, y=value, group=name, color=name)) +
geom_line(size=2) + scale_color_hue(labels=c("Acetone",
"Acetylcarnitine", "3-Hydroxybutyrate", "Dicarboxylate"))
```

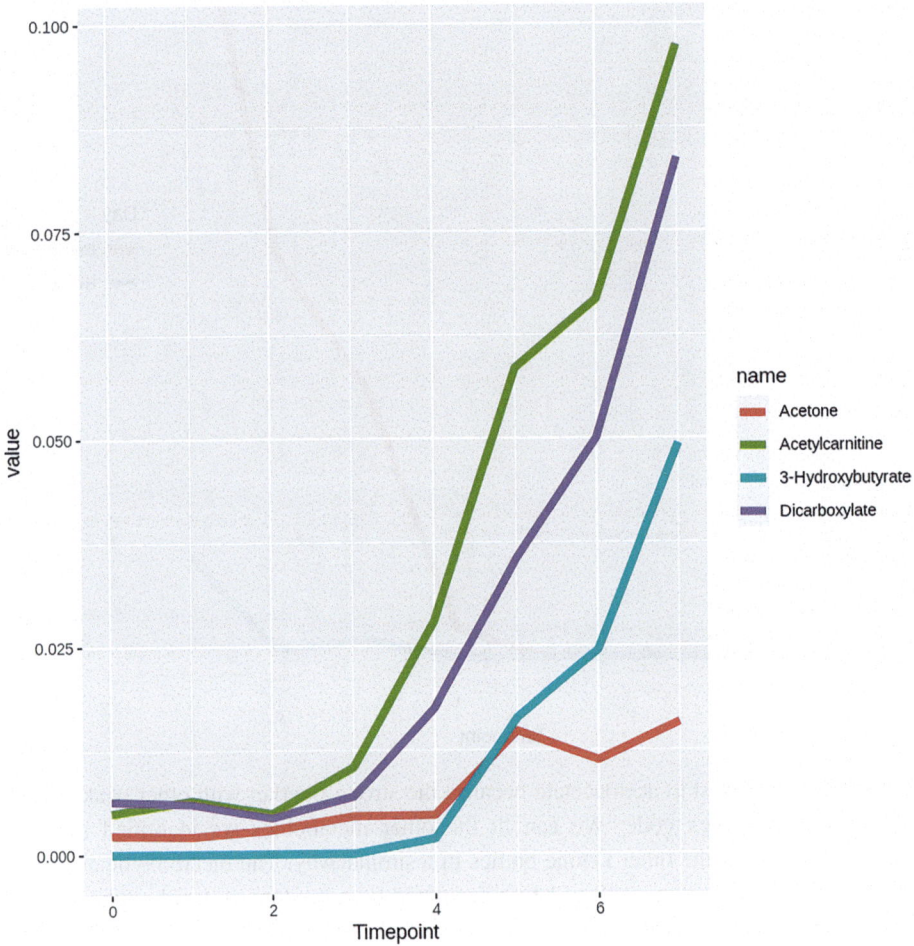

10.4   Discussion

When waking up in the morning, the body's glycogen reserves are low. When the test subject proceeds to continue fasting and simultaneously performs physical exercise, the glycogen stores eventually become depleted. The body then switches to ketosis (and gluconeogenesis) for its energy needs. This indeed happens, and we can see this by the appearance of the ketone bodies acetone, acetoacetate, and 3-HB in urine. However, the NMR spectra show the presence of many other compounds, a few of which also increase in concentration during later exercise stages, as demonstrated by multivariate analysis. The compounds were identified as acetylcarnitine and dicarboxylic acids. The appearance of these compounds tells us a lot about metabolic processes that happen in conjunction with ketosis. One group of compounds is dicarboxylic acids, which are the products of ω-oxidation of fatty acids. This is a type of oxidation that takes place at the methyl end group (ω-carbon) of the fatty acid alkyl chain after transfer of these molecules from adipose tissue to the cells where they are required. This happens in conjunction with β-oxidation, but in contrast with β-oxidation, which occurs in mitochondria and peroxisomes, ω-oxidation happens at the endoplasmic reticulum in the cytosol. A part of these dicarboxylic acids seem to easily diffuse out of the cell wall and are expelled from the body through the urine. This is not surprising, as it was previously shown that dicarboxylic acids can diffuse across membranes [23] and that orally ingested dicarboxylic acids are partly excreted by the urine [24]. Starvation is one of the conditions that induces ω-oxidation [25]. The purpose of ω-oxidation is unclear. One possibility is that it improves the solubility of fatty acids when these are present in relatively high concentrations as substrates for β-oxidation. However, this is unlikely, as maximally 10% of oxidized fatty acids undergo ω-oxidation. Its purpose could also be to steer β-oxidation away from the mitochondria, as the β-oxidation of dicarboxylic acids preferably takes place in peroxisomes [26] and the presence of dicarboxylic acids upregulates the synthesis of peroxisomal β-oxidation enzymes [27]. Another possibility is to supply the cell with succinate, which is the end product of even-chain dicarboxylic acids undergoing β-oxidation. Succinate can then be used for TCA cycle anaplerosis or as a substrate for gluconeogenesis. This also could have an antiketogenic effect [28], and hence keep ketogenesis in check. The appearance of acetylcarnitine in the urine is probably to maintain homeostasis in order to balance the supply of ketones and acetyl-CoA to the TCA cycle. Interestingly, the appearance of dicarboxylic acids at later exercise stages is also characteristic of an inborn error of metabolism called medium-chain acyl-coenzyme A dehydrogenase deficiency (MCADD), indicating that the body has reached the limit of mitochondrial β-oxidation capacity.

Thus, we have demonstrated that even with a small but well-designed experiment we can study many aspects of human metabolism simultaneously. By controlling nutrition and providing samples not only from the day of interest (the exercise while fasting day) but also from a "control day" by the same person, the amount of noise from nutrition-derived metabolites was kept at a minimum and we obtained a clear view of the human body's metabolic response to fasting and exercise.

Take Home Message

The metabolic analysis of bodily fluids results in highly complex chemical data consisting of hundreds or even thousands of signals. Moreover, relevant metabolic changes are often masked by metabolite variations caused by differences in lifestyle, diet, and medication. In turn, it is extremely important to reduce the number and influence of confounding variables and random effects as much as possible. By measuring fasted subjects at multiple timepoints, in a paired fashion with a control group (in this case, the same individual on a different day not performing physical exercise), relevant metabolites can be clearly identified among the many unchanged (irrelevant) bystanders. With the here presented example, we show that:

- Physical activity is reflected in the urinary metabolome.
- Targeted and untargeted (binning) NMR-based metabolomics can be applied to study metabolic changes in bodily fluids (urine).
- Data normalization, scaling, and multivariate analysis are key features of metabolomics analysis.
- Software libraries and databases, glued together using a modern scripting or workflow environment are key to metabolite identification and ultimately biochemical interpretation.
- When correctly applied, modern metabolomics tools have the power to confirm known and unravel novel biochemistry/biology.

References

1. Friebolin H, Becconsall JK. Basic one-and two-dimensional NMR spectroscopy. Wiley-VCH Weinheim. 2005;7
2. Bouatra S, Aziat F, Mandal R, Guo AC, Wilson MR, Knox C, Bjorndahl TC, Krishnamurthy R, Saleem F, Liu P, Dame ZT, Poelzer J, Huynh J, Yallou FS, Psychogios N, Dong E, Bogumil R, Roehring C, Wishart DS. The human urine metabolome. PLoS One. 2013;8(9):e73076.
3. De Meyer T, Sinnaeve D, Van Gasse B, Tsiporkova E, Rietzschel ER, De Buyzere ML, Gillebert TC, Bekaert S, Martins JC, Van Criekinge W. NMR-based characterization of metabolic alterations in hypertension using an adaptive, intelligent binning algorithm. Anal Chem. 2008;80(10):3783–90.
4. Kostidis S, Addie RD, Morreau H, Mayboroda OA, Giera M. Quantitative NMR analysis of intra- and extracellular metabolism of mammalian cells: a tutorial. Anal Chim Acta. 2017;980:1–24.
5. Rosen Vollmar AK, Rattray NJW, Cai Y, Santos-Neto ÁJ, Deziel NC, Jukic AMZ, Johnson CH. Normalizing untargeted periconceptional urinary metabolomics data: a comparison of approaches. Meta. 2019;9(10):198.

6. Dieterle F, Ross A, Schlotterbeck G, Senn H. Probabilistic quotient normalization as robust method to account for dilution of complex biological mixtures. Application in 1H NMR metabonomics. Anal Chem. 2006;78(13):4281–90.

7. van den Berg RA, Hoefsloot HCJ, Westerhuis JA, Smilde AK, van der Werf MJ. Centering, scaling, and transformations: improving the biological information content of metabolomics data. BMC Genomics. 2006;7(1):142.

8. Murenu E, Kostidis S, Lahiri S, Geserich AS, Imhof A, Giera M, Michalakis S. Metabolic analysis of vitreous/lens and retina in wild type and retinal degeneration mice. Int J Mol Sci. 2021;22(5):2345.

9. Gandhi S, Chinnadurai V, Bhadra K, Gupta I, Kanwar RS. Urinary metabolic modulation in human participants residing in Siachen: a 1H NMR metabolomics approach. Sci Rep. 2022;12(1): 9070.

10. Caspersen CJ, Powell KE, Christenson GM. Physical activity, exercise, and physical fitness: definitions and distinctions for health-related research. Public Health Rep. 1985;100(2):126–31.

11. Hawley JA, Hargreaves M, Joyner MJ, Zierath JR. Integrative biology of exercise. Cell. 2014;159 (4):738–49.

12. Hargreaves M, Spriet LL. Skeletal muscle energy metabolism during exercise. Nat Metab. 2020;2 (9):817–28.

13. McPherson PAC. Ketone bodies. In: Caballero B, Finglas PM, Toldrá F, editors. Encyclopedia of food and health. Oxford: Academic Press; 2016. p. 483–9.

14. Pinckaers PJM, Churchward-Venne TA, Bailey D, van Loon LJC. Ketone bodies and exercise performance: the next magic bullet or merely hype? Sports Med. 2017;47(3):383–91.

15. Johnson RH, Walton JL, Krebs HA, Williamson DH. Post-exercise ketosis. Lancet. 1969;294 (7635):1383–5.

16. Dhatariya KK, Glaser NS, Codner E, Umpierrez GE. Diabetic ketoacidosis. Nat Rev Dis Primers. 2020;6(1):40.

17. Giacomoni F, Le Corguillé G, Monsoor M, Landi M, Pericard P, Pétéra M, Duperier C, Tremblay-Franco M, Martin J-F, Jacob D, Goulitquer S, Thévenot EA, Caron C. Workflow4Metabolomics: a collaborative research infrastructure for computational metabolomics. Bioinformatics (Oxford, England). 2015;31(9):1493–5.

18. Goecks J, Nekrutenko A, Taylor J, The Galaxy T. Galaxy: a comprehensive approach for supporting accessible, reproducible, and transparent computational research in the life sciences. Genome Biol. 2010;11(8):R86.

19. Verhoeven A, Giera M, Mayboroda OA. KIMBLE: a versatile visual NMR metabolomics workbench in KNIME. Anal Chim Acta. 2018;1044:66–76.

20. Berthold MR, Cebron N, Dill F, Gabriel TR, Kötter T, Meinl T, Ohl P, Sieb C, Thiel K, Wiswedel B. KNIME: The Konstanz information miner. In: Preisach C, Burkhardt H, Schmidt-Thieme L, Decker R, editors. Data analysis, machine learning and applications. Berlin, Heidelberg: Springer; 2008. p. 319–26.

21. Chong J, Soufan O, Li C, Caraus I, Li S, Bourque G, Wishart DS, Xia J. MetaboAnalyst 4.0: towards more transparent and integrative metabolomics analysis. Nucleic Acids Res. 2018;46 (W1):W486–94.

22. Madrid-Gambin F, Oller-Moreno S, Fernandez L, Bartova S, Giner MP, Joyce C, Ferraro F, Montoliu I, Moco S, Marco S. AlpsNMR: an R package for signal processing of fully untargeted NMR-based metabolomics. Bioinformatics (Oxford, England). 2020;36(9):2943–5.

23. Liu G, Hinch B, Beavis AD. Mechanisms for the transport of α,ω-dicarboxylates through the mitochondrial inner membrane. J Biol Chem. 1996;271(41):25338–44.

24. Smith HG. The metabolism of azelaic acid. J Biol Chem. 1933;103(2):531–5.

25. Mortensen PB. C6--C10-dicarboxylic aciduria in starved, fat-fed and diabetic rats receiving decanoic acid or medium-chain triacylglycerol. An in vivo measure of the rate of beta-oxidation of fatty acids. Biochim Biophys Acta. 1981;664(2):349–55.

26. Bergseth S, Poisson J-P, Bremer J. Metabolism of dicarboxylic acids in rat hepatocytes. Biochim Biophys Acta (BBA) – Lipids Lipid Metab. 1990;1042(2):182–7.

27. Bharathi SS, Zhang Y, Gong Z, Muzumdar R, Goetzman ES. Role of mitochondrial acyl-CoA dehydrogenases in the metabolism of dicarboxylic fatty acids. Biochem Biophys Res Commun. 2020;527(1):162–6.

28. Mortensen PB. The possible antiketogenic and gluconeogenic effect of the ω-oxidation of fatty acids in rats. Biochim Biophys Acta (BBA) – Lipids Lipid Metab. 1980;620(2):177–85.

Identifying Sex-Specific Cancer Metabolites and Associations to Prognosis

11

Xinyi Shen, Shuangge Ma, Sajid A. Khan, and Caroline H. Johnson

What You Will Learn in This Chapter

In this chapter, we provide an overview of a bioinformatics workflow to identify sex-specific associations between metabolites and cancer prognosis using liquid chromatography-mass spectrometry-based metabolomics data, and clinical data acquired from cancer patients. We discuss the importance of considering sex interactions in the discovery of prognostic metabolites for cancer. We introduce the theoretical basis of data analysis approaches for survival analysis using metabolomics data with an emphasis on Cox Proportional Hazard regression analysis. Crucial issues that need to be taken into account when analyzing metabolomics data in oncology studies will be examined, such as multiple comparisons and inadequacy of analytical methods. We provide an example of survival analysis in R using data obtained from colorectal cancer (CRC) patients. R scripts are provided,

(continued)

X. Shen · C. H. Johnson (✉)
Department of Environmental Health Sciences, Yale School of Public Health, New Haven, CT, USA
e-mail: xinyi.shen@yale.edu; caroline.johnson@yale.edu

S. Ma
Department of Biostatistics, Yale School of Public Health, New Haven, CT, USA
e-mail: shuangge.ma@yale.edu

S. A. Khan
Department of Surgery, School of Medicine, Yale University, New Haven, CT, USA
e-mail: sajid.khan@yale.edu

© The Author(s), under exclusive license to Springer Nature Switzerland AG 2023
J. Ivanisevic, M. Giera (eds.), *A Practical Guide to Metabolomics Applications in Health and Disease*, Learning Materials in Biosciences,
https://doi.org/10.1007/978-3-031-44256-8_11

along with outputs that detail data processing, survival analysis, data visualization, and interpretation. Finally, an overview of limitations and potential alternative analytical approaches are provided for the identification of sex-specific prognostic cancer metabolites.

11.1 Sex Interplays with the Associations Between Cancer Prognosis and Metabolome

11.1.1 Cancer Metabolomics and Prognosis

Metabolomics techniques have been widely applied to studies in cancer metabolism. These studies have identified novel metabolic pathways and metabolic targets, revealed metabolic responses following a metabolic inhibitor treatment, and also identified metabolites associated with cancer progression and prognosis [1]. Recently, this was exemplified in a study that used liquid chromatography-mass spectrometry (LC-MS)-based metabolomics analysis to analyze serum samples from metastatic soft tissue sarcoma patients, and detected prognostic metabolites using a Cox multivariable regression model [2]. LC-MS techniques were also utilized to identify metabolomic biomarkers and create a metabolite-based risk score to improve prognosis prediction of esophageal squamous cell carcinoma [3].

11.1.2 Sex Interacts with Metabolism and Cancer Prognosis

Sex differences in cancer prognosis exist in many types of cancers. In many malignancies, males have a worse overall prognosis compared to females, including in colorectal cancer (CRC) [4]. Certain risk factors and exposures have disparate distributions by sex, which may contribute to a sex-specific prognostic outcome for cancer patients. For instance, a high inflammatory profile is associated with a greater risk of CRC in males but not in females[5]. Sex-related differences in cancer biology are also a well-discussed field of research; metabolism, sex-steroid hormones, genetic programming, epigenetics, and chromosomes have sex differences that strongly impact cancer risk and prognosis. For example, transcriptomics analysis revealed a set of genes in hepatocellular carcinoma and CRC that could be linked to sex-specific survival [6, 7]. Recently, metabolomics analysis discovered sex-specific differences in cancer, using tumor tissue samples from CRC patients [8], and the adoption of serum-based metabolic profiling that revealed sex-related differences in renal cell carcinoma [9]. Moreover, sex can interact with the association between cancer metabolism and cancer prognosis. An untargeted metabolomics study on CRC tumor tissues discovered 91 metabolites that had a unique metabolic signature showing that increased asparagine synthesis was associated with poorer prognosis in female CRC patients only [10]. In this chapter, we will show an R-based bioinformatic pipeline for the discovery of sex-specific prognostic cancer metabolites using metabolomics data.

11.2 Basics of Data Analysis Methods for Survival Data

This chapter will show survival analysis methods using R codes (version 4.0.4): Kaplan-Meier Analysis and Cox Proportional Hazard Regression analysis. This chapter also serves as an introduction to survival analysis using clinical cancer metabolomics data. For more detailed or more advanced instruction in survival analysis, please refer to *Modelling Survival Data in Medical Research* (Second or Third Edition, Chapman and Hall/CRC, 2015) by David Collett [11]. The chapter assumes readers have entry-level exposure to basic R programming using RStudio. For SAS users, *Survival Analysis Using SAS* (Second Edition, SAS Publishing, 2010) by P. Allison[12] provides additional examples.

11.2.1 Metabolomics Data Processing and Cancer Prognosis

Clinical LC-MS-based metabolomics data can be acquired from the analysis of cancer patients' serum, urine, tissues, etc. After raw data processing, statistical analysis, and metabolite identification, the protocols of which are not covered in this chapter but can be found in our previous study (Materials and Methods section) [8], a spreadsheet is obtained with the relative abundance of each metabolite annotated for each patient. A log transformation (e.g., Log_2) for each metabolite abundance makes them comparable.

Clinical and pathologic data such as age, sex, race/ethnicity, body mass index, date of cancer diagnosis, cancer stage, date of death, cause of death, and treatment history can be acquired from electronic medical records. Molecular profiling data is often available in medical records as well.

Additionally, prognostic information such as overall survival (OS), recurrence-free survival (RFS), cancer-specific survival (CSS), and stage-specific survival can be calculated from medical records by using data metrics such as date of diagnosis, date of death, and date of recurrence. These data are essential to performing survival analysis. In clinical studies, medical professionals often measure the prognosis of patients using 1-, 5-, and 10-year scales, to determine OS, RFS, and CSS depending on the aggressiveness of the cancer which is examined. Additionally, studies will often measure median survival. For example, pancreas cancer survival is aggressive and often measured in shorter intervals such as 1, 2, and 5-year OS, while less aggressive cancers such as breast cancer and melanoma are often measured in terms of 10-, 15-, and 20-year intervals with a focus on RFS. Survival responses consist of: (1) T, the survival time which is often calculated in weeks, months, or years, (2) occurrence of death, (3) occurrence of recurrence, (4) location of recurrence, and (5) cause of death (i.e., cancer vs non-cancer death). Censoring is a common phenomenon that occurs when the survival time of an individual is incomplete. Events can be left-censored, right-censored, or interval-censored. Left-censored events indicate that the event occurred prior to the documented date of entering the study. An interval-censored event happens between two observation times. Censoring becomes important to control for biases that may exist in cancer studies such as lead-time bias [13]. Most studies on cancer prognosis focus on right-censored events that occur when the failure time is greater than the last time the patient was observed, for example, a patient survives after the study ended.

Using basic functions in R software, it is possible to combine metabolomics data with clinical information and link them for each patient, creating a new dataset before moving forward to the next steps of analysis. Due to differences in data formats from different research labs, such basic data processing is not shown here. However, we will use data. frame to do this, an example is displayed in Table 11.1 (values in the table are for illustration purposes and not true values). A dataset that is in a similar format to Table 11.1 will be used for survival analysis using codes provided in Sect. 11.3. In this dataset, 91 metabolites were annotated, and their abundances were presented as numerical values. The following variables are covariates that will be used in the survival models: Sex: F, female; M, male. Clinical stage: I, II, and III. Tumor location: L, left-sided; R, right-sided. Chemo history (chemotherapy history): 0, no; 1, yes. 5-year OS is encoded by 1 for the occurrence of an event (disease relapse/death) and 0 (the subject is right-censored). Time is measured by month ≤ 5 years (60 months). A more detailed illustration of covariates can be found in Sect. 4.2. *Statistical Analysis* in the example publication [10].

For more detailed guidelines for other data processing aspects such as missing data imputation for metabolomics data, please refer to Wei et al. [14], and Faquih et al. [15].

11.2.2 Kaplan-Meier Analysis

The distribution of T can be characterized by survival function $S(t)$ and hazard function $h(t)$. The Kaplan-Meier (K-M) estimator is a non-parametric frequency-based estimator of $S(t)$. The survival time can be ordered by time of occurrence: $t_1 < t_2 \ldots < t_k$ (there are k observed event times). Determine that d_i is the number of events/failures that occur at t_i, and n_i is the number of patients at risk at t_i. Therefore, the conditional probability of surviving beyond T_i given they are at risk at the time T_i can be expressed by:

$$P(T > t_i | T > t_{i-1}) = 1 - \frac{d_i}{n_i}$$

The unconditional probability of surviving beyond t_i is the product of the conditional probabilities for the current and all previous time intervals, giving the Kaplan-Meier estimator of the survival function:

Table 11.1 An example of a dataset for survival analysis for 197 patients

ID	Sex	Age	Clinical stage	Tumor location	Chemo history	5-year OS	OS Time	Metabolite 1	Metabolite 2	...	Metabolite 91
1	F	45	II	L	0	1	35	246432	38765	...	864676
2	M	75	III	R	1	0	60	1324	77898	...	23542
...	...	...	...				...	...	...	...	...
197	M	53	I	L	1	1	59	32356	98765	...	77843

$$\widehat{S_{(t)}} = \prod_{t_i \leq t} \left(1 - \frac{d_i}{n_i}\right)$$

To test whether the survival function differs in two groups 1 and 2, Log-Rank test is used for the hypothesis testing of K-M analysis given that τ is the largest time on the study:

$$H_0 : S_1(t) = S_2(t) \forall t \in [0, \tau]$$

$$H_1 : S_1(t) \neq S_2(t) \exists t \in [0, \tau]$$

A P value less than 0.05 suggests that the null hypothesis H_0 should be rejected and thus survival significantly differs in groups 1 and 2.

The R function $\texttt{survfit()}$ in the $\texttt{survival}$ package is used to perform K-M analysis, and the function $\texttt{survdiff()}$ performs Log-Rank test.

11.2.3 Cox Proportional Hazard Regression Analysis

Cox Proportional Hazard (PH) models are used for evaluating the effects of multiple variables on survival. Variables can be continuous (such as age) and categorical (such as sex), which are not limited to two categories as K-M analysis requires. Cox PH analysis measures how factors impact the rate of a prognostic event (i.e., 5-year OS) at a particular point in time (fifth year), which is quantified as the hazard rate.

The hazard function $h(t, X)$ can be expressed in this way for a multivariable Cox PH model:

$$h(t, X) = h_0(t)\exp\left(\sum_{i=1}^{p} b_i x_i\right)$$

where, t represents the survival time, $h(t, X)$ is the hazard function determined by a set of covariates $(x_1, x_2, \ldots, x_p)$, and the coefficients $(b_1, b_2, \ldots, b_p)$, measure the effect size of covariates. h_0 is the baseline hazard when $x_1 = x_2 = \ldots = x_p = 0$, which is only dependent on t.

The proportional hazards (PH) assumption means that the ratios of hazard functions are constant over time. For example, in a univariate Cox PH model, where $h(t, X) = h_0(t) \, exp \, (X'b)$ with a continuous X as the only variable, the hazard ratio (HR) equals $exp(b)$. A $b > 0$ is equivalent to an HR >1, suggesting that the risk of failure increases as the value of variable X increases, which leads to a decreased length of survival. Whereas an HR $= 1$ suggests no effect of the variable, and an HR < 1 indicates a reduction of risk of failure.

Using R, the function `coxph()` constructs a Cox PH regression model and the output can be observed using the `summary()` function.

11.2.4 Parametric Survival Models

The Cox PH model is a semi-parametric model which makes a parametric assumption that the effect of the predictors on the hazard function is constant over time, but makes no assumption about the form of the hazard function $\lambda(t)$. In contrast, parametric survival models assume that the survival time T follows certain distributions $f(t)$ and thus $S(t)$ and $h(t)$ are also specified.

The Parametric survival model has advantages over the Cox PH model as it has a more precise estimate due to the nature of time-dependency. Although not covered in the practical example in this chapter, the proportional hazards (PH) models can also be expanded more in the parametric form.

The Exponential model assumes a constant hazard $h(t) = \lambda$, which has a stronger assumption than Cox PH assumption. Thus, the equations below characterize its survival function and the probability density function of T: $S(t) = exp\,(-\lambda t)$

$$f(t) = S(t)h(t) = \lambda\ exp(-\lambda t)$$

Weibull model assumes a monotonic hazard:

$$h(t) = \alpha t^{\alpha-1}\lambda$$

$$S(t) = exp(-\lambda t^{\alpha})$$

$$f(t) = \alpha t^{\alpha-1}\lambda\ exp(-\lambda t^{\alpha})$$

where $\lambda = exp\left(\sum_{i=1}^{p} b_i x_i\right)$ in a multivariable model, and α is the shape parameter.

Log-logistic model assumes a non-monotonic hazard that

$$h(t) = \frac{\alpha t^{\alpha-1}\lambda}{1 + t^{\alpha}\lambda}$$

$$S(t) = 1 + t^{\alpha}\lambda$$

$$f(t) = \frac{\alpha t^{\alpha-1}\lambda}{(1 + t^{\alpha}\lambda)^2}$$

Log-normal assumes that

$$f(t) = \frac{exp\left[-\frac{1}{2}\left(\frac{\ln t - \ln \lambda}{\sigma}\right)^2\right]}{t\sqrt{2\pi\sigma^2}}$$

$$S(t) = 1 - \Phi\left[\frac{\ln t - \ln \lambda}{\sigma}\right]$$

$$h(t) = \frac{f(t)}{S(t)}$$

where σ is the scale parameter, and $h(t)$ is also non-monotonic.

11.2.5 Multiple Comparisons

When analyzing regression model results from metabolomics studies, it is common to perform hundreds or thousands of hypothesis tests simultaneously. To control false discovery from multiple hypothesis testing, it is imperative to adjust the significance level (α) to reduce the chances of getting type I error (false positive) or Family-wise error rate (FWER), which is defined as the probability of getting at least one false significant result just by chance. One useful approach is Bonferroni correction which adjusts the α to α/n, where n is the number of multiple tests. Another method is to control false discovery rate (FDR), defined as the proportion of false-positive results among all significant results, proposed by Benjamini and Hochberg [16], which is less conservative than the Bonferroni correction.

The two methods are easy to apply using the R function p.adjust (p, method = p.adjust.methods, n = length(p)), where p.adjust.methods can be chosen from "bonferroni", "fdr" and other methods that cater to a specific research need. An adjusted α of 0.05 is widely adopted, but sometimes 0.1 is also used [17] to keep a larger number of significant results while still mitigating false positive levels.

11.2.6 Examination of the Interaction Term

The association between a variable and the event differs by a third variable, which can be the presence of interaction. For instance, a sex interaction occurs when the effect of a certain tumor tissue metabolite on the hazard of mortality differs by sex. In this case, the interaction presents between sex (binary: female (reference) and male) and metabolite abundance, which is continuous originally but can be categorized when needed.

The interaction analysis of biomedical data can be conducted in mainly two ways: (1) Marginal analysis that analyzes one or a small number of biomarker unit(s) at a time;

(2) Joint analysis that handles a large number of biomarkers within one model. In this chapter, we mainly describe the marginal interaction analysis that could be applied to a single metabolite or a small section of metabolites. In the interaction analysis of the exemplary publication [10], confounders (such as age, clinical stage, and tumor location) were adjusted, and the estimation was in a nonrobust manner that was based on the maximum likelihood estimation (MLE).

Mathematically, in a multivariable Cox PH model adjusted for other confounders (age, clinical stage, tumor location), the hazard of such sex interaction can be calculated with a certain metabolite M by examining the log hazard function: $\ln (h(t, M, male)) = \ln h_0(t) + b_1 M + b_2 male + b_3 M \times male + b_4 \times age + b_5 \times clinical\ stage + b_6 \times tumor\ location$.

The determination of importance of an interaction is based on the P of interaction, which is assessed by testing the significance of the sex interaction term via the Wald test, if the interaction effect includes a single coefficient b_3.

$$H_0 : b_3 = 0; H_1 : b_3 \neq 0$$

If P of interaction < 0.05, then the null hypothesis is rejected, indicating that a sex interaction is present. Using the R code `summary(coxph(Surv(time, evet) ~ M + Sex + M*Sex + age + clinical stage + tumor location, data = your data))`, the P of interaction term of `M*Sex` can be examined.

As metabolomics data has a high number of variables, a greater challenge is posed for metabolomics researchers. Therefore, before moving forward to the next section, it is recommended to perform survival analysis using R in simpler datasets. Joseph Rickert gave an introductory tutorial for the K-M analysis and Cox PH analysis using a dataset with much fewer variables [18]. Statistical tools for high-throughput data analysis (STHDS) have a basic introduction to Cox PH regression analysis using R [19].

11.3 A Practical Example of a Study for Identifying Sex-Specific Metabolites Associated with CRC Prognosis Using Cox PH Regression with R Codes

Step-by-Step Protocol

Step 1: Get familiar with the study background and research questions and by determining research goals, obtain basic knowledge about the cancer type regarding its metabolism and sex differences.

Step 2: Learn the data structure, especially the prognostic outcomes of interest, covariates, and metabolites categories, and determine the correct analytical methods before setting up a programming environment for data analysis.

(continued)

Step 3: Explore the dataset by preliminary data cleaning and performing multivariable Cox PH regression analysis on each metabolite adjusted for necessary covariates for all patients regardless of their biological sex. Conduct multiple comparisons that calculate FDR adjusted P value for the results.

Step 4: Conduct multivariable Cox PH regression analysis on females and males separately to obtain P values for each metabolite and the P values after correction for multiple comparisons.

Step 5: Calculate the P of sex interaction for metabolites that are significantly associated with either female or male prognosis.

Step 6: Conduct multivariable Cox PH analysis for a set of metabolites that are involved in specific metabolic pathways, which have important metabolic regulation in cancer progression and/or prognosis. Conduct multivariable Cox PH analysis for each pathway by sex and obtain P values of sex interaction in a method that is similar to single metabolite analysis.

Step 7: Data interpretation.

Below are the codes and outputs for the R script with annotations for instruction and explanation.

11.3.1 Code and Output (Generated by Rmarkdown)

1. Load required libraries and conduct initial data processing.

The deidentified dataset $cox.os.csv$ and a spreadsheet $name.xlsx$ containing metabolites names can be downloaded from (https://www.ebi.ac.uk/metabolights/MTBLS1129/files). Its original non-related variables were removed, and variable recoding has been partially conducted. A summary of how we treated missing values and recoded variables is provided in the *Statistical Analysis* section of the example publication [10]. Below is a brief illustration of the meaning of each column:

- Patient.id: The identification number (deidentified) of each patient, ranging from 1 to 197.
- Age: Age at diagnosis of CRC.
- OS5.status: 5-year OS status (0 or 1).
- OS5.months: Time to event (death) in months.
- RecurrenceStatus: 5-year RFS status (0 or 1).
- Recurrence5.months: Time to event (recurrence) in months.
- Sex: sex of patient (female or male).
- Side: Location of the primary tumor in the colon or rectum LCC, left-sided; RCC, right-sided.

- Stage: Pathologic stages I, II, and III.
- Chemo: Patients with adjuvant postoperative chemotherapy were coded as 1, and those without such history were coded as 0.
- Metabolite 1 to Metabolite 91: To simplify the workflow, metabolite names were replaced by an ID ranging from 1 to 91. Their full names can be found in name.xlsx as mentioned below.

The analysis of 5-year OS and 5-year RFS is almost the same, thus we only focus on the codes for OS for illustration purposes.

```
library(survival)
library(readxl)
library(jtools)
library(forestplot)
library(ggstance)
library(tab)
library(dplyr)
# set up your work directory where your data is stored
setwd("your filepath")
# read the dataset
crc<-read.csv("cox.os.csv") # The dataset is similar to Table 1 in section
2.1. The column name of the first metabolite is Metabolite1, and likewise for
all 91 metabolites.
met.name<-read_excel("name.xlsx")# read the excel file containing
metabolite names for the 91 metabolites.
met_name<-as.vector(met.name $Metabolite)
crc$stage<-ifelse(crc$stage == "I"| crc $stage=="II", 'early', 'late') #
Due to the absence of death events in female patients at clinical stage I, we
recoded the clinical stages I and II as "early stage", and III as "late stage".
#Factorize the categorial variables
crc$stage<-factor(crc$stage)
crc$sex<-factor(crc$sex)
crc$side<-factor(crc$side)
crc$Chemo<-factor(crc$Chemo)
#Log2 transformation of metabolite abundances and renaming as "mtlog"
followed by their order number.
for (i in 1:91){crc[,paste0("mtlog",i)]<-log(crc[,paste0("Metabolite",
i)],2) }
os1<-crc # Dataset for Overall survival
rec1<-crc # Dataset for Recurrence-free survival. The analysis is almost the
same with OS, thus we will not repeat the codes in this chapter.
# Divide the dataset by sex
os1.m<-subset(os1,sex=='male') # dataset for 5-yr OS in males
os1.m<-os1.m[,-1]
os1.f<-subset(os1,sex=='female') # dataset for 5-yr OS in females
os1.f<-os1.f[,-1]
```

2. Create functions that organize outputs of multivariable Cox PH regression for each metabolite by producing a summary table showing individual metabolite's name estimates of HR, *P* value, and information of 95% CI.

```
FML <-function (x, surv, varb, covariate) {
as.formula (paste0 (surv, paste ( append (paste0 (varb, x) , covariate) , collapse
= '+')))
}
#example: call FML("1") and get "Surv() ~ mtlog1 + sex + side + stage + age"
# A function to get all the results
getall<-function (i, data, num_pred) {
sum<-summary (coxph (i, data = data))
HR <- round (sum$coefficients [, 2] , 2)
P_Value <- round (sum$coefficients [, 5] , 3)
LCI <- format (round (sum$conf.int [, 3] , 2) , 2)
UCI <- format (round (sum$conf.int [, 4] , 2) , 2)
CI95 <- paste0 (LCI, '-', UCI)
#show statistical significance
sig<- rep (NA, num_pred)
for (i in 1:num_pred) {if (P_Value [i] <0.05) {sig[i]="*"}else if (P_Value
[i] <0.1) {sig[i]="."}else {sig[i]=""}}
result<-cbind (HR=HR, CI95=CI95, LCI=LCI, UCL=UCI, P_Value=P_Value,
sig=sig)
}
```

3. Set up fixed variables for further usage.

```
num_met<-91 #number of metabolites
Surv.os<-'Surv (time = OS5.months, event = OS5.status) ~'
# Survival object of 5-year OS
Surv.rfs<-'Surv (time = Recurrence5.months, event =RecurrenceStatus) ~' #
Survival object of 5-year RFS
covar0<-c ('sex', 'side', "stage", "Chemo", "age") # A full set of covariates
covar1<-c ('side', "stage", "Chemo", "age") # A set of covariates for
sex-specific analysis
```

4. Create a function that gives outputs of multivariable Cox PH analysis in a table including FDR adjusted *P* values for each metabolite. The *coxphtable* function requires a series of inputs—the number of metabolites (*num_met*), a survival object (*Surv*), a list of covariates (*covar*), and the data for analysis (*data*).

```
coxphtable<-function (num_met, Surv, covar, data) {
out_multi=data.frame ()
num_pred=1+length (covar)
```

```
for (i in 1:num_met){
out<-as.data.frame(getall(FML(i,Surv,'mtlog',covar),
data,num_pred))
out_multi<-rbind(out_multi,out)
}
#only extract rows that contain outputs of estimates of HR of metabolites:
out_multi_mtlog=data.frame()
for (i in 1:num_met){
out_multi_mtlog<-rbind(out_multi[num_pred*(num_met-i)
+1,],out_multi_mtlog)}
out_multi_mtlog$fdr.p<- round(p.adjust(as.numeric(out_multi_mtlog
$P_Value), method = "fdr", n = num_met),3)
output<-cbind(met_name,out_multi_mtlog) #bind the metabolite name in the
first column
return(output)
}
```

5. Perform multivariable Cox PH analysis for each of 91 metabolites for all patients and then by sex.

```
#OS
output<-coxphtable(91,Surv.os,covar0,os1)
head(output) # print the output
##              met_name HR   CI95 LCI UCL P_Value sig fdr.p
## mtlog1    2'-Deoxyuridine 0.88 0.74-1.06 0.74 1.06  0.191    0.846
## mtlog2 3'-O-methylguanosine 0.81 0.63-1.03 0.63 1.03   0.08  . 0.660
## mtlog3    Acetyl-lysine 0.85 0.75-0.97 0.75 0.97  0.016  * 0.528
## mtlog4         Adenine 0.99 0.70-1.40 0.70 1.40  0.955    0.980
## mtlog5       Adenosine 1.17 0.98-1.39 0.98 1.39  0.087  . 0.660
## mtlog6         ADMA_pos 0.91 0.72-1.13 0.72 1.13  0.386    0.956
#Export the summary table as a csv file
write.csv(output,"Your filepath/
os.individual.all.csv", row.names = T)
#RFS: Similarly, we conduct single metabolite analysis for 5-yr RFS
output<-coxphtable(91,Surv.rfs,covar0,rec1)
head(output)
##              met_name HR   CI95 LCI UCL P_Value sig fdr.p
## mtlog1    2'-Deoxyuridine 0.83 0.69-1.00 0.69 1.00  0.056  . 0.330
## mtlog2 3'-O-methylguanosine 0.88 0.70-1.12 0.70 1.12   0.3    0.593
## mtlog3    Acetyl-lysine 0.84 0.74-0.96 0.74 0.96  0.012  * 0.212
## mtlog4         Adenine 1.17 0.80-1.70 0.80 1.70  0.41    0.728
## mtlog5       Adenosine 0.98 0.81-1.19 0.81 1.19  0.854    0.952
## mtlog6         ADMA_pos 0.94 0.75-1.17 0.75 1.17  0.562    0.825
#Female OS
output.os.f<-coxphtable(91,Surv.os,covar1,os1.f)
head(output.os.f) #Show the first 6 rows of outputs
```

```
##             met_name  HR    CI95 LCI UCL P_Value sig fdr.p
## mtlog1    2'-Deoxyuridine 0.74 0.50-1.09 0.50 1.09  0.132    0.809
## mtlog2 3'-O-methylguanosine 0.8 0.52-1.23 0.52 1.23  0.309    0.809
## mtlog3     Acetyl-lysine 0.97 0.75-1.25 0.75 1.25  0.809    0.979
## mtlog4         Adenine 0.99 0.56-1.75 0.56 1.75  0.966    0.988
## mtlog5        Adenosine 0.9 0.69-1.17 0.69 1.17  0.421    0.910
## mtlog6        ADMA_pos 1.19 0.82-1.71 0.82 1.71  0.358    0.880
#Export the summary table as a csv file
write.csv(output.os.f,"Your filepath/
os.individual.female.csv", row.names = T)


rn.os.f<-rownames(subset(output.os.f,P_Value<0.05))
name.os.f<-subset(output.os.f,P_Value<0.05)$met_name


#Male OS
output.os.m<-coxphtable(91,Surv.os,covar1,os1.m)
head(output.os.m) #Print the first 6 rows of outputs
##             met_name  HR    CI95 LCI UCL P_Value sig fdr.p
## mtlog1    2'-Deoxyuridine 0.92 0.74-1.15 0.74 1.15  0.455    0.752
## mtlog2 3'-O-methylguanosine 0.8 0.59-1.08 0.59 1.08  0.138    0.516
## mtlog3     Acetyl-lysine 0.82 0.71-0.96 0.71 0.96  0.012  * 0.182
## mtlog4         Adenine  1 0.65-1.55 0.65 1.55  0.995    0.995
## mtlog5        Adenosine 1.29 1.03-1.62 1.03 1.62  0.026  * 0.223
## mtlog6        ADMA_pos 0.78 0.59-1.03 0.59 1.03  0.076  . 0.396
#Export the summary table as a csv file
write.csv(output.os.m,"Your filepath/
os.individual.male.csv", row.names = T)
rn.os.m<-rownames(subset(output.os.m,P_Value<0.05))
name.os.m<-subset(output.os.m,P_Value<0.05)$met_name
```

6. Interaction testing for metabolites that are significantly associated with either female or male OS, which can be acquired using $union()$ function.

```
name.os.sex<-union(name.os.f,name.os.m) #metabolite name
rn.os.sex<-union(rn.os.f,rn.os.m) #row number


cv<-paste(covar0,collapse = '+')


FML1 <-function(surv,varb) {
    as.formula(paste0(surv,paste(varb,"+","sex*",varb,   "+",cv)))
}
df.int.os=data.frame()


for (i in rn.os.sex) {
    out<-as.data.frame(getall(FML1(Surv.os,i),os1,7))
    df.int.os<-rbind(df.int.os,out)
}
```

```
out.df.int.os=data.frame()
for (i in 1:length(rn.os.sex)){
   out.df.int.os<-rbind(df.int.os[7*i,c(5,6)],        out.df.int.os)
}
rownames(out.df.int.os)<-rev(rn.os.sex)
colnames(out.df.int.os)[1]<-"P_interaction"
```

Print the P of sex interaction for each metabolite that are either
significantly associated with female or male OS. Metabolites marked with "*"
have a significant sex interaction (P<0.05); and those marked with "." have a
trend toward significance (0.05< P< 0.1).

```
out.df.int.os
##      P_interaction sig
## mtlog91      0.028  *
## mtlog87      0.024  *
## mtlog82      0.035  *
## mtlog73      0.035  *
## mtlog63      0.31
## mtlog42      0.441
## mtlog17      0.002  *
## mtlog14      0.897
## mtlog12      0.026  *
## mtlog11      0.151
## mtlog8      0.092 .
## mtlog5      0.038  *
## mtlog3      0.298
## mtlog84      0.01  *
## mtlog80      0.004  *
## mtlog75      0.073 .
## mtlog50      0.009  *
## mtlog39      0.018  *
sex_met_os<-rownames(subset(out.df.int.os,sig=="*"))
```

Number of significant metabolites with significant interaciton P

```
nrow(subset(out.df.int.os,sig=="*"))
## [1] 11
```

7. Forest plot that summarized the HR (95% CI) for each sex-specific metabolite by sex (Fig. 11.1, 11.2, 11.3).

```
library(metaviz)
library(grid)
library(ggplotify)

vec.os.sex<-rep(NA,length(rn.os.f))
vec.os.sex<-sort(as.numeric(substr(sex_met_os,6,nchar(sex_met_os))))

name_sex_met.os<-met_name[vec.os.sex]
met_name.os.sex<- c("Metabolite","",name_sex_met.os)
```

Female (OS)

Metabolite	HR (95% CI)
Adenosine	0.90 (0.69-1.17)
Asparagine	1.51 (0.89-2.57)
Citrulline	1.66 (0.97-2.83)
Glycerol 3-phosphate	3.64 (1.26-10.49)
LysoPC(16:0)	1.52 (1.03-2.25)
Serine	1.25 (0.65-2.41)
Succinate	0.35 (0.12-0.99)
Threonine	1.12 (0.67-1.87)
UDP-D-Glucose	0.81 (0.67-0.97)
Uracil	1.24 (0.56-2.76)
Xanthosine	1.22 (0.86-1.73)

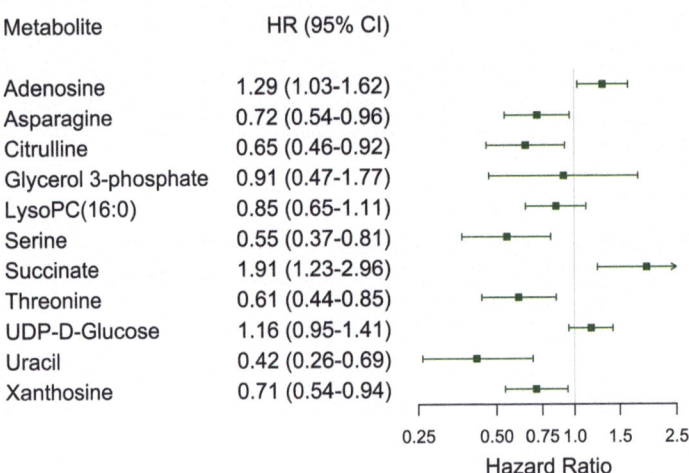

Fig. 11.1 The associations between sex-specific metabolites and female 5-year OS

Male (OS)

Metabolite	HR (95% CI)
Adenosine	1.29 (1.03-1.62)
Asparagine	0.72 (0.54-0.96)
Citrulline	0.65 (0.46-0.92)
Glycerol 3-phosphate	0.91 (0.47-1.77)
LysoPC(16:0)	0.85 (0.65-1.11)
Serine	0.55 (0.37-0.81)
Succinate	1.91 (1.23-2.96)
Threonine	0.61 (0.44-0.85)
UDP-D-Glucose	1.16 (0.95-1.41)
Uracil	0.42 (0.26-0.69)
Xanthosine	0.71 (0.54-0.94)

Fig. 11.2 The associations between sex-specific metabolites and male 5-year OS

```
tabletext.os.f<-cbind(
 met_name.os.sex,
 c("HR (95% CI)","",paste0( format(as.numeric(as.character(output.os.f$HR
[vec.os.sex]))),nsmall=2),' (',as.character(output.os.f$CI95[vec.os.
sex]),')')))
```

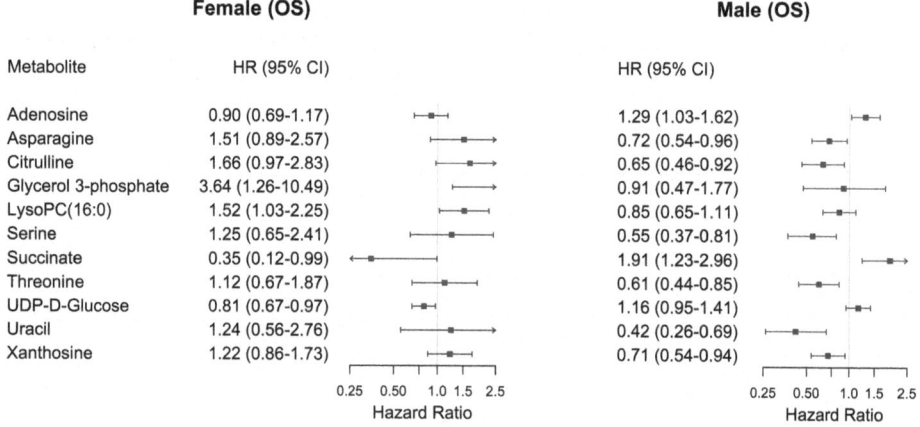

	Female (OS)		**Male (OS)**	
Metabolite	HR (95% CI)		HR (95% CI)	
Adenosine	0.90 (0.69-1.17)		1.29 (1.03-1.62)	
Asparagine	1.51 (0.89-2.57)		0.72 (0.54-0.96)	
Citrulline	1.66 (0.97-2.83)		0.65 (0.46-0.92)	
Glycerol 3-phosphate	3.64 (1.26-10.49)		0.91 (0.47-1.77)	
LysoPC(16:0)	1.52 (1.03-2.25)		0.85 (0.65-1.11)	
Serine	1.25 (0.65-2.41)		0.55 (0.37-0.81)	
Succinate	0.35 (0.12-0.99)		1.91 (1.23-2.96)	
Threonine	1.12 (0.67-1.87)		0.61 (0.44-0.85)	
UDP-D-Glucose	0.81 (0.67-0.97)		1.16 (0.95-1.41)	
Uracil	1.24 (0.56-2.76)		0.42 (0.26-0.69)	
Xanthosine	1.22 (0.86-1.73)		0.71 (0.54-0.94)	

Fig. 11.3 The associations between sex-specific metabolites and 5-year OS by sex

```
tabletext.os.m<-cbind(
 met_name.os.sex,
 c("HR (95% CI)","",paste0( format(as.numeric(as.character(output.os.m$HR
[vec.os.sex])),nsmall=2),' (',as.character(output.os.m$CI95[vec.os.
sex]),')')))


tabletext.os.m0<-cbind(
 c("HR (95% CI)","",paste0( format(as.numeric(as.character(output.os.m$HR
[vec.os.sex])),nsmall=2),' (',as.character(output.os.m$CI95[vec.os.
sex]),')')))
cochrane_from_rmeta.os.f <-
 structure(list(
  mean =c(NA,NA, as.numeric(as.character(output.os.f$HR[vec.os.sex]))),
  lower = c(NA,NA, as.numeric(as.character(output.os.f$LCI[vec.os.
sex]))),
  upper = c(NA,NA, as.numeric(as.character(output.os.f$UCL[vec.os.
sex])))),
  .Names = c("mean", "lower", "upper"),
  row.names = c(NA,length(vec.os.sex)+2),
  class = "data.frame")


cochrane_from_rmeta.os.m<-
 structure(list(
  mean =c(NA,NA, as.numeric(as.character(output.os.m$HR[vec.os.sex]))),
  lower = c(NA,NA, as.numeric(as.character(output.os.m$LCI[vec.os.
sex]))),
  upper = c(NA,NA, as.numeric(as.character(output.os.m$UCL[vec.os.
sex])))),
```

```
  .Names = c("mean", "lower", "upper"),
  row.names = c(NA, length(vec.os.sex)+2),
  class = "data.frame")

f1<-grid.grabExpr(print(forestplot(tabletext.os.f,
     cochrane_from_rmeta.os.f,
     is.summary=F,
     clip=c(0.25,2),
     xlog=TRUE,
     boxsize = 0.2,
     ci.vertices=T,
     col=fpColors(box="royalblue",line="darkblue"),
     xlab="Hazard Ratio",
     graphwidth=unit(0.3, "npc"),
     colgap=unit(0.03, "npc"),
     title="Female (OS)",txt_gp = fpTxtGp(ticks=gpar(cex=0.8),xlab=gpar
(cex=1)),
     xticks=c(0.25, 0.5, 0.75, 1.0, 1.5,2.5 ),lineheight=unit(0.7,'cm'))))
# Save the figure in a svg file
svg(filename="OS_forest_female_chapter.svg",
  width=7,
  height=5,
  pointsize=12)
forestplot(tabletext.os.f,
     cochrane_from_rmeta.os.f,
     is.summary=F,
     clip=c(0.25,2),
     xlog=TRUE,
     boxsize = 0.2,
     ci.vertices=T,
     col=fpColors(box="royalblue",line="darkblue", ),
     xlab="Hazard Ratio",
     graphwidth=unit(0.3, "npc"),
     colgap=unit(0.03, "npc"),
      title="Female (OS)",
     txt_gp = fpTxtGp(ticks=gpar(cex=0.8),xlab=gpar(cex=1)),
     xticks = c(0.25, 0.5, 0.75, 1.0, 1.5,2.5),
     lineheight=unit(0.7,'cm'))
dev.off()
# Save the figure in a svg file
svg(filename="OS_forest_male_chapter.svg",
  width=7,
  height=5,
  pointsize=12)
forestplot(tabletext.os.m,
```

```
        cochrane_from_rmeta.os.m,
        is.summary=F,
        clip=c(0.25,2),
        xlog=TRUE,
        boxsize = 0.2,
        ci.vertices=T,
        col=fpColors(box="darkgreen",line="darkgreen"),
        xlab="Hazard Ratio",
        graphwidth=unit(0.3, "npc"),
        colgap=unit(0.03, "npc"),
        title="Male (OS)",
        txt_gp = fpTxtGp(ticks=gpar(cex=0.8),xlab=gpar(cex=1)),
        xticks=c(0.25, 0.5, 0.75, 1.0, 1.5,2.5 ),
        lineheight=unit(0.7,'cm'))
dev.off()
f2<-grid.grabExpr(print(forestplot(tabletext.os.m0,
        cochrane_from_rmeta.os.m,
        is.summary=F,
        clip=c(0.25,2),
        xlog=TRUE,
        boxsize = 0.2,
        ci.vertices=T,
        col=fpColors(box="darkgreen",line="darkgreen"),
        xlab="Hazard Ratio",
        graphwidth=unit(0.3, "npc"),
        colgap=unit(0.03, "npc"),
        title="Male (OS)",
        txt_gp = fpTxtGp(ticks=gpar(cex=0.8),xlab=gpar(cex=1)),
        xticks=c(0.25, 0.5, 0.75, 1.0, 1.5,2.5 ),
        lineheight=unit(0.7,'cm'))))

svg(filename="OS_forest_merge_chapter.svg",
  width=10,
  height=5,
  pointsize=12)
gridExtra::grid.arrange(f1,f2,widths = c(2,2),ncol=2)
dev.off()
```

8. Multivariable Cox PH analysis for metabolic pathways. We created functions that generate a summary table for a certain metabolic pathways. It requires inputs of: survival object (*surv*), a list of metabolites (*mt_list*), and a list of covariates (*cov_list*).

```
fml.pw<-function(surv,mt_list,cov_list){
as.formula(paste0(surv,paste(mt_list,"+",cov_list)))
}
```

```
coxph.table<-function(fml.pw,data){
  sum<-summary(coxph(fml.pw,data=data))
  HR <- round(sum$coefficients[,2],2)
  P_Value <- round(sum$coefficients[,5],3)
  LCI <- format(round(sum$conf.int[,3],2),2)
  UCI <- format(round(sum$conf.int[,4],2),2)
  CI95 <- paste0(LCI,'-',UCI)
  result<-cbind(HR=HR,CI95=CI95,P_Value=P_Value)
}


FML.pathway.int<-function(surv,mt_list,cov_list,varb_i){
  as.formula(paste0(surv,mt_list,"+",cov_list,"+","sex*",
varb_i))
}
```

9. An example of analysis of a metabolic pathway: Asparagine synthesis pathway catalyzed by the asparagine synthetase (ASNS) enzyme. The results can be found in "*2.3. Sex-Specific Differences in CRC Prognosis Associated with Asparagine Synthesis*" in the example publication [10].

```
mt_asns<-c("mtlog12","mtlog13","mtlog35","mtlog36","mtlog9") # They
represent Asparagine, Aspartate, Glutamate, Glutamine, AMP, respectively.
mt_list<-paste(mt_asns,collapse = '+')


cov_list<-paste(covar1,collapse = '+')
cov_list.all<-paste(covar0,collapse = '+')


cox.table.os.asns.all.csv<-as.data.frame(coxph.table(fml.pw(Surv.os,
mt_list,
cov_list.all),os1))


cox.table.os.asns.f<-as.data.frame(coxph.table(fml.pw(Surv.os,mt_list,
cov_list),
os1.f))
#Summary table of the multivariable Cox PH regression model for asparagine
synthesis in females
cox.table.os.asns.f
##            HR     CI95 P_Value
## mtlog12   6.39 1.78-22.91  0.004
## mtlog13   1.47 0.64-3.40  0.365
## mtlog35   0.24 0.06-0.95  0.042
## mtlog36   0.16 0.03-1.06  0.057
## mtlog9    0.65 0.31-1.37  0.252
## sideRCC   1.63 0.47-5.65  0.438
```

```
## stagelate   14.9 2.19-101.43  0.006
## Chemo1      1.78  0.39-8.23  0.458
## age         1.09  1.01-1.19  0.037
cox.table.os.asns.m<-as.data.frame(coxph.table(fml.pw(Surv.os,mt_list,
cov_list),
os1.m))
#Summary table of the multivariable Cox PH regression model for asparagine
synthesis in males
cox.table.os.asns.m
##           HR    CI95 P_Value
## mtlog12    0.57  0.36-0.91   0.018
## mtlog13    1.32  0.75-2.32   0.339
## mtlog35    1.28  0.48-3.38   0.618
## mtlog36    1.18  0.39-3.54   0.772
## mtlog9     0.89 0.65-1.22   0.476
## sideRCC    0.8   0.31-2.10   0.653
## stagelate   6.02   1.39-26.06  0.016
## Chemo1     0.74  0.18-2.99   0.676
## age        1.15  1.07-1.23   0
num.pred<-length(mt_asns)+length(covar0)+1
df.int.os.asns=data.frame()
for (i in 1:length(mt_asns)){
 out<-as.data.frame(getall(FML.pathway.int(Surv.os,mt_list,
cov_list,mt_asns[i]),os1,num.pred))
 df.int.os.asns<-rbind(df.int.os.asns,out)
}
out.df.int.os.asns=data.frame()
for (i in 1:length(mt_asns)){
 out.df.int.os.asns<-rbind(df.int.os.asns[num.pred*i,c(5,6)],out.df.
int.os.asns)
}
rownames(out.df.int.os.asns)<-rev(mt_asns)
colnames(out.df.int.os.asns)[1]<-"P_interaction"
out.df.int.os.asns# Print the P of sex interaction for metabolites of AMP,
Glutamine, Glutamate, Aspartate, Asparagine.
##       P_interaction sig
## mtlog9     0.86
## mtlog36    0.974
## mtlog35    0.807
## mtlog13    0.639
## mtlog12    0.021 *
#Store the outputs in a list
result.asns<-list(OS.f=cox.table.os.asns.f,OS.m=cox.table.os.asns.m,
OS.interaction = out.df.int.os.asns)
```

10. Running codes for model that dichotomized asparagine levels (please refer to Table 11.2, 11.3 and Fig. 11.4, 11.5 and results for Model 2 presented in the example publication [10]).

```
#Dichotomize asparagine abundance divided by a threshold of 75% percentile:
'1' refers to low level, and '2' represents high level.
os1 <- os1 %>% mutate(asn_bi = case_when(mtlog12 <=quantile
(os1$mtlog12,0.75) ~ '1',mtlog12 > quantile(os1$mtlog12,0.75)  ~ '2'))


os1.f <- os1.f %>% mutate(asn_bi = case_when(mtlog12 <= quantile
(os1$mtlog12,0.75) ~ '1',mtlog12 > quantile(os1$mtlog12,0.75)  ~ '2'))


os1.m <- os1.m %>% mutate(asn_bi = case_when(mtlog12 <= quantile
(os1$mtlog12,0.75) ~ '1', mtlog12 > quantile(os1$mtlog12,0.75)  ~ '2'))
# Female
tabcoxph(coxph(Surv.os~asn_bi+mtlog13+mtlog35+mtlog36+mtlog9+side+stage
+Chemo+age, data=os1.f), columns = c("hr.ci", "p"), decimals = 2) #tabcoxph
() function present Cox PH model in a formatted table:
# Male
tabcoxph(coxph(Surv.os~asn_bi+mtlog13+mtlog35+mtlog36+mtlog9+side+stage
+Chemo+age, data=os1.m), columns = c("hr.ci", "p"), decimals = 2)
#P of sex interaction of asparagine (binary)
summary(coxph(Surv.os~asn_bi+mtlog13+mtlog35+mtlog36+mtlog9+side+stage
+Chemo+age+sex+asn_bi*sex, data=os1))$coefficients[11,5]
## [1] 0.05213214
```

Table 11.2 Outputs of multivariable analysis of asparagine synthesis catalyzed by asparagine synthetase (ASNS) in female 5-year OS

Variable	HR (95% CI)	P
asn_bi		
1 (ref)	–	–
	5.68 (1.06, 30.61)	0.04
mtlog13	1.21 (0.64, 2.29)	0.56
mtlog35	0.58 (0.20, 1.70)	0.32
mtlog36	0.46 (0.09, 2.28)	0.34
mtlog9	0.87 (0.45, 1.65)	0.66
Side		
LCC (ref)	–	–
RCC	1.29 (0.40, 4.14)	0.67
Stage		
Early (ref)	–	–
Late	13.44 (1.91, 94.79)	0.009
Chemo		
0 (ref)	–	–
1	0.82 (0.16, 4.13)	0.81
Age	1.09 (1.00, 1.20)	0.05

Table 11.3 Outputs of multivariable analysis of asparagine synthesis catalyzed by asparagine synthetase (ASNS) in male 5-year OS

Variable	HR (95% CI)	P
asn_bi		
1 (ref)	–	–
2	0.46 (0.11, 1.84)	0.27
mtlog13	1.29 (0.75, 2.22)	0.35
mtlog35	1.01 (0.40, 2.55)	0.99
mtlog36	0.78 (0.29, 2.10)	0.63
mtlog9	0.77 (0.58, 1.03)	0.08
side		
LCC (ref)	–	–
RCC	0.86 (0.33, 2.26)	0.76
Stage		
Early (ref)	–	–
Late	4.09 (0.99, 16.86)	0.05
Chemo		
0 (ref)	–	–
1	0.84 (0.21, 3.41)	0.80
Age	1.11 (1.05, 1.18)	<0.001

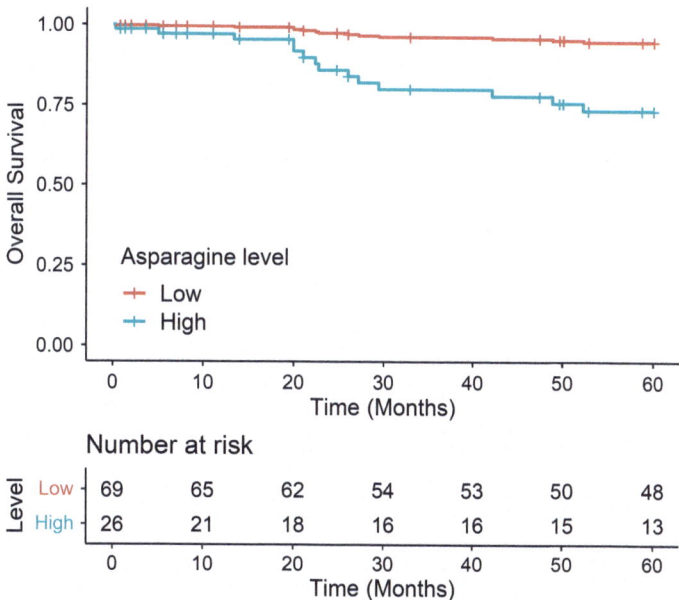

Fig. 11.4 Cox adjusted survival curves of ASNS-catalyzed asparagine for female 5-year OS

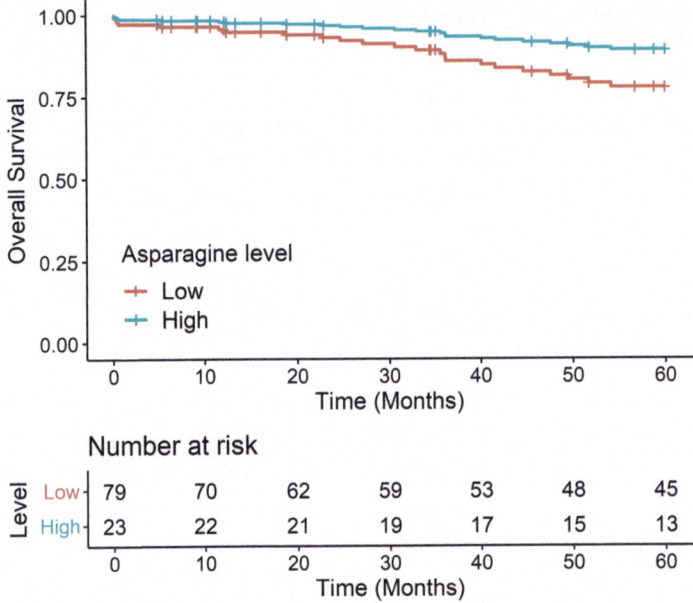

Fig. 11.5 Cox adjusted survival curves of ASNS-catalyzed asparagine for male 5-year OS

11. Adjusted Cox PH curves: Adjusted survival curves can be plotted for a Cox model adjusted for covariates using *ggsurvplot ()*.

```
# Construct new variables for side, stage, and Chemo in numerical form.
os1.f <- os1.f %>% mutate(side = case_when(side =='LCC' ~ 1, side == 'RCC' ~
2))
os1.f <- os1.f %>% mutate(stage = case_when(stage =='early' ~ 0, stage ==
'late' ~ 1))
os1.f <- os1.f %>% mutate(chemo = case_when(Chemo =='0' ~ 0, Chemo == '1' ~ 1))
# create a new data.frame with two rows, one for each value of the group with the
other covariate fixed to its average value.
new1.f.2<-data.frame(asn_bi=c('1','2'),mtlog13=rep(mean(os1.
f$mtlog13),2),mtlog35=rep(mean(os1.f$mtlog35),2),mtlog36=rep(mean(
os1.f$mtlog36),2),mtlog9=rep(mean(os1.f$mtlog9),2),side=rep(
mean(os1.f$side),2),stage=rep(mean(os1.f$stage),2),chemo=rep(mean(os1.
f$chemo),2),age=rep(mean(os1.f$age),2))


cox0.f2<-coxph(Surv(time = OS5.months, event=OS5.status)~asn_bi
+mtlog13+mtlog35+mtlog36+mtlog9+side+stage+chemo+age, data=os1.f)


pred1.f<-survfit(cox0.f2,newdata = new1.f.2,data=os1.f)
```

```
#Construct risk table from K-M curve, which is the same with the risk table of
adjusted Cox PH survival curve.
require("survminer") #Loading/Attaching the "survminer" package
km.fit <- survfit(Surv(time = OS5.months, event=OS5.status) ~ asn_bi, data =
os1.f)
km.surv <- ggsurvplot(km.fit, risk.table = TRUE,xlab="Time (Months)",
legend.title = "Level",
   legend.labs = c("Low", "High"),censor=F)


##Adjusted survival curves using coxph()
cox.surv <- ggsurvplot(pred1.f, data=os1.f,
              conf.int = FALSE,
              xlab="Time (Months)",ylab='Overall Survival',
              legend.labs = c("Low","High"),
              legend.title = "Asparagine level",
              legend=c(0.2,0.2),
              font.legend=14,
              risk.table.title="Group")
# Attach risk table below the curve
cox.surv$table <- km.surv$table
print(cox.surv, risk.table.height = 0.3)
### Male
os1.m <- os1.m %>% mutate(side = case_when(side =='LCC' ~ 1,side == 'RCC' ~
2))
os1.m <- os1.m %>% mutate(stage = case_when(stage =='early' ~ 0,stage ==
'late' ~ 1))
os1.m <- os1.m %>% mutate(chemo = case_when(Chemo =='0' ~ 0,Chemo == '1' ~ 1))


new1.m.2<-data.frame(asn_bi=c('1','2'),mtlog13=rep(mean(os1.
m$mtlog13),2),mtlog35=rep(mean(os1.m$mtlog35),2),mtlog36=rep(mean(
os1.m$mtlog36),2),mtlog9=rep(mean(os1.m$mtlog9),2),side=
rep(mean(os1.m$side),2),stage=rep(mean(os1.m$stage),2),chemo=rep(mean
(os1.m$chemo),2),age=rep(mean(os1.m$age),2))


cox0.m2<-coxph(Surv(time = OS5.months, event=OS5.status)~asn_bi
+mtlog13+mtlog35+mtlog36+mtlog9+side+stage+chemo+age, data=os1.m)


pred1.m<-survfit(cox0.m2,newdata = new1.m.2,data=os1.m)


km.fit <- survfit(Surv(time = OS5.months, event=OS5.status) ~ asn_bi, data =
os1.m)
km.surv <- ggsurvplot(km.fit, risk.table = TRUE,xlab="Time (Months)",
legend.title = "Level",
   legend.labs = c("Low", "High"),censor=F)
```

```
##Adjusted survival curve for males using coxph()
cox.surv <- ggsurvplot(pred1.m, data=os1.m,
            conf.int = FALSE,
            xlab="Time (Months)",ylab='Overall Survival',
            legend.labs = c("Low","High"),
            legend.title = "Asparagine level",
            legend=c(0.2,0.2),
            font.legend=14,
            risk.table.title="Group")
cox.surv$table <- km.surv$table
print(cox.surv, risk.table.height = 0.3)
```

A more detailed data interpretation has been presented in the Results and Discussion sections of the example publication [10].

Questions
1. A study aims to investigate sex-specific metabolites for 10-year RFS in breast cancer patients. A female patient had not experienced recurrence of breast cancer since the first date of follow-up and died in the 15th year. To code her clinical information for the study goal, what are her values of RFS (0 or 1) and time (in year)?
2. How can one view the Cox PH regression analysis output in R?
3. In a multivariable Cox PH regression model for analyzing 5-year OS of kidney cancer in females using serum LC-MS-based metabolomics data, the arginine metabolite was detected and included in the model, which is adjusted for age and BMI. The abundance of arginine has been log_2 transformed. The HR of arginine is 1.25 with a 95% CI of (1.04, 1.63). While the HR of arginine is 0.75 (95% CI: 0.70-0.91) in the corresponding model for male patients. The interaction term of sex with arginine was tested, and the P value is 0.03. How would you interpret these results?

11.4   Limitations

In this section, we discuss the limitations of Cox PH methods for detecting sex-specific prognostic metabolites from the perspective of statistical methodology.

11.4.1 Violation of PH Assumption

Log transformation can aid in normalizing metabolite abundance by forcing extremely large values into smaller ones, making the value more "linear" and enabling the comparison of different metabolites. However, the HR might not be constant over time for some metabolites. Violation of the PH assumption can limit the interpretation of model performance since the HR estimation is distorted. This is common in other -omics data such as transcriptomics but can be improved by introducing the time interaction term [20]. Some machine learning models can also overcome such a restriction, such as the Random survival forest (RSF) model [21] and least absolute shrinkage and selection operator (LASSO)-Cox regression model [22], which bring a broader scope of available analytical tools. We can "safely" explore the sex-specific metabolites of cancer prognosis by performing those models for each sex.

11.4.2 Choosing Metabolites for Analysis of Metabolic Pathways

In the example study, some metabolites were grouped into several important metabolic pathways for cancer progression, which were chosen because of their existence. As the study adopted untargeted metabolomics using limited standards for matching metabolites, there might be other metabolites that are involved in the mentioned pathways or other sex-specific metabolic pathways but were not detected or identified. Moreover, the inclusion of a set of metabolites in a pathway might be risky as they were biologically correlated to each other. Therefore, more advanced statistical methods for dimensional reduction are strongly encouraged, such as incorporating principal component analysis (PCA), LASSO-Cox, or RSF. A Cox PH model following PCA can identify a combination of metabolites associated with prognosis, however, PCA is not intrinsically clear in the identification of original metabolites or for ranking them [23]. For LASSO-Cox regression, problems can occur if a large proportion of metabolites have a strong correlation with each other [23]. RSF is effective in dealing with complex relationships among metabolites but cannot provide a measure of P value to quantify the significance of metabolites [23].

Answers
1. RFS: 0; Time: 10.
2. By using the `summary()` function on a `coxph` object.
3. Adjusted for age and BMI, one unit of increase of log2 (serum arginine abundance) increases all-cause death risk by 25% for female kidney cancer patients, while it decreases male patients' risk of all-cause death by 25%. The P of sex interaction is statistically significant, suggesting that sex significantly interacts with the association between serum arginine and 5-year OS of kidney cancer.

Take-Home Message

- Metabolite levels measured from human samples can have sex-dependent associations with prognosis, suggesting that cancer metabolism may be attributable to the sex-specific prognostic outcomes of cancer.
- The use of the Cox PH regression models stratified by sex is a useful tool to reveal sex-specific associations between metabolome and cancer prognosis after examining the sex interaction term.
- With knowledge of metabolic pathways with potential sex differences that link to cancer prognosis, a Cox PH model that includes critical metabolites involved in the pathway can give a quantitative evaluation of the sex interaction, offering insights into sex-specific and targeted cancer treatments. In conclusion, this chapter illustrates a workflow of discovering sex-specific prognostic biomarkers using metabolomics data for cancer studies based on Cox PH analysis using R software (version 4.0.4).

Acknowledgments This research was funded by NIH 1R21CA223686-01 (C.H.J., S.A.K.) and American Cancer Society research scholar grant 134273-RSG-20-065-01-TBE (C.H.J.). This publication was also made possible by CTSA grant number UL1 TR001863 from the National Center for Advancing Translational Science (NCATS), components of the National Institutes of Health (NIH), and NIH roadmap for Medical Research. Its contents are solely the responsibility of the authors and do not necessarily represent the official view of NIH. This work was also supported by the Lampman Research Fund in Yale Surgical Oncology.

References

1. Hoang G, Udupa S, Le A. Application of metabolomics technologies toward cancer prognosis and therapy. Int Rev Cell Mol Biol. 2019;347:191–223.
2. Miolo G, Di Gregorio E, Saorin A, Lombardi D, Scalone S, Buonadonna A, et al. Integration of serum metabolomics into clinical assessment to improve outcome prediction of metastatic soft tissue sarcoma patients treated with trabectedin. Cancers (Basel). 2020, 1983;12(7)
3. Chen Z, Dai Y, Huang X, Chen K, Gao Y, Li N, et al. Combined metabolomic analysis of plasma and tissue reveals a prognostic risk score system and metabolic dysregulation in esophageal squamous cell carcinoma. Front Oncol. 2020;10:1545.
4. Radkiewicz C, Johansson ALV, Dickman PW, Lambe M, Edgren G. Sex differences in cancer risk and survival: a Swedish cohort study. Eur J Cancer. 2017;84:130–40.
5. Jakszyn P, Cayssials V, Buckland G, Perez-Cornago A, Weiderpass E, Boeing H, et al. Inflammatory potential of the diet and risk of colorectal cancer in the European Prospective Investigation into Cancer and Nutrition study. Int J Cancer. 2020;147(4):1027–39.
6. Kim SY, Song HK, Lee SK, Kim SG, Woo HG, Yang J, et al. Sex-biased molecular signature for overall survival of liver cancer patients. Biomol Ther (Seoul). 2020;28(6):491–502.
7. Hases L, Ibrahim A, Chen X, Liu Y, Hartman J, Williams C. The importance of sex in the discovery of colorectal cancer prognostic biomarkers. Int J Mol Sci. 2021;22(3):1354.

8. Cai Y, Rattray NJW, Zhang Q, Mironova V, Santos-Neto A, Hsu KS, et al. Sex differences in colon cancer metabolism reveal a novel subphenotype. Sci Rep. 2020;10(1):4905.
9. Deja S, Litarski A, Mielko KA, Pudelko-Malik N, Wojtowicz W, Zabek A, et al. Gender-specific metabolomics approach to kidney cancer. Metabolites. 2021;11(11)
10. Shen X, Cai Y, Lu L, Huang H, Yan H, Paty PB, et al. Asparagine metabolism in tumors is linked to poor survival in females with colorectal cancer: a cohort study. Metabolites. 2022;12(2)
11. Collett D. Modelling survival data in medical research. 3rd ed. Boca Raton: CRC Press, Taylor & Francis Group; 2015. pxvi, 532 pages p
12. Allison P. Survival analysis using SAS. 2nd ed. SAS Publishing; 2010.
13. Rollison DE, Sabel MS. 3 - Basic epidemiologic methods for cancer investigations. In: Sabel MS, Sondak VK, Sussman JJ, editors. Essentials of Surgical Oncology. Philadelphia: Mosby; 2007. p. 21–38.
14. Wei R, Wang J, Su M, Jia E, Chen S, Chen T, et al. Missing value imputation approach for mass spectrometry-based metabolomics data. Sci Rep. 2018;8(1):663.
15. Faquih T, van Smeden M, Luo J, le Cessie S, Kastenmuller G, Krumsiek J, et al. A workflow for missing values imputation of untargeted metabolomics data. Metabolites. 2020;10(12)
16. Benjamini Y, Hochberg Y. Controlling the false discovery rate: a practical and powerful approach to multiple testing. J R Stat Soc Ser B (Methodological). 1995;57(1):289–300.
17. Kuan PF, Yang X, Kotov R, Clouston S, Bromet E, Luft BJ. Metabolomics analysis of post-traumatic stress disorder symptoms in World Trade Center responders. Transl Psychiatry. 2022;12(1):174.
18. Rickert J. Survival analysis with R 2017 [cited 2022 April 15]. Available from https://rviews.rstudio.com/2017/09/25/survival-analysis-with-r/.
19. Statistical tools for high-throughput data analysis. Cox Proportional-Hazards Model 2020 [cited 2022 April 15]. Available from http://www.sthda.com/english/wiki/cox-proportional-hazards-model.
20. Zeng Z, Gao Y, Li J, Zhang G, Sun S, Wu Q, et al. Violations of proportional hazard assumption in Cox regression model of transcriptomic data in TCGA pan-cancer cohorts. Comput Struct Biotechnol J. 2022;20:496–507.
21. Dietrich S, Floegel A, Troll M, Kuhn T, Rathmann W, Peters A, et al. Random Survival Forest in practice: a method for modelling complex metabolomics data in time to event analysis. Int J Epidemiol. 2016;45(5):1406–20.
22. Di Poto C, Ferrarini A, Zhao Y, Varghese RS, Tu C, Zuo Y, et al. Metabolomic characterization of hepatocellular carcinoma in patients with liver cirrhosis for biomarker discovery. Cancer Epidemiol Biomarkers Prev. 2017;26(5):675–83.
23. Antonelli J, Claggett BL, Henglin M, Kim A, Ovsak G, Kim N, et al. Statistical workflow for feature selection in human metabolomics data. Metabolites. 2019;9(7)

Further Reading

Johnson CH, Gonzalez FJ. Challenges and opportunities of Metabolomics. J Cell Physiol. 2012;227 (8):2975–81.
Zhang F, Zhang Y, Zhao W, Deng K, Wang Z, Yang C, et al. Metabolomics for biomarker discovery in the diagnosis, prognosis, survival and recurrence of colorectal cancer: a systematic review. Oncotarget. 2017;8(21):35460–72.

A Lipidome-Wide Association Study: Data Processing, Annotation, and Analysis Workflow Using MS-DIAL and R

12

Olivier Salamin, Justin Carrard, Tony Teav, Arno Schmidt-Trucksäss, Hector Gallart-Ayala, and Julijana Ivanisevic

> **What You Will Learn from This Chapter**
> - Investigate sex-related differences in lipid metabolism.
> - Understand the principles of mass spectrometry (MS)-based lipidomics.
> - Perform association analyses (i.e., multivariable linear/logistic regression analysis) to determine which lipid species are associated with the male/female sex while considering the effect of age.

Authors O. Salamin and J. Carrard have contributed equally to this chapter.

O. Salamin (✉) · T. Teav · H. Gallart-Ayala · J. Ivanisevic (✉)
Metabolomics Unit, Faculty of Biology and Medicine, University of Lausanne, Lausanne, Switzerlandolivier.salamin.1@gmail.com
olivier.salamin.1@gmail.com
Julijana.ivanisevic@unil.ch

J. Carrard (✉) · A. Schmidt-Trucksäss
Division of Sports and Exercise Medicine, Department of Sport, Exercise and Health, University of Basel, Basel, Switzerlandjustin.carrard@unibas.ch

12.1 Theoretical Background

This section provides the theoretical background on metabolic phenotyping and its application to human research. We highlight major determinants of the serum lipid composition and describe the principles of mass spectrometry (MS)-based lipidomics as a tool to uncover sex-related differences. Using healthy young (in their twenties) and aged (in their seventies) participants from the Complete Health cohort, we illustrate the workflow on how MS-based lipid signatures can be translated into biologically and physiologically relevant information.

12.1.1 Comprehensive Lipid Signature as a Phenotyping Tool

Detailed multiparametric lipid profiles (or signatures at the molecular species level) constitute a powerful phenotyping tool because they capture the metabolic individuality, including the response to challenges imposed by our lifestyle [1]. Lipid phenotyping at the molecular species level can be combined with standardized clinical measurements such as anthropometric parameters (e.g., body mass index or waist circumference), cardiorespiratory fitness (CRF), and routine blood biochemistry, to identify sets of molecular species associated with cardiometabolic risk factors [2]. The ultimate goal is to use these specific signatures to improve the sensitivity and accuracy of disease risk prediction and optimize disease prevention by influencing patient lifestyle [3]. Dysregulation of lipid metabolism is specifically and tightly associated with the onset and development of cardiometabolic diseases (CMDs). Currently, cholesterol and triglycerides are the main markers for assessing cardiometabolic risk in the clinical environment. However, these biomarkers, combined with clinical parameters such as blood pressure, are suboptimal for early prediction and accurate diagnostics of adverse events [4]. Thanks to technological advancements, we can now measure hundreds of lipid species from a minimal amount (~5 µL) of human blood specimen, with more specificity and sensitivity than ever before. Beyond prognostic and diagnostic value, a prospective lipid profiling of human populations will significantly enhance our understanding of lipid metabolism at the population level. Using association analysis and testing causal effects (in combination with genomic data), we can identify the metabolic pathways and processes affected by age-related changes, including hormonal changes that precede disease onset. Because the onset, progression, and manifestation of CMDs are influenced by aging and sex differences, a better understanding of sex-specific and age-related differences in healthy individuals is necessary before associating the alterations in lipid signatures with a specific disease [5, 6]. For the present study, we have selected a subset of clinically healthy participants in their twenties and seventies to investigate the sex-related differences in serum lipid composition age-dependently. Both age groups went through an extensive clinical characterization.

12.1.2 MS-Based Lipidomics

Untargeted lipidomics approaches tend to maximize the lipid characterization (including structural elucidation) by combining different chromatographic conditions such as C18 and C30 reversed-phase liquid chromatography (RPLC) and hydrophilic interaction liquid chromatography (HILIC) coupled to high-resolution mass spectrometry (HRMS) instruments (Q-TOF and Q-Orbitrap) in positive and negative ionization modes [7]. Data acquisition is performed in full scan mode, screening the entire mass range from 50 to 1700 Da. For identification purposes, pooled quality control (QC) samples, representative of the entire sample batch, are analyzed in tandem mass spectrometry (MS/MS) mode using one of the two data acquisition possibilities: data-dependent acquisition (DDA) or data-independent acquisition (DIA). Conventional DDA operates with a pre-defined intensity threshold that determines the selection of precursor ions (acquired in full scan mode at MS1 level) to be isolated for fragmentation. The fragmentation occurs in series where the most abundant ions are prioritized (top-n approach), limiting the fragmentation of low abundant ions and, therefore, the number of identified metabolites/lipids. To overcome this limitation, the iterative DDA with computer-driven exclusion was introduced to maximize the number of fragmented precursor ions. A pooled sample is analyzed in multiple consecutive injections where the precursor ions previously selected for fragmentation are excluded on a rolling basis. In DIA mode, all ions are fragmented simultaneously, including the very low abundant species. However, the link between the precursor and its product ions is lost and must be reconstructed using deconvolution algorithms. This is a challenging process due to a high number of co-eluting ions.

Lipid annotation can be achieved with varying levels of specificity (Fig. 12.1), from class and sum composition level down to fatty acyl/alkyl composition, position, and structure (i.e., double bond position and stereoisomer). The confidence and specificity of lipid annotation depend on the quality and quantity of acquired chemical information, including accurate m/z ratio, retention time (in applied analytical conditions), MS/MS spectra, ion mobility value, and additional unique fragmentation spectra (using ultraviolet photodissociation) to determine regioisomers and double bond(s) position. In untargeted analysis, a putative identification at the class sum composition and/or fatty acyl/alkyl level is usually achieved by matching experimentally acquired MS/MS data against MS/MS spectra predicted in silico or recorded on pure standards and stored in available spectral libraries. Today, over 20 spectral databases, such as METLIN [9] (an open-access database with a collection of over 942,000 spectra), HMDB [10], LipidBlast [11], mzCloud (https://www.mzcloud.org/), NIST (https://www.nist.gov/programs-projects/tandem-mass-spectral-library) and MassBank [12], are available to facilitate the identification of known and unknown metabolites and lipids. Several data processing software contain built-in spectral libraries, such as MS-DIAL [13, 14] (RIKEN) and Compound Discoverer (Thermo Scientific). MS-DIAL utilizes open-access libraries such as MassBank to provide putative hits based on accurate mass and MS/MS match. At the same time, lipidomic data can be annotated using LipidBlast, an in silico-generated spectral library [11]. Unlike other more

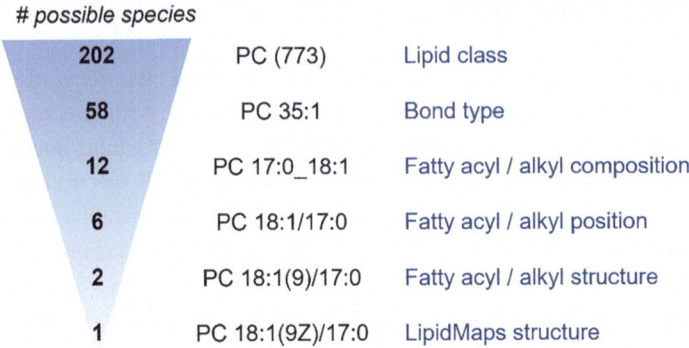

Fig. 12.1 Different levels of lipid annotation. This figure was modified from Züllig and Köfeler 2020 [8]

polar metabolites, lipids show consistent mass spectral fragmentations which can be efficiently predicted. The largest spectral library for lipids, LipidBlast, was created on a rule-based generation of characteristic fragmentations: the loss of the polar head groups, the acyl or alkyl chain losses from precursor ions (M—sn1 and M—sn2), and product ions of the fatty acid (FA) fragments (sn1 and sn2; best observed in negative ionization as [FA—H]$^-$) [11]. In the present study, we used MS-DIAL for lipid data preprocessing, signal intensity drift correction, and lipid annotation using LipidBlast.

12.1.3 Statistical Data Analysis: Multivariable Linear Regression Analysis

Multiple linear regression modeling was performed to assess associations between lipids, age, sex, and the interaction between age and sex while adjusting for the following potential clinical confounders: body fat (%), statins intake, glycated hemoglobin (HbA1c, %), daily total physical activity (min), blood sampling time, and fasting time before blood sampling [15–23]. In the present study, lipids were used as dependent variables, while the other parameters were independent explanatory variables. Lipid abundances were log_2-transformed before statistical analysis. Continuous dependent and independent variables were z-standardized before analysis [24]. Post-hoc tests, using the emmeans R-package (version 1.4.8), were calculated to determine and compare the estimated marginal means of each lipid in aged and young participants within both sexes [25]. The same was done for females and males in both age groups. The resulting β coefficients were converted to a percentage of difference for ease of interpretation [25, 26]:

$$\text{Percentage of difference} = \left(2^\beta - 1\right) \times 100$$

Finally, lipid species were ordered by descending percentage values of differences for all four age and sex groups. The normality of the residuals was checked graphically before running the above-mentioned statistical tests. For each statistical test, p-values were adjusted using the Benjamini-Hochberg method [27]. Adjusted p-values ≤ 0.05 were considered significant. Unless otherwise specified, statistical analyses were carried out using R (version 4.0.2) [28]. Rain plots were computed using a previously published R-code [29].

12.2 Research Question and Experimental Setup

12.2.1 Research Question and Hypothesis

The main objective of this healthy population study was to investigate the sex-related differences in serum lipid composition with age. For this, we have selected a subset of clinically healthy participants in their twenties and seventies from the Complete Health cohort established at the University of Basel (https://www.complete-project.ch/english/contact/) [30]. The participants underwent extensive clinical characterization (as described below), and their fasted blood sera were used for lipid profiling (Fig. 12.2). Our hypothesis was that sex and age constitute the primary determinants of serum lipid composition.

12.2.2 Experimental Design and Data Collection

The investigated subset consisted of 73 young (25.2 ± 2.6 years, 43% female) and 77 aged (73.5 ± 2.3 years, 48% female) individuals (Fig. 12.2). As reported in the study protocol, only healthy participants from the Basel area (Switzerland), who did not have exercise-limiting chronic diseases, were non-smokers, or quit at least ten years ago, were included in the COmPLETE-Health study [31]. This excluded participants with a history of coronary artery disease, stroke, heart failure, lower-extremity artery disease, any malignant tumor, diabetes, obesity, clinically apparent kidney failure, severe liver disease, chronic obstructive pulmonary disease GOLD stages two to four, arterial hypertension grades two and three, drug or alcohol abuse, exercise-limiting musculoskeletal conditions, and clinically manifest Alzheimer's disease or dementia. The study was conducted by the Declaration of Helsinki, approved by the Ethics Committee of North-Western and Central Switzerland (EKNZ 2017–01451), and registered on ClinicalTrials.gov (NCT03986892). All participants provided written informed consent.

12.2.2.1 Clinical Characterization

Data was collected between January 2018 and June 2019. Before the clinical examination, participants were instructed not to diverge from habitual eating behavior (for the previous 72 h) and to avoid exercising, drinking alcohol (for the previous 24 h), and drinking caffeinated beverages (for the previous 4 h). Participants were allocated into one of the five possible time slots (08:00, 10:00, 12:00, 14:00, and 16:00) according to their availability,

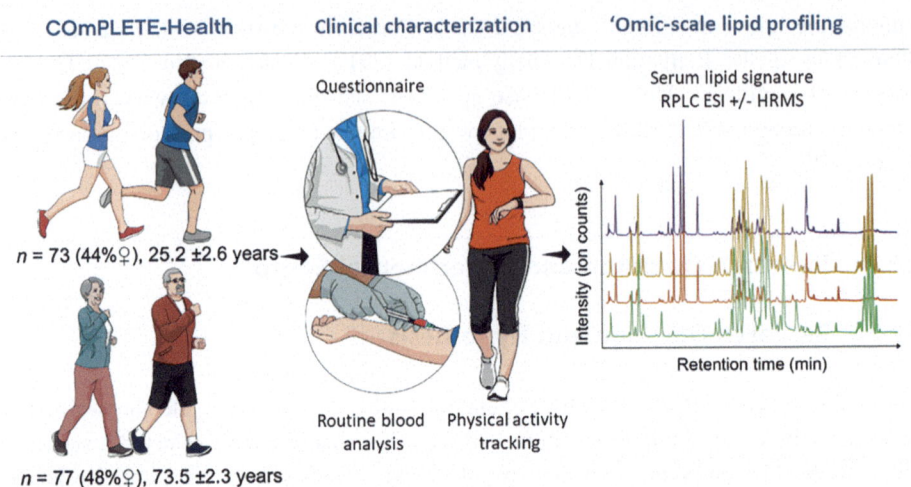

Fig. 12.2 Experimental design and collected data. A subset of the Complete Health cohort of young and aged individuals underwent clinical characterization and metabolic phenotyping. Their blood sera were used for lipid profiling. This figure was adapted from Carrard et al. [5]

and each measurement took approximately 4 hours. After an hour of measurements at rest, trained medical staff collected blood samples in a fasted state (at least 3 hours) by venipuncture of the cubital fossa (2 × 7.5 mL serum-gel, Monovette®, Sarstedt, Nümbrecht, Germany). Serum samples were gently shaken for 30 min, centrifuged (3000 rpm, 10 min; 20–23 °C), aliquoted, and frozen at −80 °C.

Smoking status was assessed by telephone interview before the examination, while physicians reviewed medical history and medications by questionnaires on site. Body fat content was quantified using a four-segment bioelectrical impedance analysis (Inbody 720, Inbody Co. Ltd., Seoul, South Korea). Physical activity was objectively monitored over the 14 consecutive days following the clinical examination using a wrist-worn triaxial accelerometer (GeneActive Activinsights Ltd., Kimbolton, UK). Data were analyzed using the validated open-source Excel macro file "General physical activity" (version 2), quantifying total and moderate-to-vigorous physical activity in minutes per day (moderate defined as 4.00–6.99 Metabolic Equivalent of Task (METS) and vigorous ≥ 7 METS) [32]. The recruitment and data collection processes have been previously described in detail in the study protocol [31].

12.2.2.2 Biochemical Analysis

Total cholesterol, LDL-C, HDL-C, and triglyceride concentrations were analyzed from serum using an Olympus AU680 automatic analyzer (Beckman Coulter, Brea, CA, USA), enzymatic reagents (DiaSys, Holzheim, Germany), and secondary standards (Roche Diagnostics, Mannheim, Germany). HbA1c was quantified from whole blood by high-pressure liquid chromatography using D-10 (Bio-Rad, Hercules, CA, USA).

12.2.2.3 Sample Preparation

As previously reported by Carrard et al. [5], lipids were extracted by adding 200 μL of n-Butanol/Methanol (1:1) solution to 40 μL of serum. Following centrifugation for 15 min at 4000 g at 4 °C, the resulting supernatants were collected and transferred to liquid chromatography–mass spectrometry vials for injection.

12.2.2.4 Liquid Chromatography–High-resolution Mass Spectrometry (LC-HRMS)

Serum lipid extracts were analyzed by reversed-phase liquid chromatography coupled to high-resolution mass spectrometry (RPLC-HRMS) instrument (Agilent 6550 iFunnel Q-TOF LC/MS, Agilent Technologies, Santa Clara, CA, USA). In both positive and negative ionization modes, the chromatographic separation was carried out on a Zorbax Eclipse Plus C18 (1.8 μm, 100 mm × 2.1 mm I.D. column) (Agilent Technologies, Santa Clara, CA, USA). The mobile phases were composed of A = 60:40 (v/v) Acetonitrile:water with 10 mM ammonium acetate and 0.1% acetic acid and B = 88:10:2 Isopropanol:acetonitrile:water with 10 mM ammonium acetate and 0.1% acetic acid. The flow rate was 600 μL/min, the column temperature was set at 60 °C, and the sample injection volume was 2 μl. Electrospray ionization source conditions were set as follows: dry gas temperature 200 °C, nebulizer 35 psi and flow 14 L/min, sheath gas temperature 300 °C and flow 11 L/min, nozzle voltage 1000 V, and capillary voltage +/− 3500 V. Full scan acquisition in the range of 100–1700 mass-to-charge ratio (m/z) was applied while iterative MS/MS data-dependent acquisition at 25 eV was used to acquire the MS/MS data on pooled quality control (QC) samples. Iterative MS/MS was performed in five consecutive injections using computer-driven exclusion. The scan rate was set at 3 spectra/s with a duration of 333.3 ms/spectrum and a narrow isolation width of 1.3 m/z. The mass error tolerance was +/− 20 ppm with a retention time exclusion tolerance of +/− 0.1 min. Precursor ions were excluded after two spectra, with a maximum of three precursors per cycle. The precursor threshold was set to an absolute threshold of 5000 counts in positive mode and 2500 counts in negative mode. Pooled QC samples (representative of the entire samples set) were analyzed every ten samples throughout the overall analytical run to correct the signal intensity drift and remove the peaks with poor reproducibility (CV > 20%). In addition, a series of diluted quality controls (dQC) were prepared by dilution with n-butanol:methanol (1:1): 100%QC, 50%QC, 25%QC, 12.5%QC, and 6.25%QC and analyzed at the beginning and the end of the sample batch. This dQC series was used as a filter to remove the features in which MS signal response was not linear (correlation with dilution factor <0.8).

12.3   Data Processing, Statistical Analysis, and Interpretation

The workflow of data preprocessing, normalization, lipid annotation, and statistical analysis is presented in Fig. 12.3. The software used is indicated in the panels on the left side.

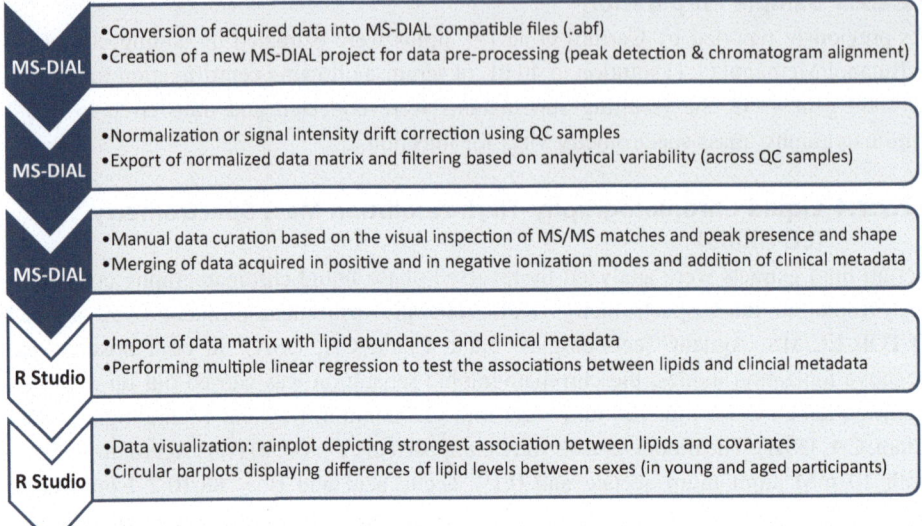

Fig. 12.3 Workflow for data preprocessing and statistical analysis

12.3.1 MS-DIAL for Data Processing and Lipid Annotation

12.3.1.1 Data Conversion

The raw data files (acquired in vendor-specific format) must be converted to an open community-driven format to analyze your data with MS-DIAL. The first step consists of downloading the Reifycs file converter at https://www.reifycs.com/AbfConverter/ and converting the raw data (*.d) to the format analysis base file (*.ABF) compatible with MS-DIAL. Specify the location for data storage and drag your raw files to the Abf window for conversion.

12.3.1.2 MS-DIAL Download

Download open-access MS-DIAL software compatible with your computer setup at http://prime.psc.riken.jp/compms/, and once the download is completed, open the program by selecting the "exe" file. The workflow described below concerns specifically lipid data acquired in DDA mode. The data acquired in different analytical conditions can also be processed; for more information, please refer to the online tutorial using https://mtbinfo-team.github.io/mtbinfo.github.io/MS-DIAL/tutorial.html

12.3.1.3 Create a New Project

Start by creating a new project (File → New project). The start-up window (Fig. 12.4) will be displayed. First, select the project file path containing the ABF files, then specify the main analytical conditions used for data acquisition: ionization, separation, MS method,

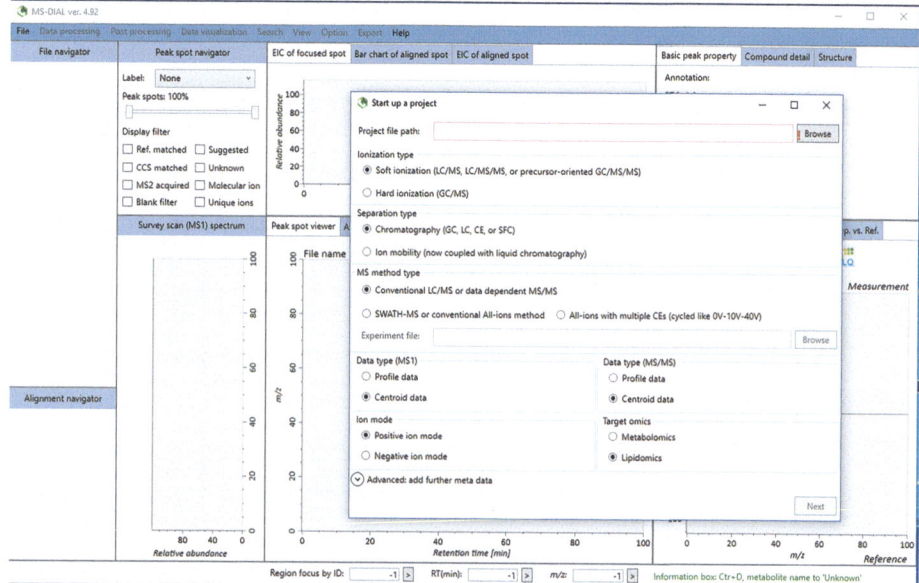

Fig. 12.4 "Start up a project" window

data type, ionization mode, and target omics (metabolomics or lipidomics). Proceed by clicking on the "next" tab.

A new project window will open for you to ***upload the ABF files*** you wish to process. Following the upload, for each file, you will need to specify its type (Sample, Standard, QC, or Blank), the class ID (for statistical analysis), the batch number if applicable, the analytical order (in which your samples were analyzed) and the injection volume (Fig. 12.5). The analytical order will be used for the signal intensity drift correction. Once finalized, you can proceed by clicking the "next" tab.

12.3.1.4 Analysis Parameter Setting

If data analysis has already been performed with a specific parameter setting, MS-DIAL allows you to load your previously saved parameter setting (bottom-left button; Fig. 12.6).

Otherwise, the analysis parameter setting begins with defining ***data collection*** parameters (Fig. 12.6). Here, you should set the mass accuracy (MS1 and MS2 tolerance depending on m/z ratio deviation across consecutive scans), and the analysis ranges for RT, MS1, and MS/MS. While focusing on small molecules ($<$2000 Da), the maximum charged number can be set to 2 for isotope recognition. The multithreading parameter depends on your computer's performance (open task manager $\rightarrow$ open resource monitor).

The minimum values for peak detection should be set for peak width and height (Fig. 12.7). These parameters are defined based on the analyst's experience using the dedicated MS spectrometer. Peaks below the specified threshold will be considered background noise. In the dataset acquired in this study, the peak height threshold was set

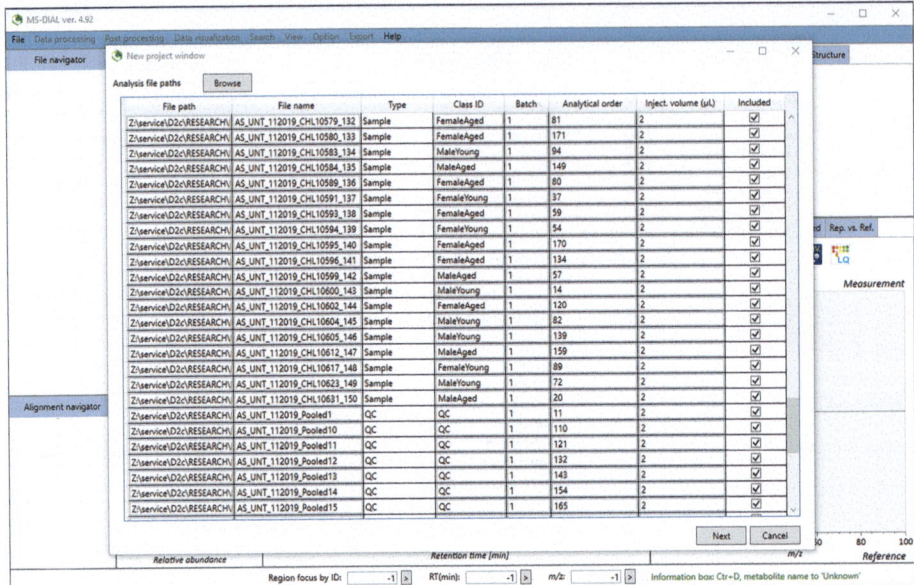

Fig. 12.5 "New project window" for the upload of ABF files

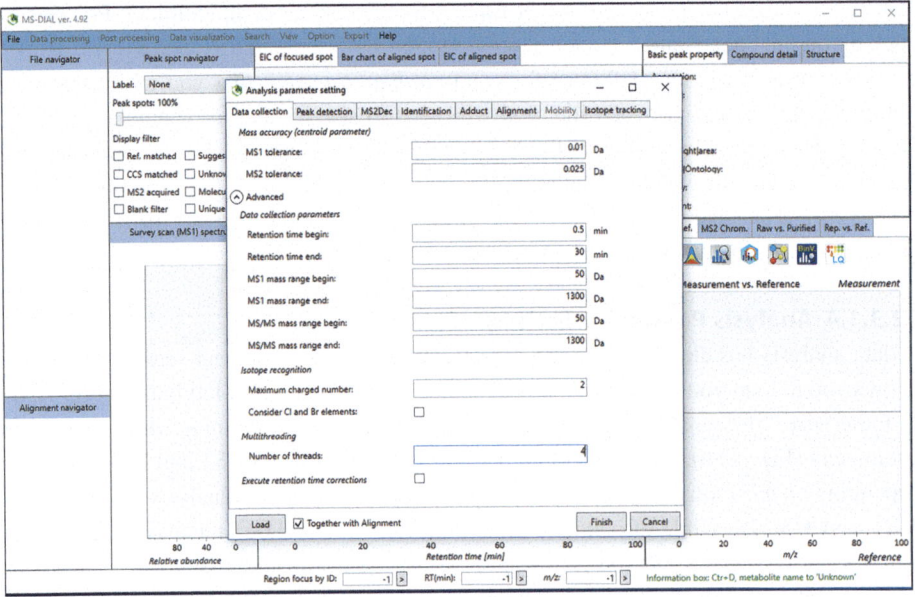

Fig. 12.6 Analysis parameter setting—Data collection

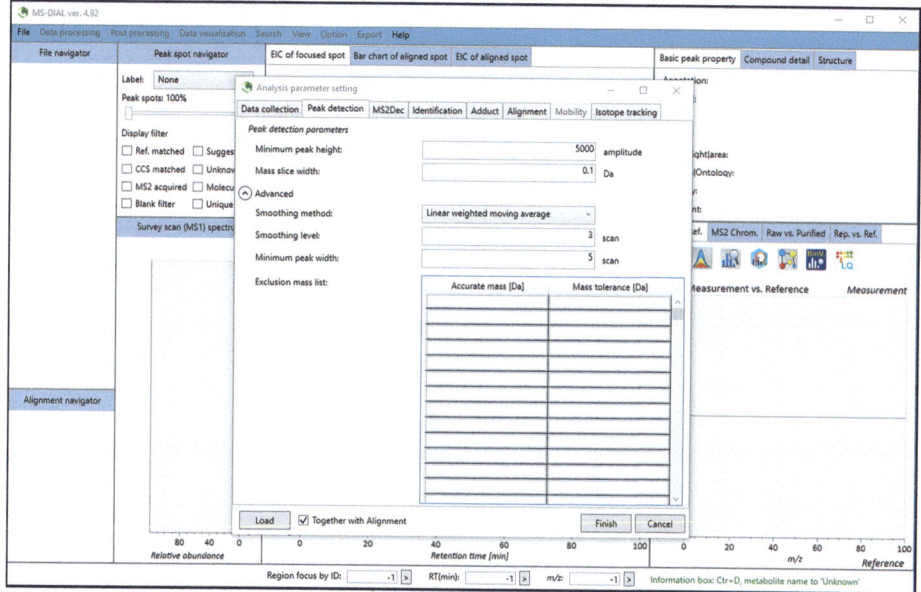

Fig. 12.7 Analysis parameter setting—Peak detection

at 100 (ion counts). The minimum peak width is set to "5 scans" or data points necessary for peak definition. The recommended smoothing level is 1–3, and the "Linear weighted moving average" smoothing method is used for peak detection by default to accurately determine the peak's left and right edges. Finally, if there are already known unwanted ions (coming from columns, vials, or solvent contaminants), adding them to the "Exclusion mass list" is possible. The contaminants are usually detected across the entire chromatographic gradient.

The MS/MS, data deconvolution parameters must be defined for data acquired in DIA mode. However, you should still define a cut-off value for MS/MS abundance to reduce the extraction of noisy spectra (Fig. 12.8). The presented dataset's MS/MS abundance cut-off was set at 100. To remove the ions after the focused precursor ion (m/z > precursor ion), which is recommended, you should check the "Exclude after precursor ion" box.

In the context of "lipidomics" projects, for *lipid annotation*, the MS/MS data will be matched, by default, against the *in silico-generated* LipidBlast library, which is integrated into the MS-DIAL lipidomics workflow. In the *lipid database setting*, you should select the lipid subclasses and corresponding adducts that can be detected in respective analytical conditions (check all for the characterization of a new matrix; Fig. 12.9). The adduct type should also be selected between formate and acetate. When RT information is unavailable (in the selected database/library), "100" should be the default value. The MS1 and MS2 tolerance depends on the instrument type (Orbitrap or Q-TOF). The identification score cut-off can be set (to 80%) to filter out the false positive matches.

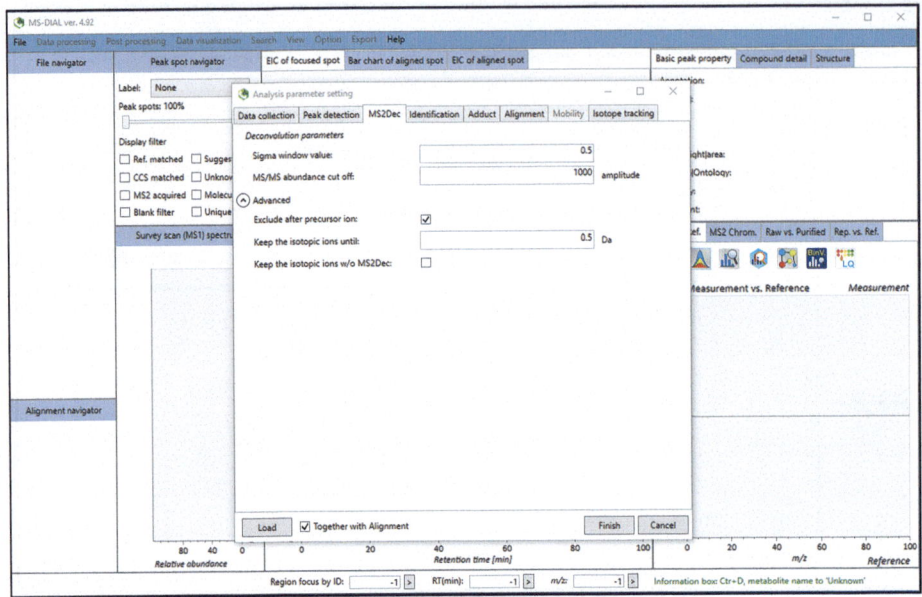

Fig. 12.8 Analysis parameter setting—MS2Dec.

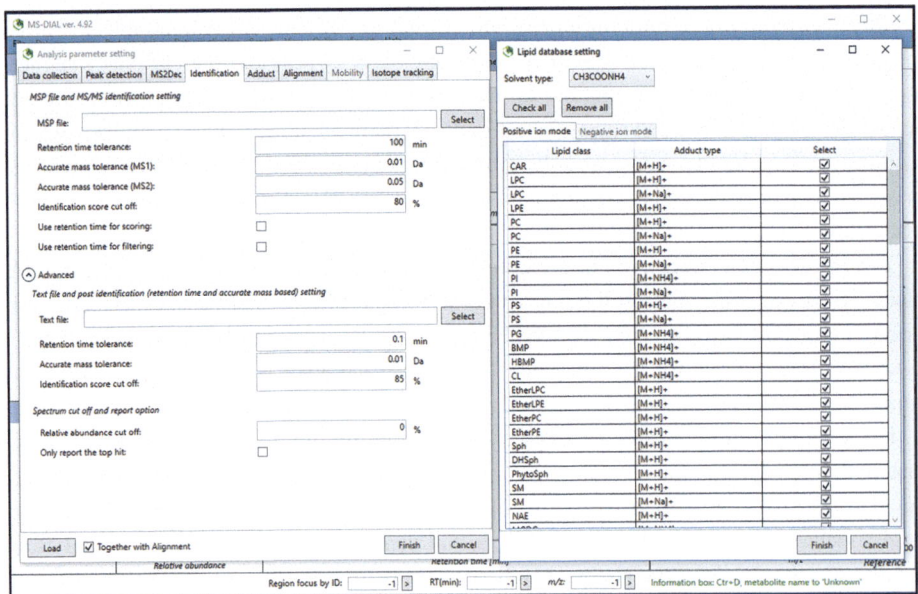

Fig. 12.9 Analysis parameter setting—Identification

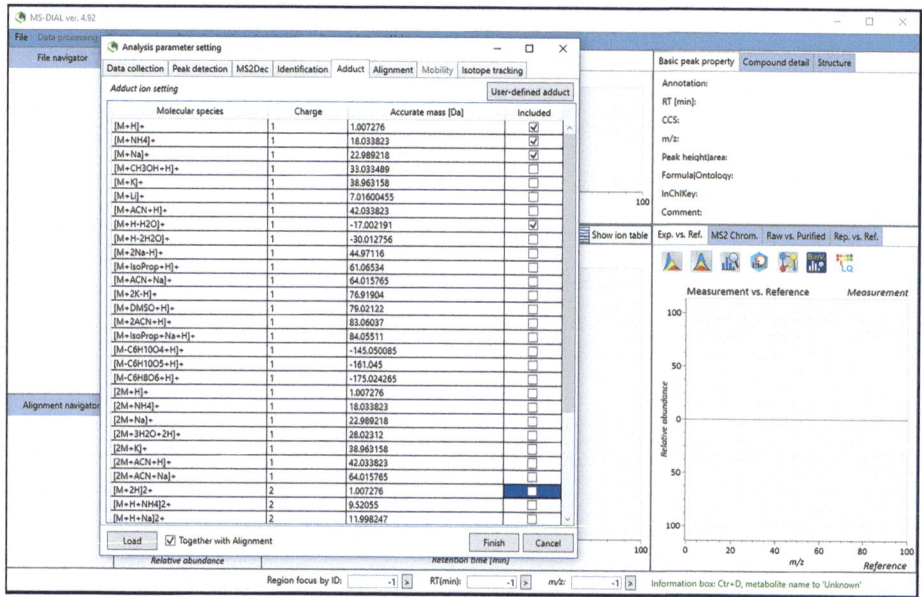

Fig. 12.10 Analysis parameter setting—Adduct

For *metabolite identification*, the MSP file referring to a specific spectral library must be selected for matching. Multiple spectral libraries (depending on the model organism or developer) can be downloaded from the RIKEN website at http://prime.psc.riken.jp/compms/msdial/main.html#MSP

When one uses an **in-house generated database** with recorded accurate mass and RT, the tolerance parameters for matching can be defined under the "Advanced" tab.

In addition, the appropriate adduct ions (depending on applied analytical conditions) can be defined in the ***Adduct panel*** ("User-defined adduct"; Fig. 12.10). In lipidomics, the adducts [M+H]+, [M+NH4]+, [M+H-H2O]+, [M+Na]+, [M-H]-, and [M+CH3COO]- are usually selected. Adduct annotation is done based on defined mass differences, retention time, and peak shape (i.e., co-elution rule).

Finally, in the ***alignment*** panel, you should select the reference file that will be used for chromatogram alignment (Fig. 12.11). Here, we usually specify one QC file against which all the other profiles will be aligned. The RT and MS1 tolerance depends on the expected retention time shift (when using specific chromatography) and *m/z* ratio deviation across consecutive runs (or samples), respectively. These parameters will be used for grouping peaks representing the same ion across all samples. Under the "advanced" tab, you can also define the minimal number of samples of the same group (usually around 50–80%) in which an ion must be detected to be considered as a "real" peak. You can also define the parameters for blank subtraction and check the gap-filling function. When finished setting up the analysis parameters, select "Finish" to launch the data processing.

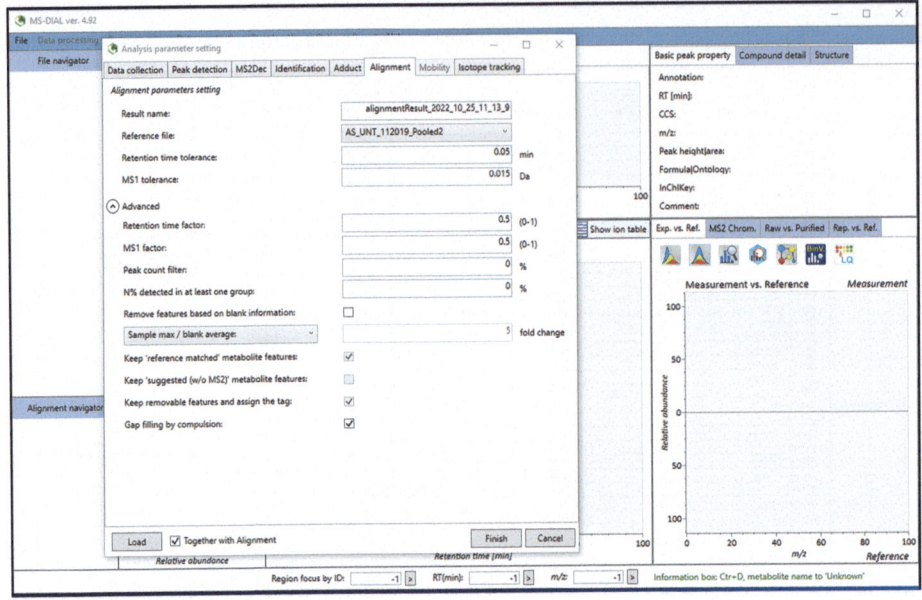

Fig. 12.11 Analysis parameter setting—Alignment

12.3.1.5 Interactive Data Exploration Using MS-DIAL Graphical Interface

Once the data processing has been finalized, you should "double click" on the alignment result in the "Alignment navigator" (bottom-left) to visualize the aligned results (Fig. 12.12). Each spot on the "Alignment spot viewer" (bottom-center) represents an aligned ion feature detected across all samples. Ions are mapped as a function of their retention time (x-axis) and m/z (y-axis).

In lipidomics projects, each lipid (sub)class is colored differently. In the "Peak spot navigator" (upper-left), the projected data can be filtered based on different matching parameters. When the "Reference matched" ions box is checked, only features with the corresponding MS/MS match will be displayed. In this case, the MS/MS matches can be visualized on the bottom right panel ("Rep. vs. Ref" where experimentally acquired MS/MS spectrum is compared to a reference MS/MS spectrum). The metadata (name, adduct type, m/z, formula, structure) for the selected annotated ion are displayed in the upper-right panel "Basic peak property," "Compound detail," and "Structure" tabs.

When the "Suggested" box is ticked among different filters, all features with accurate mass match (at MS1 level only) will be displayed. The "Unknown" box displays all features that do not match any compound in the database or library used for matching, neither at the MS/MS nor at the MS level only.

On the upper-center window, the "Bar chart of aligned spot" depicts the boxplots of normalized peak height for each pre-defined Class IDs (Fig. 12.13). With the right click, the boxplot chart can be changed to a bar plot ("Change chart type to…"), and the

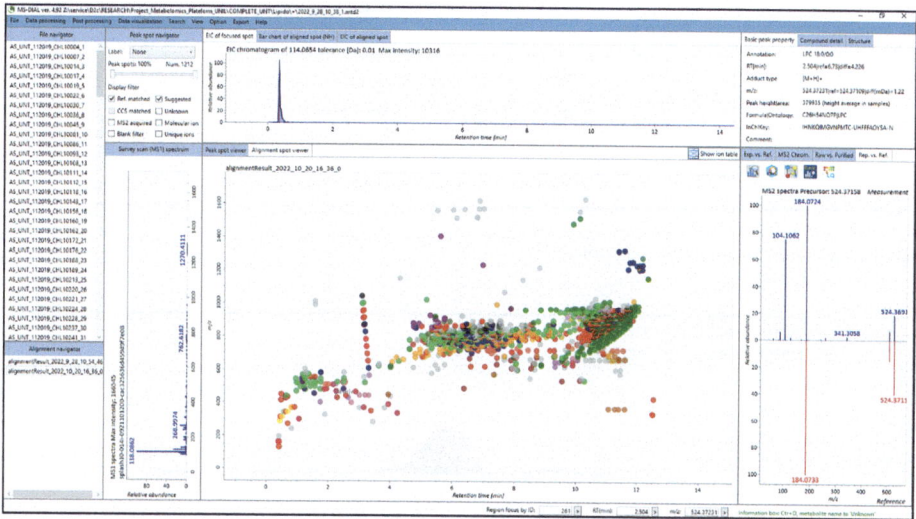

Fig. 12.12 Graphical user interface of MS-DIAL

normalized peak height (NH) can be swapped to the original peak height (OH) or peak area (OA) ("Change data source to...").

To assess the peak quality associated with the selected feature, click on the adjacent "EIC of aligned spot" tab (Fig. 12.14). You can scrutinize the individual peaks across all analyzed samples by right-clicking on the window and selecting the "Table viewer for curating each chromatogram" option.

12.3.1.6 Signal Intensity Drift Correction and Data Filtration

Data normalization or signal intensity drift correction can be performed directly in MS-DIAL or using another external statistical program (R, SIMCA, etc.). In MS-DIAL, the functionality for data normalization can be found under the "Data visualization" panel ("Normalization"). When no internal standard spike was performed, the most applied method for drift correction is *LOWESS* (locally weighted scatterplot smoothing) modeling (Fig. 12.15). LOWESS uses the QC data acquired periodically throughout the run to correct for the signal intensity drift (which is inherent to the MS detector) over time. The setting of the LOWESS span should be defined, and the optimal span can be calculated directly by selecting the "Span optimization" option. Once normalized, the data should be filtered for analytical variability before statistical analyses. This step of discarding features for which the analytical variability or CV across QC samples is higher than 20% can be performed in the exported dataset using the calculated average and standard deviation across pooled QC samples.

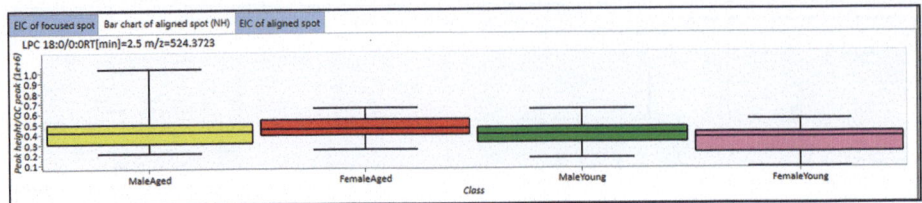

Fig. 12.13 Bar chart of aligned spot (NH)

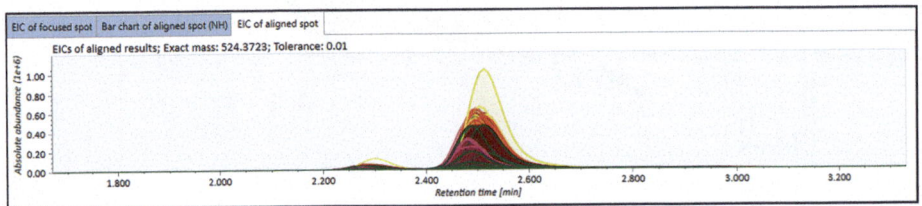

Fig. 12.14 Extracted ion chromatogram (EIC) of aligned spot

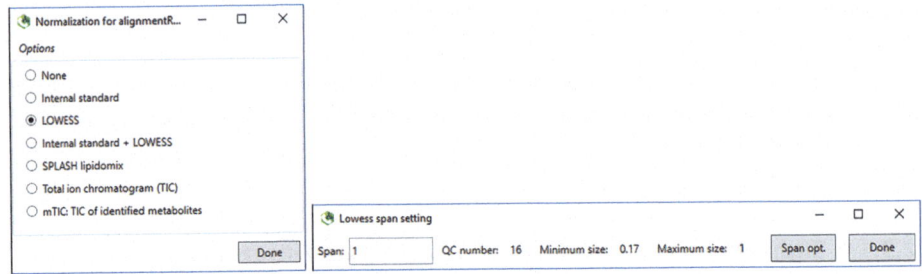

Fig. 12.15 Normalization of the alignment result using MS-DIAL

12.3.1.7 Data Curation

Automated lipid annotation through MS/MS matching against LipidBlast will generate a certain number of *false positive and redundant matches*, which must be manually curated by visual inspection to avoid misidentification and repetition. Here, we provide a roadmap for interactive data exploration and curation. It is **important** to remember that poor-quality data and redundancy (if kept) will affect statistical tests and findings. For additional insights on data curation, visit the MS-DIAL tutorial https://mtbinfo-team.github.io/mtbinfo.github.io/MS-DIAL/tutorial.html#section-5-4 or https://github.com/respiratory-immunology-lab/metabolome-lipidome-MSDIAL

First, *export the data processing results* via the "Export" tab—"Alignment result." You should define the "Directory" and the alignment file to be used and select "Normalized data matrix (Area)." In the exported data matrix (.txt file), the annotation of ions will be displayed at three different levels: "*lipid name*" (with corresponding "TRUE" MS/MS

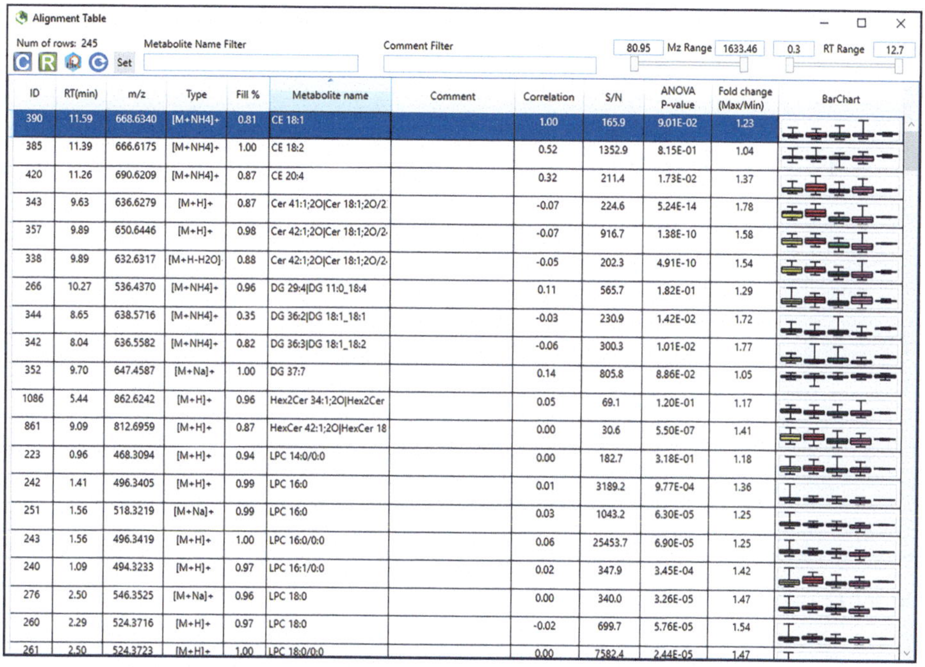

ID	RT(min)	m/z	Type	Fill %	Metabolite name	Comment	Correlation	S/N	ANOVA P-value	Fold change (Max/Min)	BarChart
390	11.59	668.6340	[M+NH4]+	0.81	CE 18:1		1.00	165.9	9.01E-02	1.23	
385	11.39	666.6175	[M+NH4]+	1.00	CE 18:2		0.52	1352.9	8.15E-01	1.04	
420	11.26	690.6209	[M+NH4]+	0.87	CE 20:4		0.32	211.4	1.73E-02	1.37	
343	9.63	636.6279	[M+H]+	0.87	Cer 41:1;2O|Cer 18:1;2O/2		-0.07	224.6	5.24E-14	1.78	
357	9.89	650.6446	[M+H]+	0.98	Cer 42:1;2O|Cer 18:1;2O/2		-0.07	916.7	1.38E-10	1.58	
338	9.89	632.6317	[M+H-H2O]	0.88	Cer 42:1;2O|Cer 18:1;2O/2		-0.05	202.3	4.91E-10	1.54	
266	10.27	536.4370	[M+NH4]+	0.96	DG 29:4|DG 11:0_18:4		0.11	565.7	1.82E-01	1.29	
344	8.65	638.5716	[M+NH4]+	0.35	DG 36:2|DG 18:1_18:1		-0.03	230.9	1.42E-02	1.72	
342	8.04	636.5582	[M+NH4]+	0.82	DG 36:3|DG 18:1_18:2		-0.06	300.3	1.01E-02	1.77	
352	9.70	647.4587	[M+Na]+	1.00	DG 37:7		0.14	805.8	8.86E-02	1.05	
1086	5.44	862.6242	[M+H]+	0.96	Hex2Cer 34:1;2O|Hex2Cer		0.05	69.1	1.20E-01	1.17	
861	9.09	812.6959	[M+H]+	0.87	HexCer 42:1;2O|HexCer 18		0.00	30.6	5.50E-07	1.41	
223	0.96	468.3094	[M+H]+	0.94	LPC 14:0/0:0		0.00	182.7	3.18E-01	1.18	
242	1.41	496.3405	[M+H]+	0.99	LPC 16:0		0.01	3189.2	9.77E-04	1.36	
251	1.56	518.3219	[M+Na]+	0.99	LPC 16:0		0.03	1043.2	6.30E-05	1.25	
243	1.56	496.3419	[M+H]+	1.00	LPC 16:0/0:0		0.06	25453.7	6.90E-05	1.25	
240	1.09	494.3233	[M+H]+	0.97	LPC 16:1/0:0		0.02	347.9	3.45E-04	1.42	
276	2.50	546.3525	[M+Na]+	0.96	LPC 18:0		0.00	340.0	3.26E-05	1.47	
260	2.29	524.3716	[M+H]+	0.97	LPC 18:0		-0.02	699.7	5.76E-05	1.54	
261	2.50	524.3723	[M+H]+	1.00	LPC 18:0/0:0		0.00	7582.4	2.44E-05	1.47	

Fig. 12.16 Alignment table listing annotated lipids using MS/MS or MS only match against LipidBlast

match), "*w/o MS2: lipid name*" (based on accurate MS1 match only) and "*Unknown*" (without match against applied database). For data curation, you may simultaneously open the exported data matrix (in Excel) and the MS-DIAL interface with the corresponding project. For the present study, we decided to consider only "MS/MS matched" hits by applying the "Reference matched" filter (in the Peak spot navigator) as described above. To facilitate the interactive exploration of spectral data, we advise you to open the alignment table in a separate window by selecting the "Show ion table" (central panel in MS-DIAL). This alignment ion table lists annotated lipids based on MS/MS or MS only matching against LipidBlast (Fig. 12.16). Each row represents one annotated peak, ion, or lipid species and displays the ID, RT (min), *m*/*z* ratio, ion form (protonated, deprotonated, adduct type), S/N ratio, and statistics (i.e., group comparison).

Now, you can sort the alignment table and the data matrix in Excel by the "Metabolite name" column and explore the MS/MS matches by visual inspection of each match between the experimentally acquired spectrum (*in blue*) and the reference spectrum (*in red*) (Fig. 12.17). By selecting the Compound search' window, you can verify each compound's list of putative hits (with corresponding scores). In the "Compound search" window, you can adjust the tolerated mass shift for MS matching, access the "Library information," and interactively explore the matches (Fig. 12.17). In the present study, on top of MS/MS confirmed hits, we also kept the relevant hits w/o MS/MS data for

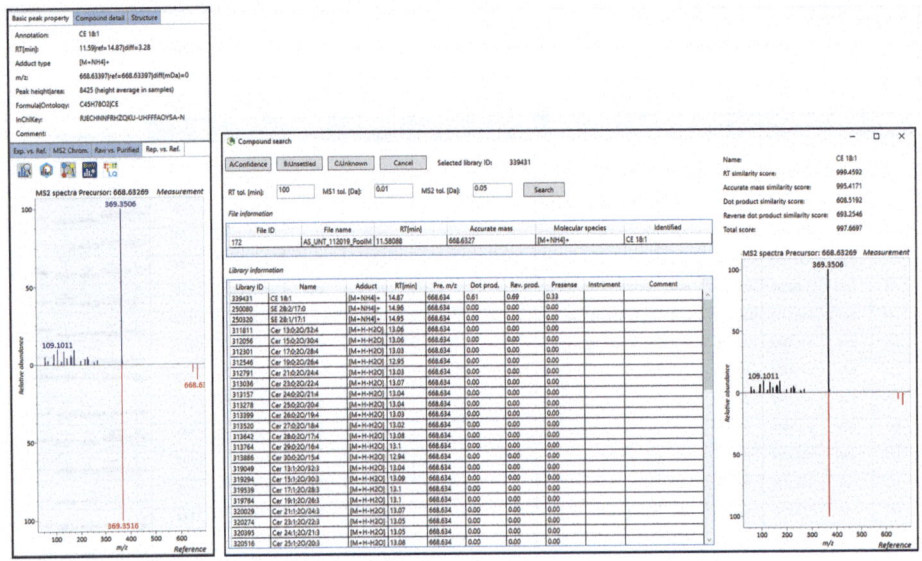

Fig. 12.17 Interactive exploration of MS/MS matched spectra

subsequent targeted MS/MS acquisition using an inclusion list. This refers to the ions of interest whose levels vary significantly between the explored classes (i.e., female vs. male and young vs. aged, in our case). Following targeted MS/MS acquisition for the ions of interest, the list of MS/MS matched compounds with a high confidence level was merged with the list of previously annotated lipids for further statistical analyses (Fig. 12.18).

12.3.1.8 Exploratory Statistical Analysis (PCA, OPLS, Pathway Map) in MS-DIAL

For data quality assessment (i.e., QC grouping) and to rapidly explore the sample clustering and potential outliers, MS-DIAL offers the possibility to perform the exploratory multivariate analysis, including the unsupervised principal component analysis (PCA), hierarchical clustering analysis (HCA) and discriminant partial least squares (PLS). The type of analysis can be selected in the "Data visualization" panel. PCA was selected in the example presented below; the "Scale method" was set to "Pareto scale," and the data were Log10 transformed. For this, only the "Ref. matched" metabolites were kept. In the PCA score plot, the sample clustering according to age can be observed (i.e., the distinction between young—green and pink, and aged—red and yellow). In the loadings plot, the contribution of the reference-matched lipids to the sample clustering is depicted (Fig. 12.19).

An additional tool for data visualization and mapping onto pathways is a "Pathway map." Three background knowledge pathway databases (animal, plant, or user defined) can be selected, and up to four aligned datasets can be loaded. The pathway mapper application will generate a map of annotated lipids in the context of interconnected biochemical pathways (Fig. 12.20). Each node represents a (sub)class of lipids with boxplots for the

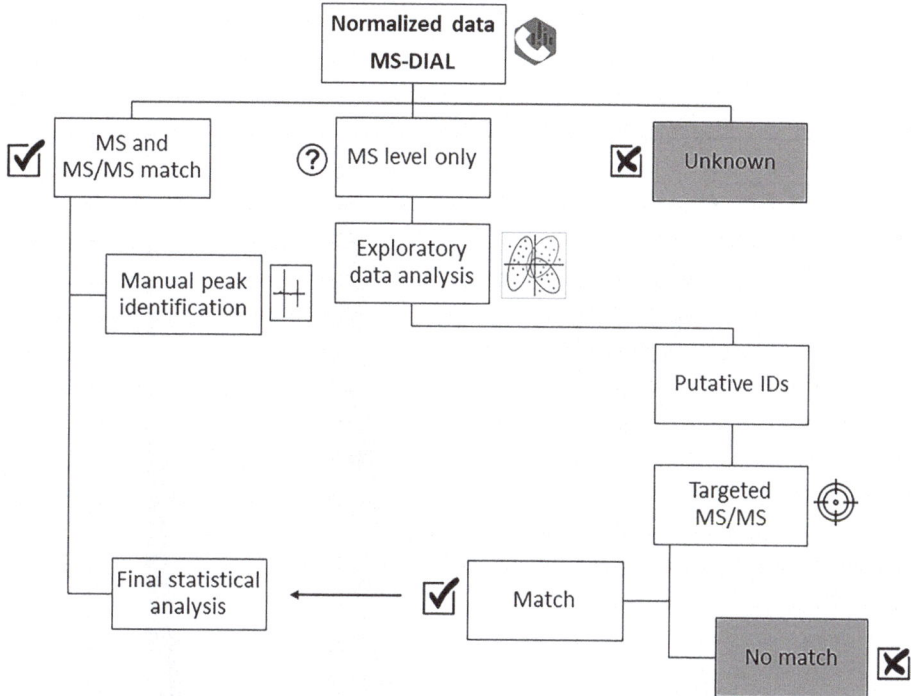

Fig. 12.18 Summary of the workflow from raw data extracted from MS-DIAL to statistical analyses

"Ref. matched." The differences in lipid levels between the study groups can be rapidly explored by selecting specific nodes.

12.3.1.9 Data Table Formatting for Downstream Statistical Analyses in R Environment

Following manual data curation, the normalized data matrices acquired in positive and negative ionization modes are merged into a single dataset for statistical analysis. The overlap between positive and negative modes should be filtered to remove the redundancy. The more selective matches with fatty acyl/alkyl composition are kept, like phospholipid annotations in negative mode. Putative hits based only on the accurate mass match, such as free fatty acids, were discarded from the present data set. Lastly, the missing values should be imputed directly in MS-DIAL (by selecting "Replace zero values with 1/10 of minimum peak height over all samples" in the "*Missing value option*" during data export) or *à posteriori* using specific imputation algorithm [33, 34]. The curated dataset of the Complete Health cohort contains 214 annotated lipid species, which were used for association analyses with clinical metadata as described in the section below.

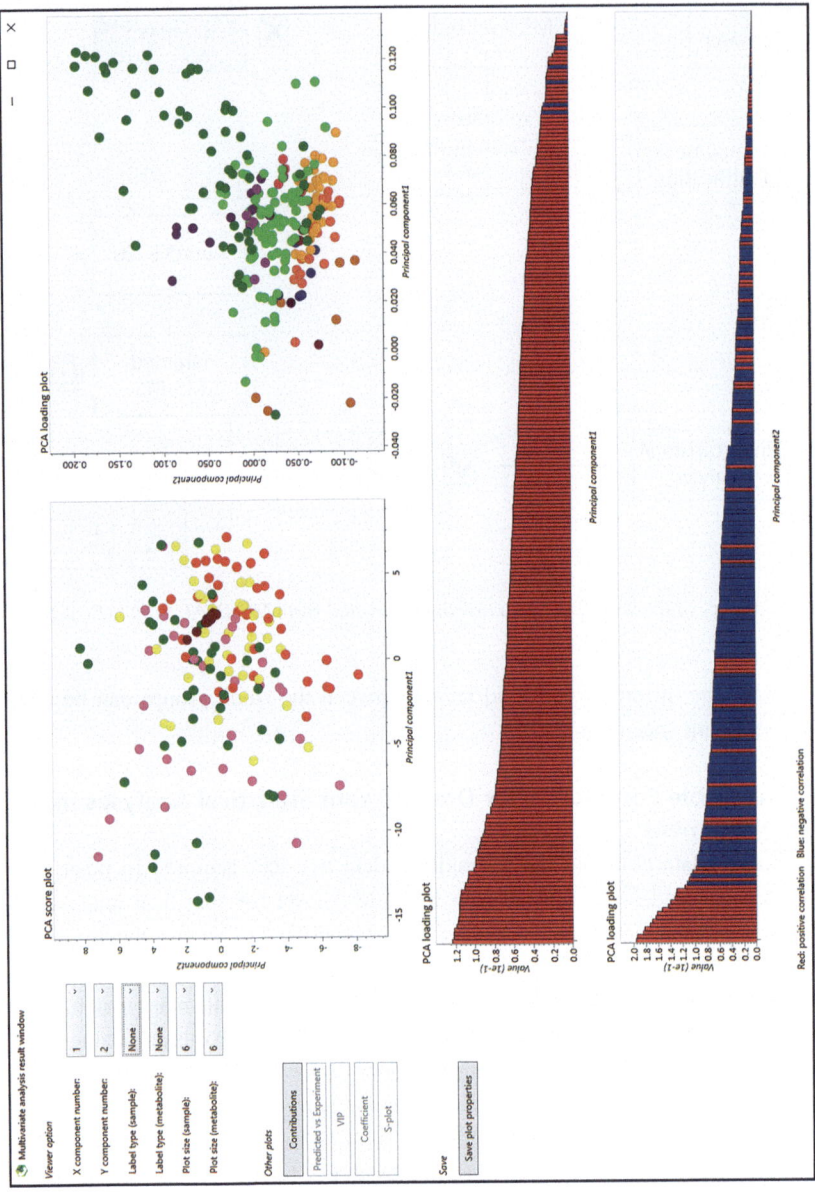

Fig. 12.19 Principal component analysis using MS-DIAL

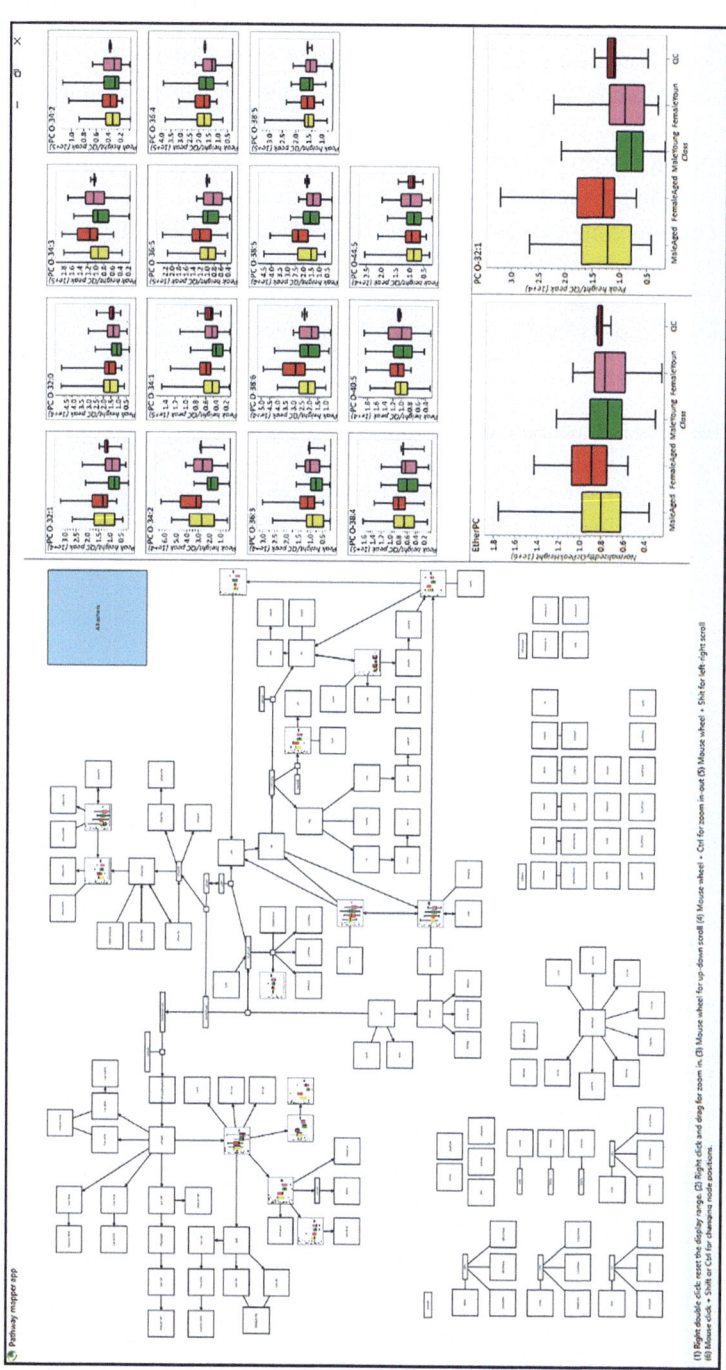

Fig. 12.20 Pathway analysis in MS-DIAL

12.3.2 R Code for Statistical Data Analysis

The R-code and related data set can be downloaded from GitHub (https://github.com/ JustinCarrard/A-lipidome-wide-association-study), which requires a suitable locally installed RStudio environment. Start by loading the packages necessary for the data analysis (see full R script). If some of these packages still need to be installed on your computer, use install.packages("") to install the missing ones.

Then, start by importing the formatted lipid dataset combined with metadata and use the function "str" to display the internal structure of the data frame:

```
dat <- as.data.frame(read_excel("data_book_chapter.xlsx"))
str(dat)
```

The data frame is first adapted concerning "Sampling time"

```
#Remove the wrong date from the sample time column
dat$"Sampling time" = gsub("1899-12-31","",dat$"Sampling time")

#Group sampling time into 5 categories
dat$Sampling_time_cat <- cut(
chron::times(dat$"Sampling time")
, breaks = chron::times(c(
"08:00:00"
, "10:00:00"
, "12:00:00"
, "14:00:00"
, "16:00:00"
, "23:59:59"
))
, labels = c("8-9.59","10-11.59","12-13.59","14-15.59","16-18.59")
, include.lowest = TRUE
, right = TRUE)

dat[, c("Sampling time", "Sampling_time_cat")]
```

For downstream analyses, lipid values need to be log2 transformed. To do so, a log2 function is applied to each lipid, and new columns "_log2" are added to the data frame:

```
variables_to_transform <- names(dat)[24:237] # Name of variables to
transform with log2
for (i in variables_to_transform) {
  dat[, paste0(i, "_log2")] <- log2(dat[, i])
}
```

The continuous and dependent variables are then z-standardized and new columns "_std" are added to the data frame:

```
variables_to_standardize <- names(dat)[c(6:23, 239:452)]
for (i in variables_to_standardize) {
  dat[, paste0(i, "_std")] <- scale(dat[, i], center = TRUE, scale = TRUE)
}
```

12.3.2.1 Rain Plot Investigating the Associations Between Lipid Species, Age, Sex, and Other Covariates

The data frame is now ready to run multiple linear regressions. The variables "Decade," "Sex," and "Statins" need to be defined as a factor as they will be used as categorical variables ("Decade"=0 corresponds to "Young," otherwise =1 corresponds to "Aged"; "Sex"=0 corresponds to female, otherwise=1 corresponds to male; "Statins"=0 corresponds to no statin used, otherwise =1 corresponds to the use of statins).

```
my_lms <- lapply(dat[,c(471:684)], function(x) lm(x ~ dat$Decade * dat$Sex +
dat$Statins + dat$PBF_std + dat$Total_PA_std + dat$HbA1c_std + dat
$Sampling_time_cat + dat$Fasting_std, data = dat))
```

The "lm" regression function testing for the interaction between age and sex-adjusted for other potential confounders (statins intake, body fat (%), glycated hemoglobin, daily physical activity) is applied for each lipid species using the "lapply" function.

Following this step, the main results of the regression (estimate, standard error, t-value, p-value, response factor) are extracted into a data frame for each covariate. For binary covariates, the "1" condition is chosen. Here, an example is shown for the results related to age:

```
#extract Decade results
res_frame_Decade <- data.frame(
  term = names(my_lms)
  , estimate = NA
  , std.error = NA
  , statistic = NA
  , p.value = NA
  , response = "NA"
)

for (i in seq_along(names(my_lms))) {

  sum_tmp <- summary(my_lms[[i]])

  res_frame_Decade$estimate[which(res_frame_Decade$term %in% names
  (my_lms)[i])] <- sum_tmp$coefficients[, "Estimate"]["dat$Decade1"]
```

```
res_frame_Decade$std.error[which(res_frame_Decade$term %in% names
(my_lms)[i])] <- sum_tmp$coefficients[, "Std. Error"]["dat$Decade1"]
res_frame_Decade$statistic[which(res_frame_Decade$term %in% names
(my_lms)[i])] <- sum_tmp$coefficients[, "t value"]["dat$Decade1"]
res_frame_Decade$p.value[which(res_frame_Decade$term %in% names
(my_lms)[i])] <- sum_tmp$coefficients[, "Pr(>|t|)"]["dat$Decade1"]
res_frame_Decade$response[which(res_frame_Decade$term %in% names
(my_lms)[i])] <- "Aged"

remove(sum_tmp)


}
```

Once this model is applied for each covariate, the data frames are merged into one file. The p-values for each association are adjusted using the Benjamini-Hochberg (BH) correction method. The adjusted p-values are then categorized into different levels depending on their significance.

```
#Combine_data_frames
Overall <- rbind(
 res_frame_Decade
 , res_frame_Sex
 , res_frame_Statins
 , res_frame_PBF
 , res_frame_HbA1c
 , res_frame_Total_PA
)


#Adjust p-values
Overall <- Overall[order(Overall$p.value),]
Overall$BH <- p.adjust(Overall$p.value, method = "BH")


#Categorise BH p-values
Overall$BH_cat <- NA
Overall$BH_cat[Overall$BH > 0.05] <- "> 0.05"
Overall$BH_cat[Overall$BH <= 0.05 & Overall$BH > 0.01] <- "≤ 0.05"
Overall$BH_cat[Overall$BH <= 0.01 & Overall$BH > 0.001] <- "≤ 0.01"
Overall$BH_cat[Overall$BH <= 0.001 & Overall$BH > 0.0001] <- "≤ 0.001"
Overall$BH_cat[Overall$BH <= 0.0001] <- "≤ 0.0001"
Overall$BH_cat <- factor(Overall$BH_cat, levels = c("> 0.05", "≤ 0.05", "≤
0.01", "≤ 0.001", "≤ 0.0001"))
```

The data frame combining all results is then ordered by decreasing "estimates" values to determine the lipid species with the biggest estimates.

```
#Order by decreasing estimates
Overall <- Overall[order(-Overall$estimate),]
Overall
#remove suffixes from lipid species' name
Overall$term=gsub("_log2_std","",Overall$term)

#extract the 16 lipid species displaying the biggest estimates
Overall_top <- subset(Overall, term == "PC 16:0_18:0" | term == "SM 33:1;2" |
term == "SM 34:2;2" | term == "LPE O-18:1" | term == "PE O-20:1_20:4" | term ==
"Cer 18:1;2/16:0" | term == "PE O-18:1_22:5" | term == "PE O-18:1_18:1" | term
== "LPC 22:5" | term == "SM 34:1;2" | term == "PC 17:0_18:1" | term == "LPC 17:
0" | term == "PC 16:0_22:5" | term == "LPI 18:0" | term == "PC O-37:5" | term ==
"PC O-38:6" ,
   select=c(term, estimate, std.error, statistic, p.value, response, BH,
BH_cat))


plot_data <- Overall_top
```

Once the lipid species displaying the most meaningful associations are extracted into a new data frame, the rain plot (Fig. 12.21) can be conceived. In the following example, the lipid species themselves are investigated. Still, the rain plot may also be constructed using lipid subclasses by adding this extra information to the "Overall" data frame.

```
# Start by defining the parameters of the theme and the palette that will be
used
## Palette
palette <-
 # Blue
 c("#053061",
 "#313695",
 "#4575b4",
 "#74add1",
 "#abd9e9",
 "#e0f3f8",
 "#fee090",
 "#fdae61",
 "#f46d43",
 "#d73027",
 "#a50026",
 '#67001f')
# Red


# Calculate symmetric limits based on the most extreme value
max_abs_estimate <- max(abs(plot_data$estimate))
```

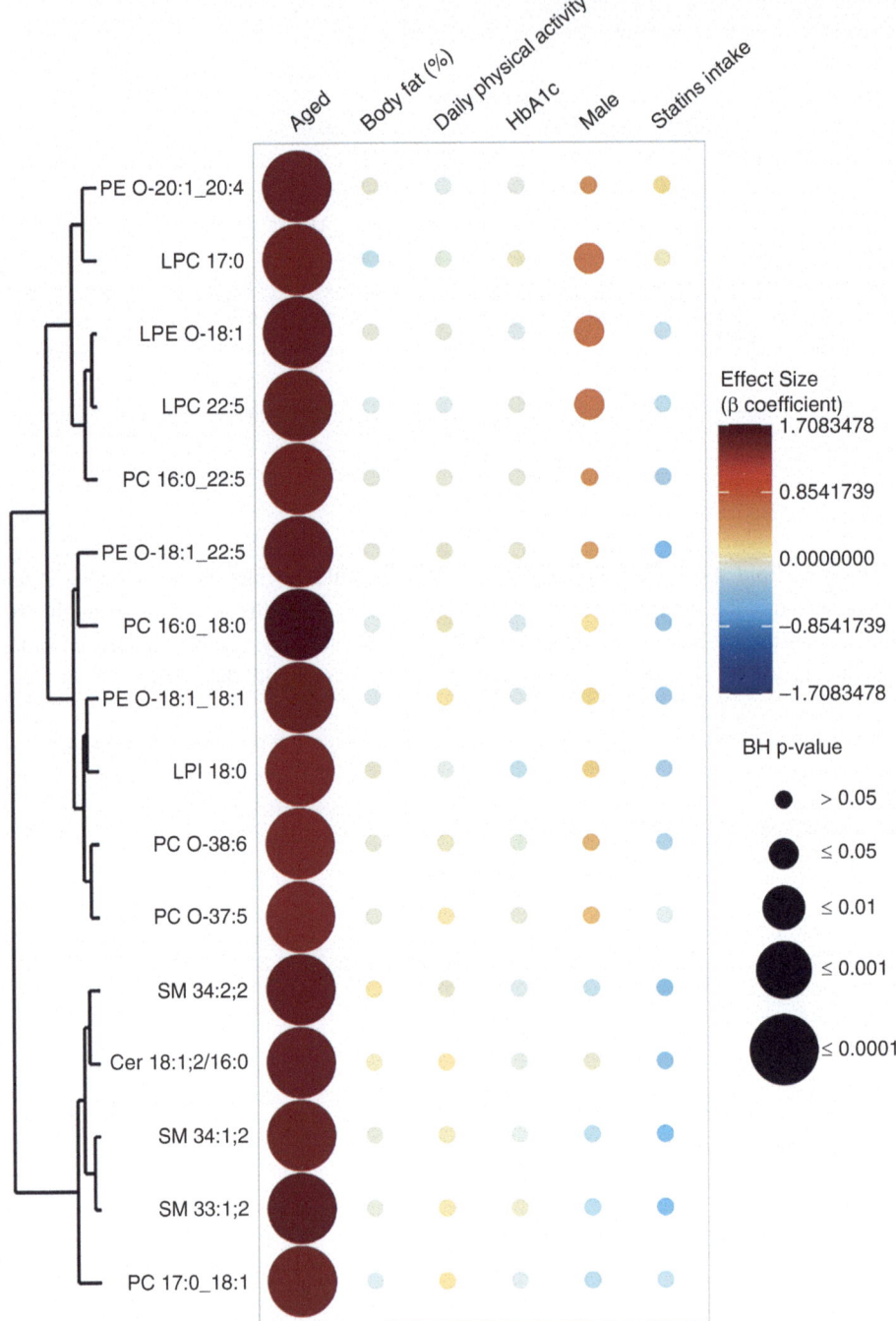

Fig. 12.21 Rain plot displaying associations between lipid species, age, sex, and clinical variables

```
max_lim <- max_abs_estimate
min_lim = -1 * max_lim


## Theme


thm <-
  # Good starting theme + set text size
  theme_light(base_size = 7) +
  theme(
  # Remove axis ticks and titles
  axis.title.x = element_blank(),
  axis.ticks.x = element_blank(),
  axis.title.y = element_blank(),
  axis.ticks.y = element_blank(),

  # Remove gridlines and boxes
  panel.grid.major = element_blank(),
  panel.grid.minor = element_blank(),
  axis.line = element_blank(),
  legend.key = element_blank(),

  # White backgrounds
  panel.background = element_rect(fill = 'white'),
  plot.background = element_rect(fill = 'white'),
  legend.background = element_rect(fill = 'white'),

  # Angle text
  axis.text.x.top = element_text(angle = 45, hjust = 0)
)


# Ordering by Cluster


# Convert to the matrix and reshape to obtain lipid species estimate values for
each response for clustering.
cluster_data <-
 plot_data %>%
 select(response, term, estimate) %>%
 spread(response, estimate)

rnms <-
 cluster_data$term

cluster_data <-
 cluster_data %>%
```

```
  select(-term) %>%
  as.matrix()

rownames(cluster_data) <- rnms

# cluster dependent variable terms
clust <- hclust(dist(cluster_data), method = 'ward.D2')
# `clust$order` orders `term` into clusters
term_order <-
  clust$labels[clust$order]

# Convert term to a factor, ordered by `term_order`
plot_data_clo <-
  plot_data %>%
  mutate(term = factor(term, levels = term_order))

# Create the rainplot using cluster ordered data
rainplot <-
  ggplot(plot_data_clo) +
  geom_point(aes(x = response, y = term, colour = estimate, size = BH_cat)) +
  scale_x_discrete(position = 'top') +
  scale_size_manual(name = expression("BH p-value"), values = c(2, 4, 6, 8,
10), drop = FALSE) +
  scale_color_gradientn(
    'Effect Size\n(Î² coefficient)',
    colors = palette,
    limits = c(min_lim, max_lim),
    breaks = c(min_lim, min_lim / 2, 0 , max_lim / 2, max_lim)
  ) +
  guides(colour = guide_colourbar(order = 1),
  size = guide_legend(order = 2)) +
  thm

# Adding dendrograms
dendro_dat <- segment(dendro_data(clust))

dendro <-
  # Empty ggplot with the same y-scale as rainplot
  ggplot() +
  geom_blank(aes(y = term), data = plot_data) +
  theme_dendro() +
  # 'expand' controls whitespace around the dendrogram. The non-zero argument
  # may need to be increased if the line thickness of the dendrogram is
  # increased to make sure the entire dendrogram is plotted
  scale_x_discrete(position = 'top', expand = c(0, 0.03, 0, 0)) +
```

```
# Draw dendrogram
geom_segment(aes(x = -y, y = x, xend = -yend, yend = xend),
       colour = 'black',
       data = dendro_dat)


p <- ggarrange(dendro, rainplot, ncol = 2, widths = c(1, 5))
p


#export rain plot as a .png
ggsave(paste('path', paste("rain_plot_top16_lipid_species.png", sep =
""), sep = "/"), p, width = 4.3*1, height = 6*1, units = "in", dpi = 300)
```

The output table featuring the association results is also saved from investigating the outcomes further.

```
#rename columns
names(Overall)[1] <- "Dependent_variable"
names(Overall)[2] <- "Î² coefficient"
names(Overall)[3] <- "standard error"
names(Overall)[5] <- "p-value"
names(Overall)[6] <- "Independent variables"
names(Overall)[7] <- "BH p-value"
names(Overall)[8] <- "Categorical BH p-value"


#export excel file
write_xlsx(Overall, "assocations.xlsx")
```

12.3.3 Circular Bar Plots Displaying the Percentage of Difference in Lipid Levels Between Females and Males

In this example, we will show how to perform circular bar plots exhibiting differences in lipid levels between two groups of individuals. In the two examples below, the sex differences in the circulatory lipidome composition were investigated age dependently. In the full R script, the sex-related age difference was also described.

ANCOVA was used while considering other variables to evaluate the differences between two (or more) groups of interest. To perform this statistical analysis, the data frame is first reshaped to obtain two columns containing the lipid species name and the corresponding abundance value in everyone.

```
# Define variables to plot
(vars_to_plot <- names(dat)[c(471:684)])
```

```
# Reshape to long
dat_long <- reshape2::melt(
 dat
 , id.vars = c("Decade", "Sex", "Statins", "PBF_std", "Total_PA_std",
"HbA1c_std", "Sampling_time_cat","Fasting_std")
 , measure.vars = vars_to_plot
)


#rename columns
names(dat_long)[9] <- "Lipid_species"
# Recode factors
dat_long$Sex <- factor(dat_long$Sex)
dat_long$Decade <- factor(dat_long$Decade)
dat_long$Sampling_time_cat <- factor(dat_long$Sampling_time_cat)
```

The ANCOVA model is then applied to each lipid species using a loop function, and a data frame is created with the extracted results, including p-values for the age, sex, and interaction effects, the mean difference between the investigated sub-groups (old, young, female, male) and their associated p-values.

```
p_val_frame <- data.frame(
 Lipid_species = character()
 , p_val_sex = numeric()
 , p_val_decade = numeric()
 , p_val_interaction = numeric()
 , mean_diff_decade_f = numeric()
 , mean_diff_decade_m = numeric()
 , mean_diff_sex_y = numeric()
 , mean_diff_sex_o = numeric()
 , p_mean_diff_decade_f = numeric()
 , p_mean_diff_decade_m = numeric()
 , p_mean_diff_sex_y = numeric()
 , p_mean_diff_sex_o = numeric()
)


combinations <- unique(dat_long$Lipid_species) #remove duplicates lipid
species

for (i in seq_along(combinations)) {

 dat_tmp <- subset(dat_long, Lipid_species %in% combinations[i])

 mod <- lm(value ~ Decade*Sex + Statins + PBF_std + Total_PA_std + HbA1c_std +
Sampling_time_cat + Fasting_std, data = dat_tmp)
```

```
ANOVA_res <- Anova(mod, type = "II")


mean_diff_decade <- as.data.frame(emmeans::contrast(emmeans(mod,
"Decade", by = "Sex"), "revpairwise"))
mean_diff_sex <- as.data.frame(emmeans::contrast(emmeans(mod, "Sex", by =
"Decade"), "pairwise"))


tmp_frame <- data.frame(
  Lipid_species = combinations[i]
  , p_val_sex = ANOVA_res$`Pr(>F)`[which(rownames(ANOVA_res) %in% "Sex")]
  , p_val_decade = ANOVA_res$`Pr(>F)`[which(rownames(ANOVA_res) %in%
"Decade")]
  , p_val_interaction = ANOVA_res$`Pr(>F)`[which(rownames(ANOVA_res) %in%
"Decade:Sex")]
  , mean_diff_decade_f = mean_diff_decade$estimate[mean_diff_decade$Sex %
in% "0"]
  , mean_diff_decade_m = mean_diff_decade$estimate[mean_diff_decade$Sex %
in% "1"]
  , mean_diff_sex_y = mean_diff_sex$estimate[mean_diff_sex$Decade %in% "0"]
  , mean_diff_sex_o = mean_diff_sex$estimate[mean_diff_sex$Decade %in% "1"]
  , p_mean_diff_decade_f = mean_diff_decade$p.value[mean_diff_decade$Sex %
in% "0"]
  , p_mean_diff_decade_m = mean_diff_decade$p.value[mean_diff_decade$Sex
%in% "1"]
  , p_mean_diff_sex_y = mean_diff_sex$p.value[mean_diff_sex$Decade %in%
"0"]
  , p_mean_diff_sex_o = mean_diff_sex$p.value[mean_diff_sex$Decade %in%
"1"]
)


p_val_frame <- rbind(p_val_frame, tmp_frame)


rm(dat_tmp, mod, ANOVA_res, tmp_frame, mean_diff_decade, mean_diff_sex)


}
```

The *p*-values are then adjusted using the BH correction method and ranked.

```
p_vals_adjusted_sex_decade <- p.adjust(c(
 p_val_frame$p_val_sex
 , p_val_frame$p_val_decade
 , p_val_frame$p_val_interaction
), method = "BH") #adjust p-value with BH correction
```

```
p_val_frame$p_val_sex_adj <- p_vals_adjusted_sex_decade[seq(1, dim
(p_val_frame)[1])]
p_val_frame$p_val_decade_adj &lt;- p_vals_adjusted_sex_decade[seq(1, dim
(p_val_frame)[1]) + dim(p_val_frame)[1]]
p_val_frame$p_val_interaction_adj &lt;- p_vals_adjusted_sex_decade[seq
(1, dim(p_val_frame)[1]) + 2*dim(p_val_frame)[1]]

p_vals_adjusted_fmyo <- p.adjust(c(
 p_val_frame$p_mean_diff_decade_f
 , p_val_frame$p_mean_diff_decade_m
 , p_val_frame$p_mean_diff_sex_y
 , p_val_frame$p_mean_diff_sex_o
), method = "BH")

p_val_frame$p_val_decade_f_adj &lt;- p_vals_adjusted_fmyo[seq(1, dim
(p_val_frame)[1]) + 0*dim(p_val_frame)[1]]
p_val_frame$p_val_decade_m_adj &lt;- p_vals_adjusted_fmyo[seq(1, dim
(p_val_frame)[1]) + 1*dim(p_val_frame)[1]]
p_val_frame$p_val_sex_y_adj &lt;- p_vals_adjusted_fmyo[seq(1, dim
(p_val_frame)[1]) + 2*dim(p_val_frame)[1]]
p_val_frame$p_val_sex_o_adj &lt;- p_vals_adjusted_fmyo[seq(1, dim
(p_val_frame)[1]) + 3*dim(p_val_frame)[1]]

p_val_frame$p_val_sex_adj_cat <- NA
p_val_frame$p_val_decade_adj_cat <- NA
p_val_frame$p_val_interaction_adj_cat <- NA

p_val_frame$p_val_sex_adj_cat[p_val_frame$p_val_sex_adj > 0.05] <- ">
0.05"
p_val_frame$p_val_sex_adj_cat[p_val_frame$p_val_sex_adj <= 0.05 &
p_val_frame$p_val_sex_adj > 0.01] <- "≤ 0.05"
p_val_frame$p_val_sex_adj_cat[p_val_frame$p_val_sex_adj <= 0.01 &
p_val_frame$p_val_sex_adj > 0.001] <- "≤ 0.01"
p_val_frame$p_val_sex_adj_cat[p_val_frame$p_val_sex_adj <= 0.001 &
p_val_frame$p_val_sex_adj > 0.0001] <- "≤ 0.001"
p_val_frame$p_val_sex_adj_cat[p_val_frame$p_val_sex_adj <= 0.0001] <- "≤
0.0001"
```

This last step is then repeated for "decade" and "interaction," and these categories are defined as factors. Furthermore, new p-value categories for differences in age within gender are generated similarly (see full R script). Finally, the percentage difference between age groups for each gender and between genders for each age group are calculated for each lipid species. These percentage differences are added to the data frame. To do so, the linear regression estimates for each condition are used and applied as exponents. The data frame containing the p-values, p-value categories, means, and percentage differences is then exported.

```
#add a percentage of the difference between aged vs. young and male vs. female
p_val_frame$Percentage_difference_aged_vs_young_female = 100*
((2^p_val_frame$mean_diff_decade_f)-1) #to calculate the percent change
between aged and young females
p_val_frame$Percentage_difference_aged_vs_young_male = 100*
((2^p_val_frame$mean_diff_decade_m)-1) #to calculate the percent change
between aged and young males
p_val_frame$Percentage_difference_female_vs_male_young = 100*
((2^p_val_frame$mean_diff_sex_y)-1) #to calculate the percent change
between young females and males
p_val_frame$Percentage_difference_female_vs_male_aged = 100*
((2^p_val_frame$mean_diff_sex_o)-1) #to calculate the percent change
between aged females and males
```

The creation of the circular bar plots is then initiated using these results. Start by importing an additional file containing subclass classification for each lipid species in the dataset.

```
dat_header <- as.data.frame(read_excel("Lipid_subclasses_1.xlsx"))
```

Herein, circular bar plot examples are displayed only for sex differences (Figs. 12.22 and 12.23). For each condition (age differences or sex differences), a new dataset needs to be created, and a subset for sub-groups (age in this example) with significant p-values ($p < 0.05$) must be defined.

```
dat_cbp_3 <- p_val_frame[, names(p_val_frame)[c(1, 25, 26, 29, 30)]]
#Extract lipid species, p-values categories, and percentage differences for
sex differences in both age groups
dat_cbp_4 <- dat_header[, names(dat_header)[c(1, 2)]]
dat_cbp_sex <- merge(dat_cbp_3, dat_cbp_4, by = "Lipid_species")

dat_cbp_sex$color <- factor(dat_cbp$Lipid_subclass, levels=c("CE", "Cer",
"DG", "GSL", "LPC", "LPE", "LPE-O", "LPI", "PC", "PC-O", "PE", "PE-O", "PE-
P", "PI", "SM", "TG"), labels=rainbow(16))

### For the 'young' group
# create a subset for young with p< 0.05 only
dat_cbp_y <- filter(dat_cbp_sex, p_val_sex_y_adj_cat == "≤ 0.05"|
p_val_sex_y_adj_cat =="≤ 0.01"|p_val_sex_y_adj_cat =="≤ 0.001"|
p_val_sex_y_adj_cat =="≤ 0.0001")

#Define Lipid_subclass as factor
dat_cbp_y$Lipid_subclass <- factor(dat_cbp_y$Lipid_subclass)
dat_cbp_y$p_val_sex_y_adj_cat <- factor(dat_cbp_y$p_val_sex_y_adj_cat)
```

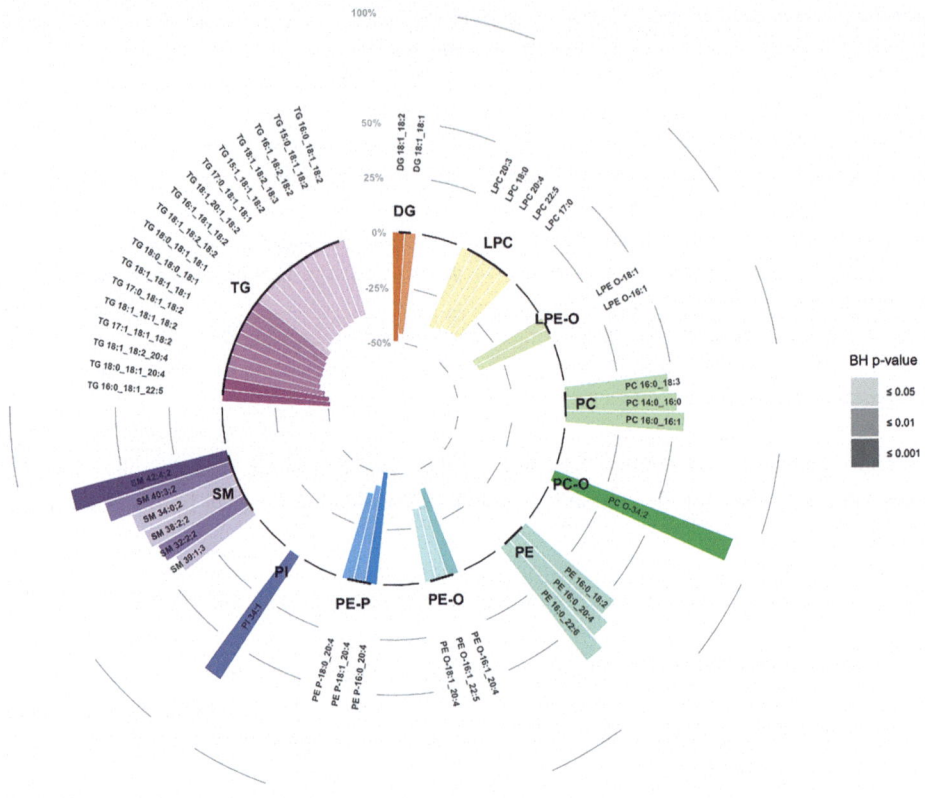

Fig. 12.22 Circular bar plot displaying percentage differences of lipid species between young females and males. BH: Benjamin-Hochberg, DG: diacylglycerol, LPC: lyso-glycerophosphocholines, LPE-O: lyso-alkyl-glycerophosphoethanolamines, PC: glycerophosphocholines, PC-O: alkyl-glycerophosphocholines, PE: glycerophosphoethanolamines, PE-O: alkyl-glycerophosphoethanolamines, PI: glycerophosphoinositol, SM: sphingomyelins and TG: triacylglycerols

```
levels(dat_cbp_y$p_val_sex_y_adj_cat) <- c("≤ 0.05", "≤ 0.01", "≤ 0.001",
"≤ 0.0001")

# Order data
dat_cbp_y = dat_cbp_y %>% arrange(Lipid_subclass,
Percentage_difference_female_vs_male_young)
```

Once the dataset is ready, the circular bar plot can be prepared by setting the number of "empty bars" at the end of each group, the y position of each label, the baselines, and the grid (scales).

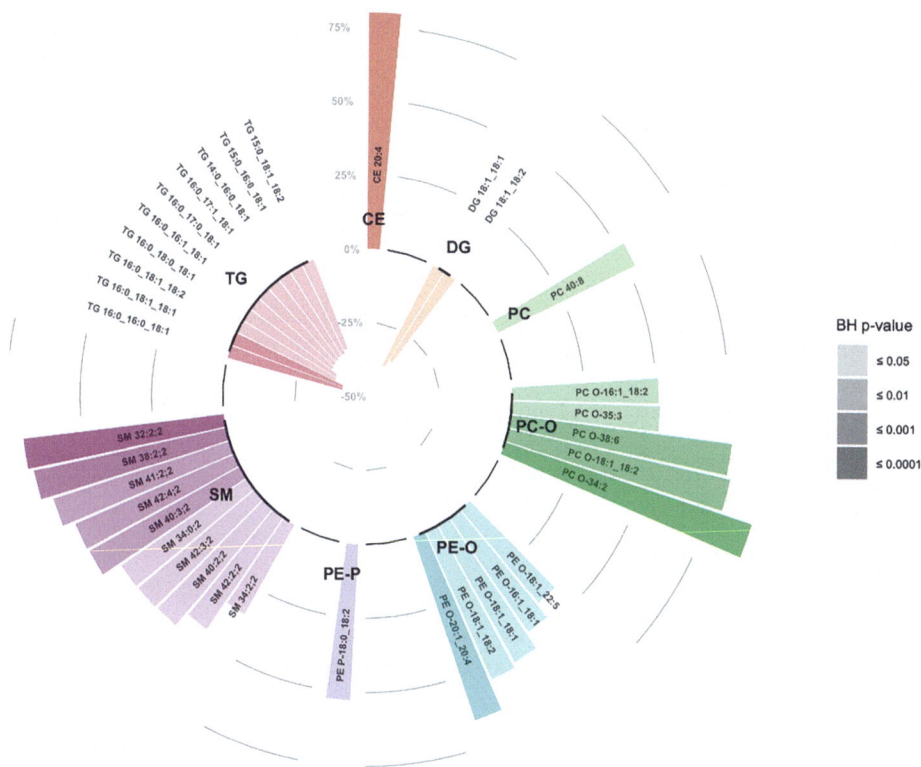

Fig. 12.23 Circular bar plot displaying the percentage difference of lipid species between aged females and males. BH: Benjamin-Hochberg, CE: cholesterol ester, DG: diacylglycerol, PC: glycerophosphocholines,   PC-O: alkyl-   glycerophosphocholines,   PE-O: alkyl- glycerophosphoethanolamines, SM: sphingomyelins and TG: triacylglycerols

```
# Set a number of 'empty bars' to add at the end of each Lipid_subclass
empty_bar <- 4
to_add <- data.frame(matrix(NA, empty_bar*nlevels(dat_cbp_y
$Lipid_subclass), ncol(dat_cbp_y)))
colnames(to_add) <- colnames(dat_cbp_y)
to_add$Lipid_subclass <- rep(levels(dat_cbp_y$Lipid_subclass),
each=empty_bar)
dat_cbp_y <- rbind(dat_cbp_y, to_add)
dat_cbp_y<- dat_cbp_y %>% arrange(Lipid_subclass)
dat_cbp_y$id <- seq(1, nrow(dat_cbp_y))


# Get the name and the y position of each label
label_data <- dat_cbp_y
number_of_bar <- nrow(label_data)
angle <- 90 - 360 * (label_data$id-0.5) /number_of_bar
```

```
label_data$hjust <- ifelse ( angle < -90, 1, 0)
label_data$angle &lt;- ifelse (angle &lt; -90, angle+180, angle)

# prepare a data frame for baselines
base_data <- dat_cbp_y %>%
 group_by (Lipid_subclass) %>%
 summarize (start=min(id), end=max(id) - empty_bar) %>%
 rowwise () %>%
 mutate (title=mean(c(start, end)))

# prepare a data frame for the grid (scales)
grid_data <- base_data
grid_data$end &lt;- grid_data$end[ c( nrow(grid_data), 1:nrow(grid_data) -
1)] + 1
grid_data$start <- grid_data$start - 1
grid_data <- grid_data[-1,]

# Make the plot
p &lt;- ggplot (dat_cbp_y, aes (x=as.factor(id),
y=Percentage_difference_female_vs_male_young, fill=color, alpha = factor
(p_val_sex_y_adj_cat))) + # Note that id is a factor. If x is numeric, there is
some space between the first bar

 geom_bar (aes (x=as.factor(id),
y=Percentage_difference_female_vs_male_young, fill=color, alpha = factor
(p_val_sex_y_adj_cat)), stat="identity") +

# Add a val=100/50/25/0/-25/-50 lines. Do it at the beginning to make sure bar
plots are OVER it.
 geom_segment (data=grid_data, aes (x = end, y = 100, xend = start, yend = 100),
colour = "grey", alpha=1, size=0.3 , inherit.aes = FALSE ) +
 geom_segment (data=grid_data, aes (x = end, y = 50, xend = start, yend = 50),
colour = "grey", alpha=1, size=0.3 , inherit.aes = FALSE ) +
 geom_segment (data=grid_data, aes (x = end, y = 25, xend = start, yend = 25),
colour = "grey", alpha=1, size=0.3 , inherit.aes = FALSE ) +
 geom_segment (data=grid_data, aes (x = end, y = 0, xend = start, yend = 0),
colour = "black", alpha=1, size=0.3 , inherit.aes = FALSE ) +
 geom_segment (data=grid_data, aes (x = end, y = -25, xend = start, yend = -25),
colour = "grey", alpha=1, size=0.3 , inherit.aes = FALSE ) +
 geom_segment (data=grid_data, aes (x = end, y = -50, xend = start, yend = -50),
colour = "grey", alpha=1, size=0.3 , inherit.aes = FALSE ) +

# Add text showing the value of each 100/50/25/0/-25/-50 lines
```

```
 annotate ("text", x = rep (max (dat_cbp_y$id) ,6) , y = c (-50, -25, 0, 25,
50, 100), label = c ("-50%", "-25%", "0%", "25%", "50%", "100%") ,
color="grey", size=2 , angle=0, fontface="bold", hjust=1) +

 geom_bar (aes (x=as.factor (id) ,
y=Percentage_difference_female_vs_male_young, fill=color, alpha = factor
(p_val_sex_y_adj_cat)), stat="identity") +
 scale_alpha_discrete (name = expression ("BH p-value") , range=c (0.10,
0.40, 0.70, 1)) +
 ylim (-80,100) +
 theme_minimal () +
 theme (
   legend.position = "none",
   axis.text = element_blank (),
   axis.title = element_blank (),
   panel.grid = element_blank (),
   plot.margin = unit (rep (-1,4) , "cm")
 ) +
coord_polar () +
theme (
   legend.position = "right",
   legend.text=element_text (size=6) ,
   legend.title = element_text (size = 8) ,) +
 scale_fill_manual (values = c ("#FF6D00", "#FFDB00", "#B6FF00", "#49FF00",
"#00FF24", "#00FF92", "#00FFFF", "#0092FF", "#4900FF", "#B600FF",
"#FF00DB")) +
 guides (fill = "none") +
 geom_text (data=label_data, aes (x=id, y=28, label=Lipid_species,
hjust=hjust) , color="black", fontface="bold",alpha=0.6, size=1.8,
angle= label_data$angle, inherit.aes = FALSE ) +

 # Add baseline information
 geom_segment (data=base_data, aes (x = start, y = 0, xend = end, yend = 0) ,
colour = "black", alpha=0.8, size=0.6 , inherit.aes = FALSE ) +
 geom_text (data=base_data, aes (x = title, y = 10, label=Lipid_subclass) ,
hjust=c (0.55, 0.55, 0.7, 0.55, 0.55, 0.7, 0.55, 0.55, 0.55, 0.3, 0.55), colour
= "black", alpha=0.8, size=3, fontface="bold", inherit.aes = FALSE)
p
```

For sex differences in the aged group, the same R-code is applied. A subset of the data frame must first be created for this specific group.

```
### For the 'aged' group
# create a subset for aged with p< 0.05 only
```

```
dat_cbp_o <- filter(dat_cbp_sex, p_val_sex_o_adj_cat == "≤ 0.05" |
p_val_sex_o_adj_cat =="≤ 0.01" |p_val_sex_o_adj_cat =="≤ 0.001" |
p_val_sex_o_adj_cat =="≤ 0.0001")


#Define Lipid_subclass as factor
dat_cbp_o$Lipid_subclass <- factor(dat_cbp_o$Lipid_subclass)
dat_cbp_o$p_val_sex_o_adj_cat <- factor(dat_cbp_o$p_val_sex_o_adj_cat)
levels(dat_cbp_o$p_val_sex_o_adj_cat) <- c("≤ 0.05", "≤ 0.01", "≤ 0.001",
"≤ 0.0001")


# Order data
dat_cbp_o = dat_cbp_o %>% arrange(Lipid_subclass,
Percentage_difference_female_vs_male_aged)
```

12.4 Discussion

12.4.1 Unraveling Age- and Sex-Associated Lipid Signature

All participants of the present study were free of exercise-limiting chronic diseases and fulfilled the WHO recommendations for daily physical activity. In addition, participants were randomly selected from rural and urban areas and not as a random sample of hospital admissions without pathologic findings. Thus, the investigated lipidome is thought to reflect clinically healthy phenotypes [35]. In this way, we could identify age- and sex-associated differences in the serum lipid profile, avoiding the confounding effects of symptomatic cardiometabolic diseases on lipid metabolism. Thereby, this study confirmed that age and sex determine the composition of the serum lipidome [36, 37]. Association analysis, after adjusting for body fat, daily physical activity, sex, statin intake, and HbA1c, showed that the major portion (>80 %) of annotated lipid species exhibited significantly higher levels in aged participants: 12 of 16 lipid subclasses were significantly and positively associated with age [5]. Notably, out of this pool of annotated lipids, no lipid class was observed to be negatively associated with age or to decrease in the aged population. The strongest positive associations with age were observed for ether-linked phospholipids (alkyl–phosphatidylethanolamines), phosphatidylinositols, ceramides, and sphingomyelins. If we take a closer look into these age-associated differences, we can observe that many differences are sex dependent, with more lipid species showing significantly higher levels in aged women compared to aged men (Fig. 12.23). In brief, many distinct species belonging to two specific classes of ether-phospholipids (PE-O & PC-O) and of lysophospholipids (LPC & LPE) were positively associated with age in females only. Beyond these age-related differences, the serum lipidome composition appears sex-specific regardless of age. In females, we could observe significantly higher levels of sphingomyelins (SMs) and lower levels of triacylglycerols (TAGs) regardless of age (Figs. 12.22 and 12.23). Several hypotheses to explain these sex differences focus mainly on sex hormone-driven regulation of enzyme activity. The differential, sex-specific activity of enzymes implicated in the synthesis and catabolism of specific lipid classes might be responsible for the differential regulation of specific pathways in women compared to men [6]. The differences in dietary regimen might also contribute to the observed differences in lipid levels, specifically concerning species containing polyunsaturated fatty acids.

12.4.2 Lipidomics Entering the Clinics?

In clinical practice, lipid measurements are often limited to HDL-C, LDL-C, total cholesterol, and triglycerides. Although these parameters have been proven effective in evaluating cardiovascular risk, recent data demonstrated that specific ceramide species predict cardiovascular risk beyond them, calling for detailed lipid analysis at the molecular species level [38–41]. The results of the present study support this call considering that, in several cases, associations could be observed only when zooming into species diversity within the same lipid subclass. Different species within the same subclass have distinct biological roles, as illustrated by the cardiometabolic favorable and deleterious ceramide species [40, 42–44]. Undoubtedly, lipidomic studies will provide new biomarkers to improve risk assessment, patient stratification, diagnosis, and follow-up of patients suffering from cardiometabolic diseases [45]. Importantly, this pilot study on clinically healthy individuals allowed for the elucidation of sex-specific age-associated changes in lipid metabolism: and these findings suggest a differential impact of aging on lipid metabolism in men compared to women. These findings also strongly indicate the need to establish sex-specific biological reference intervals for disease risk assessment.

12.4.3 Limitations

The cross-sectional nature of this study allows only for the establishment of associations, and not causality, between clinical and lipid phenotypes [46]. Importantly, the discrepancies in the selectivity of lipid annotation complicate data interpretation and comparison across different studies. Lastly, all participants lived in a small geographical area in Switzerland; therefore, results might not be generalizable to populations living in other regions of the world.

References

1. Medina J, et al. Omic-scale high-throughput quantitative LC–MS/MS approach for circulatory lipid phenotyping in clinical research. Anal Chem. 2023;95:3168–79.

2. Huynh K, et al. High-throughput plasma lipidomics: detailed mapping of the associations with cardiometabolic risk factors. Cell Chem Biol. 2019;26:71–84.e74.

3. Després J-P. Predicting longevity using metabolomics: a novel tool for precision lifestyle medicine? Nat Rev Cardiol. 2020;17:67–8.

4. Selby K, et al. Low statin use in adults hospitalized with acute coronary syndrome. Prevent Med. 2015;77:131–6.

5. Carrard J, et al. Metabolic view on human healthspan: a lipidome-wide association study. Metabolites. 2021;11:287.

6. Beyene HB, et al. High-coverage plasma lipidomics reveals novel sex-specific lipidomic fingerprints of age and BMI: evidence from two large population cohort studies. PLoS Biol. 2020;18:e3000870.

7. Lange M, et al. AdipoAtlas: a reference lipidome for human white adipose tissue. Cell Rep Med. 2021;2:100407.

8. Züllig T, Köfeler HC. High resolution mass spectrometry in lipidomics. Mass Spectrom Rev. 2021;40:162–76.

9. Xue J, Guijas C, Benton HP, Warth B, Siuzdak G. METLIN MS(2) molecular standards database: a broad chemical and biological resource. Nat Methods. 2020;17:953–4.

10. Wishart DS, et al. HMDB 5.0: the Human Metabolome Database for 2022. Nucl Acids Res. 2022;50:D622–d631.

11. Kind T, et al. LipidBlast in silico tandem mass spectrometry database for lipid identification. Nat Methods. 2013;10:755–8.

12. Horai H, et al. MassBank: a public repository for sharing mass spectral data for life sciences. J Mass Spectrom. 2010;45:703–14.

13. Tsugawa H, et al. MS-DIAL: data-independent MS/MS deconvolution for comprehensive metabolome analysis. Nat Methods. 2015;12:523–6.

14. Tsugawa H, et al. A lipidome atlas in MS-DIAL 4. Nat Biotechnol. 2020;38:1159–63.

15. Gerl MJ, et al. Machine learning of human plasma lipidomes for obesity estimation in a large population cohort. PLoS Biol. 2019;17:e3000443.

16. Chapman MJ, et al. LDL subclass lipidomics in atherogenic dyslipidemia: effect of statin therapy on bioactive lipids and dense LDL. J Lipid Res. 2020;61:911–32.

17. Takeda H, et al. Lipid profiling of serum and lipoprotein fractions in response to pitavastatin using an animal model of familial hypercholesterolemia. J Proteome Res. 2020;19:1100–8.

18. Lu J, et al. High-coverage targeted lipidomics reveals novel serum lipid predictors and lipid pathway dysregulation antecedent to type 2 diabetes onset in normoglycemic chinese adults. Diabetes Care. 2019;42:2117–26.

19. Pino MF, et al. Endurance training remodels skeletal muscle phospholipid composition and increases intrinsic mitochondrial respiration in men with Type 2 diabetes. Physiol Genomics. 2019;51:586–95.

20. Fikenzer K, Fikenzer S, Laufs U, Werner C. Effects of endurance training on serum lipids. Vascul Pharmacol. 2018;101:9–20.

21. Chua EC, et al. Extensive diversity in circadian regulation of plasma lipids and evidence for different circadian metabolic phenotypes in humans. Proc Natl Acad Sci USA. 2013;110:14468–73.

22. Gooley JJ. Circadian regulation of lipid metabolism. Proc Nutr Soc. 2016;75:440–50.

23. Hyötyläinen T, Orešič M. Optimizing the lipidomics workflow for clinical studies—practical considerations. Anal Bioanal Chem. 2015;407:4973–93.

24. Schielzeth H. Simple means to improve the interpretability of regression coefficients. Methods Ecol Evol. 2010;1:103–13.

25. Searle SR, Speed FM, Milliken GA. Population marginal means in the linear model: an alternative to least squares means. Am Stat. 1980;34:216–21.
26. Introduction to SAS. UCLA: Statistical Consulting Group. Accessed September 16, 2020. https://stats.idre.ucla.edu/other/mult-pkg/faq/general/faqhow-do-i-interpret-a-regression-model-when-some-variables-are-log-transformed/.
27. Benjamini Y, Hochberg Y. Controlling the false discovery rate: a practical and powerful approach to multiple testing. J R Stat Soc Ser B (Methodologic). 1995;57:289–300.
28. R Core Team. R: a language and environment for statistical computing. Vienna, Austria: R Foundation for Statistical Computing; 2020. https://www.R-project.org/
29. Henglin M, et al. A single visualization technique for displaying multiple metabolite-phenotype associations. Metabolites. 2019;9
30. Wagner J, et al. Functional aging in health and heart failure: the COmPLETE Study. BMC Cardiovasc Disorders. 2019;19:180.
31. Wagner J, et al. Functional aging in health and heart failure: the COmPLETE Study. BMC Cardiovasc Disord. 2019;19:180.
32. Esliger DW, et al. Validation of the GENEA Accelerometer. Med Sci Sports Exer. 2011;43:1085–93.
33. Dekermanjian JP, Shaddox E, Nandy D, Ghosh D, Kechris K. Mechanism-aware imputation: a two-step approach in handling missing values in metabolomics. BMC Bioinformatics. 2022;23:179.
34. Wei R, et al. Missing value imputation approach for mass spectrometry-based metabolomics data. Scientific Reports. 2018;8:663.
35. Bull FC, et al. World Health Organization 2020 guidelines on physical activity and sedentary behaviour. Br J Sports Med. 2020;54:1451–62.
36. Gonzalez-Covarrubias V, et al. Lipidomics of familial longevity. Aging Cell. 2013;12:426–34.
37. Johnson LC, et al. The plasma metabolome as a predictor of biological aging in humans. GeroScience. 2019;41:895–906.
38. Mach F, et al. 2019 ESC/EAS Guidelines for the management of dyslipidaemias: lipid modification to reduce cardiovascular risk: The Task Force for the management of dyslipidaemias of the European Society of Cardiology (ESC) and European Atherosclerosis Society (EAS). European heart journal. 2019;41:111–88.
39. Miller M, et al. Triglycerides and cardiovascular disease. Circulation. 2011;123:2292–333.
40. Laaksonen R, et al. Plasma ceramides predict cardiovascular death in patients with stable coronary artery disease and acute coronary syndromes beyond LDL-cholesterol. Eur Heart J. 2016;37:1967–76.
41. Hilvo M, et al. Development and validation of a ceramide- and phospholipid-based cardiovascular risk estimation score for coronary artery disease patients. Eur Heart J. 2019;41:371–80.
42. Havulinna AS, et al. Circulating ceramides predict cardiovascular outcomes in the population-based FINRISK 2002 cohort. Arteriosc Thromb Vasc Biol. 2016;36:2424–30.
43. Meeusen JW, et al. Plasma ceramides. Arteriosc Thromb Vasc Biol. 2018;38:1933–9.
44. Chew WS, et al. Large-scale lipidomics identifies associations between plasma sphingolipids and T2DM incidence. JCI Insight. 2019;5
45. Quehenberger O, Dennis EA. The human plasma lipidome. N Engl J Med. 2011;365:1812–23.
46. Chu SH, et al. Integration of metabolomic and other omics data in population-based study designs: an epidemiological perspective. Metabolites. 2019;9:117.

Quantitative Analysis of Eicosanoids and Other Oxylipins

13

Katharina M. Rund and Nils Helge Schebb

Abbreviations

4-HNE	4-Hydroxy-2-nonenal
ARA	Arachidonic acid (C20:4 n6)
BCA	Bicinchoninic acid
BHT	Butylated hydroxytoluene
CE	Collision energy
COX	Cyclooxygenase
CXP	Collision cell exit potential
CYP	Cytochrome P450 monooxygenase
DHA	Docosahexaenoic acid (C22:6 n3)
DiHETE	Dihydroxy-eicosatetraenoic acid
DiHETrE	Dihydroxy-eicosatrienoic acid
DMEM	Dulbecco's modified Eagle medium
DMSO	Dimethyl sulfoxide

Supplementary Information The online version contains supplementary material available at https://doi.org/10.1007/978-3-031-44256-8_13.

K. M. Rund · N. H. Schebb (✉)
Chair of Food Chemistry, Faculty of Mathematics and Natural Sciences, University of Wuppertal, Wuppertal, Germany
e-mail: katharina.rund@schebb-web.de; nils@schebb-web.de

DP	Declustering potential
EDTA	Ethylenediaminetetraacetic acid
EP	Entrance potential
EPA	Eicosapentaenoic acid (C20:5 n3)
EpETrE	Epoxy-eicosatrienoic acid
ESI	Electrospray ionization
EtOH	Ethanol
FCS	Fetal calf serum
GIT	Gastrointestinal tract
HAc	Acetic acid
HETE	Hydroxy-eicosatetraenoic acid
HHT/HHTrE	Hydroxy-heptadecatrienoic acid
HpETE	Hydroperoxy-eicosatetraenoic acid; hydroperoxy-ARA
IC_{50}	Half maximal inhibitory concentration
IS	Internal standard
IsoP	Isoprostane
KOH	Potassium hydroxide
LC	Liquid chromatography
LLOQ	Lower limit of quantification
LOD	Limit of detection
LOX	Lipoxygenase
LPS	Lipopolysaccharide
LT	Leukotriene
LTA_4	5S-$Trans$-5,6-oxido-7E,9E,11Z,14Z-eicosatetraenoic acid
LTAH	Leukotriene A_4 hydrolase
LTB_4	5S,12R-Dihydroxy-6Z,8E,10E,14Z-eicosatetraenoic acid
m/z	Mass-to-charge ratio
MDA	Malondialdehyde
MeOH	Methanol
MRM	Multiple reaction monitoring
MS	Mass spectrometry
MS/MS	Tandem mass spectrometry
n3-PUFA	Omega-3 polyunsaturated fatty acid
n6-PUFA	Omega-6 polyunsaturated fatty acid
NSAID	Non-steroidal anti-inflammatory drug
PBS	Phosphate buffered saline
PCR	Polymerase chain reaction
PG	Prostaglandin
$PGF_{2\alpha}$	9α,11α,15S-Trihydroxy-prosta-5Z,13E-dien-1-oic acid
PUFA	Polyunsaturated fatty acid
ROOH	Organic hydroperoxide

RP	Reversed-phase
rpm	Rounds per minute
RPMI	Roswell Park Memorial Institute (cell culture medium)
RSD	Relative standard deviation
S/N	Signal-to-noise ratio
SD	Standard deviation
SPE	Solid phase extraction
SRM	Selected reaction monitoring
t-AUCB	*Trans*-4-[4-(3-adamantan-1-yl-ureido)-cyclohexyloxy]-benzoic acid
t-BOOH	*Tert*-butyl hydroperoxide
TGF-β1	Transforming growth factor-β1
t_R	Retention time
TX	Thromboxane
ULOQ	Upper limit of quantification
UPLC	Ultra-high-performance liquid chromatography

What You Will Learn in This Chapter?

- Application of a quantitative targeted metabolomics approach to lipid mediators as an important complement to non-targeted metabolomics and semiquantitative methods.
- Concept of quantitative LC-MS analysis using external calibration and internal standards.
- Strategies for the analysis of low-abundant endogenous compounds/lipids.
- Basic cell culture strategies.
- Evaluation of oxidative stress based on lipid peroxidation products.
- Key lipid mediators formed during inflammation in the cyclooxygenase and lipoxygenase pathways of the arachidonic acid cascade and their pharmacological modulation.
- Lipid mediators are functional markers with dual function reflecting both non-enzymatic formation by autoxidation and enzymatic formation as cellular response.

13.1 Introduction

Polyunsaturated fatty acids (PUFA) occur in all biological systems and are major constituents of the polar lipids in the cell membrane. These PUFA act as precursors of signaling molecules — following a concept widely used by biological systems: The generation of signaling molecules from an abundant species of a major class of biomolecules such as, e.g., amino acids (serotonin, epinephrine, histamine, thyroxine, etc.) or cholesterol (cortisol, testosterone, estradiol, vitamin D, etc.). The signaling molecules resulting from PUFA — lipid mediators — comprise a multitude of different oxygenated PUFA. A challenge for analytical methods is that the concentrations of the signaling molecules are low, and orders of magnitude lower than those of their precursors. Targeted analysis of lipid mediators is currently carried out by LC-MS/MS as summarized in [1, 2], allowing the selective and sensitive detection and quantification of multiple analytes in parallel.

Lipid mediators are formed from PUFA by oxidation leading to eicosanoids (oxygenated C20 PUFA) and other oxylipins (all oxygenated PUFA). Because arachidonic acid (ARA, C20:4 n6) is the major n6-PUFA occurring in mammalian cells, it plays a key role in the formation of highly biologically active lipid mediators [3]. Though this chapter focuses on ARA-derived eicosanoids (which are best investigated), it should be noted that oxylipins can be formed from all PUFA. This is why the diet, i.e., the intake of long-chain n3-PUFA, such as EPA and DHA, directly modulates the oxylipin formation and pattern in the organism [4].

PUFA bearing one or more 1,4-pentadiene systems (two double bounds with a methylene group (CH_2) in between) can easily be chemically oxidized by free radicals and singlet oxygen [5, 6]. Initially hydroperoxides (ROOH) are formed reacting further to a plethora of oxidation products. A break in the carbon chain leads to aldehydes, which are the cause of rancidity in food (plant oils, potato chips, etc.) or reactive markers (e.g., malondialdehyde (MDA), 4-hydroxy-2-nonenal (4-HNE)) of oxidative stress. This oxidative stress is characterized by a misbalance of radical production and removal processes in cells and is associated with several diseases [7–9] — even though it is often unclear whether it is the cause or the consequence. In order to investigate oxidative stress in cells, tissues, and the whole organism stable products are needed.

Oxylipins can serve here as ideal biomarkers: Reduction of initially formed hydroperoxides (ROOH, Hp) — which takes place rapidly in the cell through the action of glutathione peroxidases — leads to hydroxy-PUFA, in case of ARA particularly 5-, 9-, 11-, and 15-hydroxy-eicosatetraenoic acid (HETE) (Fig. 13.1), of which those regioisomers that are not enzymatically formed can be used as marker of oxidative stress. Specifically, stable cyclic reaction products formed *via* bicycloendoperoxide intermediates from the initial peroxy-fatty acids — so-called isoprostanes (IsoP) — are established biomarkers of oxidative stress, induced, e.g., by smoking, cardiovascular or neurological diseases [10, 11].

In the first experiment of this chapter, oxidative stress is induced in a cell line by incubation with *tert*-butyl hydroperoxide (*t*-BOOH), directly inducing lipid peroxidation (Fig. 13.2). This leads to a time- and dose-dependent increase of isoprostanes demonstrating their applicability to monitor oxidative stress in biological systems. Analysis by LC-MS/MS using the targeted oxylipin metabolomics approach additionally enables the simultaneous quantification of hydroxy-PUFA, i.e., HETE derived from hydroperoxy-PUFA (HpETE), which can be formed by autoxidation as well as by enzymes (see

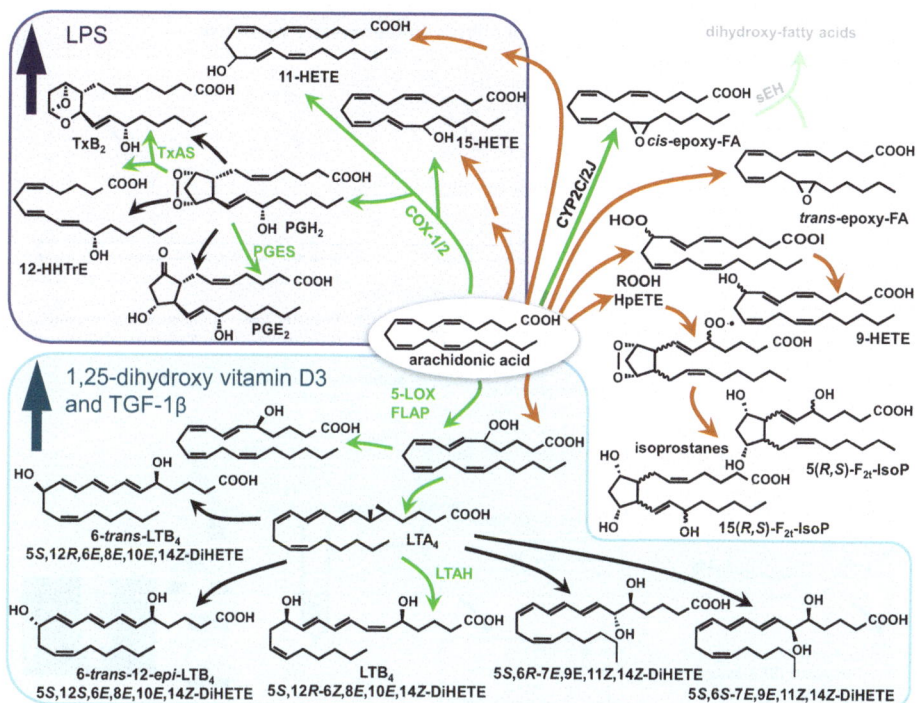

Fig. 13.1 Simplified overview of oxylipin formation. Shown are the main pathways of arachidonic acid-derived eicosanoid formation by COX, LOX, and CYP as well as non-enzymatic autoxidative formation. Formation of enzymatic products is illustrated by green arrows, autoxidation by orange arrows, black arrows depict chemical non-enzymatic breakdown. Furthermore, modulation of eicosanoids by the differentiation and stimulation used in the experiment to investigate the inflammatory response (THP-1 assay, 13.2) is indicated

below). Moreover, the effect of oxidative stress on epoxy-PUFA, i.e., EpETrE, is explored. Though epoxidation of double bounds is not a dominating reaction in the course of autoxidation, we recently uncovered that the ratio of the stereoisomers (*cis*- and *trans*-epoxy-PUFA) is a new marker of oxidative stress [12]. Based on the evaluation of these three classes of oxylipins the power of quantitative oxylipin metabolomics to assess oxidative stress is demonstrated. Correlating the different results enables to learn about the biology of (non-enzymatic) oxylipin formation.

Conversion of PUFA by the enzymes of the ARA-cascade leads to similar oxygenation products. However — as for all enzymatic catalysis — the reactions are much more selective and give rise to specific products [3]. Though some of these products can also be formed by non-enzymatic reactions, the product resulting from enzymatic conversion is just one of a large number of possible regio- and stereoisomers (e.g., prostaglandin $F_{2\alpha}$ ($PGF_{2\alpha}$)) or even not formed without a specific enzyme which is highlighted in the second experiment for leukotriene B_4 (LTB_4).

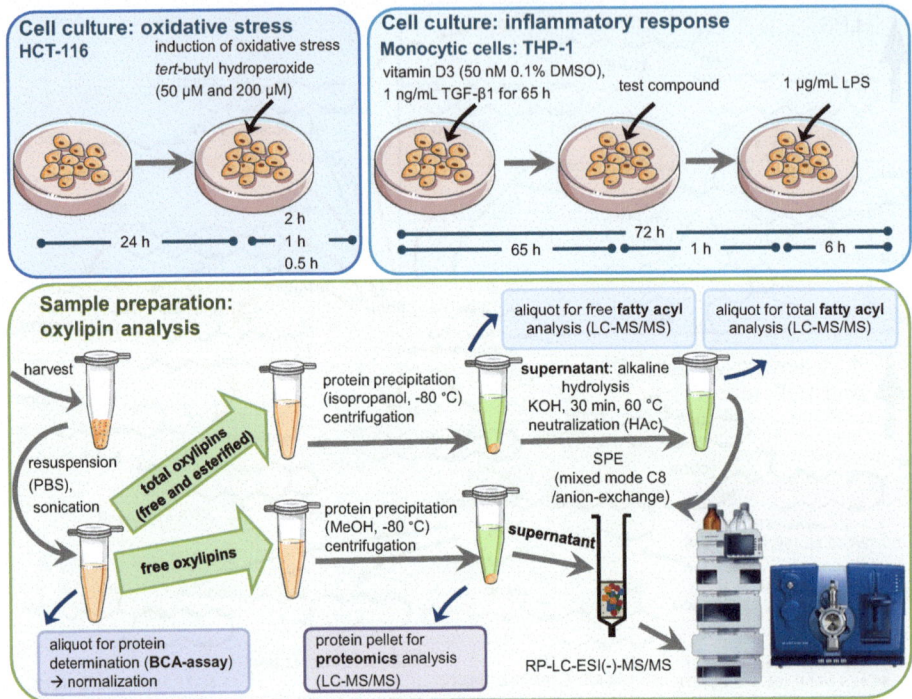

Fig. 13.2 Workflow of cell incubation and sample preparation. Shown is the incubation strategy used in the experiments in this chapter: Left: Induction of oxidative stress by *t*-BOOH in adherent HCT-116 cells; Right: Differentiation and compound testing in the context of inflammation induced by LPS using the monocytic cells THP-1. Bottom: Sample preparation strategy for the analysis of free and esterified oxylipins. The possible parallel analysis of protein levels, as well as fatty acyls is highlighted

Three (super)families of enzymes catalyze the formation of eicosanoids and other oxylipins. By far the best investigated is the cyclooxygenase (COX) pathway. COX generates an unstable bicycloendoperoxide-peroxy product (PGG$_2$) from ARA, which is reduced to the unstable PGH$_2$ which can be further converted by specific synthases (and in case of PGE$_2$, PGD$_2$ and TXB$_2$ also non-enzymatically) to prostaglandins (PG), thromboxane (TX) and prostacyclin (Fig. 13.1) [10, 13]. The latter are regulators of, e.g., pain, inflammation, fever, and blood coagulation. Thus, COX is a major target of widely used pharmaceuticals including the best-selling over-the-counter drugs aspirin, diclofenac, and ibuprofen, summarized as non-steroidal anti-inflammatory drugs (NSAIDs). Two isoforms of COX exist: COX-1 being constitutively expressed, e.g., in the gastrointestinal tract (GIT), and the inducible COX-2 playing a key role in inflammation. In the second experiment a monocytic cell line (THP-1), which is differentiated to a macrophage-like phenotype is used. Here, an inflammatory response, and thus induction of COX is elicited by lipopolysaccharide (LPS), a constituent of the cell wall of Gram-negative bacteria

(Fig. 13.2). This is assessed based on a dramatic increase in the production of PGE_2, a major prostanoid formed during acute inflammation, as well as two non-enzymatically formed PGH_2 breakdown products: 12-HHTrE and TXB_2 (Fig. 13.1) [14]. The experimental setup also allows the testing of common drugs, enabling to investigate hands-on the mechanism and biological effect of drugs known from daily life.

Different hydroxy-PUFA are formed (through reduction of initially generated hydroperoxides) as side products of COX activity, i.e., 11-HETE and 15-HETE (Fig. 13.1) [15, 16]. Again, it is a key learning for the interpretation of data from metabolomics to evaluate and interpret the levels of these oxylipins, which can be formed *via* different enzymatic as well as non-enzymatic pathways. A highly interesting extension of the experiment could be the investigation of aspirin, irreversibly blocking prostanoid formation, while COX-2 is still giving rise to 15-H(p)ETE [15, 17, 18].

The second pathway of the ARA-cascade is catalyzed by several lipoxygenases (LOX), forming in the first step positional and stereospecific hydroperoxides, which can be reduced to hydroxy-PUFA, i.e., HETE in the case of ARA. The nomenclature of the enzymes originates from the position where the hydro(peroxy) group is inserted within the ARA molecule. Two 12-LOX, two 15-LOX, and the 5-LOX are found in humans [19].

In the second experiment, only the 5-LOX pathway is investigated giving rise to physiologically active leukotrienes: The initially formed 5-hydroperoxy-fatty acid is converted by 5-LOX to the unstable epoxide LTA_4 (Fig. 13.1). Similar to PGH_2 in the prostanoid formation pathway, this intermediate is then transformed by specific enzymes. The LTC_4 synthase reduces the epoxy group forming a glutathione conjugate, which causes muscle contraction of the smooth muscles playing a key role in the regulation of the lung function (not shown). Hydrolysis by leukotriene A_4 hydrolase (LTAH) leads to LTB_4, a highly potent chemoattractant for neutrophils. This means, if this compound is released by a cell, it attracts neutrophils to come to the site thereby increasing the (local) inflammatory response. In this experiment, the formation of LTB_4 is analyzed in THP-1 cells. Moreover, in parallel also the isomers of LTB_4 (*trans*- and *epi*-isomers of LTB_4 as well as 5,6-DiHETE isomers) are monitored which are chemical breakdown products of the unstable LTA_4 formed when the short lived (about 20 s in a biological setting) LTA_4 is not timely converted by LTAH (Fig. 13.1).

The third pathway of the ARA-cascade is catalyzed by cytochrome P450 monooxygenase (CYP) enzymes giving also rise to highly potent lipid mediators [20, 21]. Terminal hydroxylation results in the formation of 20-HETE — a hydroxy-PUFA that cannot be generated by autoxidation — playing a key role in the regulation of blood pressure. Moreover, epoxy-PUFA (Fig. 13.1) are formed acting as potent anti-inflammatory and vasodilatory mediators. Investigating the formation of these mediators is beyond the scope of the current educational experiments (for the investigation of the modulation of the CYP pathway by phytochemicals see [22]).

It should be noted that the experiments carried out in this chapter only highlight a few of the large number of oxylipins formed from ARA and all other PUFA, which can be in parallel investigated using the described methodology (for the methods please refer to [14, 23–25]).

With the application of quantitative targeted oxylipin metabolomics for the evaluation of oxidative stress, the first experiment demonstrates how oxylipins can be used as markers for this (patho)physiological condition.

In the second set of experiments, oxylipins are monitored as active mediators of inflammation, generated by COX-2 and 5-LOX. It is demonstrated how these oxylipins can be modulated by stimuli and how their formation can be blocked by drugs. Here, targeted metabolomics serves as an indispensable tool to monitor the target engagement of the drugs (Fig. 13.2).

13.2   Chemicals and Materials

All chemicals and materials needed for the experiments are listed in the supplementary material (Table S1). Preparation of solutions for cell culture, solid phase extraction (SPE) and LC-MS/MS analysis is described in the supplementary material.

13.3   Step-By-Step Protocol

13.3.1 Cell Cultivation: Maintaining Cells in Culture

Thawing of cells stored in cryo tubes in liquid nitrogen:

a. Gently thaw the cell pellet in the cryo tube in a water bath (37 °C, approx. 2 min).
b. Transfer the cells under sterile conditions to a 15-mL conical centrifuge tube containing 10 mL warmed medium (see Sects. 13.3.2 and 13.3.3 for medium composition).
c. Centrifuge (5 min, room temperature, 300–500 × g).
d. Remove the medium.
e. Add 7 mL (for HCT-116) or 5 mL (for THP-1) fresh medium and resuspend the cells thoroughly.
f. Seed/transfer the cells:
   i. For HCT-116 (adherent cells): Seed in a T25 flask (25 cm^2) for adherent cells.
   ii. For THP-1 (suspension cells): Seed in a 22.1 cm^2 dish for suspension cells.
g. Check growth and morphology of the cells under the microscope on the following days, replace the medium with fresh if needed, and transfer the cells to bigger dishes when almost confluent.

NOTE: *Use the cells only for at most 10 passages after defrosting, enabling the replication of experiments from frozen cell stocks (at comparable passage number).*

Check absence of mycoplasmas (every second month): These are small bacteria that can contaminate the cell culture and alter the biology of the test system. To verify their absence in the cell culture, collect an aliquot of the cell culture medium from nearly confluent cells (right before transfer) and determine mycoplasmas using, e.g., a commercial mycoplasma detection kit for conventional PCR.

Generation of a lab stock of frozen cells:

a. Transfer the cells from the dish to a conical centrifuge tube.
   i. For HCT-116 (adherent cells): Detach the cells from the dish using trypsin (as described in Sect. 13.3.2, Step 3. a. – g).
   ii. For THP-1 (suspension cells): Gently resuspend the cells before transfer.
b. Determine the cell number in the suspension, e.g., with a Neubauer chamber.
c. Centrifuge (5 min, room temperature, $600 \times g$).
d. Remove the medium and resuspend the cells in cold fetal calf serum (FCS) containing 10% DMSO leading to 2×10^6 cells/mL (HCT-116) or 10×10^6 cells/mL (THP-1).
e. Transfer 1 mL in cryo tubes and freeze at $-80\ °C$ overnight using a cell freezing container (e.g., "Mr. Frosty") allowing a freezing rate of 1 °C/min, which is the optimal condition to conserve cells.
f. Transfer the cryo tubes to liquid nitrogen storage.

NOTE: Check the viability of the cells in an aliquot of the generated frozen cell pellets by performing the thawing steps described under Sect. 13.3.1 a–g.

13.3.2 Cell Culture Assay: Oxidative Stress in HCT-116 Cells

1. The adherent human colorectal carcinoma cell line HCT-116 (obtained from the German Collection of Microorganisms and Cell Cultures GmbH (DSMZ, Braunschweig, Germany)) is cultivated in Dulbecco's modified Eagle medium (DMEM) supplemented with 10% FCS, 100 U/mL penicillin, 100 µg/mL streptomycin (P/S, 2%) and 2 mM L-glutamine (1%) in a humidified incubator at 37 °C and 5% CO_2 in 10 cm dishes (60.1 cm^2) for adherent cells.
   Maintaining the cell culture: Transfer cells every 2–3 days:
2. Replace medium every 2 days (if no cell passaging is carried out, otherwise see Step 3):
   a. Carefully remove the medium (e.g., using a membrane pump and pipette).
   b. Replace with fresh warmed medium.
3. Passaging every 2–3 days (Monday, Wednesday, Friday):
   a. Carefully remove the medium.
   b. To detach the cells, add 1.5 mL trypsin (0.25% trypsin in PBS-EDTA) to the cells, swivel the dish to wet the surface completely, and remove the trypsin directly.
   c. Add 1.5 mL trypsin (0.25% trypsin in PBS-EDTA), let it act while swiveling the dish for 30 s, and it remove again.
   d. Incubate the cells on the dish without liquid for 1.5 min at 37 °C.

e. Tap the dish, e.g., against the palm of the hand or the bench, and observe if the cells are detaching and moving from the surface of the dish. If the cells do not move, incubate for another 10 s at 37 °C and check again. Repeat in steps of 10 s until the cells are detaching.

NOTE: If protein expression levels are intended to be determined the use of trypsin for detachment of the cells should be omitted to avoid premature protein digestion. For this, scraping is recommended to detach the cells from the surface of the dish (see Sect. 13.3.3, Step 4. a).

f. Add 10 mL fresh warmed medium and flush the surface of the dish thoroughly to detach and remove all cells from the surface of the dish.
g. Transfer the cells and medium to a conical centrifuge tube.
h. Determine the cell number.
i. Seed 1×10^6 (on Monday and Wednesday) or 0.8×10^6 (on Friday) cells in a total of 10 mL fresh medium in new dishes (60.1 cm^2) for adherent cells.

4. Incubation with *tert*-butyl hydroperoxide (*t*-BOOH):
   a. Seed 2×10^6 cells in a total of 10 mL fresh medium in new dishes (60.1 cm^2) for adherent cells.
   b. After 24 h of growth remove the medium, add fresh medium (without FCS), and add 10 μL *t*-BOOH (in water). For dose- and time-dependent investigation of the effects recommended final concentrations in the dishes (60.1 cm^2 for adherent cells, 10 mL medium) are 50 μM and 200 μM *t*-BOOH for an incubation time of 30 min, 1 h and 2 h.

5. Harvest the cells using trypsin as described in Step 3. a–e.
   a. Add 5 mL cold PBS + 5% FCS to the detached cells in suspension, and transfer to a 15-mL conical centrifuge tube.
   b. Centrifuge (5 min, 4 °C, 200–600 × g).
   c. Remove the supernatant liquid.
   d. Resuspend the cell pellet in 1 mL cold PBS and transfer to a 1.5-mL reagent tube yielding a pellet of approx. $4–8 \times 10^6$ cells. The 15-mL tube should be rinsed with additional ~300 μL cold PBS, which are transferred to the same 1.5-mL reagent tube.
   e. Centrifuge (5 min, 4 °C, 600–1000 × g) and remove the liquid.

6. Freeze the cell pellet at −80 °C until analysis.
7. Exclude cytotoxic effects of *t*-BOOH at the used concentrations and incubation times, e.g., by resazurin (Alamar Blue) assay [26] and lactate dehydrogenase assay [27, 28].

Question (1)

Why is it important to determine the cytotoxicity of the test compound at the used concentration? Which parameters for different cytotoxic endpoints can be assessed?

13.3.3 Cell Culture Assay: Inflammatory Response in THP-1 Cells

1. The monocytic cell line THP-1 (obtained from the German Collection of Microorganisms and Cell Cultures GmbH (DSMZ, Braunschweig, Germany)) is cultivated in suspension in bicarbonate buffered RPMI 1640 medium supplemented with 10% FCS, 100 U/mL penicillin, 100 µg/mL streptomycin (P/S, 2%), and 2 mM L-glutamine (1%) in a humidified incubator at 37 °C and 5% CO_2 in 10 cm dishes (58.8 cm^2) for suspension cells.

2. *Maintaining the cell culture:* Transfer cells every 4–5 days (Monday and Friday):
   a. Gently resuspend the cells in the dish and transfer cells and medium to a conical centrifuge tube.
   b. Determine the cell number.
   c. Seed 1×10^6 cells (on Monday and let them grow till Friday) or 1.8×10^6 cells (on Friday and let them grow till Monday) in a total of 10 mL fresh medium in new 10 cm dishes (58.8 cm^2) for suspension cells.

3. Differentiation and incubation with test compounds:
   a. For differentiation of cells, prepare medium by adding 10 µL 50 µM vitamin D3 (in DMSO) and 10 µL 1 µg/mL TGF-β1 (in PBS) to 10 mL RPMI 1640 medium resulting in 50 nM vitamin D3 (0.1% DMSO) and 1 ng/mL TGF-β1. Seed cells at densities of 0.125×10^6 cells/mL in 10 mL of this medium in 10 cm dishes (60.1 cm^2) for adherent cells as cells will become partially adherent during differentiation. Allow the cells to differentiate for 72 h.
   b. For incubation with test compounds, replace the cell culture medium 7 h before the end of the differentiation (65 h after the start of the differentiation) with serum-free 50 mM TRIS buffered RPMI medium (2% P/S, 1% L-glutamine):
      i. Transfer non-attached cells and medium to a conical centrifuge tube and add directly 5 mL serum-free 50 mM TRIS buffered RPMI medium (2% P/S, 1% L-glutamine) to the dish.
      ii. Centrifuge the tube (5 min, room temperature, $500 \times g$) and remove the medium.
      iii. Add 5 mL serum-free 50 mM TRIS buffered RPMI medium (2% P/S, 1% L-glutamine) to the cell pellet, resuspend the cells and transfer back to the dish.
   c. Add 10 µL of the test compound (pharmacological drug/inhibitor) in DMSO (e.g., 100 µM indomethacin yielding a final concentration of 100 nM) or DMSO (0.1%) as control. Suggestions for possible test compounds are summarized in Table 13.1.
   d. After 1 h of preincubation, add 10 µL 1 mg/mL LPS (in PBS) to the medium resulting in 1 µg/mL LPS (for control add 10 µL PBS) for 6 h.

Table 13.1 Suggested test compounds and concentrations for the incubation to assess the modulation of the inflammatory response in THP-1 cells

Test compound	Concentration range
Dexamethasone	10 pM–1 µM
Indomethacin	1 nM–10 µM
Celecoxib	1 nM–10 µM

Question (2)

Which control incubations should be carried out to allow meaningful interpretation? Which aspects should be considered when performing control incubations?

4. Harvest all adherent and non-adherent cells:
   a. Scrape the cells from the dish in the culture medium with a cell scraper, and thoroughly flush the dish with the suspension.
   b. Transfer cells and medium to a 15-mL conical centrifuge tube and place on ice.
   c. Centrifuge (5 min, 4 °C, 200–600 × g).
   d. Remove the medium.
   e. Add 5 mL PBS and resuspend thoroughly.
   f. Centrifuge and remove the supernatant again.
   g. Resuspend the cell pellet in 1 mL PBS (containing 1% protease inhibitor mixture with AEBSF, Aprotinin, Bestatin, E-64, Leupeptin, and Pepstatin A, if proteomics analysis of the cell pellet is intended to be carried out).
   h. Transfer to a 1.5-mL reagent tube yielding a pellet of approx. $6-9 \times 10^6$ cells.
   i. Centrifuge (5 min, 4 °C, 500 × g) and remove the supernatant.
   j. Freeze the cell pellet at −80 °C until analysis.
5. Exclude cytotoxic effects of the test compounds at the used concentrations, eg., by resazurin (Alamar Blue) assay [26] and lactate dehydrogenase assay [27, 28].

13.3.4 Sample Preparation

NOTE: Work on ice, and store samples on ice during sample preparation to minimize artificial oxylipin formation.

Free and total oxylipins, protein expression levels as well as fatty acyls can be determined from a single cell pellet (Fig. 13.2).

1. Resuspend the cell pellets in an exact volume (300–500 µL) of MeOH/H$_2$O (50/50, *v/v*) (containing 1% protease inhibitor mixture with AEBSF, Aprotinin, Bestatin, E-64, Leupeptin, and Pepstatin A, if proteomics analysis of the cell pellet is intended to be carried out), and add 10 µL inhibitor/antioxidant solution (0.2 mg/mL BHT, 100 µM of the COX inhibitor indomethacin, 100 µM of the soluble epoxide hydrolase inhibitor *trans*-4-[4-(3-adamantan-1-yl-**u**reido)-**c**yclohexyloxy]-**b**enzoic acid (*t*-AUCB) and 100 µM of the 15-LOX inhibitor BLX3887 in methanol (MeOH)).
2. Sonicate (keep the tube in ice during the sonication):
   a. Small ultrasonic tip (1 mm): amplitude 100%, unpulsed, cycle 1, 8–10 s.
   b. Ultrasonic tip (3 mm): level 2, 20% output, unpulsed, 8–10 s.

Repeat the sonication for another 10 s if the cell suspension does not look clear when viewed against the light.

3. Determine the protein content in an aliquot of the sonicate (use an exact volume of 10–30 µL for an appropriate dilution) *via* the bicinchoninic acid (BCA) assay [29] allowing normalization of the oxylipin concentration to the protein content. (Expected protein concentration in the sonicate: for HCT-116 cells 3–6 mg/mL, for THP-1 cells 0.4–2 mg/mL). If the cell number of the pellet is known the oxylipin concentration can alternatively be normalized to the cell number.

   For an assay as carried out in our laboratory with a working range of 0.1–1 mg/mL BSA the following dilutions are suggested for the BCA assay when the cell pellet is resuspended in 500 µL MeOH/H_2O (with inhibitors) for sonication:

   a. For HCT-116 cells an 1 + 4 dilution.

   b. For THP-1 cells an 1 + 2 dilution.

4. The homogenous cell lysate can be split to analyze free and total (the sum of free and esterified) oxylipins from the same cell pellet depending on the scientific question. Based on that the volume of this step should be decided, e.g., whether a higher sensitivity is needed for free or total oxylipins.

   If both free and total oxylipins are analyzed it is suggested to use 350 µL of the sonicate for the determination of free oxylipins, and 100 µL of the sonicate for the determination of total oxylipins, and the rest for protein determination *via* the BCA assay. Otherwise, select volumes to achieve best sensitivity for the scientific question.

> **Question (3)**
> With the described methodology it is possible to analyze free and total (the sum of free and esterified) oxylipins. Can you give examples in which experimental setups the analysis of free or total oxylipins makes more sense and why? For the two experiments described here, is the analysis of free or total oxylipins more meaningful to characterize the resulting biological effects?

a. For free oxylipins: Add the 2.8-fold volume of ice-cold MeOH and 10 µL internal standards (IS) (100 nM of $^2H_4$–PGE$_2$, $^2H_4$–TXB$_2$, $^2H_4$–LTB$_4$, $^2H_8$–5-HETE, $^2H_8$–12-HETE, $^2H_8$–15-HETE, $^2H_{11}$–11,12-DiHETrE, $^2H_{11}$–8(9)-EpETrE, $^2H_{11}$–14(15)-EpETrE) using a repeating syringe dispenser ("Hamilton-Repeater") (500 µL syringe).

b. For total oxylipins: Add 400 µL ice-cold isopropanol and 10 µL internal standards (IS) (100 nM of $^2H_4$–15-F$_{2t}$-IsoP, $^2H_{11}$–5-(R,S)-5-F$_{2t}$-IsoP, $^2H_8$–5-HETE, $^2H_8$–12-HETE, $^2H_8$–15-HETE, $^2H_{11}$–11,12-DiHETrE, $^2H_{11}$–8(9)-EpETrE, $^2H_{11}$–14(15)-EpETrE) using a repeatingsyringe dispenser (500 µL syringe).

5. Vortex samples thoroughly, and precipitate proteins by freezing at −80 °C for at least 30 min.

   Possible break for hours up to several days.

6. When samples are taken from $-80\ °C$ freezer, leave them at room temperature for 1–2 min, then vortex briefly.
7. Centrifuge (10 min, 4 °C, 20,000 x g).

    OPTION: The pellet is discarded or can be used to determine protein levels using targeted LC-MS/MS-based proteomics analysis as described in [14]. Therefore, we recommend to centrifuge with a slightly lower acceleration and to store the pellet at −80 °C until proteomics analysis.

8. The supernatant is used for oxylipin analysis:
    a. For free oxylipins, the supernatant can be loaded directly onto the prepared solid phase extraction (SPE) cartridge (see Sect. 13.3.5, Step 3).
    b. For total oxylipins, a hydrolysis step is required.
9. Hydrolysis for total oxylipins:
    a. Transfer the supernatant to a 1.5-mL reagent tube.
    b. Add 100 µL 0.6 M KOH in $MeOH/H_2O$ (75/25, *v/v*).
    c. Vortex.
    d. Hydrolyze the sample for 30 min at 60 °C using a pre-heated shaker (500 rpm).
    e. After hydrolysis, cool the sample immediately on ice and neutralize it by adding 20 µL acetic acid (HAc, 25% in water).
    f. Vortex.
    g. Centrifuge the sample very briefly in order to collect the liquid on the bottom of the tube.

    OPTION: Fatty acyls can be determined in an aliquot of thesample according to [30]. For analysis of free fatty acyls use an appropriate aliquot of the sample before hydrolysis (e.g., dilute 50 µL + 50 µL EtOH). Foranalysis of total fatty acyls use an appropriate aliquot of the hydrolyzed sample (e.g., dilute 20 µL + 180 µL EtOH for low-abundant fatty acyls, and 10 µL + 490 µL EtOH for high-abundant fatty acyls).

13.3.5 Solid Phase Extraction

1. Prepare solid phase extraction (SPE) cartridges (mixed-mode anion exchange/C8, Bond Elut Certify II, 200 mg, 3 mL, Agilent) by washing with:
    a. One column volume of ethyl acetate/*n*-hexane (75/25, *v/v*) containing 1% HAc.
    b. One column volume of MeOH.
    c. One column volume of 0.1 M disodium hydrogen phosphate (Na_2HPO_4) buffer in $H_2O/MeOH$ (95/5, *v/v*) (adjusted to pH 6.0 with HAc). Close the valve when the solution is 2–3 mm above the stationary phase.
2. Add 2.0 mL 0.1 M Na_2HPO_4 buffer in H_2O (adjusted to pH 6.0 with HAc) to the cartridges.
3. Load samples with a glass Pasteur pipette and mix thoroughly with the buffer.

NOTE: The content of organic solvent on the cartridge should be kept below 16% to prevent breakthrough/elution of the analytes during the loading step. If necessary, the volume of the buffer needs to be adjusted or the sample needs to be evaporated appropriately.

4. Check pH using pH strips (5.1–7.2 scale); only if necessary, carefully adjust pH to 6.0 with diluted HAc (if HAc has to be added, only a few μL are needed).
5. Open valves and let samples run by gravity until completely sunk into the stationary phase.
6. Wash with:
   a. One column volume of H_2O.
   b. One column volume of MeOH/H_2O (50/50, *v/v*).
7. Dry samples with vacuum:
   a. Close valves of all cartridges and create a stable −200 mbar negative pressure within the manifold.
   b. Open valves of two or three samples for drying the cartridges.
      NOTE: Drying of the cartridges can be verified by putting a pipette cone on top of the cartridges (cone should tighten).
   c. Close valves after 30 s (it is not critical if samples dry a few seconds longer; however, do not dry them longer than 1 min).
   d. Repeat drying step for all samples in pairs of three or two.
8. Elute analytes with 2.0 mL ethyl acetate/*n*-hexane (75/25, *v/v*) with 1% HAc by gravity in glass tubes containing 6 μL 30% glycerol in MeOH (a dispenser resistant to organic solvents can be used to measure the eluent volume). Remove last drops of eluent from the stationary phase by applying positive pressure with a pipette cone at the top of the cartridge.
9. Evaporate samples to dryness using a vacuum centrifuge (1 mbar, 30 °C, ~60 min).
10. Reconstitute samples in 50 μL MeOH using a repeating syringe dispenser (2.5 mL syringe) and dissolve samples by sonication and vortexing.

NOTE: Inclusion of one or more secondary internal standards in the reconstitution solvent enables to determine the extraction efficiency of the sample preparation (see Sect. 13.4.1); (for examples of secondary internal standards see [25]).

Question (4)
How would you select a secondary internal standard? What characteristics need to be considered?

11. Transfer samples completely into 1.5-mL reagent tubes.
12. Freeze samples at −80 °C for at least 30 min.

Possible break: Reconstituted samples can be stored for at least 2 months at − 80 °C with only slight changes in the oxylipin pattern (<20% for most analytes, CAVE: quantification of isoprostanes might be impaired).

13. Centrifuge (10 min, 4 °C, 20,000 x g).
14. Transfer clear (!) supernatant into vial with insert. Centrifuge samples again if supernatant is not completely clear.

13.3.6 LC-ESI(−)-MS/MS Analysis

The analysis is carried out using reversed-phase (RP) LC-MS/MS on a triple quadrupole instrument operated in selected reaction monitoring mode (SRM, frequently termed MRM) following negative electrospray ionization (ESI(−)).

The selection of transitions, as well as electronic parameters and source settings, requires extensive optimization, which is described in detail in [23, 24].

An excellent chromatographic separation is also mandatory for oxylipin analysis because several oxylipins cannot be separated by MS/MS (a list of typical critical separation pairs can be found in the supplementary information of [31]).

Here, we describe an optimized method using a liquid chromatography system composed of a 1290 Infinity LC system (Agilent, Waldbronn, Germany) with autosampler, binary pump, and column oven coupled to a QTRAP mass spectrometer (Sciex, Darmstadt, Germany). For data acquisition and instrument control Analyst Software, and for integration and quantification Multiquant Software is used.

However, when using (slightly) different instrumentation, the LC-MS/MS parameters need to be adapted to the used system.

1. Inject (5 μL) the samples into the LC-MS/MS system (keep samples in a 4 °C cooled autosampler rack until injection).
2. For liquid chromatography, an UPLC system is needed as a backpressure of about 600 bar is reached during the analysis. Separate analytes on a Zorbax Eclipse Plus C 18 column (2.1 × 150 mm, particle size 1.8 μm; RRHD; Agilent) equipped upstream with an in-line filter (0.3 μm, 2 mm ID, 1290 infinity II in-line filter; Agilent) and a SecurityGuard Ultra C18 cartridge as pre-column (2.1 × 2 mm; Phenomenex) at 40 °C.
3. For chromatographic separation use the following binary gradient at a flow rate of 0.3 mL/min with solvent A 0.1% HAc/solvent B (95/5, v/v), and solvent B acetonitrile/ MeOH/HAc (800/150/1, v/v/v): 0–1.0 min isocratic 21% B, 1.0–1.5 min linear from 21% B to 26% B, 1.5–10.0 min linear from 26% B to 51% B, 10.0–19.0 min linear from 51% B to 66% B, 19.0–25.1 min linear from 66% B to 98% B, 25.1–27.6 min isocratic 98% B, 27.6–27.7 min linear from 98% B to 21% B, followed by reconditioning for 3.8 min. During the first 2 min and the last 6 min of each run the LC flow is directed to waste using the 2-position-6-port valve integrated in the MS.

4. For mass spectrometric detection use negative electrospray ionization (ESI(−)) with the following source settings: ion-spray voltage: −4500 V, curtain gas (N$_2$): 35 psi, nebulizer gas (gas 1, zero air generated with a zero air generator): 60 psi, and drying gas (gas 2, zero air): 60 psi at a temperature of 475 °C. The offset of the sprayer is 0.250 cm for the vertical axis, and 0.550 cm for the horizontal axis, the electrode protrusion is approx. 1 mm.

5. For detection of the analytes use scheduled SRM with nitrogen as collision gas (set to "high", 12 psi), a detection window of ± 22.5 s around the expected retention time, and a cycle time of 0.4 s.

6. For each analyte use the optimized compound-specific parameters. Optimized parameters for the selected analytes evaluated within the experiments in this chapter are summarized in Table 13.2. A list of more oxylipins covered by this methodology and their mass spectrometric parameters can be found in [14].

7. For calibration, mix stock solutions of the individual authentic standards (in MeOH) and dilute in glass volumetric flasks (5–100 mL) with MeOH at 10 concentration levels (0.1, 0.25, 0.5, 1, 2, 5, 10, 20, 100, and 500 nM), each with 20 nM of the internal standards. An appropriate procedure is described in detail in [14, 25].

   NOTE: For evaluation of extraction efficiency of the sample preparation based on second internal standards a secondary calibration curve should be prepared (covering 20–120% recovery of IS1).

8. Analyze calibration standards with the same method as samples.

13.4   Data Analysis/Interpretation

In contrast to non-targeted metabolomics, with targeted metabolomics the metabolites in the biological samples are quantified. Thus, the key step is the calculation of concentrations of the analyzed compounds. Similar to all other chromatography-mass spectrometry-based quantitative methods used in food, pharmaceutical, forensic or environmental chemistry, and other fields the concentration is calculated based on external calibration. In order to compensate for losses occurring by sample preparation, matrix effects and instability of the MS-signal IS are used. Thus, instead of the peak area of the analytes, the peak area ratio of the analyte to its assigned IS is used.

> **Question (5)**
> What is ion suppression occurring in ESI-MS and how do you detect it? Why are isotopically labeled standards the only way to enable a robust quantification of analytes in biological samples with varying matrix (matrix = all other compounds in the sample)?

Table 13.2 Parameters for the targeted LC-ESI(−)-MS/MS analysis of selected oxylipins covered in the experiments

Analyte	Mass transition		Electronic parameters				Internal standard	tR	Calibration range		
	Q1	Q3	DP [V]	EP [V]	CE [V]	CXP [V]		[min]	LOD [nM]	LLOQ [nM]	ULOQ [nM]
15-F$_{2t}$-IsoP (8-iso-PGF$_{2\alpha}$)	353.1	193.1	−95	−10	−34	−8	^{2}H$_4$-15-F$_{2t}$-IsoP	7.58	0.10	0.25	500
TXB$_2$	369.2	169.1	−80	−10	−24	−7	^{2}H$_4$-TXB$_2$	7.68	0.25	0.50	1000
5(R,S)-5-F$_{2t}$-IsoP (5-iPF$_{2\alpha}$-VI)	353.2	114.8	−85	−10	−27	−8	^{2}H$_{11}$-5-(R,S)-5-F$_{2t}$-IsoP	8.07	0.1	0.25	500
PGE$_2$	351.2	271.3	−80	−10	−23	−6	^{2}H$_4$-PGE$_2$	8.91	0.25	0.50	750
6-$trans$-LTB$_4$	335.2	195.1	−80	−10	−22	−9	^{2}H$_4$-LTB$_4$	13.36	0.1	0.25	250
6-$trans$-12-epi-LTB$_4$	335.2	195.1	−85	−10	−20	−9	^{2}H$_4$-LTB$_4$	13.51	0.1	0.25	500
LTB$_4$	335.2	195.1	−80	−10	−22	−9	^{2}H$_4$-LTB$_4$	13.83	0.05	0.1	500
5(S),12(S)-DiHETE	335.2	195.1	−80	−10	−21	−8	^{2}H$_4$-LTB$_4$	14.40	0.025	0.05	500
12-HHTrE	279.1	179.0	−70	−10	−15	−8	^{2}H$_{11}$-11,12-DiHETrE	15.62	0.25	0.5	500
5(S),6(R)-DiHETE (ARA)	335.2	115.1	−70	−10	−20	−8	^{2}H$_{11}$-11,12-DiHETrE	17.33	0.020	0.039	390
5(S),6(S)-DiHETE (ARA)	335.2	115.1	−70	−10	−20	−8	^{2}H$_{11}$-11,12-DiHETrE	17.80	0.022	0.045	223
15-HETE	319.2	219.2	−80	−10	−18	−8	^{2}H$_8$-15-HETE	20.08	0.11	0.22	220
11-HETE	319.2	167.2	−80	−10	−21	−7	^{2}H$_8$-12-HETE	20.68	0.022	0.044	219
9-HETE	319.2	167.2	−80	−10	−21	−7	^{2}H$_8$-5-HETE	21.45	0.27	0.4	265
5-HETE	319.2	115.2	−80	−10	−19	−7	^{2}H$_8$-5-HETE	21.74	0.018	0.035	350

(continued)

Table 13.2 (continued)

14(15)-EpETrE	319.2	219.2	−90	−10	−15	−4	$^2H_{11}$–14(15)-EpETrE	22.45	0.1	0.25	500
trans-14(15)-EpETrE	319.2	219.2	−90	−10	−15	−4	$^2H_{11}$–14(15)-EpETrE	22.60		Relative quantification based on 14(15)-EpETrE	
11(12)-EpETrE	319.2	167.2	−85	−10	−16	−7	$^2H_{11}$–8(9)-EpETrE	22.98	0.05	0.1	500
trans-11(12)-EpETrE	319.2	167.2	−85	−10	−16	−7	$^2H_{11}$–8(9)-EpETrE	23.13		Relative quantification based on 11(12)-EpETrE	
8(9)-EpETrE	319.2	155.2	−90	−10	−16	−6	$^2H_{11}$–8(9)-EpETrE	23.16	0.25	0.5	500
trans-8(9)-EpETrE	319.2	155.2	−90	−10	−16	−6	$^2H_{11}$–8(9)-EpETrE	23.31		Relative quantification based on 8(9)-EpETrE	
$^2H_4$–15-F_{2t}-IsoP	357.2	196.8	−75	−10	−34	−8	*Internal standard*	7.55			
$^2H_4$-TXB$_2$	373.3	173.2	−85	−10	−23	−8	*Internal standard*	7.66			
$^2H_{11}$–5(R,S)-5-F-IsoP	364.2	115.2	−70	−10	−30	−10	*Internal standard*	8.04			
$^2H_4$-PGE$_2$	355.2	275.3	−80	−10	−24	−6	*Internal standard*	8.88			

(continued)

Table 13.2 (continued)

Analyte	Mass transition		Electronic parameters				Internal standard	t_R	Calibration range		
	Q1	Q3	DP [V]	EP [V]	CE [V]	CXP [V]		[min]	LOD [nM]	LLOQ [nM]	ULOQ [nM]
$^2H_4$-LTB$_4$	339.2	197.2	−80	−10	−22	−9	Internal standard	13.76			
$^2H_{11}$–11,12-DiHETrE	348.2	167.2	−85	−10	−26	−8	Internal standard	16.31			
$^2H_8$–15-HETE	327.2	226.0	−90	−10	−18	−8	Internal standard	19.88			
$^2H_8$–12-HETE	327.2	184.2	−85	−10	−20	−8	Internal standard	20.93			
$^2H_8$–5-HETE	327.2	116.1	−80	−10	−19	−8	Internal standard	21.60			
$^2H_{11}$–14(15)-EpETrE	330.2	219.3	−90	−10	−16	−4	internal standard	22.32			
$^2H_{11}$–8(9)-EpETrE	330.2	155.0	−80	−10	−16	−7	Internal standard	23.05			
Aleuritic acid	303.1	268.8	−75	−10	−41	−10	Secondary internal standard	5.39			
(1-(1-(Ethyl-sulfonyl)piperidin-4-yl)-3-(4-(trifluoromethoxy)phenyl)urea)	394.0	176.0	−100	−10	−21	−10	Secondary internal standard	10.69			

(continued)

Table 13.2 (continued)

12-oxo Phytodienoic acid (OPDA)	291.1	165.0	−90	−10	−27	−10	*Secondary internal standard*	15.56
12-[[(tricyclo[3.3.1.13,7] dec-1-ylamino)carbonyl] amino]-dodecanoic acid (AUDA)	391.0	240.1	−100	−10	−25	−10	*Secondary internal standard*	19.66

Shown are the mass transitions for quantification in scheduled SRM mode (m/z of precursor (Q1) and characteristic fragment (Q3) ion). The instrument settings (declustering potential (DP), entrance potential (EP), collision energy (CE), and collision cell exit potential (CXP)) are shown for a Sciex 5500 QTRAP instrument. Moreover, the assigned internal standard (IS) for each analyte, as well as the retention time (tR), the limit of detection (LOD), the calibration range (lower limit of quantification (LLOQ), upper limit of quantification (ULOQ)) of our method are provided as orientation for method development

13.4.1 Quantification by External Calibration Using Internal Standards

1. Integrate peak areas of analytes and internal standards.
2. Determine areas of the analytes and corresponding internal standards (Table 13.2), and calculate area ratios.
3. Determine the limit of detection (LOD; S/N $\geq$ 3) and the lower limit of quantification (LLOQ; S/N $\geq$ 5 and accuracy of the calibration level 80–120%, see Step 4c) based on the signal-to-noise ratio (S/N) of the peak.
4. Determine the calibration curve:
   a. Plot the peak area ratio at the individual calibration level against the respective concentration, and determine the calibration curve using linear least square regression (weighting: $1/x$ or $1/x^2$).
   b. Verify the absence of signal saturation: If the signal of the highest calibration level is below the linear calibration curve, this concentration is not within the linear range, but above the upper limit of quantification (ULOQ), and should be removed from the calibration. Otherwise, the highest injected calibration level limits the upper linear range; do not extrapolate.
   c. Use the calibration curve to calculate the concentrations of the calibration levels. Evaluate the quality of the calibration curve by determining the accuracy of the determined $vs.$ the theoretical concentration at each calibration level. For all calibration levels, the accuracy should be within 100 $\pm$ 15%, for the LLOQ within 100 $\pm$ 20%. Otherwise repeat the calibration.

   NOTE: Quantification can also be performed without weighting. However, using the suggested $1/x$ or $1/x^2$ weighting more accurate results will be obtained at low concentrations.

   OPTION: For evaluation of sample preparation — more specifically of the extraction efficiency — determine a second calibration curve using the area ratio of internal standards added at the beginning of sample preparation (IS1) to assigned secondary internal standards (IS2) used for sample reconstitution. Only samples with a sufficient predefined extraction efficiency should be further evaluated.

5. For quantification, calculate the oxylipin concentration in the vial based on the analyte to corresponding IS area ratio in the sample using the external calibration of the respective authentic standards (see Step 4). Only determine concentrations between the LLOQ and the highest calibration level/ULOQ to ensure reliable quantification (as indicated in [32]). Suggested unit: nmol/L ($=$ nM) or pmol/mL.
6. Taking the determined protein concentration, the volumes used for homogenization and oxylipin analysis, and the reconstitution volume after SPE into account calculate the concentration of oxylipins in the cells, e.g., in pmol/mg protein [14]. Alternatively, the oxylipin concentration can be calculated based on the number of cells, e.g., in fmol/10^6 cells [12].

7. For epoxy-fatty acid regioisomers, two peaks can be detected with *trans*-epoxy-PUFA isomers eluting 0.14–0.3 min after their corresponding *cis*-isomers.
   a. Characterize *trans*-epoxy-PUFA based on retention time and identical MS-fragmentation pattern as described [12, 33–35].
   b. For the individual regioisomers determine the *trans/cis*-epoxy-PUFA ratio.

13.4.2 Evaluation of the Quantitative Data

1. Keep in mind what is your/the hypothesis of the experiment: Which oxylipins do you expect to be up/downregulated by the stimuli used? Which experimental groups are you going to select as controls to investigate the impact of the stimuli?

   NOTE: Further data evaluation and statistical analysis can be performed using, e.g., GraphPad Prism, R, origin, SPSS, or a similar software. Data evaluation can also be carried out using Microsoft Excel, however, it has limited functions for statistical analysis.

2. For each set of samples (i.e., replicates of the same treatment) calculate the mean and the standard deviation (SD) and relative standard deviation (RSD).

   NOTE: For specific pairs of oxylipins where a ratio is biologically meaningful to be determined (e.g., the ratio of cis- and trans-epoxy-PUFA), calculate the ratio for each sample. Then the mean and SD are calculated leading to a higher precision.

3. Plot the results in diagrams allowing to compare the concentration of individual oxylipins or ratio of oxylipin pairs of the treated samples with the controls (mean ± SD) (examples can be found in [12, 14]).

- HCT-116/*t*-BOOH: Compare the incubations with different *t*-BOOH concentrations with the control incubation without *t*-BOOH. Which oxylipins are elevated, which are not changed? Is it possible to deduce a dose- and time-dependent effect of *t*-BOOH incubation? Correlate different oxylipins which may result from non-enzymatic conversion against each other.
- THP-1/LPS: Which oxylipins are not affected and which are elevated in incubations with LPS (positive control) *vs.* without LPS (negative control)? Can the test compounds block the LPS-induced formation of oxylipins? Are there concentration-dependent effects of the test compounds?

13.4.3 Interpretation and Learnings

Oxidative stress leads to the formation of reactive oxygen species. Among other cellular biomolecules, lipids are oxidized by radical chain reactions.

Both in living cells as well as in foods (e.g., plant oils) particularly unsaturated fatty acids are prone to oxidation, a process called aut(o)oxidation. Here, initially hydroperoxy radicals are formed which further react to volatile aldehydes, such as MDA, but also to the oxylipins measured in this experiment. The formation of hydroxy-PUFA from hydroperoxy-PUFA occurs by reduction, e.g., catalyzed by cellular glutathione peroxidases. Thus, the level of hydroxy-PUFA should be analyzed as potential marker of oxidative stress. Besides hydroxy-PUFA, other oxylipins are also elevated by oxidative stress. With the targeted oxylipin metabolomics approach several different oxylipin classes are analyzed in parallel.

Learnings and questions from this experiment: Evaluate which oxylipin class, and which isomers within the classes are strongest elevated by *t*-BOOH induced oxidative stress. Based on the literature [10–12, 23], understand how prostanoid-like isoprostanes are formed during oxidative stress, and suggest potential mechanisms for the formation of *trans*-epoxy-PUFA can be formed. Evaluate which oxylipin class, and which isomers within this class are strongest elevated by *t*-BOOH-induced oxidative stress. Suggest which (set of) oxylipins could be used to assess autoxidation in diseases associated with oxidative stress, and how these experiments/sampling could be carried out.

The first set of experiments demonstrates that PUFA are non-enzymatically oxidized to oxylipins.

However, oxylipins are also formed by specific enzymes giving rise to highly potent lipid mediators. In the experiments with THP-1/LPS, the focus is set on the 5-LOX- and COX-derived oxylipins, which are formed in macrophage-like cells upon an inflammatory stimulus.

Question (6)
What is LPS and how does it elicit an inflammatory stimulus in mammalian cells?

Learnings and questions from this experiment: Evaluate the 5-LOX- and COX-derived oxylipins and their modulation by LPS treatment. Is there an overlap with oxylipins formed by autoxidation?

5-LOX Which oxylipins can be used to monitor 5-LOX activity? What is the difference between the enzymatic and the non-enzymatic LTA$_4$ products (Fig. 13.1)? How can the level of the different LTB$_4$ isomers be interpreted with respect to the LTAH activity in the cells?

COX Inflammation is a major target of today's pharmaceuticals. Among the pharmaceuticals sold over-the-counter, NSAIDs (ibuprofen, indomethacin, diclofenac) are under the top-sellers in pharmacies in Europe and worldwide. These compounds directly target COX enzymes and elicit their effects by reducing the formation of pain-mediating and fever causing prostanoids. However, these compounds block both COX-1 and COX-2 activity, causing severe side effects, such as ulcers in the gastrointestinal tract, by COX-1 inhibition. In order to selectively block COX-2, selective inhibitors have been developed, such as celecoxib. The strongest agents to dampen inflammation are steroids, such as the synthetic glucocorticoid dexamethasone.

In the second experiment, these two classes of drugs are tested for their effects on the inflammatory response: What are the similarities, what are the differences in the effects of the compounds (classes) on the oxylipin formation particularly based on COX activity in the employed model of human macrophages? Compare the concentration range in which the compounds are active. If it is possible to analyze a sufficient number of different concentrations, calculate dose–response curves and compare the IC_{50} values.

Based on the potency and selectivity of the modulation of 5-LOX, COX(s), and autoxidation derived products, this assay also allows to evaluate the inhibitory activity of new compounds.

Take Home Message
- Oxylipins are formed non-enzymatically by (lipid) autoxidation, as well as by specific enzymatic reactions. In mammals, enzymatic formation comprises three main pathways catalyzed by cyclooxygenases (COX), lipoxygenases (LOX), and cytochrome P450 monooxygenases (CYP). The initial products are further converted by several other enzymes forming specific lipid mediators with distinct biological activity.
- Multiple oxylipins can be formed by several pathways. For example, 15-HETE can be formed by LOX, as a side product by COX, and also as a minor product by CYP. Moreover, 15-HETE is a major product of autoxidation.
- Autoxidation leads to a multitude of different oxylipins: Hydroxy-fatty acids such as 15-HETE, *trans*-epoxy-fatty acids, as well as isoprostanes serve as markers of oxidative stress. Thus, the parallel analysis of different oxylipins can serve together as markers of oxidative stress (see [12] for further information).
- Several oxylipins are highly potent lipid mediators, playing a key role in the regulation of pain, fever, and inflammation. Thus, oxylipin formation is a major drug target, e.g., as shown for COX-inhibiting non-steroidal anti-inflammatory drugs (NSAIDs) in the inflammaton assay.
- Oxylipins occur in biological samples in low concentrations, e.g. in cells a range of fmol to pmol per mg protein can be expected. Targeted LC-MS/MS is currently the best-suited approach to quantify the low-abundant lipid mediators. Quantification is carried out by external calibration with internal standards. The experiments clearly show that basic method validation is indispensable including the definition of upper and lower limits of quantification, accuracy, and recovery rates.

References

1. Gladine C, Ostermann AI, Newman JW, Schebb NH. MS-based targeted metabolomics of eicosanoids and other oxylipins: analytical and inter-individual variabilities. Free Radic Biol Med. 2019;144:72–89.
2. Willenberg I, Ostermann AI, Schebb NH. Targeted metabolomics of the arachidonic acid cascade: current state and challenges of LC-MS analysis of oxylipins. Anal Bioanal Chem. 2015;407(10):2675–83.
3. Buczynski MW, Dumlao DS, Dennis EA. Thematic review series: proteomics – an integrated omics analysis of eicosanoid biology. J Lipid Res. 2009;50(6):1015–38.
4. Ostermann AI, West AL, Schoenfeld K, Browning LM, Walker CG, Jebb SA, et al. Plasma oxylipins respond in a linear dose–response manner with increased intake of EPA and DHA: results from a randomized controlled trial in healthy humans. Am J Clin Nutr. 2019;109(5):1251–63.
5. Frankel EN. Lipid Oxidation. 2nd ed. Bridgwater: The Oily Press an Imprint of PJ Barnes & Associates; 2005.
6. Yin H, Xu L, Porter NA. Free radical lipid peroxidation: mechanisms and analysis. Chem Rev. 2011;111(10):5944–72.
7. Sies H, Jones D. Oxidative stress. Encyclopedia of stress. 2nd ed. New York: Academic Press; 2007. p. 45–8.
8. Sies H, Berndt C, Jones DP. Oxidative stress. Annu Rev Biochem. 2017;86(1):715–48.
9. Forman HJ, Zhang H. Targeting oxidative stress in disease: promise and limitations of antioxidant therapy. Nat Rev Drug Discov. 2021;20(9):689–709.
10. Jahn U, Galano JM, Durand T. Beyond prostaglandins - Chemistry and biology of cyclic oxygenated metabolites formed by free-radical pathways from polyunsaturated fatty acids. Angew Chem. 2008;47(32):5894–955.
11. Milne GL, Yin H, Hardy KD, Davies SS, Roberts LJ II. Isoprostane generation and function. Chem Rev. 2011;111(10):5973–96.
12. Rund KM, Heylmann D, Seiwert N, Wecklein S, Oger C, Galano J-M, et al. Formation of trans-epoxy fatty acids correlates with formation of isoprostanes and could serve as biomarker of oxidative stress. Prostaglandins Other Lipid Mediat. 2019:144.
13. Smith WL, Urade Y, Jakobsson P-J. Enzymes of the cyclooxygenase pathways of prostanoid biosynthesis. Chem Rev. 2011;111(10):5821–65.
14. Hartung NM, Mainka M, Pfaff R, Kuhn M, Biernacki S, Zinnert L, et al. Development of a quantitative proteomics approach for cyclooxygenases and lipoxygenases in parallel to quantitative oxylipin analysis allowing the comprehensive investigation of the arachidonic acid cascade. Anal Bioanal Chem. 2023;515(5):913–33.
15. Smith WL, DeWitt DL, Garavito RM. Cyclooxygenases: structural, cellular, and molecular biology. Annu Rev Biochem. 2000;69(1):145–82.
16. Powell WS, Rokach J. Biosynthesis, biological effects, and receptors of hydroxyeicosatetraenoic acids (HETEs) and oxoeicosatetraenoic acids (oxo-ETEs) derived from arachidonic acid. Biochim Biophys Acta (BBA) – Mol Cell Biol Lipids. 2015;1851(4):340–55.
17. Gottschall H, Schmöcker C, Hartmann D, Rohwer N, Rund K, Kutzner L, et al. Aspirin alone and combined with a statin suppresses eicosanoid formation in human colon tissue. J Lipid Res. 2018;59(5):864–71.
18. Vane JR, Botting RM. The mechanism of action of aspirin. Thromb Res. 2003;110(5):255–8.
19. Haeggstrom JZ, Funk CD. Lipoxygenase and leukotriene pathways: biochemistry, biology, and roles in disease. Chem Rev. 2011;111(10):5866–98.

20. Westphal C, Konkel A, Schunck W-H. Cytochrome P450 enzymes in the bioactivation of polyunsaturated fatty acids and their role in cardiovascular disease. In: Hrycay EG, Bandiera SM, editors. Monooxygenase, peroxidase and peroxygenase properties and mechanisms of cytochrome P450. Cham: Springer; 2015. p. 151–87.

21. Morisseau C, Hammock BD. Impact of soluble epoxide hydrolase and epoxyeicosanoids on human health. Annu Rev Pharmacol Toxicol. 2013;53(1):37–58.

22. Kampschulte N, Alasmer A, Empl MT, Krohn M, Steinberg P, Schebb NH. Dietary polyphenols inhibit the cytochrome P450 monooxygenase branch of the arachidonic acid cascade with remarkable structure-dependent selectivity and potency. J Agric Food Chem. 2020;68(34): 9235–44.

23. Rund KM, Ostermann AI, Kutzner L, Galano J-M, Oger C, Vigor C, et al. Development of an LC-ESI(−)-MS/MS method for the simultaneous quantification of 35 isoprostanes and isofurans derived from the major n3- and n6-PUFAs. Anal Chim Acta. 2018;1037:63–74.

24. Kutzner L, Rund KM, Ostermann AI, Hartung NM, Galano J-M, Balas L, et al. Development of an optimized LC-MS method for the detection of specialized pro-resolving mediators in biological samples. Front Pharmacol. 2019:10.

25. Koch E, Mainka M, Dalle C, Ostermann AI, Rund KM, Kutzner L, et al. Stability of oxylipins during plasma generation and long-term storage. Talanta. 2020;217:121074.

26. O'Brien J, Wilson I, Orton T, Pognan F. Investigation of the Alamar Blue (resazurin) fluorescent dye for the assessment of mammalian cell cytotoxicity. Eur J Biochem. 2000;267(17):5421–6.

27. Mulac D, Lepski S, Ebert F, Schwerdtle T, Humpf H-U. Cytotoxicity and fluorescence visualization of ergot alkaloids in human cell lines. J Agric Food Chem. 2013;61(2):462–71.

28. Kamiloglu S, Sari G, Ozdal T, Capanoglu E. Guidelines for cell viability assays. Food Frontiers. 2020;1(3):332–49.

29. Smith PK, Krohn RI, Hermanson GT, Mallia AK, Gartner FH, Provenzano MD, et al. Measurement of protein using bicinchoninic acid. Anal Biochem. 1985;150(1):76–85.

30. Koch E, Wiebel M, Hopmann C, Kampschulte N, Schebb NH. Rapid quantification of fatty acids in plant oils and biological samples by LC-MS. Anal Bioanal Chem. 2021;413(21):5439–51.

31. Mainka M, Dalle C, Pétéra M, Dalloux-Chioccioli J, Kampschulte N, Ostermann AI, et al. Harmonized procedures lead to comparable quantification of total oxylipins across laboratories. J Lipid Res. 2020;61(11):1424–36.

32. Schebb NH, Kühn H, Kahnt AS, Rund KM, O'Donnell VB, Flamand N, et al. Formation, signaling and occurrence of specialized pro-resolving lipid mediators—what is the evidence so far? Front Pharmacol. 2022;13

33. Jiang H, McGiff JC, Quilley J, Sacerdoti D, Reddy LM, Falck JR, et al. Identification of 5,6-trans-epoxyeicosatrienoic acid in the phospholipids of red blood cells. J Biol Chem. 2004;279(35): 36412–8.

34. Jiang H, Quilley J, Reddy LM, Falck JR, Wong PY, McGiff JC. Red blood cells: reservoirs of cis- and trans-epoxyeicosatrienoic acids. Prostaglandins Other Lipid Mediat. 2005;75(1–4):65–78.

35. Aliwarga T, Raccor BS, Lemaitre RN, Sotoodehnia N, Gharib SA, Xu L, et al. Enzymatic and free radical formation of cis- and trans-epoxyeicosatrienoic acids in vitro and in vivo. Free Radic Biol Med. 2017;112:131–40.

Part VI

Spatial Metabolite Analysis Using MS-Assisted Tissue Imaging

Comprehensive Spatial Lipidomics of Formalin-Fixed Paraffin-Embedded Tissue Guided by Mass Spectrometry-Imaging

14

Vanna Denti, Martin Piazza, Andrew Smith, and Giuseppe Paglia

What You Will Learn in This Chapter

This chapter aims to provide a practical guide to performing comprehensive spatial lipidomics of formalin-fixed paraffin-embedded (FFPE) tissue guided by Matrix Assisted Laser Desorption/Ionisation-mass spectrometry-imaging (MALDI-MSI), presenting an overview of the key methodological aspects as well as the type of data that can be obtained when using this approach. Moreover, it also aims to highlight the more extensive and reliable lipid identifications that can be obtained when an additional trapped ion mobility spectrometry (TIMS) dimension is employed.

In the data presented in this chapter, the number of annotated lipids was enhanced by combining lipidomics MS-imaging with LC-MS-based analysis of cells obtained through laser-capture microdissection (LMD), overcoming some of the challenges related to lipid identification that are often faced when using MALDI-MSI alone. Moreover, given the specificity of the isolated tissue regions, it also helps to overcome certain issues encountered when performing bulk lipidomics analysis from whole-tissue sections where key lipidomic information from smaller regions of interest may be diluted or even lost.

V. Denti · M. Piazza · A. Smith · G. Paglia (✉)
University of Milano-Bicocca, School of Medicine and Surgery, Vedano al Lambro, Italy
e-mail: vanna.denti@unimib.it; m.piazza30@campus.unimib.it; andrew.smith@unimib.it; giuseppe.paglia@unimib.it

© The Author(s), under exclusive license to Springer Nature Switzerland AG 2023
J. Ivanisevic, M. Giera (eds.), *A Practical Guide to Metabolomics Applications in Health and Disease*, Learning Materials in Biosciences,
https://doi.org/10.1007/978-3-031-44256-8_14

14.1   Introduction

The study of lipids in their native spatial context has rapidly developed in recent years and has taken a prominent position at the forefront of lipidomics research in the context of human disease [1]. This is particularly true in the field of oncology where "lipid reprogramming" or imbalance of lipid metabolism are associated with aberrant tumorigenic processes that can occur heterogeneously within the cellular network [2, 3]. Spatial lipidomics approach can help to capture this complexity that characterises numerous diseases and enable the detection of lipidomic alterations within small cell subpopulations, or tissue regions, that may otherwise get diluted, or missed, if a whole-tissue lipidomics approach was employed [4].

In particular, this progress has been facilitated by recent advancements in MSI instrumentation which now offer near single-cell lateral resolutions with acquisition speeds upwards of 20 pixels per second [5] whilst maintaining the sensitivity necessary to detect endogenous metabolites and lipids present in the micromolar concentration range [5, 6]. Moreover, the ability to map the distribution of these relatively low abundant lipids with a low ionisation efficiency has been further enhanced with the introduction of laser-induced postionisation (MALDI-2). Postionisation (PI) has been shown to produce a marked increase in the number of lipid signals detected in tissue [5–7], enabling a more extensive portion of the lipidome to be accessed and exploited for improved mechanistic insights on different pathological processes. It may also further push the boundaries when aiming to explore the lipidome at a single-cell level. To accompany this, TIMS has also been recently integrated within high-throughput MALDI-MSI instrumentation [8] and represents a further technical advancement that has the potential to revolutionise this field considering the resolving power associated with this form of IMS and enables isobaric, in addition to isomeric, lipids to be separated and mapped within pathological tissue.

Until now, the majority of clinical MALDI-MSI studies have used snap-frozen tissue for spatial lipidomic investigations, however, most clinical biopsies are stored as formalin-fixed paraffin-embedded (FFPE) tissue, the gold standard for specimen preservation in pathology units [9]. This form of preservation initiates the process of amine–thiol cross-linking and the resulting methylene bridges inactivate enzymatic activity, stabilising the biomolecules within the tissue [10]. Contrary to the spatial mapping of proteins and N-glycans, for example this renders spatial lipidomics challenging given that many lipid species are depleted from the tissue by the use of paraffin wax and organic solvents during tissue processing. Notwithstanding, recent studies have highlighted that, despite this depletion, certain solvent-resistant lipid species are maintained in FFPE tissues and can be exploited for spatial lipidomics studies within a clinical context.

In particular, Denti et al. recently described a protocol that incorporated an antigen retrieval (AR) step in order to liberate solvent-resistant lipids that were trapped within the network of cross-linked proteins generated by formalin fixation [11]. Doing so increases the ion yield for lipid species in clear cell renal cell carcinoma (ccRCC) FFPE tissue when working in a positive-ion modality, including phosphocholine (PC), lipopolysaccharide (LPS), sphingomyelin (SM) and phosphatidylserine (PS) species. Moreover, the same protocol was also applied to a patient-derived xenograft model of breast cancer where the robust technical reproducibility of the protocol was highlighted along with its potential to highlight alterations within the spatial lipidome of the tumours that were induced by the administration of a glutaminase inhibitor [12].

In addition, in a negative ion modality, the detection of phospholipid species such as phosphatidylinositols (PI) and phosphatidic acids (PA) is also favoured. Denti et al. assessed the tumour-immune landscape in colorectal cancer and found that the relative abundance of a number of lipids, in particular PI(O-40:3) and PS(44:1), were correlated with the degree of CD3+ and CD8+ immune infiltrates [13].

Even though MALDI-MSI alone can enable a certain degree of characterisation regarding the spatial lipidome in FFPE tissue, it may not always provide sufficient molecular depth to obtain more detailed biological insights. Combining MALDI-MSI with a more traditional LC-MS-based lipidomics approach will result in a more comprehensive view of the altered lipidome in pathological tissue samples. In other terms, MALDI-MSI can provide spatial context and help to individuate molecularly distinct cellular regions that can then be better characterised by LC-MS-based lipidomics.

14.2 Practical Guide

14.2.1 Biological Context

Considering the inherent cellular heterogeneity of many solid tumours, omics techniques that allow us to investigate how spatially structured communities of individual cells act and interact in the context of their networked environment are particularly valuable. At the level of the lipidome, it is known that lipid species such as phosphatidylinositol and phosphatidylserine may contribute to inflammatory processes as well as innate tumour immunity [14, 15], therefore having a downstream effect on their immunogenic nature and, consequently, prognostic outlook. One such tumour is represented by clear cell renal cell carcinoma (ccRCC) and has therefore been used as a proof-of-concept.

In this section, we present a practical guide that outlines an approach utilised to perform detailed lipidomic mapping of ccRCC FFPE tissue. This approach can also be applied to other disease contexts where maintaining spatial structure is fundamental.

14.2.2 Tissue Treatment

For many years, fresh frozen tissue was the unique option for performing spatial lipidomics studies with MALDI-MSI due to the wide range of cell membrane lipids that could be detected. However, recently developed protocols have also rendered spatial lipidomics studies using archived formalin-fixed paraffin-embedded specimens feasible, extracting clinically relevant molecular information from pathological tissue. As mentioned previously, one such protocol involves a tissue treatment workflow that is similar to that employed for spatial proteomics of FFPE tissue, requiring deparaffinisation, achieved by melting of the paraffin wax at 60 °C followed by washes in toluene, gradual rehydration and an antigen retrieval step using a 10 mM citric acid buffer at 97 °C. This antigen retrieval step is employed in order to help liberate cross-linked lipids that become trapped during the formalin fixation process, in particular phosphatidylethanolamine (PE) and PS, which ionise readily in a positive-ion modality, and whose amine headgroups form amine-thiol cross-links with other amine containing proteins or lipids [11]. Naturally, one would also expect that the yield of cross-linked lipids liberated may also be impacted by the antigen retrieval buffer, and pH, given that it is well known that proteins with a diverse tertiary structure, and therefore physicochemical properties, can be more efficiently unfolded in specific pH conditions. However, this has yet to be investigated and represents a further avenue of research that can be exploited to increase the lipid coverage achievable when analysing FFPE tissues.

In the workflow presented in this chapter, we performed a tissue treatment that requires only deparaffinisation. This approach, with the aid of the appropriate MALDI matrices discussed in Sect. 2.3, allows for improved visualisation of phospholipid species which deprotonate efficiently in negative ion mode, such as PI and PA, and has shown promising results in oncological tissue readily infiltrated by immune cells [13]. In addition to visualising a complementary portion of the spatial lipidome, this approach offers several advantages: it does not only facilitate the detection of small metabolites, along with the polar lipids, as highlighted in Buck et al. [16], but it is also more compatible with additional spatial omics such as N-glycomics and proteomics. Whilst the concept of spatial multiomics may be outside the scope of this work, such an approach may provide complementary molecular information that strengthens the insights into lipid metabolism gained by this workflow.

14.2.3 Matrix Selection and Deposition

In this work, as described in the previous section, we have employed a MALDI matrix that allows an efficient deprotonation of phospholipids and small molecules for analysis in negative ion mode. For this purpose, a solution of 10 mg/mL 9-aminoacridine (9-AA) dissolved in a 70% methanol solution was deposited using the HTX TM-Sprayer™ (HTX Technologies, LLC, Chapel Hill, NC, USA) with the following parameters: temperature 85 °C; number of passes 6; flow rate 0.2 mL/min; velocity 1100 mm/min; track spacing 2 mm; pressure 10 psi.

14.2.4 MALDI-MS Imaging

MALDI-MSI analysis can be performed at different m/z ranges with different lateral resolutions. The newest laser technologies, together with more reproducible matrix deposition techniques, allow for improved imaging down to a lateral resolution of 5 μm. However, performing MALDI-MSI analysis at such high spatial resolution is time consuming and leads to the generation of high-dimension data files that require powerful machines for the data deconvolution. Moreover, the sensitivity of the acquisition is affected by a small diameter laser beam and more energy or more laser shots should be applied to improve the S/N of the acquired signal. Nevertheless, in some cases a high image resolution is needed to obtain specific data on single cell type or small structure within the tissue. Another point to be considered is that novel technologies applied to MALDI source, such as MALDI-2 postionisation, significantly improve the sensitivity of the analysis without losing the high spatial resolution [17].

Therefore, before performing a MALDI-MSI analysis, it is crucial to select the lateral resolution necessary to answer the specific biological question.

In the case reported, all imaging analyses were performed using a timsTOF fleX mass spectrometer (Bruker Daltonics, Bremen, Germany) equipped with a Smartbeam™ 3D laser. The acquisition was performed in negative ion mode, in the m/z range of 300–1200. External calibration was performed using red phosphorus clusters in the m/z range of 0–2000. The number of shots accumulated for each spectrum depends on the amount of matrix deposited on the tissue and the laser scan range. The optimal number of shots (to be accumulated) is a stepwise process: one can progressively apply the increase of 50 shots in each step to determine the best signal-to-noise ratio for the analytes of interest.

For the matrix deposition described in the previuous paragraph, the optimal number of shots accumulation for each spectrum was 200. The measurement regions were rastered at a lateral resolution of 20 × 20 (x, y) μm with a laser scan range of 16 μm per pixel.

14.2.5 MALDI-MSI Data Processing and Deconvolution

Each MALDI-MSI dataset can be represented by a three-dimensional data cube where the x- and y-axes comprise the intensity images for each m/z feature, whilst the z-axis contains the mass spectral information. Consequently, processing and deconvolution strategies are critical prerequisites to interpreting highly complex MALDI-MSI datasets. Generally, MSI data pre-processing involves six principal steps: normalisation, baseline subtraction, smoothing, peak picking, peak alignment and spatial denoising [18].

Nowadays, many software packages and data analysis platforms for data processing are available, such as BASIS, Cardinal R package, MSIReader, MZmine3, mmass and all open-source, along with the commercially available SCiLS Lab (https://scils.de/). The last has become widely used due to the possibility to handle vendor-neutral file formats and to its intuitive user interface [19–21].

For the present workflow, SCiLS Lab 2023a was used to elaborate the MALDI-Q-TOF-MSI obtained as described in Sect. 2.4. In this case, no baseline correction or smoothing was performed during the data import. Peak picking and the dataset normalisation were performed using the Root Mean Square algorithm. MALDI-MSI data and the obtained spatial segmentation can be co-registered with the histological image directly within SCiLS Lab software using .svs and high-resolution .jpeg images. This tool allows for the association of molecular changes measured with MALDI-MSI to the histopathological changes observed in the histological image.

To spatially resolve lipidomic alterations in the tissue, an unsupervised segmentation analysis, which clusters spectra based on similarity patterns, can be performed. In this case, Bisecting k-Means algorithm and Correlation Distance metric with weak spatial denoising were used. The tissue clustering, overlaid with the histological image, can be visualised in Fig. 14.1 on the right. Three main regions were found, highlighted in pink, orange and green, and associated with the annotations performed by the pathologist on the FFPE section. Each region, identified by MALDI-MSI, was then used to guide the laser-capture microdissection (LMD) described in Sect. 2.6. Three specific regions were dissected: the haemorrhagic nodule (HN), tumour (T) and cortex (C).

14.2.6 MSI-Guided Laser-Capture Microdissection

Laser-capture microdissection (LMD) is a laser-based dissection technique that allows dissecting specific cells or tumour regions under microscopic guidance [22].

In the workflow reported in this chapter, MALDI-MSI tissue segmentation was used to guide the dissection. As exemplified in Fig. 14.1, three regions of interest (ROI) of 6.5 mm^2 were identified with the comparative pattern analysis using SCiLS Lab.

Briefly, once the MALDI-MSI acquisition was finished, the ITO slide analysed was washed 2 times with ethanol 100% for 30 seconds to remove the 9AA matrix. The slide was then inserted in a Laser Microdissection Microscope (LeicaLMD7) stage and microdissection was performed using the following parameters:

Laser Power 40, Aperture 10, Speed 20, Specimen Balance 2, Line Spacing For Draw +Scan 20, Head Current 80%, and Pulse Frequency 119.

Three collectors were selected, each containing an Eppendorf tube and named after the annotation assigned to each region obtained from the segmentation. The ROIs were imported into the software and automatic cut was performed using the "Draw+Scan" option that pulverises the tissue within the imported regions. Finally, three Eppendorf tubes containing the same amount of pulverised tissue from the regions HN, T and C were obtained, as shown in Fig. 14.1. Considering that the FFPE section used for the analysis was 5 μm thick, the estimated number of cells collected in each Eppendorf tube corresponds to 25,000 cells.

These samples were then used to perform region-specific lipids extraction.

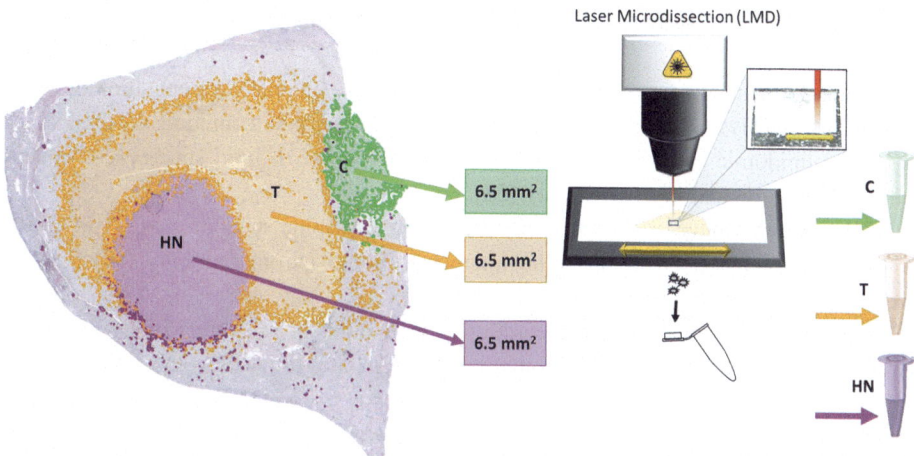

Fig. 14.1 MALDI-MSI-guided LMD. On the left, a histological image co-registered to the unsupervised segmented regions of interest (ROIs). Each region was named following the annotation obtained by a pathologist. HN, in pink corresponds to Haemorragic Nodule (tumour with blood deposits); T for the tumour region in orange; C, in green, for the cortex (non-tumoral) region. Three ROIs (corresponding to HN, T and C) of 6.5 mm^2 were dissected with LMD from the same tissue analysed with MALDI-MSI. The pulverised tissue was collected in an Eppendorf tube cup. On the right, the samples obtained were then processed for lipids extraction as described below

14.2.7 Lipids Extraction of LMD Tissue

The extraction process employed in traditional lipidomics workflows usually involves homogenisation steps that result in the loss of information regarding the tissue or cells of origin. That can lead to the dilution of molecular information deriving from a small number of cells and to a consequent loss of relevant biological insights. The possibility of extracting the lipidome content from specific tissue regions or cells using LMD can overcome this limit, enriching lipidomic results with the spatial information specific to this region.

Lipid extraction of LMD microdissected regions was carried out using the same solution used to dilute the 9AA matrix (2.3). Therefore, 20 µL of a 70% MeOH solution was added to each Eppendorf tube, vortex mixed for 30 s and incubated in a thermomixer at 60 °C for 30 min, mimicking the extraction conditions that occurred during the 9AA matrix spray. Finally, following the centrifugation at 15,000 rpm for 15 min at 4 °C the supernatants were transferred to glass vials for LC-MS analysis.

14.2.8 UHPLC-Ion Mobility-MS Lipidomic Analysis

Ultra High-performance liquid chromatography (UHPLC) enables the separation of lipids prior to MS analysis. Lately, UHPLC-MS has become the technique of choice for untargeted lipidomic experiments due to its broad coverage of lipid species, convenient sample preparation and high sensitivity [23].

The two most widely used approaches for chromatographic separation in UHPLC-MS for lipidomic analyses are hydrophilic interaction liquid chromatography (HILIC) and reversed-phase chromatography (RP).

Whilst HILIC separates lipids according to their polar head groups (which result in distinct lipid class separation), RP chromatography separates lipids by composition of their fatty acyl chains [24].

In RP separation lipids retention time increases with an increasing fatty acyl carbon number and decreases with an increased number of double bonds. Therefore, it is possible to separate different lipid species from the same class by their cumulative double bond index.

Recently, the advantages of coupling classical LC-MS-based lipidomics approaches with ion mobility spectrometry have been demonstrated [1]. IM-MS provides a novel analytical platform that can separate ions beyond mass-to-charge ratio (*m/z*), exploiting the different mobility of ions in a chamber filled with a buffer gas and subjected to an electrical field. The mobility of an ion relates directly to its shape, size, and charge as well as to the nature of the buffer gas and can be used to resolve isobaric and isomeric compounds. Trapped ion mobility (TIMS) was first introduced in 2011 by Fernandez Lima et al. It reverses the concept of the classical drift tube IMS. [25] Indeed, TIMS maintains the ions in a stationary position, against a gas flow, and then releases them according to their mobility. In particular, TIMS enables work to be performed in Parallel Accumulation Serial Fragmentation (PASEF®) operation mode coupled with the traditional LC-MS, and this approach has been proven to enhance the lipidome coverage [26, 27].

For the work presented in this chapter, the extracted lipidome was analysed using an LC-MS platform which includes an Elute UHPLC system coupled to a timsTOF fleX spectrometer. RP chromatographic separation was achieved using an ACQUITY UPLC CSH C18 Column.

Mobile phase A was 10 mM ammonium acetate: acetonitrile (40:60 v:v) with 0.1% formic acid and B was Isopropanol: Phase A (90:10 v:v).

A 20-minute gradient followed by one minute of equilibration time was performed as follows: 0 min 99% A, 1 min 99% A, 1.10 min 60% A, 5 min 20% A, 11 min 20% A, 12 min 1% A, 18 min 1% A, 18.10 min 60% A, 20 min 99%. The flow rate was 0.25 ml/min and the column temperature was 55 °C.

Data-dependent parallel accumulation serial fragmentation (PASEF) was performed, operating in negative ionisation mode in the full scan range of 100–1350 *m/z* and in the mobility range 0.55–1.90 V·s/cm^2.

External m/z calibration was performed using a Sodium Formate: Tuning Mix (MMI-L Low Concentration Tuning Mix, Agilent Technologies, USA) (1:1 v/v) solution. External mobility calibration was performed using a Tuningmix solution. Additionally, a calibration segment containing the infused Sodium Formate: Tuning Mix (1:1 v/v) solution was included in a calibration segment from 0 to 0.3 min of each analysis run for internal calibration. The injection volume was 5 µL and each sample (HN, T and C) was analysed in duplicates.

14.2.9 Feature Detection and Identification

During the data-dependent (DDA) LC-IMS-MS analysis, as described above, retention time (RT), m/z and collision cross-section (CCS) of the precursor ions were acquired.

Therefore, the matrix of data obtained from this analytical technique is very complex and needs appropriate software for pre-processing, feature extraction, MS/MS fragment association to the precursor ion and, finally, identification.

Also in this case, some open-source software and plugins are available, such as MZmine3 along with commercial solutions such as Metaboscape® and the vendor-neutral Lipostar [28] (https://www.moldiscovery.com/software/lipostar/).

Metabolite annotation and/or identification can be classified into four levels of confidence [29]:

– Identified (level 1)
– Putatively annotated compounds (level 2)
– Putatively characterised compound classes (level 3)
– Unknown compounds (level 4)

Level 1 identification requires that 2 or more orthogonal properties of an authentic chemical standard analysed in the same laboratory are compared to experimental data acquired with the same analytical methods. In level 2, the annotation is verified using MS/MS libraries and does not require matching with data obtained from standards analysed in the same laboratory.

For the data presented in this chapter, feature detection and identification were performed with Metaboscape® software both for LC-IMS-MS and for MALDI-QTOF imaging data.

For LC-IMS-MS analysis, a feature detection range of m/z 100–1300 and RT of 0–20 min were set. Annotation of negatively charged ions was performed including the following species and adducts: $[M-H]^-$, $[M-H-H_2O]^-$, $[M + CH_3COO]^-$ and $[M + Cl]^-$. For MALDI-MSI data, feature detection range m/z 300–1200 and RT 0–20 min were set. Feature annotation in negative ionisation mode analysis was performed including the following ions: $[M-H]^-$ and $[M + Cl]^-$.

Metabolite identification was carried out considering m/z accuracy, RT, MS/MS score, isotopic pattern and CCS compared to an *in-house* lipids' library and to software-included spectral libraries, HMDB and METLIN. A total of 116 lipids were identified and for 86 of these lipids MS/MS spectra were acquired and used to confirm their identity. MALDI-MSI features were putatively assigned considering m/z accuracy and isotopic pattern compared to an *in-house* lipids library and to software-included libraries, HMDB and METLIN.

As presented in Fig. 14.2, with MALDI-MSI operating in negative ion mode it was possible to annotate mostly PI, ceramides and PE, whilst annotation of LC-IMS-MS in negative ion mode allowed for the detection of additional lipid species such as SM, PS and PC, in addition to the species previously detected with MALDI-MSI. This result can be explained by the different types of ionisation occurring in the two types of analysis. With MALDI in negative ionisation mode, it was possible to observe mainly [M-H]- and [M +Cl]- ions, whereas in LC-MS analysis with mobile phase assisting the ionisation, [M +CH$_3$COO]- ions could also be detected. The presence of acetate adducts allows for the detection of lipid species that are otherwise more easily detected as positive ions, such as PC and SM. In fact, as can be observed in Fig. 14.2 these classes of lipids were not annotated in MALDI analysis. However, it is worth mentioning that the process of removing the MALDI matrix (9-AA) with methanol prior to LC-IMS-MS analysis can lead to loss of some lipid species (see questions and answers section).

Nevertheless, the combination of both techniques provided a wide coverage of tissue lipidome, whilst keeping the spatial information and avoiding the dilution of low abundant cell lipidome. In Fig. 14.2a, the relative ion distribution of 3 annotated lipids, specifically localised in the 3 ROIs are shown. On the right side of the same image, one can appreciate the example of lipid identified considering the MS/MS spectrum in addition, and the boxplot reporting the intensity of that ion across three different ROIs, obtained from LMD.

14.2.10 Lipidomic Data Integration

Many biostatistical and bioinformatics tools are currently offering novel approaches for the integration of spatial-omic data [30–32], which may represent a bottleneck with regards to future application. A critical point to consider when integrating these data is that they should derive from the very same tissue section [33], which alleviates said integration issues. Combining the spatial and molecular information from the same section might help to improve the discrimination of tissue from a molecular standpoint, ensuring congruence between molecular information and the cell type of origin.

In the work presented in this chapter, the annotations obtained from both datasets were imported into SCiLS Lab and used to perform a partially supervised analysis of the MALDI-MSI dataset created as described in sect. 14.2.5. The annotated feature list was used to perform first a segmentation with Bisecting k-Means algorithm and Correlation Distance metric, with weak spatial denoising. In Fig. 14.3a, a panel of the obtained clusters could be observed. In this case, it was possible to segment the sample into five classes based on the molecular similarity (i.e. pattern analysis) of each cluster. On the top of the

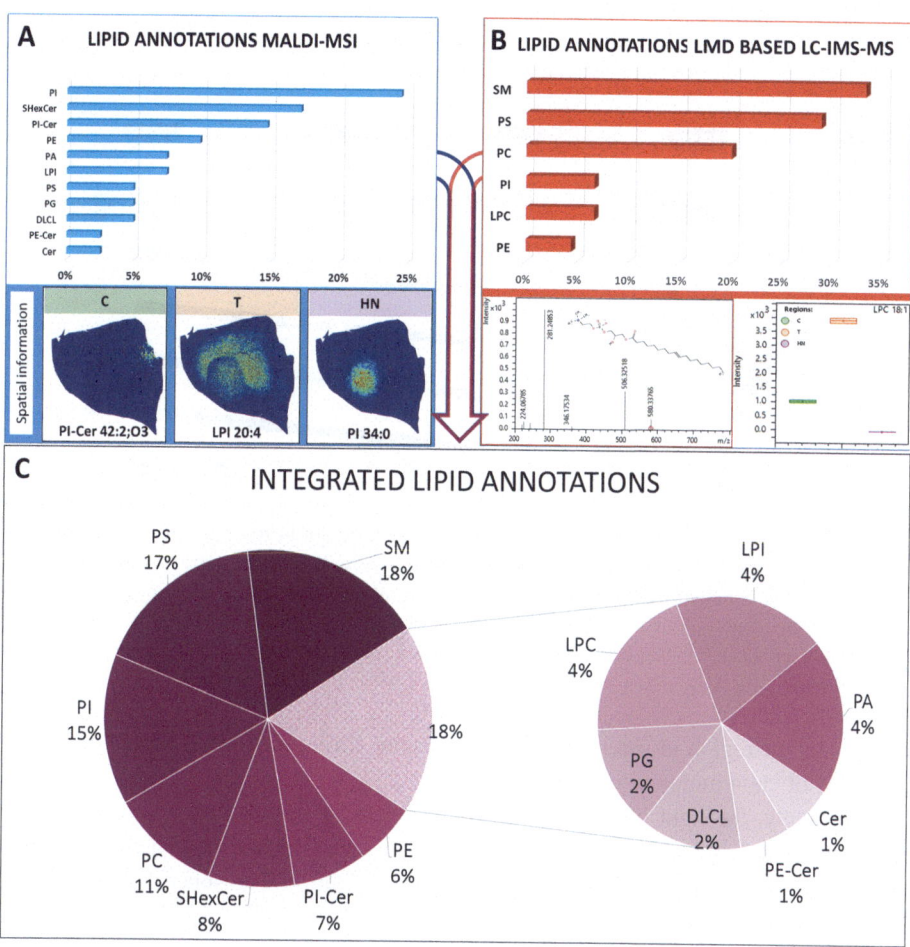

Fig. 14.2 Lipid annotation using MALDI-MSI and LC-IMS-MS. (**a** and **b**) The percentage of lipids belonging to a specific lipid class is reported for the features annotated from the MALDI-MSI and from the LC-IMS-MS dataset, respectively. As shown, PS, PE and PI were found in both cases, whilst other lipid classes were specifically found in one or the other dataset. As shown in the bottom of panels **a** and **b**, additional information could be obtained from the two types of analyses: Spatial localisation and relative intensity of lipids plotted as MALDI-MS images (in panel **a**), MS/MS spectra and relative intensity of lipid features (in HN, T and C) from the UHPLC-IMS-MS analysis. (**c**) the percentage of each lipid class in the integrated dataset is reported. As shown, the most abundant lipid classes annotated are SM, PS and PI

figure, a 10X magnification of the histological image corresponding to each region is reported.

These regions could be selected as ROIs and used to perform a Principal Component Analysis (PCA) (Fig. 14.3b) that clustered each MALDI-MSI pixel (each point in the PCA) based on the variance associated with each pixel in the three components shown. We can

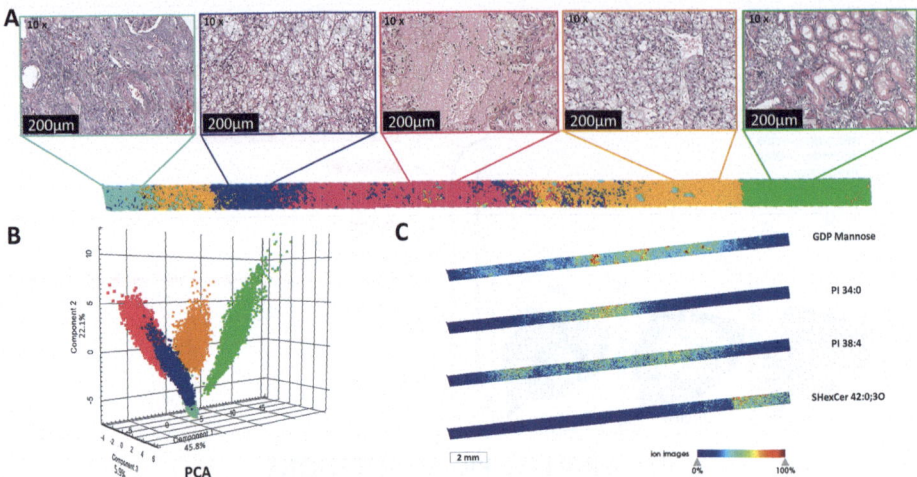

Fig. 14.3 Example of data integration results. (**a**) 10X magnification of histological images corresponding to the regions obtained by the partially supervised clustering. (**b**) A principal component analysis was obtained from the annotated and integrated feature list. Each point corresponds to one pixel of the same colour ROI (top). (**c**) MALDI-MS images of GDPMannose, PI 34:0, PI 38: 0 and SHexCer 42:0:3O, respectively. A colour-coded scale indicating the relative intensity of annotated ions

observe that the region in green represents a separate cluster that matches the histological annotation of "non tumoural cortex".

Interestingly, the light blue cluster on the left also clustered apart from all the other regions and was associated with a "fibrotic" region. The dark pink coloured region was associated with the haemorrhagic nodule, as seen in the first unsupervised clustering. Interestingly, the region indicated in sect. 14.2.5 as Tumor (Fig. 14.1, orange region) could be divided in two distinct clusters when using the integrated dataset: the blue and the orange ones (Fig. 14.3a and b). Both were assigned as tumors (ccRCC) based on the histological presentation (Fig. 14.3a). However, using imaging data at the molecular level, two distinct regions could be identified within the tumor (blue and orange regions Fig. 14.3a and b). They form separate clusters as illustrated in the PCA (Fig. 14.3b). Moreover, looking at the clusters, small circular regions in light blue can be observed in Fig. 14.3a. Interestingly, the distribution of the ion annotated as GDPMannose correlated with the presence of these specific circular regions (Fig. 14.3c). Finally, regarding the histological images, they also appear to be colocalised with blood vessels.

The other ions reported in Fig. 14.3c were annotated as PI 34:0, PI 38:0 and SHexCer 42:0:3O and colocalised with HN nodule, Tumour and non-tumour regions, respectively. Interestingly, the annotated hexosylceramide was previously observed to be altered in the same type of tumour, both in tissue and in urine samples [34].

Questions

1. Is it possible to use this spatial lipidomics approach on fresh frozen tissue?
2. What kind of lipid information can be obtained from the analysis in positive ionisation mode?
3. Is it possible to perform laser-capture microdissection (LMD) without removing the MALDI matrix with a polar solvent beforehand?
4. Do the processing steps with organic solvents such as toluene affect the lipid composition?
5. How large does the microdissected tissue region need to be to obtain a sufficient amount of material for lipidomics analysis by LC-MS?
6. Can we guide laser microdissection using histological images?

Answers

1. Yes, it is also possible to apply the same principles to fresh frozen tissue. Naturally, given that this tissue source is not subjected to processing with various organic solvents and paraffin wax, the lipid content will also be greater. However, tissue-specific adjustments would need to be made in terms of MALDI matrix deposition [35] as well as the cutting parameters for LMD to obtain optimal results.
2. In general, in positive mode, the ionisation of phospholipid species including phosphocholine (PC), lipopolysaccharide (LPS), sphingomyelin (SM) and phosphatidylserine (PS) tends to be favoured. Conversely, in negative ionisation mode, phospholipid species such as phosphatidylinositols (PI) and phosphatidic acids (PA) are favoured.
3. In this workflow, the MALDi matrix (9-AA) was removed with methanol to avoid interference with the cutting parameters of the laser in LMD. Moreover, the deposition of matrix crystal residue within the LC column should be avoided. Alternatively, the first minute of analysis can be diverted to waste (i.e. not injected to MS). However, as described in this chapter, the application of alternative matrices followed by an extensive cleanup protocol to wash away the matrix deposits, will also allow for the LMD of matrix-coated tissue.
4. Yes, unfortunately, tissue processing with paraffin wax and organic solvents will cause the depletion of many lipid species. In particular, class-specific depletion of amine-containing lipids upon formalin fixation has been reported along with an increased degree of lipid hydrolysis [36]. Nevertheless, recent studies employing MALDI-MSI and FTIR spectroscopy suggest that the solvent-resistant lipid

(continued)

species are preserved in FFPE tissues and may provide diagnostically relevant information.

5. In this workflow, regions of 6.5 mm^2 were microdissected from a 5-μm thick FFPE tissue section and subjected to LC-MS lipidomics. Considering the average volume of a cell, approximately 10 μm^3, this equates to 25,000 cells and results in sufficient sensitivity for the detection and annotation of a broad range of lipid species. Naturally, in lipid-rich tissue, only a few thousand cells can result in solid lipid detection. The number of cells obtained by LMD can be increased, but the process is time consuming. For example, the LMD of three regions of 6.5 mm^2 required approximately 1 hour and 45 min.

6. It is possible to perform histology-guided microdissection and lipidomics of pathological tissue, however, this should be done by employing a serial section. When haematoxylin and eosin staining is performed, many of the non-polar solvent-resistant lipids are further depleted due to multiple washes with polar solvents which will limit the lipid coverage. Using a serial tissue section for histological staining can overcome this issue to a certain extent. However, the serial sections may not necessarily have the same spatial composition which can lead to imprecise LMD. Moreover, it has been well documented that molecular alterations may underline greater tissue heterogeneity than routine histology and employing a molecular guide will thus provide greater biological insights.

Take-Home Message
- MALDI-MS imaging of FFPE tissue samples allows for the collection of spatial information on lipid distribution across different regions and provides guidance for LMD of specific regions of interest for LC-MS lipidomics.
- The LC-TIMS-MS approach working in PASEF data-dependent mode allowed to obtain high quantity and quality of MS/MS data (77% of identified lipids), despite the low number of cells used per sample (25,000 cells).
- The use of PASEF-TIMS coupled with LC-MS improves both sensitivity and quality of the lipid analysis by improving the acquisition of MS/MS spectra. MALDI-MSI analysis can also be coupled with IMS. However, when working with MALDI-IM-MSI, with FFPE samples, a compromise between sensitivity, time of analysis and isomer separation should be reached.

Acknowledgements The work presented in this chapter was facilitated by Fondazione Gigi & Pupa Ferrari Onlus and Regione Lombardia: programma degli interventi per la ripresa economica: sviluppo di nuovi accordi di collaborazione con le università per la ricerca, l'innovazione e il trasferimento tecnologico.

References

1. Paglia G, Smith AJ, Astarita G. Ion mobility mass spectrometry in the omics era: challenges and opportunities for metabolomics and lipidomics. Mass Spectrom Rev. 2022;41:722–65.
2. Sgobba E, Daguerre Y, Giampà M. Unravel the local complexity of biological environments by MALDI mass spectrometry imaging. Int J Mol Sci. 2021;22:12393.
3. Salita T, Rustam YH, Mouradov D, Sieber OM, Reid GE. Reprogrammed lipid metabolism and the lipid-associated hallmarks of colorectal cancer. Cancers. 2022;14:3714.
4. Dewez F, et al. MS imaging-guided microproteomics for spatial omics on a single instrument. Proteomics. 2020;20:e1900369.
5. Ogrinc Potočnik N, Porta T, Becker M, Heeren RMA, Ellis SR. Use of advantageous, volatile matrices enabled by next-generation high-speed matrix-assisted laser desorption/ionization time-of-flight imaging employing a scanning laser beam. Rapid Commun Mass Spectrom. 2015;29:2195–203.
6. Swales JG, et al. Spatial quantitation of drugs in tissues using liquid extraction surface analysis mass spectrometry imaging. Sci Rep. 2016;6:37648.
7. Soltwisch J, et al. MALDI-2 on a trapped ion mobility quadrupole time-of-flight instrument for rapid mass spectrometry imaging and ion mobility separation of complex lipid profiles. Anal Chem. 2020;92:8697–703.
8. Spraggins JM, et al. High-performance molecular imaging with MALDI trapped ion-mobility time-of-flight (timsTOF) mass spectrometry. Anal Chem. 2019;91:14552–60.
9. Smith A, et al. Molecular signatures of medullary thyroid carcinoma by matrix-assisted laser desorption/ionisation mass spectrometry imaging. J Proteome. 2019;191:114–23.
10. Metz B, et al. Identification of formaldehyde-induced modifications in proteins: reactions with model peptides. J Biol Chem. 2004;279:6235–43.
11. Denti V, et al. Antigen retrieval and its effect on the MALDI-MSI of lipids in formalin-fixed paraffin-embedded tissue. J Am Soc Mass Spectrom. 2020;31:1619–24.
12. Denti V, et al. Reproducible lipid alterations in patient-derived breast cancer xenograft FFPE tissue identified with MALDI MSI for pre-clinical and clinical application. Meta. 2021;11:577.
13. Denti V, et al. Lipidomic typing of colorectal cancer tissue containing tumour-infiltrating lymphocytes by MALDI mass spectrometry imaging. Meta. 2021;11:599.
14. Gil-de-Gómez L, et al. A phosphatidylinositol species acutely generated by activated macrophages regulates innate immune responses. J Immunol. 2013;190:5169–77.
15. Smith CM, Li A, Krishnamurthy N, Lemmon MA. Phosphatidylserine binding directly regulates TIM-3 function. Biochem J. 2021;478:3331–49.
16. Buck A, et al. High-resolution MALDI-FT-ICR MS imaging for the analysis of metabolites from formalin-fixed, paraffin-embedded clinical tissue samples. J Pathol. 2015;237:123–32.
17. Barré FPY, et al. Enhanced sensitivity using MALDI imaging coupled with laser postionization (MALDI-2) for pharmaceutical research. Anal Chem. 2019;91:10840–8.
18. Smith A, Piga I, Denti V, Chinello C, Magni F. Elaboration pipeline for the management of MALDI-MS imaging datasets. Methods Mol Biol. 2021;2361:129–42.
19. Pluskal T, Castillo S, Villar-Briones A, Oresic M. MZmine 2: modular framework for processing, visualizing, and analyzing mass spectrometry-based molecular profile data. BMC Bioinformatics. 2010;11:395.

20. Strohalm M, Hassman M, Košata B, Kodíček M. *mMass* data miner: an open source alternative for mass spectrometric data analysis. Rapid Commun Mass Spectrom. 2008;22:905–8. https://doi.org/10.1002/rcm.3444.

21. Andersen MK, et al. Simultaneous detection of zinc and its pathway metabolites using MALDI MS imaging of prostate tissue. Anal Chem. 2020;92:3171–9.

22. Mahalingam M. Laser capture microdissection: insights into methods and applications. Methods Mol Biol. 2018;1–17. https://doi.org/10.1007/978-1-4939-7558-7_1.

23. Smirnov D, Mazin P, Osetrova M, Stekolshchikova E, Khrameeva E. The Hitchhiker's guide to untargeted lipidomics analysis: practical guidelines. Metabolites. 2021;11:713. https://doi.org/10.3390/metabo11110713.

24. Züllig T, Köfeler HC. High resolution mass spectrometry in lipidomics. Mass Spectrom Rev. 2021;40:162–76. https://doi.org/10.1002/mas.21627.

25. Fernandez-Lima F, Kaplan DA, Suetering J, Park MA. Gas-phase separation using a trapped ion mobility spectrometer. Int J Ion Mobil Spectrom. 2011;14:93–8. https://doi.org/10.1007/s12127-011-0067-8.

26. Chen X, et al. Trapped ion mobility spectrometry-mass spectrometry improves the coverage and accuracy of four-dimensional untargeted lipidomics. Anal Chim Acta. 2022;1210:339886.

27. Vasilopoulou CG, et al. Trapped ion mobility spectrometry and PASEF enable in-depth lipidomics from minimal sample amounts. Nat Commun. 2020;11:331.

28. Goracci L, et al. Lipostar, a comprehensive platform-neutral cheminformatics tool for lipidomics. (2017) https://doi.org/10.1021/acs.analchem.7b01259.

29. Sumner LW, et al. Proposed minimum reporting standards for chemical analysis. Metabolomics. 2007;3:211–21. https://doi.org/10.1007/s11306-007-0082-2.

30. Smets T, De Keyser T, Tousseyn T, Waelkens E, De Moor B. Correspondence-aware manifold learning for microscopic and spatial omics imaging: a novel data fusion method bringing mass spectrometry imaging to a cellular resolution. Anal Chem. 2021;93:3452–60.

31. Ji AL, et al. Multimodal analysis of composition and spatial architecture in human squamous cell carcinoma. Cell. 2020;182:1661–2. https://doi.org/10.1016/j.cell.2020.08.043.

32. Alexandrov T. Spatial metabolomics and imaging mass spectrometry in the age of artificial intelligence. Ann Rev Biomed Data Sci. 2020;3:61–87. https://doi.org/10.1146/annurev-biodatasci-011420-031537.

33. Hériché J-K, Alexander S, Ellenberg J. Integrating imaging and omics: computational methods and challenges. Ann Rev Biomed Data Sci. 2019;2:175–97. https://doi.org/10.1146/annurev-biodatasci-080917-013328.

34. Jirásko R, et al. Altered plasma, urine, and tissue profiles of sulfatides and sphingomyelins in patients with renal cell carcinoma. Cancers. 2022;14:4622.

35. Wang H-YJ. Matrix-assisted laser desorption/ionization-mass spectrometry imaging of lipids in the ischemic rat brain section: a practical approach. Methods Mol Biol. 2021;2306:299–311.

36. Dannhorn A, et al. Evaluation of formalin-fixed and FFPE tissues for spatially resolved metabolomics and drug distribution studies. Pharmaceuticals. 2022;15:1307.

Investigating the Warburg Effect in Renal Cell Carcinoma Using Spatial DYnamic MetabolOmics

15

Rosalie Rietjens, Gangqi Wang, and Bram Heijs

What You Will Learn from This Chapter
- Spatial lipidomics and metabolomics using mass spectrometry imaging
- Isotope tracing in biological tissues
- Visualizing the Warburg effect in cancer tissues

15.1 Theoretical Background

This chapter focuses on the application of stable-isotope tracing in mass spectrometry imaging to unravel changes in the metabolic profile of renal cell carcinoma. Using a pre-recorded dataset, we will illustrate how to use spatial lipidomics data for spatial

R. Rietjens · G. Wang
Department of Internal Medicine (Nephrology) and Einthoven Laboratory of Vascular and Regenerative Medicine, Leiden University Medical Center, Leiden, The Netherlands

The Novo Nordisk Foundation Center for Stem Cell Medicine (reNEW), Leiden University Medical Center, Leiden, The Netherlands

B. Heijs (✉)
The Novo Nordisk Foundation Center for Stem Cell Medicine (reNEW), Leiden University Medical Center, Leiden, The Netherlands

Center for Proteomics and Metabolomics, Leiden University Medical Center, Leiden, The Netherlands

SCiLS Lab, Bruker Daltonics GmbH, Bremen, Germanyb.p.a.m.heijs@lumc.nl
bram.heijs@bruker.com

segmentation of the tissue, and subsequently explain the data analysis strategies for dynamic metabolic measurements. Essential insights into the theory of mass spectrometry (imaging), stable-isotope tracing, as well as the human kidney, its metabolism and expected changes upon renal cell carcinoma will be elaborated.

15.1.1 Spatial Metabolomics and Lipidomics Using Mass Spectrometry Imaging

First, you will find a short introduction into the basic principles underlying spatial metabolomics/lipidomics analysis using matrix-assisted laser desorption/ionization time-of-flight mass spectrometry imaging (MALDI-TOF-MSI). By the end of this section, you will understand the concepts behind this technique and its use in metabolomics/lipidomics, how samples should be treated for MSI, and some of the pitfalls to avoid when designing a MALDI-MSI experiment.

15.1.1.1 Mass Spectrometry

Many analytical chemistry approaches are based on molecular detection, identification, and quantification by mass spectrometry (MS). Although MS encompasses a large variety of technologies, virtually all commercially available MS instruments share the same basic layout, consisting of an ionization source, one or more mass analyzers, a detector, a vacuum system, and a computer for instrument setup and data acquisition. There is a large variety in each of these main components, and different combinations will determine important practical and analytical characteristics of specific MS instruments, such as the use of solid *vs.* liquid samples, fragmentation type, resolving power, mass accuracy, and sensitivity. These considerations consequently affect the applications for which a specific mass spectrometer can be used. It is beyond the scope of this introduction to go in-depth into all different MS platforms, but a few mass spectrometers essential for MSI will be explained in following sections.

Mass Spectra

The output of a mass spectrometer is commonly represented as a mass spectrum; the value on the horizontal axis of a mass spectrum reflects the mass-to-charge ratio (m/z) of the detected gas-phase analyte ions, and the value on the vertical axis represents the intensity which is a measure of the abundance of the analyte ion in the analyzed sample. One can make a broad distinction between two types of mass spectra; i) MS1 spectra and ii) fragmentation spectra, or MSn spectra, in which n represents the number of subsequent fragmentations. MS1 spectra provide a broad overview of all detected compounds in an analyzed sample and their relative quantity compared to each other. An MS1 spectrum can only link an m/z feature to a chemical composition (i.e. $C_8H_{15}NO_6$ for *N*-acetylhexosamine). MSn spectra provide structural information on selected, isolated, and fragmented compounds, which can be used to identify the molecular structure (e.g., the distinction between the isomeric *N*-acetylgalactosamine and *N*-acetylglucosamine) [1].

Mass Spectrometer Performance

Several important analytical characteristics can be determined from a mass spectrum; (i) mass accuracy: the difference between the measured mass of a compound and the theoretical mass derived from its chemical formula usually expressed in absolute numbers (10^{-3} u, or mDa), or relative numbers (parts per million, or ppm); (ii) resolving power: the ability of a mass spectrometer to distinguish two peaks of equal height with a slightly different m/z:

$$R = \frac{m/z}{\delta m/z} = \text{resolving power}$$

The smallest peak separation ($\delta m/z$) at which the two peaks can be separated is called resolution, and is defined as the width of a peak, at 50% of its maximum peak height, or full-width-at-half-maximum (FWHM); (iii) sensitivity: the response of the recorded signal to a change in concentration of the measured analyte. Inherently, mass spectrometers produce and record noise coming from both electrical and chemical interferences. The presence of noise calls for a threshold to distinguish true signals from background noise, the signal-to-noise ratio (S/N).

Since the further focus of this chapter will be the use of MALDI-TOF-MSI, a brief introduction will follow on the MALDI mechanism and the principles of TOF mass spectrometry. For more comprehensive information and background reading into other ionization mechanisms, mass analyzers, fragmentation, please refer to the **Further reading** section towards the end of the chapter for literature suggestions.

15.1.1.2 Matrix-Assisted Laser Desorption/Ionization

The main function of the ionization source is to convert from the solid or liquid molecular analytes contained by the sample into gas-phase ions; cations in positive ion-mode and anions in negative ion-mode. MSI requires an ionization source that can directly probe and produce ions from a solid sample [2]. One of the most common ionization methods able to directly convert analytes from solid phase molecules to gas-phase ions is MALDI. This ionization strategy is based on the illumination of a matrix-doped sample with a pulsed UV laser (Fig. 15.1a). The chemical matrix used is typically a small organic molecule dissolved in organic solvent and has to be a strong absorber of UV light at the wavelength of the laser (Fig. 15.1b). During the evaporation of the organic solvent, the matrix crystallizes and molecular analytes are embedded and co-crystallized with the matrix. Upon illumination of the matrix with the laser, rapid super heating causes both desorption of single surface (matrix) ions, as well as an explosive phase transition (called ablation) creating larger clusters of neutral and charged matrix and analyte molecules, the combination of which is referred to as the MALDI plume. In the MALDI plume, charges will be transferred mainly through the addition or removal of protons (H^+), or the addition of metal ions (Na^+, K^+, Li^+, Ag^+) or halogens (Cl^-) resulting in either cations or anions which are accelerated towards

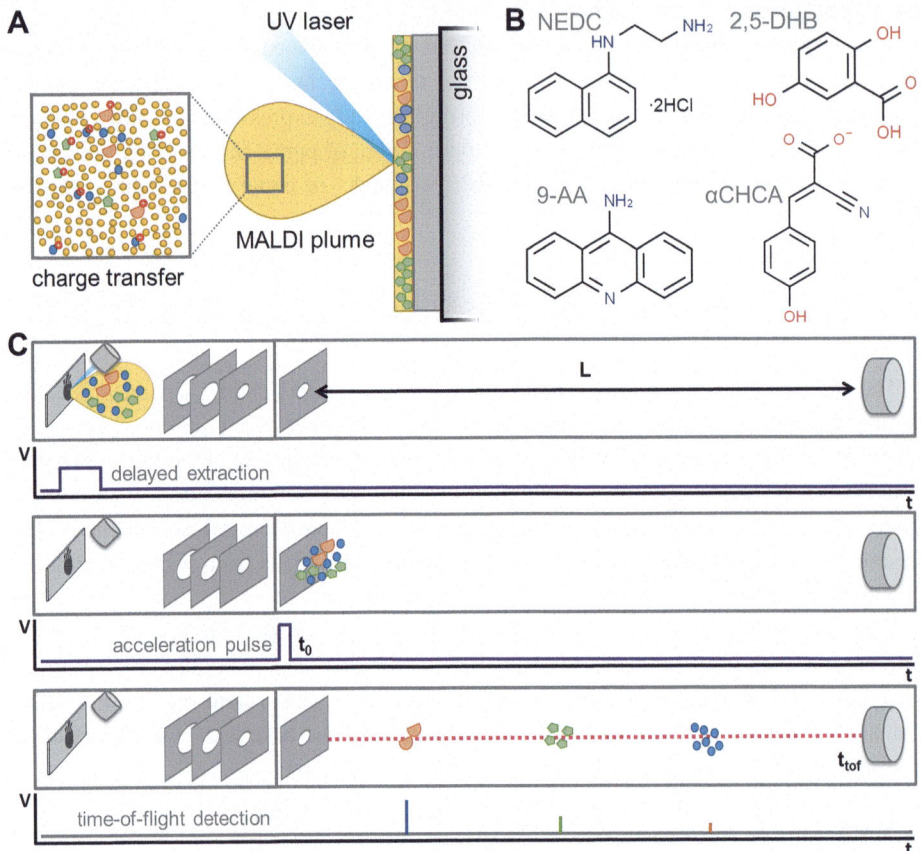

Fig. 15.1 (**a**) Schematic of the MALDI process. In MALDI-MS, the analyte molecules are mixed with a chemical matrix and illuminated with a UV laser. In the resulting gas-phase ion cloud, the MALDI plume, charge transfer, and secondary ionization processes take place, creating gas-phase analyte ions. (**b**) Common MALDI matrices used for negative ion-mode metabolomic MALDI-MSI N-(1-Naphthyl)ethylenediamine dihydrochloride (NEDC), and 9-aminoacridine (9-AA), and positive ion-mode 2,5-dihydroxybenzoic acid (2,5-DHB), and α-cyano-4-hydroxycinnamic acid (αCHCA). (**c**) Schematic of TOF-MS. In a TOF mass analyzer, the ions are transferred from the ionization source to a vacuum drift tube. At t_0, the ions are exposed to an electrostatic pulse, accelerating them towards the detector (at distance L). A difference in their resulting velocity separates the ions in space and time. The m/z can be calculated for each analyte with a differential t_{tof}

the mass analyzer, which will be discussed in the next section. It is important to note that the majority of ions generated in during the MALDI process only carry single charges [3].

15.1.1.3 Time-of-flight Mass Spectrometry

One of the simpler mass analyzers to comprehend is the axial time-of-flight, or TOF, mass spectrometer (Fig. 15.1c). TOF mass analyzers are pulsed systems, and therefore perfectly compatible with MALDI-based ion generation. The ions produced in the MALDI process are transferred from the sample target in the ionization source to the ion optics by means of a strong electric field between the sample target and the first counter electrode in the optics. The ions are accelerated into a drift tube, which they enter all having the same kinetic energy. The time-of-flight (t_{tof}) can be defined as the time interval between the MALDI laser pulse and the impact of the ion on the detector. The m/z for each ion can be calculated using the following formula:

$$\frac{m}{z} = \left(\frac{L^2}{2eV}\right)t_{tof}^2$$

In which the constants L, the length of the ion flight path, and eV, the electrostatic potential of the accelerating pulse are within brackets, and the t_{tof} is the measured time-of-flight. It can easily be deduced that the higher the molecular weight of the analyte, the longer its time-of-flight [4].

15.1.1.4 Mass Spectrometry Imaging

MSI is based on the acquisition of spatially correlated mass spectra from discrete positions in a Cartesian coordinate system virtually projected on a sample surface. Each recorded spectrum is barcoded with an *XY*-coordinate and placed in a virtual data cube, in which the *X* and *Y* axes represent the X and Y coordinates, and the *Z*-axis represents the m/z axis of the mass spectrum. Each individual voxel, or 3D pixel, in this data cube contains the intensity of a single m/z feature at the given *XY*-coordinate. Through the selection of a specific m/z feature, representing an analyte, one can visualize the intensity distribution of the analyte over the sample surface (Fig. 15.2) [5].

15.1.1.5 Sample Preparation for *in situ* Metabolomics Using MALDI-MSI

Direct molecular imaging by MALDI-MSI is one of the most common tools for *in situ* metabolomics. While there are many applications beyond metabolomics (i.e., proteomics and glycomics) and clinical research (i.e., food, insect, and plant biochemistry), the majority of applications focuses on the analysis of mammalian tissues. Metabolomics by MALDI-MSI is commonly applied to thin sections obtained from fresh frozen tissue material, although analysis of metabolites from formalin-fixed and paraffin-embedded material has also been reported [6]. This immediately poses the two main challenges in studying the metabolome in its spatial context; (i) post-mortem degradation and (ii) molecular delocalization through lateral diffusion [7, 8].

Post-Mortem Degradation

The metabolome is extremely dynamic and alters rapidly upon changes in the environment. Resecting a tissue specimen (i.e., an organ or tumor) from a human or animal body inevitably requires the disconnection of that tissue from the blood flow, causing nutrient

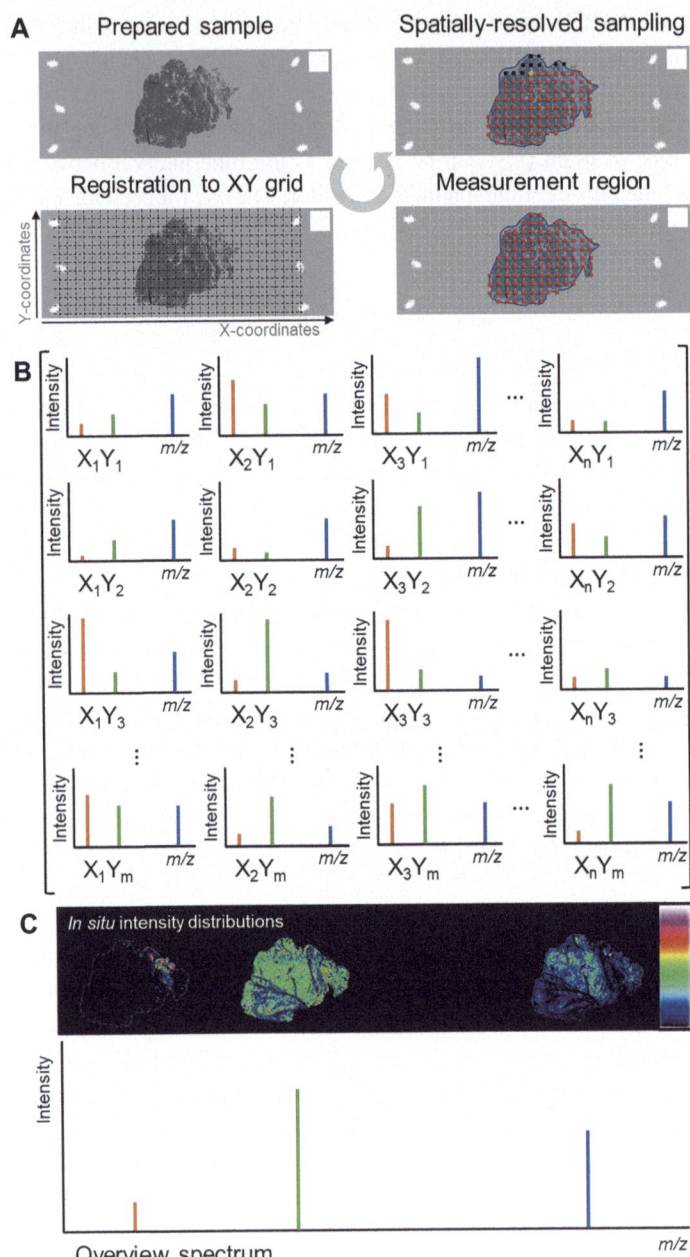

Fig. 15.2 Schematic representation of the mass spectrometry imaging (MSI) principle. (**a**) Registration of the prepared sample to a Cartesian grid with predefined raster width. Followed by definition of the measurement area and spatially-resolved sampling. (**b**) Construction of a data cube from the collection of single mass spectra recorded for every pixel coordinate in the defined measurement area. (**c**) Calculation of a representative overview spectrum, and visualization of intensity distributions for *m/z* features of interest. The color scale represents the relative intensity differences of the selected *m/z* feature between the measured pixels

and oxygen depletion which immediately start to affect the metabolome [9]. Especially the intracorporeal ischemia time, during which the tissue is still inside the body but disconnected from blood flow, is problematic due to the tissue being present at the optimal working temperature of all endogenous enzymes. Naturally, for a representative metabolomics study, this time should be minimized upon sample collection. Once resected, a tissue should be handled swiftly and is typically flash frozen in seconds using liquid nitrogen-cooled isopentane.

Molecular Delocalization

MSI is used to analyze molecules in the spatial context of the tissue, thus the localization of the metabolites is of utmost importance. Most metabolites are polar molecules and dissolve easily in water, which makes avoiding condensation one of the primary objectives during the entire MALDI-MSI sample preparation [8]. Tissues, stored at -80°C, should be transferred on dry ice at all times, and equilibrated to room temperature using a vacuum freeze-drier prior to MALDI matrix application. Once at room temperature, the slide-mounted sections should be handled swiftly. After taking a pre-MSI optical scan of the glass slide, required for setting up the virtual Cartesian coordinate system, the sample preparation for MALDI-MSI typically only involves applying the MALDI matrix.

MALDI Matrix Application

The application of the MALDI matrix is generally done in one of two ways: spray-based (Fig. 15.3a) or sublimation-based (Fig. 15.3b) matrix application. For the spray-based matrix application, the MALDI matrix should be dissolved. For metabolomics approaches, this is typically done in a high-organic solvent, minimizing the amount of water to limit delocalization. The preparation of the high-organic solvent is a balancing act, since there should still be some water present to extract the polar metabolites from the tissue. For example, for the measurements described below, the N-(1-naphthyl) ethylenediamine dihydrochloride (NEDC) matrix was dissolved in methanol:acetonitrile:water (70:25:5 % v/v/v). The dissolved matrix is then homogeneously sprayed over the tissue using a robot. The nebulization of the matrix solution into a fine spray is achieved pneumatically, using ultrasound or electrospray. The fine matrix droplets that land on the tissue allow analytes to extract from the tissue, and upon drying, incorporate into the matrix crystal. It is important here that the resulting matrix crystals should not exceed the dimensions of the desired pixel size, as it is impossible to determine the exact location of origin from an analyte within the confines of a single matrix crystal [8]. The use of solvents makes that the spray-based approach comes with the risk of slight delocalization of the analyte molecules.

In sublimation systems, the solid MALDI matrix and sample are brought into a vacuum chamber. The matrix is heated under vacuum, causing it to sublimate. The sample, placed above the matrix, is cooled, causing the gas-phase matrix to condensate onto the sample surface, allowing surface molecules to co-crystalize with the matrix. The sublimation method has some advantages over the spray-based method; i) it is a solvent-free approach, limiting analyte delocalization, and ii) during the procedure the sample is cooled and stored

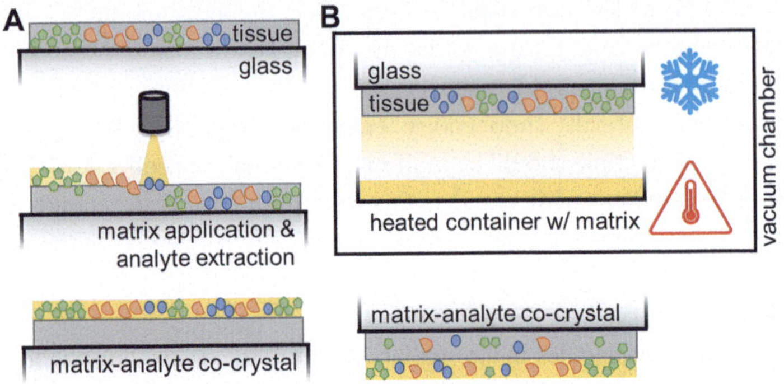

Fig. 15.3 (**a**) Spray-based matrix application of MALDI matrix is commonly performed using a pneumatic or ultrasound nebulizer mounted on a X–Y–Z-stage robot. The dissolved matrix is sprayed homogeneously onto the tissue section, allowing for maximum extraction of analytes, although at the risk of delocalization. (**b**) Sublimation-based application of MALDI matrix is a solvent-free matrix application approach performed under vacuum conditions. Through heating the matrix, it sublimates into the vacuum chamber. Upon touching the cooled sample, the matrix condensates and forms a very fine crystal layer on top of the tissue. Although the cooling and vacuum conditions aid in minimizing delocalization and post-mortem degradation, molecular extraction using sublimation-based matrix application is compromised leading to lower sensitivity for some analytes

in vacuum, limiting post-mortem degradation. Naturally, sublimation-based matrix application also has its limitations. The most obvious one is the limited extraction of analytes into the matrix which affects the measurement sensitivity of certain analytes.

15.1.1.6 MALDI-MSI Measurement Setup

Once the MALDI matrix has been applied, the tissue will be transferred to the MALDI-MS system to set up the MSI measurement. Depending on the instrument vendor, the slide with sections is usually mounted in a target carrier that is positioned in an *XY* robotic MALDI stage. The laser is focused on a fixed position, and between different pixels the target carrier with the sample is moved in the *X* and *Y* plane. Upon loading the target in the MS system, the first thing to do is to register the pre-MSI optical scan to the acquisition software of the MS system, essentially linking the MALDI stage *XY* motor positions to specific pixels in the pre-MSI optical scan image. This should be done as accurately as possible [10]. Once the registration of the pre-MSI optical image to the acquisition software is performed, the MS method can be optimized. This consists of five steps. (i) Setting up the desired *m/z* range; for metabolites and lipids this is typically *m/z* 80–1500. (ii) Setting up the desired ablation field size and matching laser focus. Note that the ablation field size should not exceed the desired spatial resolution of the MSI analysis, as it leads to undesired oversampling. (iii) Determining the optimal laser energy; too low of a laser energy results in insufficient ionization, and consequently produces poor spectra that will translate into "dead pixels". Too high laser energies result in extensive matrix cluster formation and analyte fragmentation, as well as an increase of the effective laser spot diameter which might compromise spatial resolution. (iv) Determining the optimal number of laser shots per pixel; based on the tradeoff between sensitivity (high number of shots per pixel) and throughput (low number of shots per pixel) one can select the optimal number of laser shots per pixel for the experiment. (v) External mass calibration; using a known compound mixture to calibrate the instrument mass response.

Once the MS method is optimized and calibrated, the measurement areas on the tissues can be indicated and the spatially correlated data acquisition can be started.

15.1.1.7 MSI Data Pre-processing and Feature Extraction

After the MSI data acquisition, the data needs pre-processing prior to data analysis. Typical steps in pre-processing are:

(i) Baseline subtraction. Setting the noise level of the recorded single spectra to zero to ultimately enhance the signal-to-noise ratio (S/N) [11].
(ii) Normalization. Multiplying mass spectra with an intensity-scaling factor to correct for and minimize the effect of systematic errors introduced during the MALDI-MSI analysis [12].
(iii) Feature selection, or peak picking. Defining true m/z features from noise in the representative (i.e., average, sum, or base peak) spectrum using a predefined S/N cutoff value [13].
(iv) Feature extraction. Extracting the intensity information for each m/z feature defined in the feature selection from each of the single pixel spectra [14].

The result of the pre-processing is a workable peak matrix with the per-pixel intensity information for all selected m/z features. This pre-processed peak matrix is the starting point for the data analysis procedure described in the Exercises below.

15.1.2 Stable-Isotope Tracing in (Pre)clinical Tissue Specimens

Cell metabolism is a dynamic process characterized by parameters such as cellular metabolite levels, metabolic flux, and nutrient contributions to different metabolic pathways [15]. Mass spectrometry-based metabolomics has a central role in measuring metabolite levels in both physiological and pathological conditions. Changes in metabolite levels indicate altered cellular metabolic states and are related to processes such as biosynthesis, energy metabolism, and catabolism. However, metabolite levels per se do not directly reflect the metabolic rates, or fluxes, of the pathways, nor do they reflect the origin sources of the measured metabolites. Think of it as a bank account, both a rich and a poor person can have the same amount of money on the bank. Despite having the same balance, the rich person likely has a much higher in- and outflow of money and can thus afford a different lifestyle. Stable-isotope tracing is a common tool to get insight into the fluxes of metabolism, as well as nutrient partitioning, which will be explained in more detail later. In stable isotope tracing experiments, the so-called heavy nutrients (e.g., $^{13}C_6$-glucose or $^{13}C_5$-, $^{15}N_2$-glutamine) are introduced into a biological system, and the incorporation and enrichment of the stable isotopes (^{13}C or ^{15}N) into downstream metabolic compounds is assessed [16]. Measuring both metabolite levels and metabolic fluxes at a single timepoint usually requires a metabolic steady or pseudo-steady state. This is characterized by constant or minimal changes in metabolite levels or metabolic fluxes during the time course of the isotope tracing experiment. Sometimes steady state cannot be achieved in a natural biological system, which emphasizes the need for time course experiments, and dynamic labeling calculations, as well as non-stationary flux analysis, such as acute signaling events or nutrient modulations [15]. The duration of the time course may vary depending on the research question; e.g., glycolysis reaches metabolic steady state within approximately 10 min, whereas for the TCA cycle this often takes several hours.

Different stable isotope-labeled nutrients can be used to target different metabolic pathways. For example, uniformly labeled glucose (U-$^{13}C_6$-glucose, also noted as M+6 glucose) is the most commonly used nutrient to trace glycolysis (Fig. 15.4a), the TCA cycle (Fig. 15.4b), as well as other metabolic pathways related to glucose metabolism. However, to measure flux of the oxidative pentose phosphate pathway (PPP), 1,2-$^{13}C_2$-glucose is more common [16]. Uniformly labeled glutamine (U-$^{13}C_5$-glutamine; M+5 glutamine) is often used for TCA cycle flux estimation, as it results in highly abundant labeling of TCA cycle intermediates. Its conversion, through α-ketoglutarate via reductive carboxylation, results in the production of M+5 labeled citrate, which means U-$^{13}C_5$-glutamine can also be used to elucidate the contribution of glutamine to lipogenesis via the reductive carboxylation pathway—which is the reversed direction of the TCA cycle. Alternatively, when labeled glutamine enters the oxidative TCA cycle, it will result in M+4 labeled succinate, and malate, as well as M+3 labeled α-ketoglutarate for the second cycle through the TCA cycle [17]. To measure the direct contribution of different nutrients to metabolic pathways, it is necessary to conduct tracer experiments with all circulating nutrients of interest, which can be determined by a straightforward matrix calculation [18].

Stable-isotope tracing lends itself perfectly for *in vitro* studies, although *in vivo* experiments have also been performed and applied in cancer patients to study tumor cell metabolism via either bolus injection or constant infusion. Nutrient partitioning has proven important for tumor cell survival and the function of immune cells in the tumor microenvironment [19]. Unfortunately, to study nutrient partitioning of tumor cells directly in patients, multi-tracer analyses would be required, and these are not feasible using either bolus injection or constant infusion. *Ex vivo* culturing of human tissue, following vibratome slicing, has provided a promising strategy that allows multi-tracer experiments using single tissue samples [20]. This approach can be combined with the parallelized introduction of various stable isotope-labeled nutrients to the incubation medium, which allows for an efficient and biochemically meaningful labeling of metabolically active cells [21]. Since the metabolic labeling by incubation of vibratome-sectioned tissue slices with

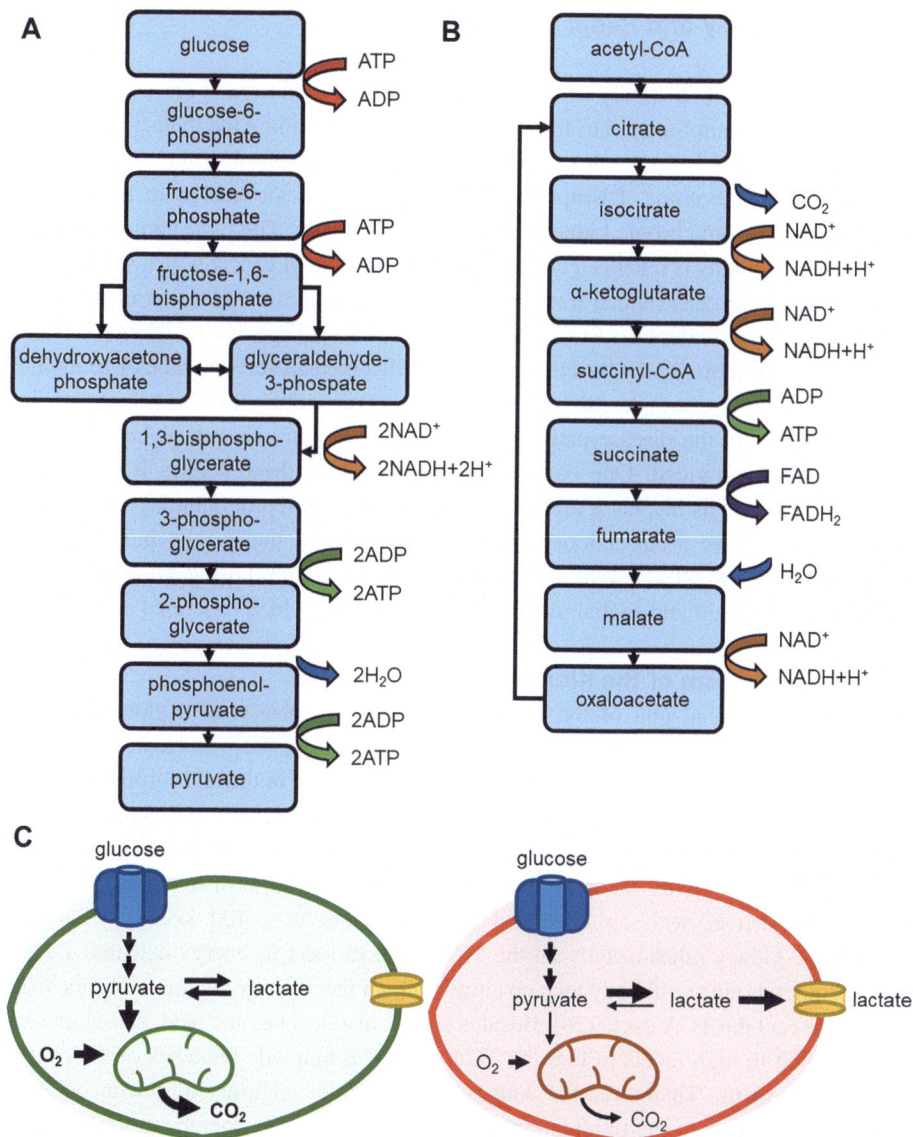

Fig. 15.4 Metabolism and the Warburg effect. (**a**) Schematic overview of glycolysis. (**b**) Schematic overview of the tricarboxylic acid (TCA) cycle. (**c**) Schematic representation of the Warburg effect

stable isotope-labeled tracers takes place in situ, this strategy is perfectly compatible with spatial metabolomics tools such as MALDI-MSI.

15.1.3 The Kidney and Renal Cell Carcinoma

The human kidney is a highly complex organ, with up to twenty known cell types contributing to its main function of filtering our blood. Maintaining the molecular integrity of all these cell types is a complex process, requiring strict control of the transcriptome, proteome, and metabolome. Disruption of these processes can emerge in a variety of diseases, ranging from chronic kidney disease to renal cancer. The most common type of kidney cancer in adults is renal cell carcinoma (RCC), a type of cancer which originates in the proximal tubule cells, which transport the primary urine after filtration of the blood [22]. The 5-year survival rate of RCC patients is around 50–70%, however when the cancer metastasizes the prognosis is substantially worse, with a median survival time of 13 months and a 5-year survival rate under 10% [22]. Giving its severe nature, RCC is heavily studied to better understand the disease pathogenesis and progression, as well as how effective treatment can be provided. One common finding amongst these studies is the role of metabolism; RCC cells display a grade-dependent metabolic reprogramming [23]. In this chapter, we will have a closer look into the metabolism of the healthy human kidney (Fig. 15.5a) as well as RCC and its surrounding tissue (Fig. 15.5b) using the *in situ* stable-isotope tracing method and spatial metabolomics by MALDI-MSI described above.

15.1.3.1 Metabolism of the Kidney and RCC

The primary functional unit of the kidney, the nephron, consists of a glomerulus and Bowman's capsule, connected serially to a proximal tubule, loop of Henle, and distal convoluted tubule. The various tubules play an important role in the reabsorption of water and salts from the filtrate originating from the glomerulus. These reabsorption processes are mostly mediated by active ion transport channels, making the kidney one of the most energy demanding organs of our body. This makes that the human kidney is highly metabolically active, with an estimated metabolic rate of >400 kcal/kg tissue/day [24, 25]. The kidney relies heavily on the TCA cycle to meet its energy demand. To this end, the kidney is able to directly take up citrate, one of the TCA cycle intermediates, from the blood to fuel the TCA cycle [26]. Besides citrate, also lactate, uric acid, and glutamine are reabsorbed in high levels to fuel the TCA cycle through side branches of the central carbon metabolism. Therefore, the kidney also portrays significant gluconeogenetic capabilities, resulting in net-oxidation of lactate into pyruvate, thereby contributing to maintaining the circulating redox homeostasis.

RCC originates in the high energy demanding proximal tubule cells. In case of RCC, the activity of the TCA cycle and susequent oxidative phosphorylation (OxPhos) are decreased, whereas anaerobic glycolysis and the PPP activity are increased. The metabolic swtich from OxPhos to anaerobic glycolysis is a distinct feature of cancer cells, and is also known as the Warburg effect (Fig. 15.4c) [27]. Even though enough oxygen is available for

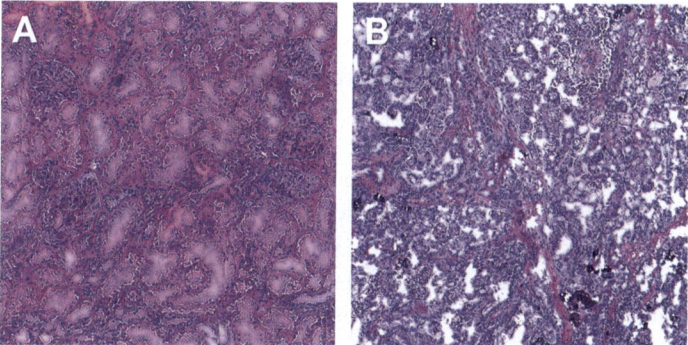

Fig. 15.5 Kidney and renal cell carcinoma histology. (**a**) Representative image of H&E stained normal human kidney. (**b**) Representative image of H&E stained human RCC

OxPhos, cancer cells preferentially use glycolysis for energy production. To still provide sufficient ATP, the cell has to drastically increase its glycolytic flux since the ATP yield of anaerobic glycolysis (2 mol ATP/mol glucose) is much lower compared to OxPhos (~38 mol ATP/mol glucose). This results in a net increase of lactate production, which subsequently can be used for biomass incorporation and cell proliferation; highly beneficial for the fast-dividing cancer cell. Another phenomenon that can be attributed to the Warburg effect is the decreased glucose contribution to the TCA cycle. Of course, these two phenomena go hand in hand and are both indicative of the metabolic shift resulting from the Warburg effect.

15.2 Research Aim

To establish an experiment that allows the metabolic differentiation between healthy proximal tubular cells and RCC cells, by visualizing dynamic differences in glucose metabolism within the tissue.

15.3 Hypothesis and Experimental Setup

As the healthy human kidney relies predominantly on the TCA cycle for energy production and RCC relies on glycolysis, an *in situ* dynamic metabolic tracing experiment of glycolysis activity will allow the distinction between healthy proximal tubule cells and RCC tumor cells, and visualization of the Warburg effect.

15.3.1 Experimental Setup

15.3.1.1 Tissue Preparation, *In Situ* Isotope Incubation, MALDI-MSI, and Staining

A patient with RCC underwent surgical resection to remove the cancer. A biopsy from the RCC tissue was taken and preserved for metabolomics purposes. Besides the RCC tissue, the surrounding healthy tissue was sampled serving as control. Tissues were sliced using a vibratome (Fig. 15.6a). Since for this particular *ex vivo* experiment we are interested in visualizing the dynamics of the Warburg effect, U-^{13}C$_6$-glucose was used as metabolic tracer of the glycolysis. Tissue slices were incubated for 2 hours after which they were quenched using liquid nitrogen. Different tissue slices underwent 0 (control), 15, 30, 60, and 120 minutes of incubation with labeled glucose (Fig. 15.6b). After snap freezing, the tissues were further prepared for MALDI-MSI analysis. First, 10 μm thick tissue sections were sectioned using a cryotome and thaw-mounted on indium-tin-oxide (ITO)-coated glass slides. NEDC matrix was dissolved at 7 mg/mL in a mixture of solvents (70: 25:5 methanol:acetonitrile:deionized water (% v/v/v)) and applied to the tissue section

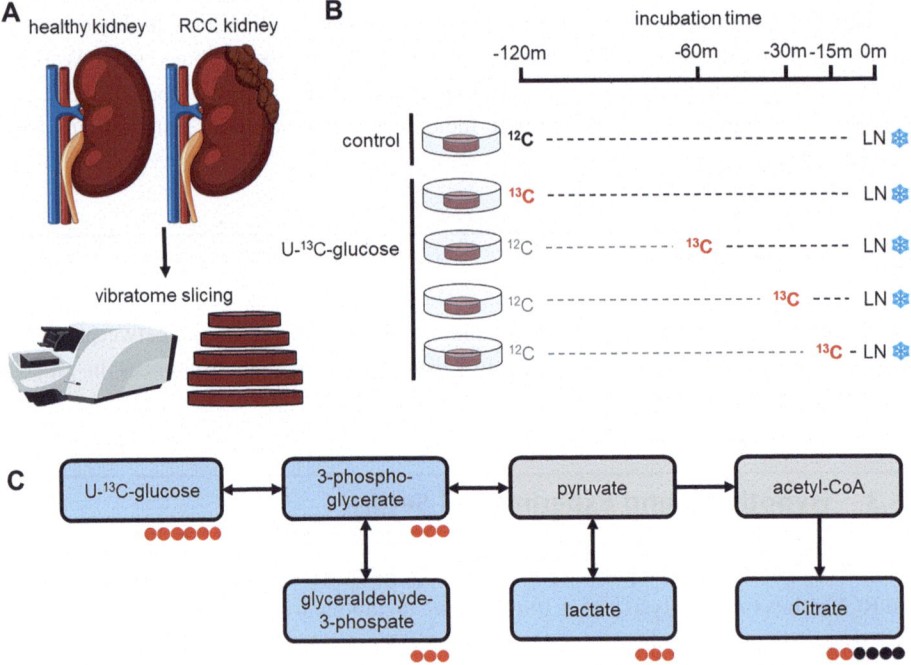

Fig. 15.6 Stable-isotope tracing in tissue culture. (**a**) Schematic overview of the vibratome slicing procedure. (**b**) Overview of the stable-isotope tracing tissue culture time course experiment. At the indicated time points, the label with regular glucose was exchanged with medium containing U-^{13}C-glucose. At time point 0m, the tissues were quenched using liquid nitrogen. (**c**) Detected metabolites with the number of incorporated stable isotope-labeled carbon atoms in red

using a pneumatic sprayer. Then, negative ion-mode MALDI-TOF-MSI analysis of the sections was performed using a Bruker Daltonics rapifleX system at a 5×5 µm^2 spatial resolution. During this analysis, anions within a m/z range of 60–1000 were recorded. After MALDI-MSI data acquisition, the remaining MALDI matrix was removed from the MSI-analyzed tissue by various organic solvent washing steps. The remaining tissue was then stained with several immunofluorescence markers (Lotus Tetragonolobus Lectin (LTL) for proximal tubular cells, E-Cadherin (ECAD) for distal tubular cells and collecting duct, and Nephrin 1 (NPHS1) for podocytes), which allowed us to identify the different epithelial cell types in the tissue.

15.3.2 MSI Data Pre-processing

Following the MALDI-MSI data acquisition, the data for every incubation time point was loaded into a proprietary software package provided by the instrument vendor (SCiLS Lab PRO, v2023a) with baseline correction using a convolution algorithm. The dataset was normalized to the total ion count (TIC). Spectral recalibration and a two-step peak picking on the average spectrum were performed in mMass [28]; (i) untargeted peak picking (S/N > 3) was performed on the m/z range between 450 and 1000 Th), and (ii) targeted m/z feature selection was performed on specific metabolites and isotopologues expected to derive from the stable-isotope tracing experiment and based on the theoretical m/z values (Fig. 15.6c). The peak list was imported into SCiLS Lab, which was used for per-pixel feature extraction and data exporting.

15.3.3 Necessary Software and Exemplary Dataset

Since the SCiLS Lab software is not freely available, we provide the pre-processed datasets (Table 15.1). The associated datasets, IsoCorrectoR file templates (Tables 15.2 and 15.3) and the R scripts (in Rmarkdown) which are referred to throughout the chapter, are available for downloading from OSF.io: https://osf.io/juc8k/?view_only= bfa29e828318419e95bb83f1ec149b99. The programming language R was used for most data analysis steps, e.g. data transformation, spatial segmentation, data integration, metabolite intensity imputation and visualization 15.3. A list of required R packages is provided below (Table 15.4).

15.3.4 Research Questions and Exercises

Using your newly acquired knowledge from the introduction, as well as the provided code and the example datasets, you can train yourself to perform the data transformations and

Table 15.1 Overview with datasets containing the per-pixel intensity information for all selected *m/z* features

Timepoint	Dataset
Control	Kidney_RCC_lipids.csv
15 min	Kidney_RCC_13C_15min.csv
30 min	Kidney_RCC_13C_30min.csv
60 min	Kidney_RCC_13C_60min.csv
120 min	Kidney_RCC_13C_120min.csv
Additional files	
Pixel ID and coordinates	Kidney_RCC_coordinate.csv

Table 15.2 Files needed for the isotope correction package IsoCorrectoR

File	Contains
ElementFile	Information on the elements important for the isotope correction process
MoleculeFile	Information on the molecules to be corrected for natural isotope abundance/tracer purity
MeasurementFile	The measured data that needs to be corrected

Table 15.3 Overview of datasets containing the isotope corrected values for enrichment visualization.

Timepoint	Dataset
15 min	CorrectedFractions15.csv
30 min	CorrectedFractions30.csv
60 min	CorrectedFractions60.csv
120 min	CorrectedFractions120.csv

steps to perform *in situ* metabolic dynamics analysis. In this chapter, we aim to answer the following central research questions:

1. Can we differentiate RCC from healthy kidney tissue on the basis of their metabolic histology? In other words, can we use unsupervised multivariate statistical approaches to isolate pixels obtained from a cancerous tissue from those obtained from a healthy kidney tissue?
2. Can we differentiate RCC from healthy kidney by visualizing the Warburg effect using *in situ* dynamic metabolomics? In other words, can we find differences in the contribution of U-^{13}C$_6$-glucose to glycolysis and TCA cycle between cancerous tissue and healthy kidney?

Table 15.4 Overview of R packages required for the data analysis strategies described below

Package	Purpose
ggplot2	Visualization
ggrepel	Visualization; repel overlaying labels in ggplot2
ggcorrplot	Visualization; easy correlation matrix
pheatmap	Visualization; drawing clustered heatmaps
patchwork	Visualization; combining plots
viridis	Visualization; colors suited for black-white and color
IsoCorrectoR	Isotope abundance correction
Seurat	Clustering, data integration, metabolite imputation
dplyr	Data manipulation
Tidyverse	Data manipulation
reshape2	Data manipulation

15.4 Exercises

Throughout the remainder of this chapter, you will be guided through the workflow outlined in Fig. 15.7. This will be a good starting point for any *in situ* dynamic metabolomics study. Obviously, the options to expand on this analysis pipeline are endless and will not be within the scope of this chapter.

15.4.1 Dimensionality Reduction and Spatial Segmentation of MSI Data

The analysis of a tissue section by MSI generates very high dimensional data. At a 5×5 μm^2 spatial resolution, a measurement area of 1 mm^2 contains 40,000 pixels. For each pixel, a mass spectrum is recorded each consisting of typically 250.000 datapoints, resulting in a total of 10×10^9 datapoints per square millimeter analyzed. In order to create interpretable data, one needs to reduce the data complexity. Following complexity reduction that is achieved through peak picking and feature selection (Sect. 15.1.1.7), the next step in dimensionality reduction is achieved through spatial segmentation of the data. Here, multivariate statistical tools (i.e., principal component analysis (PCA), t-distributed stochastic neighbor embedding (tSNE), uniform manifold approximation and projection (UMAP), etc.) are used to calculate groups of pixels that are highly similar, which in this case means they have mass spectra with comparable peak intensity profiles. By color-coding the clusters, and plotting the clusters using the pixel *XY* coordinates, one can construct an image that shows the spatial distribution of metabolically similar pixels [29].

During the first exercise in our dynamic metabolism data analysis pipeline, you will perform a metabolome-driven spatial segmentation of the healthy kidney control and RCC tissue to identify groups of pixels with similar metabolic profiles. The clustering algorithm

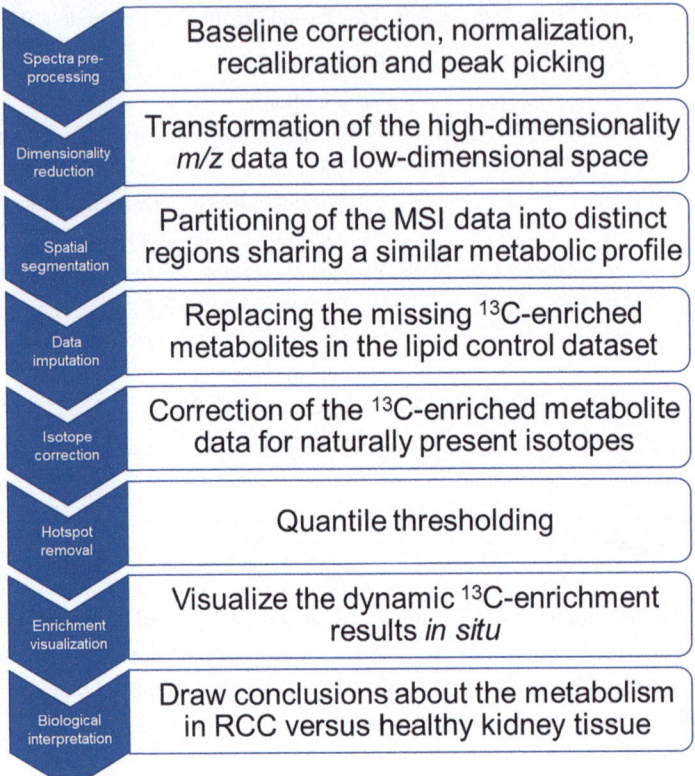

Fig. 15.7 Schematic of the data analysis workflow utilized for the in situ spatial metabolomics analysis

chosen for the dimensionality reduction is UMAP [30]. In the final UMAP plot, pixels that have a similar metabolic profiles will end up in close together and consequently will be assigned to the same cluster. Subsequently, you will reconstruct the cluster image, resulting in a chemically segmented visualization of the tissue. This image will be referred to as the "metabolic histology image" of the tissue.

15.4.1.1 R Code and Explanations

The first goal of this exercise is to perform the UMAP-based dimensionality reduction; you can use the R Script "**Reduction_Segmentation**" with corresponding .csv files "Kidney_RCC_lipids.csv" and "Kidney_RCC_coordinate.csv" for this.

We start by loading the exemplary lipid MSI dataset into our R working environment. Since the lipid profile is highly cell-type specific and stable throughout the isotope-labeling experiment, we can use these features for spatial segmentation, cell-type identification, and anchor-based data integration [31]. The file we have provided you with is the combined data of both the healthy kidney and the RCC tissue. Throughout the exercise, it will appear that these two tissues are metabolically indeed very distinct from one another.

```
# Load in the lipidome dataset
MSIref <- read.csv(file = 'Kidney_RCC_lipids.csv', row.names = 1, header =
TRUE, sep = ",")
# Transform the countmatrix into dataframe suitable for Seurat
MSIref <- MSIref * 100 %>%
 round(digits = 0)
MSIref <- as.data.frame(t(MSIref))
```

After data transformation, the data is now in a suitable format to load it into the *Seurat* package using the following code:

```
MSIdata <- CreateSeuratObject(counts = MSIref, project = "RCC")
```

To put the dataset to a common scale, without distorting the relative differences in ranges of intensity values, the data needs to be normalized and scaled. After data transformation, a PCA will be performed which determines the neighbors of each pixel. The results will later be used by the UMAP algorithm.

```
MSIdata <- SCTransform(MSIdata, verbose = F)
MSIdata <- RunPCA(MSIdata, assay = "SCT", verbose = FALSE)
# Perform Elbowplot to assess suitable number of PC's for subsequent
analysis
ElbowPlot(MSIdata)
```

The elbow plot (Fig. 15.8) shows the standard deviation for each of the calculated principle components (PCs) and is a useful tool to determine how many PCs should be selected to represent the majority of the variation held within the dataset. Based on visual inspection of the plot, you determine the PC at which the change in standard deviation starts to taper off. Despite being subjective, it is a quick and efficient method for choosing the number of PCs to use.

Based on the elbow plot in Fig. 15.8, we choose to work with the first eight PCs to run the UMAP algorithm. There are two important, the so-called, hyper parameters that have a significant effect on the results: resolution and min.dist. You can play around yourself with these parameters to see how they affect the resulting UMAP embedding.

```
# Look for pixels that overlap in the PCA space
MSIdata <- FindNeighbors(MSIdata, dims = 1:8)
# Iteratively groups cells together to a certain optimal point
MSIdata <- FindClusters(MSIdata, resolution = 0.5)
```

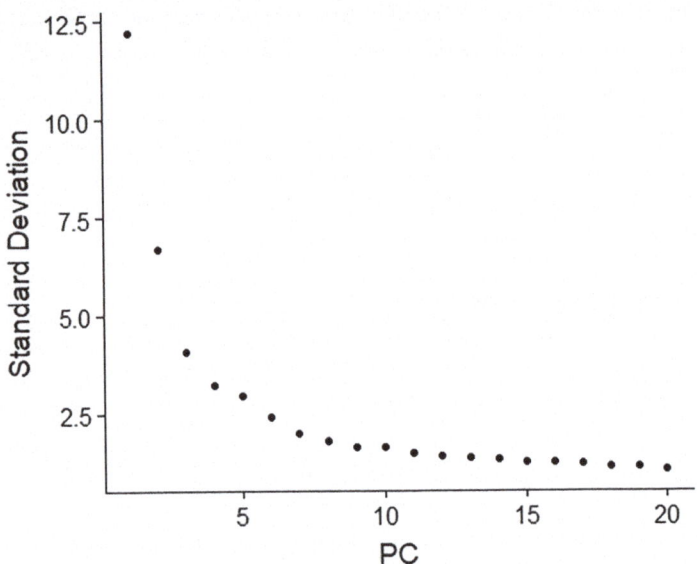

Fig. 15.8 Elbow plot displaying the standard deviation for each calculated principle component

```
# The RunUMAP function learns the underlying manifold of the data in order to
place similar cells together in a low-dimensional space
MSIdata <- RunUMAP(object = MSIdata, dims = 1:8, n.neighbors = 15L,
min.dist = 0.05, check_duplicates = FALSE)
DimPlot(object = MSIdata, reduction = 'umap', label = TRUE, pt.size = 1,
label.size = 5)
DimPlot(object = MSIdata, reduction = 'umap', label = TRUE, pt.size = 1,
label.size = 5, group.by = "orig.ident")
```

After the dimensionality reduction, the clusters are named using meaningless integers which do not give any information about the biological meaning of these clusters (Fig. 15.9a). Since in our experimental setup, we expect to find a mixture of cells, including a variety of healthy kidney cells as well as cancer cells, we are interested in assigning the different cell types to the different clusters. To achieve this, we need to compare the spatial representation of the dimensionality reduction—the "metabolic histology"—with the cell-type information obtained from the immunofluorescence microscopy images. This allows us to determine which clusters are positive for which cell-type markers.

The second part of the exercise is to reconstruct the segmentation cluster distributions and generate the metabolic histology image. To recreate the cluster images, we first need to load the *XY* coordinates for each of the analyzed MSI pixels in R.

```
# Create a dataframe with the xy coordinates from the imaging run
xycoord <- read.csv(file = 'Kidney_RCC_coordinate.csv', row.names = 1,
```

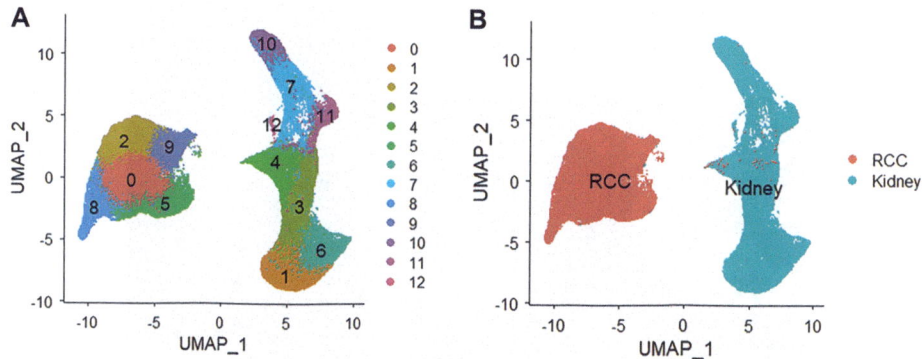

Fig. 15.9 Two-dimensional scatterplot visualization of the UMAP embedding with color-coding representing (**a**) the clusters identities and (**b**) the original pixel identities ("orig.ident")

```
header = TRUE, sep = ",")
xycoord$y1 <- xycoord$y * -1
```

The next step is to associate the *XY* coordinates to the clustered pixels in the Seurat object, and visualize the individual cluster images. An example for cluster 1 is shown below (Fig. 15.10).

```
# Extract the cluster information from the dimensionality reduction
cluster <- as.data.frame(as.matrix(MSIdata@active.ident))
# Select which cluster you want to visualize, by setting this to 1 and all
others to 0
cluster_int <- "1"
cluster$V1 <- replace(cluster$V1, cluster$V1 != cluster_int, 0)
cluster$V1 <- replace(cluster$V1, cluster$V1 == cluster_int, 1)
# Merge the cluster information with the xy coordinate system
dataframe <- merge(cluster, xycoord, by = 'row.names')
# Transform data for pheatmap
dataframe$V1 <- as.numeric(dataframe$V1)
a = dcast(dataframe, y~x, value.var = "V1")
row.names(a) <- a[,1]
data <- as.matrix(subset(a, select = -c(y)))
# Visualize the spatial distribution of cluster of interest
pheatmap(data, scale = "none", cellwidth = 0.6, cellheight = 0.6,
cluster_rows = FALSE, cluster_cols = FALSE, legend = T, show_rownames = F,
show_colnames = F, border_color = FALSE, fontsize = 10,
color = viridis(250), na_col = "WHITE", breaks = NA, main = cluster_int)
```

To assign cell-type information to the clusters, we compare the visualizations of each of the clusters with the immunofluorescence (IF) images we have of the post-MSI analyzed tissue

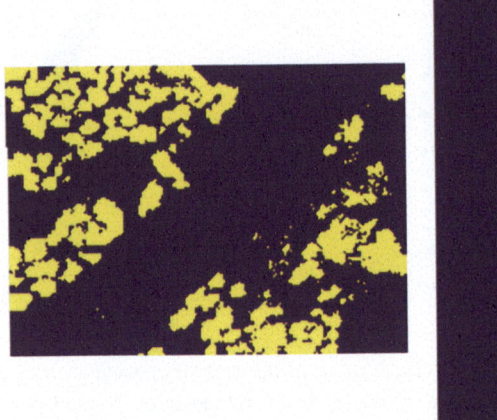

Fig. 15.10 Visualization of the spatial distribution of cluster 1. The left panel is the healthy kidney tissue, the right panel represents the RCC tissue

(Fig. 15.11). From these IF images, it becomes apparent immediately that major histological transformations have occurred in the RCC sample compared to the healthy kidney tissue. The glomerular and tubular structures have mostly disappeared, leaving a dedifferentiated and unstructured tissue which mainly consists of cancer cells and stroma tissue. Although the RCC tissue is quite heterogeneous, both on the histological level evidenced by the IF staining and the lipidomic level evidenced by the presence of several RCC clusters in the UMAP embedding, it is beyond the scope of this chapter to go into the details of intratumor heterogeneity. Therefore, in further processing steps, the RCC tissue as a whole will be regarded as a single group.

For this chapter, we have provided you with a vector with the cell types of interest (new.cluster.ids) which you can use to assign the cluster identities.

```
# Change the cluster names to the newly identified cell type names
new.cluster.ids <- c("RCC", "LTL_1", "RCC", "Unidentified_1","Glom/
Vessel"                              , "RCC", "LTL_2",
"Unidentified_2", "RCC", "RCC",
                                     "ECAD_1", "ECAD_2", "Unidentified_3")
names(new.cluster.ids) <- levels(MSIdata)
MSIdata <- RenameIdents(MSIdata, new.cluster.ids)
DimPlot(object = MSIdata, reduction = "umap", label = TRUE, pt.size = 1,
            label.size = 5)
```

As final part of this exercise, you can now visualize the metabolic histology of the tissue. Once again, a data transformation step is required to enable the spatial visualization of the metabolic histology using ggplot.

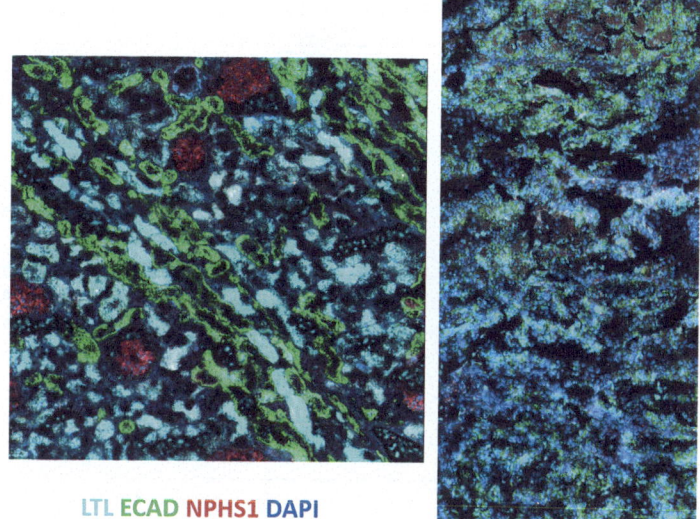

LTL ECAD NPHS1 DAPI

Fig. 15.11 Immunofluorescence staining. The left panel represents the healthy kidney tissue, the right panel represents the RCC tissue—LTL (turquoise) for proximal tubular cells, ECAD (green) for distal tubular cells and collecting duct, NPHS1 (red) for podocytes, and DAPI (blue) for cell nuclei

```
# Data extraction out of the Seurat object
embeddings <- as.data.frame(MSIdata@reductions[["umap"]]@cell.
embeddings)
ident <- as.data.frame(MSIdata@active.ident)
# Data transformation
vector <- row.names(xycoord)
xycoord$pixID <- vector
vector <- row.names(embeddings)
embeddings$pixID <- vector
vector <- row.names(ident)
ident$pixID <- vector
names(ident)[1] <- "Ident"
# Merging everything into 1 dataset
spat_UMAP_Kidney <- merge(xycoord, embeddings, by = 'pixID') %>%
merge(ident, by = 'pixID')
# Using ggplot to visualize the metabolic histology
ggplot(spat_UMAP_Kidney, aes(x = x, y = y)) +
geom_tile(aes(fill = Ident)) +
coord_fixed() +
theme_void()
```

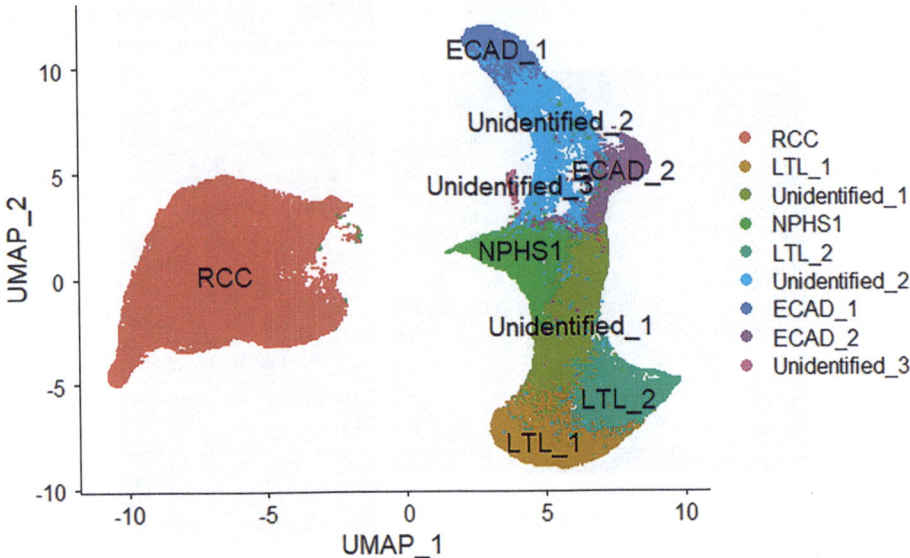

Fig. 15.12 Two-dimensional scatterplot visualization of the UMAP embedding with annotated cluster identities. The color-coding representing the clusters identities

15.4.1.2 Description of the Results

Based on the selected parameters for the dimensionality reduction using the UMAP algorithm, we find a coarse split of the data into two large clusters, each subdivided into several smaller subclusters (Fig. 15.9a). Based on the visualization of the origin of the pixel (orig.ident), it appears that the largest contributor to variation within the tissue is whether the pixel comes from a healthy or a RCC kidney (Fig. 15.9b). The subclustering of the healthy kidney, and comparison of the cluster distributions to the IF images, additionally shows that the individual cell types within the healthy kidney each have their own metabolic profile. For example, we were able to assign various clusters to cells which were positive for LTL, indicating various proximal tubular cells in the healthy kidney tissue (Figs. 15.11, 15.12, and 15.13).

15.4.2 Metabolite Abundance Imputation

To perform the dynamic metabolic measurements, we have performed a time course of *in situ* isotope tracing experiments with U-^{13}C-glucose in both the RCC tissue and its surrounding healthy kidney tissue. Since the different timepoints are represented by different tissue slices taken from the same tissue, the following MALDI measurements were performed on physically different pieces of tissue. This effectively means we now have five metabolic snapshots of the healthy kidney and the RCC tissue. The control tissues (timepoint 0 min) will not contain any ^{13}C-enriched isotopologues, while the other

RCC

LTL_1

Unidentified_1

Glom/Vessel

LTL2

Unidentified_2

ECAD_1

ECAD_2

Unidentified_3

Fig. 15.13 Metabolic histology visualization of the UMAP embeddings. The color-coding represents the different cluster identities in the spatial context of the tissue

timepoints (15, 30, 60, and 120 min) will contain different levels of the ^{13}C-enriched downstream metabolic intermediates. To efficiently and properly evaluate the dynamics of the cellular metabolism in the different cell types, we need to impute the ^{13}C-enriched metabolic snapshots in our control dataset. This allows the direct comparison of the different timepoints using the same tissue, and consequently the direct comparison of the metabolic dynamics of specific cell types. To achieve this, we make use of a data enhancing strategy. This strategy takes a query dataset (lipid control dataset) which lacks intensity information on features of interest (the ^{13}C-enriched metabolite isotopologues and downstream metabolites), and a reference dataset that does contain the intensity information of the features of interest (the ^{13}C-datasets of different timepoints). Based on the intensity profiles of the so-called common features, shared between the query and reference datasets (the lipid m/z features), a k-nearest neighbor (kNN) analysis determines the most similar pixels between the query and reference datasets. It then takes the pixel-specific intensity of the features of interest from the reference dataset and imputes them in the most similar pixel in the query dataset. By doing this for each time point, we end up with five datasets in which we can calculate the ^{13}C-enrichment in various cell types for different metabolic intermediates. The differences in metabolic flux of the different cell types will influence the dynamics of ^{13}C-label incorporation, which we can now visualize over time, and in context of the morphology of the analyzed tissues.

The exercises below will take you through the process of imputing the different ^{13}C-labeling timepoints to the control dataset. Following data imputation, an isotope correction step is performed. This is necessary since isotopes are not only introduced with the labeling experiment, these isotopes are also naturally abundant. This natural abundance of isotopes leads to convoluted signals in the MSI dataset, which could lead to distorted biological findings. To correct for the natural abundance of isotopes, we therefore perform an isotope correction step using the IsoCorrectoR package. As this correction takes a significant amount of time and requires additional manual data transformation in Excel, the ocde and resulting data frames are provided. It is not part of the exercise to perform this yourself.

15.4.2.1 R Code and Explanations

The goal of this exercise is to perform data imputation, for which you can use the R Script "**Data_imputation**" with corresponding .csv files "Kidney_RCC_13C_15min.csv" and "Kidney_RCC_lipids.csv" for this. In the chapter, only the example for the 15 minute timepoint will be shown, you can perform the other timepoints in a similar way yourself.

Start by importing in the ^{13}C-enriched processed MALDI-MSI dataset, which will be used to impute ^{13}C isotopologue data to the lipid dataset.

```
# Load reference dataset
MSIref <- read.csv(file = 'Kidney_RCC_13C_15min.csv', row.names = 1,header
= TRUE, sep = ",")
# Reshape data into suitable format for subsequent analysis
MSIref <- MSIref * 100 %>%
 round(digits = 0)MSIref <- as.data.frame(t(MSIref))
# Create Seurat objects
MSIdata_ref <- CreateSeuratObject(counts = MSIref, project = "RCC")
# Data normalization - ref
MSIdata_ref <- SCTransform(MSIdata_ref, verbose = FALSE)
MSIdata_ref <- RunPCA(MSIdata_ref, assay = "SCT", verbose = FALSE)
ElbowPlot(MSIdata_ref)
```

Based on the elbow plot (Fig. 15.14), the first eight PCs were used to run the UMAP algorithm.

```
# UMAP analysis - ref
MSIdata_ref <- RunUMAP(MSIdata_ref, dims = 1:8)
```

Then, the lipid control dataset can be loaded into the R environment and processed in a similar way as the ^{13}C isotopologue data.

```
# Load query dataset and reshape accordingly
MSIque <- read.csv(file = 'Kidney_RCC_lipids.csv', row.names = 1,
 header = TRUE, sep=",")MSIque <- MSIque * 100 %>% round(digits = 0)MSIque <-
as.data.frame(t(MSIque))
# Create Seurat object
MSIdata_que <- CreateSeuratObject(counts = MSIque, project = "RCC")
# Data normalization - que
MSIdata_que <- SCTransform(MSIdata_que, verbose = FALSE)
```

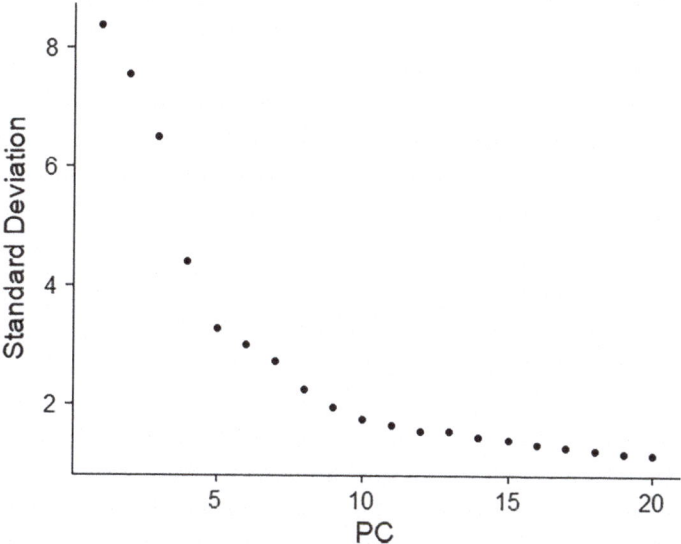

Fig. 15.14 Elbow plot displaying the standard deviation for each calculated principle component

```
MSIdata_que <- RunPCA(MSIdata_que, assay = "SCT", verbose = FALSE)
```

After processing of both the reference and query datasets, we can continue with the data imputation.

```
# Find the common features between reference and query dataset
anchors <- FindTransferAnchors(reference = MSIdata_ref, query =
MSIdata_que, normalization.method = "SCT")
# Filling in the labelling data in the control dataset based on KNN
predictions.assay <- TransferData(anchorset = anchors, refdata =
GetAssayData(MSIdata_ref[['RNA']]), prediction.assay = T,
weight.reduction = MSIdata_que[["pca"]], dims = 1:8)
# Write out csv files for further processing in Excel
data_to_write_out <- as.data.frame(as.matrix(predictions.assay@data))
data_to_write_out <- as.data.frame(t(data_to_write_out))
# Select only the 13C enriched metabolites for visualization
data_to_write_out <- data_to_write_out[,1:15]
write.csv(x = data_to_write_out, row.names = T, file = "15min.csv")
```

The resulting .csv file has the following format:

	A	B	C	D	E	F	G	H	I	J	K	L	M	N	O	P
1		X89.05	X92.04	X146.07	X147.06	X148.07	X149.09	X185.01	X188.02	X191.04	X192.04	X193.05	X215.07	X221.06	X265	X268.01
2	Kidney_1	164.2049	78.60585	3170.251	493.208	349.3059	90.98812	999.3672	688.9032	2601.084	398.3326	614.0775	3958.361	3196.907	517.793	571.1652
3	Kidney_2	131.0715	65.20852	3102.507	518.2193	330.5496	154.996	824.6879	745.2096	2528.768	579.7447	775.9839	3999.402	3337.128	566.1404	670.5507
4	Kidney_3	208.5753	189.01	6768.506	809	428.2251	258.9036	1142.644	1040.014	3750.223	689.4207	946.8252	8188.514	5513.889	648.4032	490.0168
5	Kidney_4	143.8233	97.82871	3847.535	561.4539	272.4467	199.8724	979.172	1048.885	3133.234	635.8555	839.0305	5057.1	3818.871	743.6843	598.5983
6	Kidney_5	269.3624	127.382	5101.758	956.4896	567.3086	178.4838	1153.071	1115.954	4130.564	801.2893	1136.845	6062.964	5591.37	697.8259	628.7008
7	Kidney_6	149.8921	133.5138	3759.849	565.6191	310.5893	155.7036	805.2148	872.2529	2883.201	542.5479	750.4825	4174.72	3880.181	667.6901	813.5342
8	Kidney_7	245.1582	111.4794	4346.197	490.9912	333.5009	140.0484	921.652	1091.595	3160.725	545.0753	875.3529	5181.107	4460.052	410.5579	649.548
9	Kidney_8	241.6892	92.76664	4327.685	572.2231	330.8739	201.0429	1006.773	1125.495	3187.372	708.5589	815.7928	4865.497	4251.21	662.7086	557.124
10	Kidney_9	176.681	82.74212	4045.127	524.0093	392.7269	128.7123	1062.987	914.3264	3304.156	505.4223	764.0083	5159.441	4274.472	813.3812	608.3343
11	Kidney_10	192.2751	86.38057	4397.66	529.3585	465.6634	287.0805	964.8196	930.5935	3581.457	710.9646	814.2171	5439.056	4445.235	810.1312	901.0591

Each row represents a pixel from the lipid control tissue and each column contains imputed intensities for the ^{13}C isotopologues represented here by their m/z value. In this experiment, we only introduced U-^{13}C$_6$-glucose, therefore the number of ^{13}C isotopologues from the downstream metabolites was limited to the ones represented in the example format.

The next step of the procedure is the isotope correction using the IsoCorrectoR package [32]. The package requires the input of three files which are necessary for proper isotope correction: the ElementFile, MeasurementFile, and MoleculeFile. These three files all have a set layout that should be used when importing your own data. For more information about the package and these files, see the reference listed in **Further reading**. A short reminder of the fact that running the isotope correction for these datasets took over 48 hours. The output files of the isotope correction have been provided for a smooth continuation of the workflow.

```
# Load query dataset and reshape accordingly
MeasurementFile <- read.csv(file = '15min.csv', row.names = 1, sep = ";")
a = MeasurementFile
a = data.frame(t(a))
b = data.frame(row.names(a))
# Rename file to "Measurements/Samples"
fix(b)
c = cbind(b,a)
write.csv(c, quote = F, row.names = F, file = "File_path.csv")
# Get path of IsoCorrectoR files
path.molecule <- system.file("data", "MoleculeFile.csv",
 package = "IsoCorrectoR", mustWork = TRUE);
path.element <- system.file("data", "ElementFile.csv",
 package = "IsoCorrectoR", mustWork = TRUE);
path.measurement <- system.file("data", "MeasurementFile15.csv",
 package = "IsoCorrectoR", mustWork = TRUE);
# Run correction algorithm and save results in new variable
correctionResults <- IsoCorrection(MeasurementFile = path.measurement,
 ElementFile = path.element,
 MoleculeFile = path.molecule)
```

For the enrichment visualization, we use the "IsoCorrectoR_result_CorrectedFractions" file. This file contains the corrected measurement data as fractions of the total abundance of a specific metabolite. The output format is shown in the example below:

	A	B	C	D	E	F	G	H	I	J	K
1		Kidney_1	Kidney_2	Kidney_3	Kidney_4	Kidney_5	Kidney_6	Kidney_7	Kidney_8	Kidney_9	Kidney_10
2	Lactate_0	0.683064801	0.67466836	0.532385099	0.602666919	0.685699512	0.536667229	0.694084977	0.728848771	0.687797054	0.696648258
3	Lactate_1	0	0	0	0	0	0	0	0	0	0
4	Lactate_2	0	0	0	0	0	0	0	0	0	0
5	Lactate_3	0.316935199	0.32533164	0.467614901	0.397333081	0.314300488	0.463332771	0.305915023	0.271151229	0.312202946	0.303351742
6	Glutamate_0	0.814854513	0.796687867	0.863668963	0.831067397	0.790829455	0.827455296	0.86313783	0.840111403	0.838113844	0.816200256
7	Glutamate_1	0.081818866	0.089026013	0.055918456	0.075514291	0.104370744	0.078878797	0.050287717	0.064948208	0.062569222	0.053530309
8	Glutamate_2	0.083436971	0.078352393	0.050097036	0.053412666	0.080695045	0.062575415	0.061681887	0.05912233	0.076000162	0.081361744
9	Glutamate_3	0.01988965	0.035933726	0.030315545	0.040005645	0.024104756	0.031090492	0.024892566	0.03581806	0.023316772	0.048907691
10	Glutamate_4	0	0	0	0	0	0	0	0	0	0
11	Glutamate_5	0	0	0	0	0	0	0	0	0	0
12	X3PG_0	0.599466781	0.533091921	0.531292374		0.4906127	0.515979805	0.487815273	0.465553126	0.479947787	0.545343138	0.51682877
13	X3PG_1	0	0	0	0	0	0	0	0	0	0
14	X3PG_2	0	0	0	0	0	0	0	0	0	0
15	X3PG_3	0.400533219	0.466908079	0.468707626		0.5093873	0.484020195	0.512184727	0.534446874	0.520052213	0.454656862	0.48317123
16	Citrate_0	0.755579175	0.682700839	0.730686784	0.713450956	0.714060671	0.724461814	0.72373226	0.71002588	0.758396476	0.736427914
17	Citrate_1	0.074052538	0.1183213	0.09380059	0.10507245	0.098840788	0.096112782	0.084758849	0.118168885	0.074196989	0.105230839
18	Citrate_2	0.170368287	0.198977861	0.175512626	0.181476595	0.187098541	0.179425404	0.191508891	0.171805235	0.167406535	0.158341248
19	Citrate_3	0	0	0	0	0	0	0	0	0	0
20	Citrate_4	0	0	0	0	0	0	0	0	0	0
21	Citrate_5	0	0	0	0	0	0	0	0	0	0
22	Glucose_0	0.56807141	0.560052411		0.61201798	0.584476828	0.535270067	0.533326963	0.552357064	0.548671807	0.561809231	0.565155709
23	Glucose_1	0	0	0	0	0	0	0	0	0	0
24	Glucose_2	0	0	0	0	0	0	0	0	0	0
25	Glucose_3	0	0	0	0	0	0	0	0	0	0
26	Glucose_4	0	0	0	0	0	0	0	0	0	0
27	Glucose_5	0	0	0	0	0	0	0	0	0	0
28	Glucose_6	0.43192859	0.439947589		0.38798202	0.415523172	0.464729933	0.466673037	0.447642936	0.451328193	0.438190769	0.434844291
29	X13BPG_0	0.483285929	0.465545798	0.577201429	0.561745202	0.533832269	0.458510079	0.394714295	0.551014282	0.579737899	0.481220583
30	X13BPG_1	0	0	0	0	0	0	0	0	0	0
31	X13BPG_2	0	0	0	0	0	0	0	0	0	0
32	X13BPG_3	0.516714071	0.534454202	0.422798571	0.438254798	0.466167731	0.541489921	0.605285705	0.448985718	0.420262101	0.518779417

15.4.3 Metabolic Dynamics Calculations and Visualization

As the title of this chapter states, the aim is to perform *in situ* analysis of the dynamic metabolism. The datasets resulting from the exercise in Sect. 15.4.2 contain information on the ^{13}C-enrichment for each of the pixels in the lipid control dataset. This means that similar to what was done before for constructing the metabolic histology, it is now possible to create pseudo-images containing ^{13}C-enrichment of isotopologues of various metabolites over the 120 min time course experiment. A first step in creating these pseudo-images is to perform hotspot removal on the images. This is merely an image processing step where for each image the intensity hotspots are removed, which results in higher contrast images. The hotspot removal is based on a quantile thresholding approach in which for each image, the intensities of the pixels in the highest quantile (1%) are set to the 99th quantile. The data after hotspot removal can be used to reconstruct the tissue morphology using the *XY* coordinates, and, since the coordinates were previously also linked to the UMAP clusters, data can be extracted to directly compare various cell types with one another. The exercises below will allow you to perform enrichment visualization after hotspot removal. Note that the graphs for the ^{13}C-enrichment over time were not prepared using R, and as such there is no code provided for these figures. The results with corresponding biological interpretation will however be discussed later in this section.

15.4.3.1 R Code and Explanations

This final exercise is focused on the visualization of the ^{13}C-enrichment results. These visualizations show the metabolic dynamics in the context of the tissue. To continue with this exercise, you can use the provided output .csv files from the IsotopeCorrectoR package "CorrectedFractions15.csv," "CorrectedFractions30.csv," "CorrectedFractions60.csv," and "CorrectedFractions120.csv" which contain the isotope corrected data as described in Sect. 15.4.2. The script you can use for the exercise is **"Hotspot_Enrichment."** This exercise will focus on the 15 minutes timepoint, but the same R code can be used for the other timepoints.

We start once more by loading the correct dataset into the R environment.

```
# Loading the results from IsoCorrectoR
df <- read.csv("CorrectedFractions15.csv", row.names = 1)
```

Data transformation is necessary to be able to perform visualization of ^{13}C-enrichment. This is similar to what you have seen in a previous exercise.

```
# Hotspot removal, automatically working through all columns
for(i in 1:ncol(df)){
q <- quantile(df[,i], c(.25, .50, .75, .90, .99))
max <- q[5]
df[,i][which(df[,i]>max)] <- max
}
```

After hotspot removal, we can bring in the *XY* coordinates again to make a spatial representation of the metabolite-specific ^{13}C enrichment onto our tissues of interest. We first have to combine the data frame containing the ^{13}C enrichment information with the *XY* coordinates.

```
# Combine the processed isotope corrected data with XY coordinates
xy <- read.csv(file = 'Kidney_RCC_coordinate.csv', row.names = 1,
 header = T, sep = ",")
merged_df <- merge(xy, df, by = 0)
rownames(merged_df) <- merged_df$Row.names
merged_df <- merged_df[, -1 ]
```

Since we know which column contains which metabolite, we can generate heatmaps for all our metabolites of interest.

```
# Extract the metabolites of interest
lactate3 <- merged_df[,c(1,2,6)]
glutamate2 <- merged_df[,c(1,2,9)]
x3pg3 <- merged_df[,c(1,2,16)]
citrate2 <- merged_df[,c(1,2,19)]
# Make a list with all metabolites of interest
metabolites <- list(lactate3, glutamate2, x3pg3, citrate2)
# Write a pdf file in a desired file path containing all enrichment
  visualizations
pdf("File_path.pdf") for (n in metabolites) { a = dcast(n, y~x) data <- as.
matrix(a) data <- data[,-1] print(pheatmap(data, scale= "none", cellwidth =
0.6, cellheight = 0.6, cluster_rows = F, cluster_cols = F, legend = T,
show_rownames = F, show_colnames = F, border_color= F, color=viridis(250),
na_col = "WHITE"))
}
```

In the end, these heatmaps can be put together to get a first insight into the changes in metabolic dynamics which occur between the healthy kidney and RCC tissues (Fig. 15.15).

15.4.3.2 Description of the Results

From the ^{13}C-enrichment visualizations in Fig. 15.15 and the graph representations of the results in Fig. 15.16, it becomes apparent that there is a substantial decrease of the TCA cycle-derived ^{13}C isotopomer enrichments when comparing the RCC tissue to the healthy kidney. The decrease of the citrate M+2 and glutamate M+2 fractions indicates a reduced contribution of U-^{13}C$_6$-glucose to the synthesis of these metabolites, and therefore eludes to the finding that the TCA cycle activity is lower in the RCC tissue compared to the healthy kidney. Additionally, looking at the fractions of 3PG M+3 and lactate M+3, there does not seem to be much difference between the RCC tissue and the proximal tubules in the healthy kidney.

15.5 Discussion

Besides providing the reader with an opportunity to learn the ins and outs of spatial dynamic metabolomics, we wanted to test the hypothesis that the combination of *in situ* stable-isotope tracing and spatial metabolomics using MALDI-MSI was able to visualize the metabolic anomalies introduced by Warburg effect in RCC.

Since RCC is a tumor that derives from the proximal tubuli in the healthy kidney, we wanted to perform the metabolic comparison between tumor cells and proximal tubule cells. Using a combination of the metabolic histology, based on cell-type specific lipid profiles resulting from the MALDI-MSI analysis (Fig. 15.13), and post-MSI IF staining of several kidney-specific cell types (Fig. 15.11), we were able to isolate the pixels associated to the proximal tubules (PTs; pixels positive for LTL staining). Here we performed the IF-

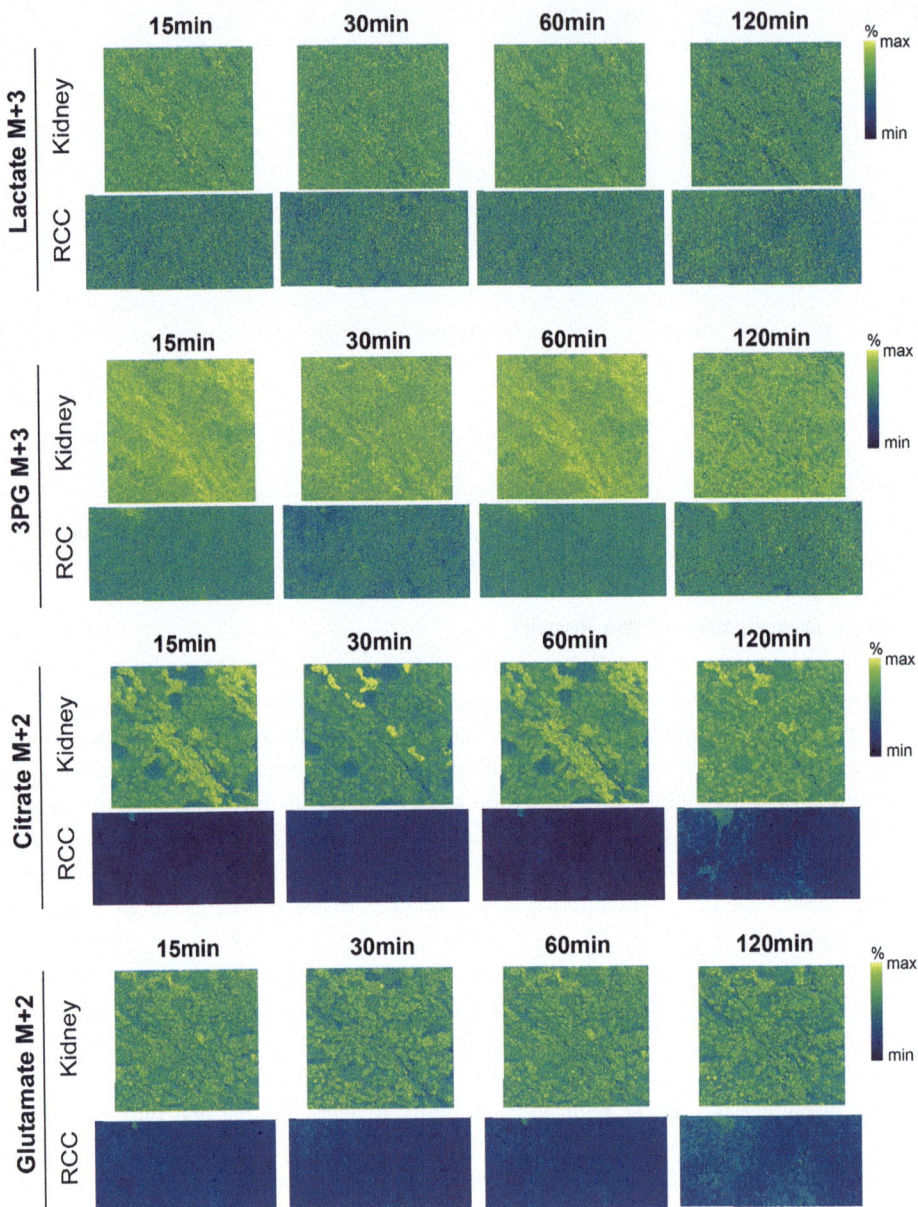

Fig. 15.15 Pseudo-images showing ^{13}C-enrichment of various glycolysis and TCA cycle intermediates over a 120 minute incubation with U-^{13}C$_6$-glucose. The color scale represents the relative ^{13}C-enrichment

based cluster identity assignment based on visual inspection, which is effective, but not the most accurate approach. To achieve the highest level of integration between IF images and

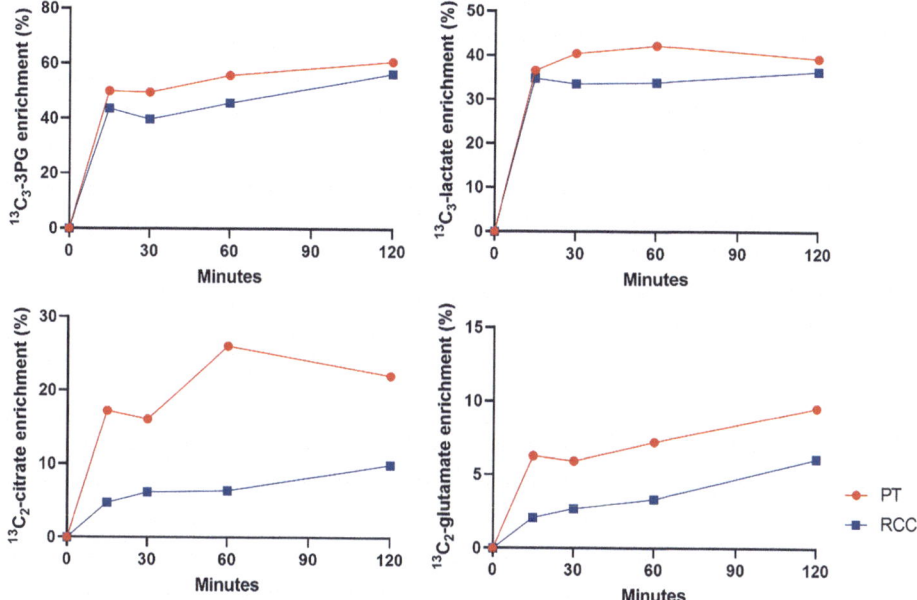

Fig. 15.16 Metabolite-specific ^{13}C enrichment over a 120 minute incubation with U-^{13}C$_6$-glucose

MALDI-MSI data one would need to resort to image co-registration strategies, this is however, beyond the scope of this chapter [10].

The Warburg effect encompasses a metabolic shift from aerobic glycolysis and OxPhos to anaerobic glycolysis and loss of OxPhos activity, even in the presence of sufficient oxygen [27]. Based on this, we expected to find a reduced contribution of stable ^{13}C-isotope-labeled glucose to the intermediates of the TCA cycle, as well as increased levels of glycolysis intermediates in the RCC tissues, with an ultimate increase in lactate production. We were indeed able to show the reduced contribution of U-^{13}C$_6$-glucose to citrate M+2, which enters the TCA cycle through the production of pyruvate M+3. Additionally, we see the reduced contribution of labeled glucose to glutamate M+2, which is the glutamate isotopomer specific to the first pass through the TCA cycle, originating from pyruvate M +3, without the interference of other metabolic pathways [15]. These findings point towards the expected decrease of TCA cycle activity in the RCC tissue in comparison to the PTs in the healthy kidney tissue. Despite this finding, however, we could not find the expected changes in dynamics of the glycolysis. Whereas we expected a RCC-specific increase in lactate M+3 enrichment, this remained nearly identical to the PTs of the healthy kidney. This could have several reasons. First, one of the inevitable effects of excessive lactate production upon the Warburg effect is lactic acidosis, which has shown to result in lower glucose consumption, and a reduction in lactate production in favor of a more oxidative and sustainable non-glycolytic phenotype [33]. To get a conclusive answer to this, we would need to quantify the total lactate pool and show a comparison of the total

amount of lactate produced in the healthy kidney compared to RCC. The absolute quantification of metabolites using MALDI-MSI is in theory possible, but would require a
completely different experimental setup than the one presented here and was outside of
the scope of this chapter [34]. Alternatively, a tissue homogenate-based technique such as
nuclear magnetic resonance (NMR) spectroscopy could be used, which is capable of
measuring absolute quantities of lactate [35]. Another contribution to the observation
could be related to the post-translational modification of proteins, like for example histone
lactylation. When cells are exposed to high levels of lactate, for example as a consequence
of the Warburg effect, they will start to utilize lactate as substrate for the post-translational
modification of histones, and as such use it as an epigenetic modulator of gene transcription
[36]. Lactate-driven alteration of gene expression has recently shown important in the
regulation of immune cells and is associated to the modulation of disease-specific immunity status [37]. To exclude this contributor would require an additional quantitative
epi-proteomics study to quantify the differential lactylation in histones originating from
RCC compared to healthy kidney, this was also deemed beyond the scope of this book
chapter.

Ultimately, we have been able to show that the metabolic histology can be used to
differentiate RCC from healthy kidney, and that it is able to reveal intratumor heterogeneity, as well as the cellular heterogeneity in the healthy kidney. Furthermore, using our
spatial dynamic metabolomics platform, we have been able to establish that RCC tissue
indeed has reduced TCA cycle activity, and that these tumor cells predominantly rely on
glycolysis for energy production. In the end, mass spectrometry imaging in combination
with stable-isotope tracing provides a powerful platform to show the *in situ* metabolic
dynamics of tissues.

Take-Home Messages

Mass spectrometry imaging can be used to gain molecular insight into complex tissue
   architecture.

Mass spectrometry imaging is able to provide additional and comprehensive mor-
   phological information to conventional histopathological staining.

The combination of spatial metabolomics and stable-isotope tracing provides a
   powerful platform to show cell-type-specific differences in the dynamics of
   metabolism.

References

1. Halim A, Westerlind U, Pett C, Schorlemer M, Rüetschi U, Brinkmalm G, Sihlbom C,
   Lengqvist J, Larson G, Nilsson J. Assignment of saccharide identities through analysis of
   oxonium ion fragmentation profiles in LC–MS/MS of glycopeptides. J Proteome Res. 2014;13
   (12):6024–32. https://doi.org/10.1021/pr500898r.

2. Ma X, Fernández FM. Advances in mass spectrometry imaging for spatial cancer metabolomics. Mass Spectrom Rev. 2022:e21804. https://doi.org/10.1002/mas.21804.
3. Dreisewerd K. The desorption process in MALDI. Chem Rev. 2003;103(2):395–426. https://doi.org/10.1021/cr010375i.
4. Boesl U. Time-of-flight mass spectrometry: introduction to the basics. Mass Spectrom Rev. 2017;36(1):86–109. https://doi.org/10.1002/mas.21520.
5. McDonnell LA, Heeren R. Imaging mass spectrometry. Mass Spectrom Rev. 2007;26(4):606–43. https://doi.org/10.1002/mas.20124.
6. Ly A, Buck A, Balluff B, Sun N, Gorzolka K, Feuchtinger A, Janssen K-P, Kuppen PJ, van de Velde CJ, Weirich G, Erlmeier F, Langer R, Aubele M, Zitzelsberger H, McDonnell L, Aichler M, Walch A. High-mass-resolution MALDI mass spectrometry imaging of metabolites from formalin-fixed paraffin-embedded tissue. Nat Protoc. 2016;11(8):1428–43. https://doi.org/10.1038/nprot.2016.081.
7. Goodwin RJ. Sample preparation for mass spectrometry imaging: small mistakes can lead to big consequences. J Proteomics. 2012;75(16):4893–911. https://doi.org/10.1016/j.jprot.2012.04.012.
8. Ščupáková K, Balluff B, Tressler C, Adelaja T, Heeren RMA, Glunde K, Ertaylan G. Cellular resolution in clinical MALDI mass spectrometry imaging: the latest advancements and current challenges. Clin Chem Laboratory Med. 2020;58(6):914–29. https://doi.org/10.1515/cclm-2019-0858.
9. Hattori K, Kajimura M, Hishiki T, Nakanishi T, Kubo A, Nagahata Y, Ohmura M, Yachie-Kinoshita A, Matsuura T, Morikawa T, Nakamura T, Setou M, Suematsu M. Paradoxical ATP elevation in ischemic penumbra revealed by quantitative imaging mass spectrometry. Antioxid Redox Signal. 2010;13(8):1157–67. https://doi.org/10.1089/ars.2010.3290.
10. Balluff B, Heeren RMA, Race AM. An Overview of image registration for aligning mass spectrometry imaging with clinically relevant imaging modalities. J Mass Spectrom Adv Clin Lab. 2022;23:26–38. https://doi.org/10.1016/j.jmsacl.2021.12.006.
11. Williams B, Cornett S, Dawant B, Crecelius A, Bodenheimer B, Caprioli R. An algorithm for baseline correction of MALDI mass spectra. Proc 43rd Annu Southeast Regional Conf - Acm-se. 2005;43:137–42. https://doi.org/10.1145/1167350.1167394
12. Deininger S-O, Cornett DS, Paape R, Becker M, Pineau C, Rauser S, Walch A, Wolski E. Normalization in MALDI-TOF imaging datasets of proteins: practical considerations. Anal Bioanal Chem. 2011;401(1):167–81. https://doi.org/10.1007/s00216-011-4929-z.
13. Abdelmoula WM, Lopez BG-C, Randall EC, Kapur T, Sarkaria JN, White FM, Agar JN, Wells WM, Agar NYR. msiPL: non-linear manifold and peak learning of mass spectrometry imaging data using artificial neural networks. Biorxiv 2020, 2020.08.13.250142. https://doi.org/10.1101/2020.08.13.250142.
14. Ràfols P, Heijs B, del Castillo E, Yanes O, McDonnell LA, Brezmes J, Pérez-Taboada I, Vallejo M, García-Altares M, Correig X. rMSIproc: an R package for mass spectrometry imaging data processing. Bioinformatics. 2020;36(11):3618–9. https://doi.org/10.1093/bioinformatics/btaa142.
15. Buescher JM, Antoniewicz MR, Boros LG, Burgess SC, Brunengraber H, Clish CB, DeBerardinis RJ, Feron O, Frezza C, Ghesquiere B, Gottlieb E, Hiller K, Jones RG, Kamphorst JJ, Kibbey RG, Kimmelman AC, Locasale JW, Lunt SY, Maddocks OD, Malloy C, Metallo CM, Meuillet EJ, Munger J, Nöh K, Rabinowitz JD, Ralser M, Sauer U, Stephanopoulos G, St-Pierre J, Tennant DA, Wittmann C, Heiden MGV, Vazquez A, Vousden K, Young JD, Zamboni N, Fendt S-M. A roadmap for interpreting 13C metabolite labeling patterns from cells. Curr Opin Biotechnol. 2015;34:189–201. https://doi.org/10.1016/j.copbio.2015.02.003.
16. Jang C, Chen L, Rabinowitz JD. Metabolomics and isotope tracing. Cell. 2018;173(4):822–37. https://doi.org/10.1016/j.cell.2018.03.055.

17. Antoniewicz MR. A guide to 13C metabolic flux analysis for the cancer biologist. Exp Mol Med. 2018;50(4):1–13. https://doi.org/10.1038/s12276-018-0060-y.

18. Hui S, Cowan AJ, Zeng X, Yang L, TeSlaa T, Li X, Bartman C, Zhang Z, Jang C, Wang L, Lu W, Rojas J, Baur J, Rabinowitz JD. Quantitative fluxomics of circulating metabolites. Cell Metab. 2020;32(4):676–688.e4. https://doi.org/10.1016/j.cmet.2020.07.013.

19. Reinfeld BI, Madden MZ, Wolf MM, Chytil A, Bader JE, Patterson AR, Sugiura A, Cohen AS, Ali A, Do BT, Muir A, Lewis CA, Hongo RA, Young KL, Brown RE, Todd VM, Huffstater T, Abraham A, O'Neil RT, Wilson MH, Xin F, Tantawy MN, Merryman WD, Johnson RW, Williams CS, Mason EF, Mason FM, Beckermann KE, Heiden MGV, Manning HC, Rathmell JC, Rathmell WK. Cell-programmed nutrient partitioning in the tumour microenvironment. Nature. 2021;593(7858):282–8. https://doi.org/10.1038/s41586-021-03442-1.

20. Roelants C, Pillet C, Franquet Q, Sarrazin C, Peilleron N, Giacosa S, Guyon L, Fontanell A, Fiard G, Long J-A, Descotes J-L, Cochet C, Filhol O. Ex-vivo treatment of tumor tissue slices as a predictive preclinical method to evaluate targeted therapies for patients with renal carcinoma. Cancer. 2020;12(1):232. https://doi.org/10.3390/cancers12010232.

21. Fan T, Lane A, Higashi R. Stable isotope resolved metabolomics studies in ex vivo tissue slices. Bio-protocol. 2016;6(3) https://doi.org/10.21769/bioprotoc.1730.

22. Padala SA, Kallam A. Clear cell renal carcinoma. [Updated 2022 May 24]. In: StatPearls [Internet]. Treasure Island (FL): StatPearls Publishing; 2022 Jan. Available from https://www.ncbi.nlm.nih.gov/books/NBK557644/

23. Bianchi C, Meregalli C, Bombelli S, Stefano VD, Salerno F, Torsello B, Marco SD, Bovo G, Cifola I, Mangano E, Battaglia C, Strada G, Lucarelli G, Weiss RH, Perego RA. The glucose and lipid metabolism reprogramming is grade-dependent in clear cell renal cell carcinoma primary cultures and is targetable to modulate cell viability and proliferation. Oncotarget. 2017;8(69): 113502–15. https://doi.org/10.18632/oncotarget.23056.

24. Tian Z, Liang M. Renal metabolism and hypertension. Nat Commun. 2021;12(1):963. https://doi.org/10.1038/s41467-021-21301-5.

25. Elia M. Organ and tissue contribution to metabolic rate. In: Kinney JM, Tucker HN, editors. Energy metabolism: tissue determinants and cellular corollaries. New York: Raven Press; 1992. p. 61–79.

26. Jang C, Hui S, Zeng X, Cowan AJ, Wang L, Chen L, Morscher RJ, Reyes J, Frezza C, Hwang HY, Imai A, Saito Y, Okamoto K, Vaspoli C, Kasprenski L, Zsido GA, Gorman JH, Gorman RC, Rabinowitz JD. Metabolite exchange between mammalian organs quantified in pigs. Cell Metab. 2019;30(3):594–606.e3. https://doi.org/10.1016/j.cmet.2019.06.002.

27. Liberti MV, Locasale JW. The warburg effect: how does it benefit cancer cells? Trends Biochem Sci. 2016;41(3):211–8. https://doi.org/10.1016/j.tibs.2015.12.001.

28. Strohalm M, Kavan D, Novák P, Volný M, Havlíček V. mMass 3: a cross-platform software environment for precise analysis of mass spectrometric data. Anal Chem. 2010;82(11):4648–51. https://doi.org/10.1021/ac100818g.

29. Alexandrov T, Becker M, Deininger S-O, Ernst G, Wehder L, Grasmair M, von Eggeling F, Thiele H, Maass P. Spatial segmentation of imaging mass spectrometry data with edge-preserving image denoising and clustering. J Proteome Res. 2010;9(12):6535–46. https://doi.org/10.1021/pr100734z.

30. Smets T, Verbeeck N, Claesen M, Asperger A, Griffioen G, Tousseyn T, Waelput W, Waelkens E, Moor BD. Evaluation of distance metrics and spatial autocorrelation in uniform manifold approximation and projection applied to mass spectrometry imaging data. Anal Chem. 2019;91(9):5706–14. https://doi.org/10.1021/acs.analchem.8b05827.

31. Wang G, Heijs B, Kostidis S, Mahfouz A, Rietjens RGJ, Bijkerk R, Koudijs A, van der Pluijm LAK, van den Berg CW, Dumas SJ, Carmeliet P, Giera M, van den Berg BM, Rabelink

TJ. Analyzing cell-type-specific dynamics of metabolism in kidney repair. Nat Metab. 2022;4(9): 1109–18. https://doi.org/10.1038/s42255-022-00615-8.

32. Paul Heinrick 2022. Introduction to IsoCorrectoR. Accessed 30 November 2022, https://bioconductor.org/packages/devel/bioc/vignettes/IsoCorrectoRGUI/inst/doc/IsoCorrectoRGUI.html

33. Xie J, Wu H, Dai C, Pan Q, Ding Z, Hu D, Ji B, Luo Y, Hu X. Beyond Warburg effect – dual metabolic nature of cancer cells. Sci Rep-UK. 2014;4(1):4927. https://doi.org/10.1038/srep04927.

34. Tobias F, Hummon AB. Considerations for MALDI-based quantitative mass spectrometry imaging studies. J Proteome Res. 2020;19(9):3620–30. https://doi.org/10.1021/acs.jproteome.0c00443.

35. Kostidis S, Addie RD, Morreau H, Mayboroda OA, Giera M. Quantitative NMR analysis of intra- and extracellular metabolism of mammalian cells: a tutorial. Anal Chim Acta. 2017;980:1–24. https://doi.org/10.1016/j.aca.2017.05.011.

36. Zhang D, Tang Z, Huang H, Zhou G, Cui C, Weng Y, Liu W, Kim S, Lee S, Perez-Neut M, Ding J, Czyz D, Hu R, Ye Z, He M, Zheng YG, Shuman HA, Dai L, Ren B, Roeder RG, Becker L, Zhao Y. Metabolic regulation of gene expression by histone lactylation. Nature. 2019;574(7779):575–80. https://doi.org/10.1038/s41586-019-1678-1.

37. Chen A-N, Luo Y, Yang Y-H, Fu J-T, Geng X-M, Shi J-P, Yang J. Lactylation, a novel metabolic reprogramming code: current status and prospects. Front Immunol. 2021;12:688910. https://doi.org/10.3389/fimmu.2021.688910.

Further Reading

Mass Spectrometry: Principles and Applications
Edmond de Hoffmann, Vincent Stroobant 2007, Wiley, West Sussex
Correcting for natural isotope abundance and tracer impurity in MS-, MS/MS- and high-resolution-multiple-tracer-data from stable isotope labeling experiments with IsoCorrectoR.
Heinrich P, Kohler C, Ellmann L, Kuerner P, Spang R, Oefner P, Dettmer K. Sci Rep. 2018;8:17910.
Getting started with Seurat
Satija Lab 2022. Accessed 30 November 2022, https://satijalab.org/seurat/articles/get_started.html
R for Data Science
Hadley Wickham, Garrett Grolemund 2017. Accessed 30 November 2022, https://r4ds.had.co.nz/

Quantitative Imaging Using SIMS

16

Cécile Becquart, Elias Ranjbari, and Michael E. Kurczy

What You Will Learn in This Chapter

In this chapter, you will learn that a biological sample prepared for NanoSIMS analysis can be an accurate representation of the subcellular chemical environment and the practical principles needed to carry out a quantitative subcellular measurement. This will be illustrated using the uptake of a drug which becomes sequestered in the subcellular environment.

C. Becquart
Drug Metabolism and Pharmacokinetics, Research and Early Development, Cardiovascular, Renal and Metabolism, BioPharmaceuticals R&D, Gothenburg, Sweden

Department of Chemistry and Molecular Biology, University of Gothenburg, Gothenburg, Sweden
e-mail: cecile.becquart@astrazeneca.com

E. Ranjbari
Department of Chemistry and Molecular Biology, University of Gothenburg, Gothenburg, Sweden

The Chemical Imaging Infrastructure, Gothenburg, Sweden
e-mail: elias.ranjbari.2@gu.se

M. E. Kurczy (✉)
Drug Metabolism and Pharmacokinetics, Research and Early Development, Cardiovascular, Renal and Metabolism, BioPharmaceuticals R&D, Gothenburg, Sweden
e-mail: michael.kurczy@astrazeneca.com

16.1   Quantitative Imaging Using NanoSIMS

While the cell is the basic unit of life, metabolic and disease-associated pathways are regulated at the subcellular level. Thus, it is important to measure the concentrations of biologically active molecules in subcellular compartments in order to better understand and to modulate the biochemical environment of the intact cell [1–4]. The principal read-outs in the fields of metabolomics and pharmacokinetics are the concentrations of metabolites and drugs in a specific biological environment. The typical environments that can be measured are limited to biofluids, tissue, and cell lysates. Unfortunately, this level of detail is not sufficient to further our understanding of biochemistry, particularly at the level of the single cell or organelle [5–7]. Nanoscale secondary ion mass spectrometry (NanoSIMS) is an ion-microscopic technique which not only provides high spatial resolution to unveil the intracellular microstructures, but it allows quantification of labeled target molecules inside subcellular organelles with high sensitivity (<1 ppm). Moreover, thanks to the high mass resolution of NanoSIMS, it is possible to take advantage of non-toxic isotopic labels. These advantages in NanoSIMS provide the required merits for both relative and absolute quantitative imaging of drugs and metabolites in both *in vivo* and *in vitro* specimens at the subcellular level.

NanoSIMS utilizes an ionizing primary ion beam (Cs^+ or O^-) to scan across the sample surface in a raster pattern. The primary ion beam sputters secondary ion species at discrete positions which are represented as pixels. A magnetic sector mass spectrometer separates individual ions, detectable by one of seven electron multiplier detectors. If the density of an ionized species is different at various regions across the sample surface, a unique signal intensity will be recorded in each pixel to reflect the pattern of the species in the sample. By registering each pixel in an *x,y* coordinate, a mass spectrometric image is generated. Although this image reveals the spatial distribution of elements and molecules, these qualitative data can be translated to quantitative data based on the signal intensity per region of interest at each mass.

NanoSIMS imaging with its incredible capabilities for imaging subcellular structures has had a significant impact on the analysis of pharmaceuticals and metabolites in biological samples. A NanoSIMS image initially represents a kind of quantitative information on the relative distribution of the imaged secondary ion species. However, the signal intensity is a function of not only the concentration of the species of interest but also the ionization efficiency and matrix effect on the extractability of secondary ions. The inherent heterogeneity of biological samples can cause a misinterpretation of the relative distribution of the analyte. To eliminate these uncontrollable variables on the image, normalization to an appropriate signal which can act as an internal standard (usually a high abundant element with homogenous spread on the surface) is critically needed to have a correct interpretation of NanoSIMS imaging data. Due to the high mass resolution, a minor isotope of an element of interest can be used for the normalization. These strategies were used by several groups to relatively quantify the drugs and metabolites in intracellular compartments.

The development of NanoSIMS for organo-metallic drugs paved the way for exploring the distribution of biologically active molecules at the subcellular level. Legin et al. combined NanoSIMS with fluorescence confocal laser scanning microscopy to characterize the subcellular distribution of ^{15}N isotopically labeled cisplatin in human colorectal cancer cells [8]. Analyzing the NanoSIMS images depicted platinum co-accumulation in sulfur-rich structures and phosphorus-rich chromatin regions. Determination of the relative intensities of platinum and nitrogen associated to secondary ion signals in different cellular compartments in NanoSIMS images suggested partial detachment of amine ligands from platinum during the accumulation process, in particular within nucleoli at elevated cisplatin concentrations. On the contrary, in research done by Lee et al., the NanoSIMS analysis of a ruthenium-based drug, RAPTA, with NanoSIMS showed co-accumulation of Ru and enriched ^{15}N, indicating the preserved coordination of PTA ligand and ruthenium [9]. This compound is a ruthenium arene complexed to a 1,3,5-triaza-7-phosphaadamantane (RAPTA) [9]. These results highlight one of the key strengths of NanoSIMS for the detection of isotopically labeled metal-based drugs to probe the ligand exchange reactions within intracellular compartments. NanoSIMS is capable of localizing a drug in specific organelles at a single-cell level without the need of combining with a high-resolution microscopy technique such as scanning electron microcopy (SEM), transmission electron microscopy (TEM), fluorescence microscopy. Neither does it require co-localization studies employing specific markers such as Fibrillarin as a nucleolar marker. Wedlock et al. compared the uptake rate and targeting of mononuclear cisplatin with a polynuclear platinum-based drug, TriplatinNC [10]. Interpretation of the NanoSIMS images of treated MCF7 cells with these drugs revealed a significantly quicker uptake and more specific localization for TriplatinNC in comparison to cisplatin.

Relative quantification of non-metallic drugs and metabolites has been widely reported in NanoSIMS imaging for comparison between two or more analytes or sample types or regions of interest in intracellular compartments. To improve identification and characterization of the cellular substructures accentuated in elemental distribution patterns, combinations of NanoSIMS with complementary analytical and imaging techniques are invaluable. For instance, Kay et al. showed that the relative amounts of both components of a peptide antisense oligonucleotide drug conjugate could be compared in the endosomal space of individual cells. Using a combination of transmission electron microscopy (TEM) and NanoSIMS, they found that the more stable linker caused the oligonucleotide to remain trapped in the endosomal space with the peptide. This enhanced entrapment was found to reduce the efficacy of the oligonucleotide [11]. He et al. also visualized the distribution of therapeutic antisense oligonucleotides labeled with bromine (Br-ASO) in some varieties of cultured cells and importantly mouse tissues (heart, kidney, and Liver) using NanoSIMS data combined with backscattered electron microscopy [12]. They demonstrated that phosphorothioate ASOs associate with filopodia and the inner nuclear membrane of cells. They also documented essential cellular and subcellular heterogeneity in ASO distribution in the mouse tissues. Lovric et al. employed the combination of NanoSIMS and TEM to show the distribution of dopamine across individual vesicles in the rat endocrine cell line, pheochromocytoma (PC12) [13]. They incubated PC12 cells with ^{13}C labeled Levo-dopa (L-DOPA, a precursor of dopamine) to visualize vesicles loaded by ^{13}C-dopamine. A significantly higher signal intensity of ^{13}C^{14}N$^-$ normalized to ^{12}C^{14}N$^-$ was observed in the incubated cells in comparison to the control cells confirming NanoSIMS' potential for the analysis of dopamine stored in a single vesicle. In a correlative NanoSIMS and scanning electron microscopy study, Jiang et al. illustrated the internalization of amiodarone into the lysosomes of macrophages [14]. Amiodarone is an antiarrhythmic medication and its molecular structure contains two Iodine atoms (^{127}I). Iodine shows a high affinity for making negative ions without isobaric interferences. The Iodine features of the amiodarone molecule produce reasonable secondary ions in NanoSIMS, that can be tracked with high sensitivity at the subcellular level without any additional labeling. Iodine and phosphorous imaging along with plotting the intensity of ^{127}I$^-$ vs ^{31}P$^-$ indicated a linear relationship between the amount of iodine and phospholipids. These results disclose that building up phospholipids in multilamellar lysosomes is linearly proportional to the drug accumulation process within the lysosomes, providing evidence of amiodarone-induced phospholipidosis. Furthermore, the relative amounts of exogenous isotopically labeled cholesterol and fatty acids have been demonstrated both *in vivo* and *in vitro* [15–18]. Although relative quantification of drugs and metabolites has been widely reported in NanoSIMS, less literature have addressed the absolute quantification of drugs and metabolites for accurate measurement of an analyte at the subcellular level.

To gain more insight into the behavior of biologically active molecules in the subcellular environment, it is clearly very important to be able to report absolute concentrations. To move the NanoSIMS measurements beyond the concept of relative quantification, Thomen et al. introduced a template for absolute quantitative NanoSIMS bioimaging for the first time [19]. They used this approach for the quantification of dopamine in the vesicular nanostructures of PC12 cells after incubation with ^{13}C labeled L-DOPA, the most potent medication for Parkinson's disease. To have a homogenous ionization efficacy throughout the sample surface, they embedded the cells in Agar 100 resin which is a type of artificial organic matrix that was well matched to the cell biomass in terms of carbon concentration. With this strategy, the proportion of each material did not greatly affect the concentration of carbon in the region of interest for the analysis, so the concentration measured in the resin-embedded cell material is an accurate representation of what is found in cell substructures before the sample preparation. To verify this approach, an electrochemical measurement was performed to measure the concentration of dopamine in single vesicles of PC12 cells. The electrochemical results showed the average dopamine concentration (60 mM) in the concentration range obtained by quantitative NanoSIMS imaging (~40–70 mM). Although ^{13}C in the resin embedding material attenuates the limit of detection to 6 mM (or 1 mM for dopamine labeled with 6 atoms of ^{13}C), however this sensitivity limitation is not problematic when the analyte is highly enriched in the cellular microstructures. While, in a homogeneous carbon material, like the embedding epoxy, sputtering away the $^{13}C^-$ ions occurs at a constant rate, this is not the case for less abundant and more heterogeneously distributed elements, like sulfur, making quantification more complicated. Becquart et al. used NanoSIMS to measure the subcellular concentration of antisense oligonucleotides (ASOs) labeled with 15 ^{34}S ($^{34}S_{15}$-ASO) in primary human hepatocytes (PHHs) [20]. ASO is a new therapeutic tool for regulation of gene expression to treat or manage a wide range of rare diseases. In order to correct the measured ion ratios and convert them to concentration, they employed a relative sensitivity factor (RSF). The RSF is the slope of the calibration curve of Agar 100 resin spiked with known amounts of ^{32}S and ^{34}S. As expected, despite using dimethyl sulfoxide (DMSO) as a standard for ^{32}S and using ^{34}S-labeled omeprazole as a standard for ^{34}S, the same RSF was obtained for both isotopes, since isotopes show the same ionization probability and transmission. The method was validated by measuring ASOs labeled with two ^{127}I ($^{127}I_2$-ASOs) in PHHs and comparison of the obtained results with the same concentration of ^{34}S-ASOs. To reach an RSF for iodine, a blank Agar 100 resin was spiked with different concentration of a ^{13}C labeled amiodarone (^{13}C-amiodarone) spanning two iodine in its molecular structure, representing a plot of $^{127}I^-/^{13}C^{12}C^-$ against ^{127}I concentration. As ^{127}I and ^{13}C are in the same molecule, the matrix effect was eliminated. Eventually, these approaches established limits of detection (LODs) of 5 and 2 µM for ^{34}S and ^{127}I, respectively, and were employed independently for measuring ASOs intracellular distribution. The matching results of these two independent labeling strategies confirmed that the method can be used for absolute quantification of ASOs in PHHs. Comparison of a naked $^{34}S_{15}$-ASO with an ASO conjugated to an uptake facilitating functionality called *N*-Acetylgalactosamine or GalNAc, (GalNAc-$^{34}S_{15}$-ASO) showed the accumulation of both species in cytosolic microstructures identified as lysosomes. Furthermore, it showed enhanced and faster uptake of conjugated-ASOs as well as a concentration-dependent uptake compared to the naked ASO.

16.2 Research Question

What is the local concentration of the compound in the subcellular environment? Here we will use all the concepts presented in this chapter to quantify the internalization of amiodarone. This antiarrhythmic drug is well known to accumulate in lysosomes and serves as an ideal system to exhibit this method.

16.3 Hypothesis and Experimental Setup

The central concept that we present in this chapter is that a biological sample prepared for NanoSIMS analysis can be viewed as a material. This change in perspective allows the opportunity to use concepts from material science to quantify biological molecules at the subcellular level. However, we need to stipulate that a resin-embedded cell or tissue is an accurate representation of the subcellular chemical environment. It is important to note that the sample will be fixed and that the water contained in the sample will have been replaced with epoxy as it is embedded in resin. By defining this as a resin-embedded biological material, a biologist can utilize the full set of concepts that material scientists have developed in the pursuit of the characterization of materials using SIMS. In addition to the ability to quantify the local concentrations of exogenous drugs and metabolites, it also makes it possible to generate quantitative standards.

There are two major ideas that should be established before we can translate material characterization for biological measurements. The first is the language we use to describe the amount of material we measure in a given volume. A material scientist may describe the number of atoms per cubic centimeter, while a biologist will report the number of moles in a liter. Here the translation is fairly simple. By way of unit conversion where atoms become moles and cubic centimeters is converted to liters, we can now easily communicate between these two disciplines. However, for this translation to be accurate, it is also critical that the volume of the live cell is a fair approximation of the volume of the cell embedded in the epoxy resin. We tested this assertion using rat pheochromocytoma (PC12) cells which have been shown to take up a well-characterized concentration of dopamine. This experiment compared the concentration of the ^{13}C labeled dopamine after the embedding to the concentration of dopamine measured by electrochemistry in live PC12 cells. The concentration of the dopamine was not greatly altered by the embedding process as the number of ^{13}C atoms per cm^3 emitted by the dopamine was equivalent to the moles per liters of dopamine contained in the live cell [19].

With this framework established, this chapter will serve to highlight the key practical steps needed to carry out quantitative subcellular measurements using the NanoSIMS. The following is a brief introduction to concepts that are not necessarily well known in the metabolomic and pharmaceutical fields but we feel will be readily accessible to researchers in these fields.

The first consideration is the **target molecule**, or the molecule that you wish to quantify. The main aspects to understand are the **expected concentration** and the **fixability** of the molecule. The expected concentration is key when choosing a **labeling strategy.** The fixability of the molecule is also key as an unfixed molecule will be washed out of the sample during sample processing. The first measurement parameter, called **ion dose** is the total number of ions that strike the sample surface over a defined area. Typically, we use the Cs^+ primary ion source equipped to the NanoSIMS, thus this is reported as Cs^+/cm^2. Monitoring this parameter is central to all the concepts that will follow. The ion dose is determined from the primary ion current, the time of the experiment, and size of the area analyzed. It is important to know what ion dose is required to reach **steady state** of secondary ion emission. This can be done by carrying out a **beam stabilization** measurement on an **epoxy standard** at a reasonably high primary ion beam current to profile through the entire sample while monitoring the ions of interest. The secondary ion counts will gradually increase, usually at different rates, as the total ion dose increases. The point where all ion counts remain constant is the steady state. As the ion dose will continue to increase over the time of the experiment, eventually the steady state will be broken due to the degradation of the sample which we call burn through. The range of ion doses between the onset of the steady state and burn through we call the **analysis window**. To make stable and comparable measurements, it is critical to make the measurement within the analysis window. In our work, we typically find the steady state at an ion dose of 1×10^{17} Cs$^+$/cm^2 while burn through varies based on the thickness of the sample. Mixing can also occur in the depth, here the **beam stabilization** measurement may also be useful as an estimate of **sputter yield** where the dose required to fully erode a sample of known thickness can be used to approximate the depth scale. As a point of reference, we have determined a depth scale of 1.6 nm per 1×10^{15} Cs$^+$/cm^2 for the epoxy we typically use for sample embedding (Agar 100). We also note that **precision** is directly related to **sensitivity**, and that precision increases with the volume of the specimen sampled, where the upper limit is defined by the analysis window. Thus, to maximize sensitivity, one should minimize the analysis area and maximize the primary ion current. That noted it is important to be conscious of the **size of the intended organelle**, a high primary ion current will have a larger beam profile and thus the image pixels may represent a mixture of the labeled molecule of interest containing organelle and the surrounding area, leading to an underestimate of the concentration. The best practice is to find the balance between the spatial resolution needed, the sensitivity and the time required to take the image.

16.3.1 Choosing the Target Molecule

When planning a quantitative NanoSIMS measurement, it is critical to become familiar with the structure of the target molecule. You want to be able to both identify positions that can be labeled and to assess fixability. To date, we have developed three labeling strategies each with advantages and disadvantages. First, the use of an isotopically labeled element which is homogeneously distributed through the sample meaning that you expect a consistent signal across the image, namely carbon. In the epoxy-embedded biological samples used for NanoSIMS analysis, the carbon content is so well matched to the carbon content of the biomaterial that ^{13}C enrichment due to the labeled compound can be directly converted to a concentration. In addition to the very straightforward data handling, this approach is a mainstay in metabolomic and medicinal chemistry labs. The drawback is that the abundance of carbon in the sample requires a high concentration of the drug or the metabolite to perturb the isotopic ratio sufficiently to detect an enrichment. Here when selecting a target molecule and using this strategy, one needs to consider the expected concentration and the availability of carbon position that can be exchanged for ^{13}C. As a rule of thumb, a small molecule such as dopamine with 6 ^{13}C labels has a limit of detection of 1 mM. Potentially a large molecule with the ability to accommodate 60 ^{13}C labels could reduce the limit of detection to the 100 μM level. Thus, it is important to have an estimate of the expected concentration of the target drug or metabolite. We have built an app to estimate the approximate sensitivity of a given molecule using the ^{13}C strategy calculation: http:/molcat.it.gu.se

The second approach uses a rare atomic label, we have focused on halogens, specifically I and Br which have a very low abundance in both the epoxy and the biological material and a high ion yield. We have also made some progress using F, however the background for F is generally quite high and difficult to predict as it is a common contamination in biological material. In practice, this makes the sensitivity of this F comparable to ^{13}C. Quantification using halogen labels requires the use of a relative sensitivity factor (RSF) which is generated by spiking a known concentration of the analyte label into an epoxy preparation. Because the embedding epoxy matches the biological material in terms of carbon concentration, we use the epoxy in a similar way as one would use blank plasma for a calibration curve for LC-MS. Meaning that the RSF with respect to $^{12}C_2$ or $^{13}C^{12}C$ in the standard will be applicable to the embedded cell. Halogen labels are very promising as the combination of low native abundance in the sample and the high ion yield produces low limits of detection potentially as low as 1 μM. While this type of labeling is not as prevalent as ^{13}C labeling, there are readily available options for brominated lipids as well as halogenated amino acids and nucleotides. While the sensitivity gained opens more possibilities, one needs to consider that the addition of halogens may alter the properties of the molecule.

The third strategy is a combination of the two above, where we use an isotopic label of a heterogeneous biologically abundant atom, for example ^{34}S which is at very low levels in the epoxy but varies across the biological material of the cell. Here the RSF is used to determine the local concentration of the endogenous sulfur which can then be used to convert the isotopic enrichment to concentration. This strategy is advantageous because the endogenous concentration of ^{34}S is approximately 100 times lower than ^{13}C in a typical sample. The result is that ^{34}S is roughly 100 times more sensitive than ^{13}C labeling [20]. However, using this strategy requires the molecule to have sulfur atoms that can be substituted for ^{34}S. Thus, the structure of the target molecule can constrain the strategy available for a planned study.

An equally important consideration when picking a target molecule is the fixability. The sample preparation procedure requires a dehydration step where the water will be replaced with the embedding epoxy. It follows that any water-soluble molecule can be translocated or removed if it is not properly fixed. The standard fixation is typically a mixture of formaldehyde and paraformaldehyde, meaning that molecules containing a primary amine are ideal. If the target molecule does not contain this function, it could be advantageous to add or expose a primary amine already in the structure, with the understanding that this may alter the physiochemical properties. Additionally, it is a standard practice to use tannic acid which stabilizes lipid membranes and has also been shown to assist in retaining molecules which lack a primary amine in subcellular structures [14].

In our experiment, we used amiodarone (Fig. 16.1) which endogenously contains two iodine atoms, so it is essentially a labeled molecule without further modification. The molecule does not however contain a primary amine thus tannic acid should be used to help keep the drug sequestered during the fixation process. In terms of the expected concentration, these cells have a large capacity for this molecule. Typical incubation conditions yield

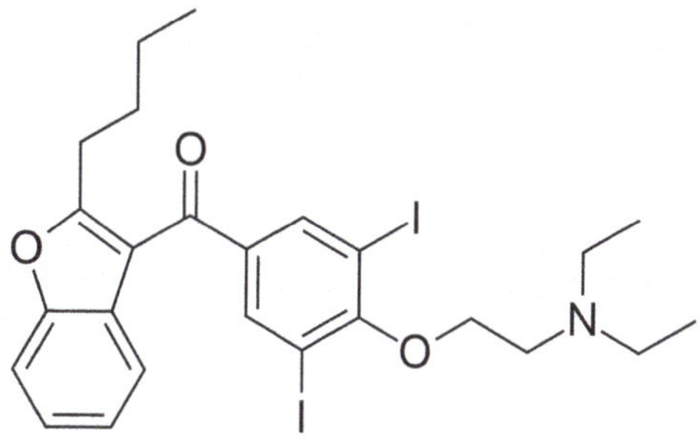

Fig. 16.1 Structure of
Amiodarone

approximately 3.5 fmol per cell [21]. Assuming the cell volume is roughly 4 pL, the total cell concentration will be on the order of 1 mM. Furthermore, the molecule is expected to be sequestered in the endosomal space which can be as small as 1% of the cell meaning the sequestered drug could be concentrated as high as 100 mM. Thus, this molecule is a sufficient target molecule fulfilling the basic criteria.

16.3.2 Standard Preparation

As our target molecule amiodarone contains iodine, we are using the second quantification strategy. Both the second and third strategies require calibration standards while quantification using ^{13}C labeling can be carried out without a standard. Data handling for ^{13}C quantification will be explained in a later section. Here we give a general protocol for preparing standards for NanoSIMS quantification.

Step-by-Step Protocol

1. Prepare a stock solution containing the target molecule; we have found that dimethyl sulfoxide (DMSO) and dimethylacetamide (DMA) solvents work well with the epoxy. When preparing the dilution series, one must consider that the ability of the epoxy to cure will be affected by the addition of the solvent, thus the solvent should be less than 5% of the total mixture, for example we typically use less than 50 μL when preparing 1 mL epoxy standards.
2. Prepare embedding resin according to instructions, in general it should be the same material used for embedding the specimen.
3. Each point of the serial dilution should be mixed into individual Eppendorf tube containing 1 mL of the epoxy. The viscous nature of the epoxy requires manual mixing typically the wooden handle of a cotton swab can be used to agitate the mixture.

(continued)

4. The samples are next cured according to the manufacturer instructions, this may require a ventilated oven.
5. Once the epoxy is cured, it should be sectioned using an ultramicrotome (300–1000 nm) and placed on a support, for example a 5 × 5 mm Si substrate (Here we suggest working with a facility that specializes in electron microscopy).

16.3.3 Sample Preparation

Sample preparation is a very important aspect of these experiments. The goal is to maintain the structure of the specimen while avoiding migration of the target molecule. As this step is critical, we suggest that the majority of this work should be carried out by labs with experience in electron microscopy where exact protocols may vary. The following is a generalized protocol.

Step-by-Step Protocol
1. Following treatment with the target molecule, the tissue or cell specimen is fixed using formaldehyde and glutaraldehyde. We typically use Karnovsky's solution (2.5% glutaraldehyde and 2% formaldehyde) prepared in 0.05 M Na-cacodylate buffer. This is typically done for 1 h at room temperature but depending on the size of sample this time may need to be adjusted.
2. A second fixation step using 1% osmium tetroxide prepared in 0.1 M Na-cacodylate buffer for 1 h at 4 °C. Again, the duration may be adjusted depending on the size of the sample.
3. A third fixation using 1% tannic acid prepared in deionized water is also done at room temperature and in the dark for 20 min.
4. If electron microscopy is required for the study, counterstaining will also be required. Here 1% uranyl acetate for 40 min also in the dark.
5. The fixed and stained samples are next dehydrated in ethanol series (30, 50, 70, 85, 95, and 100%).
6. Once the sample is in 100% ethanol, infiltration of the epoxy is facilitated by sequentially introducing the sample to 25%, 50%, and 75% epoxy in ethanol and finally the pure epoxy.
7. The epoxy sample is then cured according to the manufacturer's instructions and sectioned using an ultramicrotome (300–1000 nm). Here we reiterate that it is best to work with an EM facility that will be able to develop a protocol that best suits the intended specimen.

16.3.4 Steady State and Analysis Window Determination

To make a quantitative SIMS measurement, it is very important to keep track of the ion dose. The ion dose can be determined using a simple equation that relates the primary ion current to the time of the measurement and the area of the measurement. The primary ion current is measured with a Faraday cup which is located at the sample position called FcO. There is a set of diaphragms which are used to reduce the primary ion current called D1 1–5. Before analysis, the primary ion current corresponding to the diaphragm used for analysis as well as the diaphragm and lens one (L1) preset used for implantation. In a typical situation, we will measure 20 nA at FcP, this Faraday cup is in the central column and measures the primary ion current before it passes through a diaphragm. By switching to FcO, we can measure the current with respect to the selected diaphragm roughly we expect 15 pA for D1–1, 2 pA for D1–2, and 1 pA for D1–3. To set the implantation preset measure FcO using D1–1 and manually increase L1 until a sufficiently high current is reached, 300 pA for example, this value will be on the order of 20,000 digits and should be saved to the high current preset.

To calculate the ion dose, we suggest using a spreadsheet (Fig. 16.2) to keep track of both the implantation dose and the dose during analysis. For implantation, one should enter the target dose the field of view and the primary ion current measured for the implantation preset using the appropriate diaphragm, typically D1–1. The field of view in μm and the target dose in Cs^+/cm^2. To calculate the implantation time, choose a cell and multiply the target dose by the square of the field of view and convert to cm_2 by multiplying by 1×10^{-8}, multiply by the elementary charge (1.6×10^{-19}) to convert to coulombs, and multiply the inverse of the primary ion current being sure to convert from pA to amps (1×10^{-12} A/pA). This will report the time in seconds. If using excel, the time can be reported in minutes and seconds by dividing the result by (60*60*24) and setting the cell to the custom number format tt:mm:ss. The final equation will be in this form: = (ion dose)*(FoV μm^2) *0,00000001*1,6E-19*(1/(Primary ion current)*0,000000000001)))/(60*60*24). If you populate the sheet with a typical implantation primary ion current (300 pA) and a typical FoV 22 (μm) and set the target dose to 1×10^{17} Cs/cm$_2$, you will find that you will need to implant your sample for 4 min and 18 s.

The target dose corresponds to the steady state, the ion dose where the secondary ion emission no longer increases in response to the primary ion implantation. The easiest way to determine this dose is to carry out a beam stabilization (beam stab) measurement. This type of measurement can be selected in the tuning window. Here you can measure the secondary ion current as a function of time and using the primary ion current measured at FcO, this can easily be converted to ion dose. The beam stab function has a maximum acquisition time of 3000 s (50 min) so the implantation spreadsheet is useful when designing this type of measurement. Since we have an idea of the expected steady state, we can use the spreadsheet to choose experimental conditions where we can monitor the secondary ion current with respect to the primary ion current with the goal of exceeding the

Implatation			Analysis	
FcO pA	300		FcO pA	2
FoV um	22		FoV um	20
target dose Cs/cm2	1,00E+17		dwell time ms	1
Implatation time	00:04:18		cycles	10
			FoV pixels	256
depth implant nm	160		analysis time	00:10:55
analysis depth nm	3,2768		analysis dose	2,048E+15
total depth nm	163,2768			
(Cs/Cm2)/s	3,87E+14			

Fig. 16.2 Dose calculation spreadsheet

steady state dose within the 50 min time limit. In the case of the Agar100 epoxy, we typically find a steady state at approximately 1×10^{17} Cs^+/cm^2. Using the high current preset and the D1–1 diaphragm (300 pA primary ion current) and a FoV of 22 um, we know that this dose will be reached in 4 min and 18 s. However, it is not advisable to expose the EM detectors to very high secondary ion current so it is more appropriate to maximize the duration of the measurement to keep the secondary ion current manageable. As a rule of thumb, you can expect 100–200 k cps per pA when measuring C_2 from an epoxy sample at steady state ideally, we like to keep the counts on the EM detector below 500 k cps so here we can use D1–2 which gives 2 pA. When entered into the spreadsheet, we find using 2 pA will take 10 h. and 45 min to reach 1×10^{17} Cs^+/cm^2. Because dose is related to area, we can greatly reduce the time needed by using a smaller FoV. By entering 3 μm into the FoV field, we find that it will take 12 min to reach a dose of 1×10^{17}, and if we increase the target dose to 3×10^{17}, we find that the measurement will take 36 min which is within the 50 min time limit built into the software.

Once you have the parameters, it is just a matter of taking the measurement using a standard, which will be discussed below, that contains the ions of interest. It is possible that different ions will reach steady state at different ion doses, so in principle we pick the implantation dose to where all ions are at steady state. Once the dose is determined, we can use the spreadsheet to determine the sputtering parameters for each analysis. The following is a step-by-step protocol.

Step-by-Step Protocol

1. To reduce the risk of charging the samples should first be sputter coated with several nanometers of a conductive material; we typically use gold, however carbon and palladium also work well for this purpose.

(continued)

2. Load a standard which has been mounted on a 5 × 5 Si wafer, typically we use the "Harvard" sample holder which can accommodate 16 wafers; in practice, one should load all the samples and standards for the session at the same time.

3. Measure the primary ion current using the object Faraday cup (FcO) for the diaphragms which will be used for the measurement, typically D1–2 or D1–3, and measure the current using an implant preset using D1–1. The preset will have an elevated value for lens 1 (L1) typically above 20,000 digits and a deflector voltage (40 digits on C4Y).

4. Implant the sample to remove the gold and to reach steady state typically above 1×10^{17} Cs$^+$/cm^2, using the spreadsheet to find the appropriate implantation time for the field of view for the primary ion current measured at FcO using the implantation preset and D1–1.

5. Use the implanted standard to tune the mass table and select the appropriate entrance and aperture slits as well as the stage height (z), optimizing lens values (electrode of secondary: EOS and electrode of primary: EOP), and centering the secondary ion beam (Cy, P2P3). Also, take note of the secondary ion current measured for each Diaphragm used, considering that the secondary ion count rate at a should be kept at a reasonable level (we suggest below 500,000 cps).

6. Use the spreadsheet to determine the field of view that will allow a significantly high dose over a duration of 50 min. For example, if D1–2 was measured at 2 pA the spreadsheet a field of view of 3 μm will reach a dose of 3×10^{17} in 36 min.

7. Move the stage to a fresh area on the sample surface and run the beam stab function from the tuning window. Be sure to set the field of view determined and the appropriate diaphragm in the tuning window. In this example, 3 μm and D1–2.

8. This will generate a plot like the one shown in Fig. 16.3 with time in the x axis, which can be scaled to ion dose using the spreadsheet.

The beam stab plot shows the intensity of the ions of interest with respect to ion dose. The first observation is that the implantation greatly increases the secondary ion emission. The second observation is that the steady state is achieved for all the ions at the expected value of $1 \times 10e^{17}$ Cs$^+$/cm^2, particularly the ^{12}C^{12}C$^-$ signal and the ^{127}I$^-$ which we will use to quantify the amiodarone. It is also noteworthy that the ^{12}C^{14}N$^-$ reaches steady state at a lower ion dose reinforcing the idea that it is important to be in the steady state region to make a quantitative measurement.

The process of implantation is very important for reproducibility, but it is a destructive process which means that a significant amount of the sample will be eroded in order to reach the steady state. This should be accounted for as the sections used for these analyses are very thin usually below 500 nm thick. It is then quite useful to determine an erosion rate

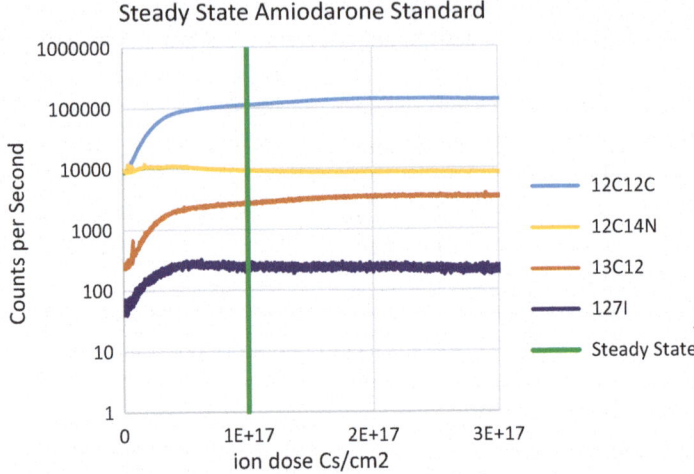

Fig. 16.3 Beam Stab plot for Amiodarone standard. This plot shows secondary ion counts per second for 4 ions of interest $^{12}C^{12}C$, $^{12}C^{14}N$, $^{13}C^{12}C$, and ^{127}I with respect to the primary ion dose. The apparent steady state is marked with a green line

for your sample. This can be done by measuring the volume of material excavated from your sample following a known ion dose using atomic force microscopy. Having done this for the Agar100 epoxy that we typically use, we find that 1.6 nm of material is removed for an ion dose of 1×10^{15} Cs^+/cm^2 or 1×10^{-15} nm/(Cs^+/cm^2), thus we add a field in the spreadsheet to multiply the input target dose by 1.6×10^{-15} nm/(Cs^+/cm^2) finding a depth of 160 nm if the target is set to 1×10^{17} Cs^+/cm^2.

It is also important to keep track of the depth during analysis as well so in the spreadsheet you should add entries for the primary ion current you will use for analysis, the FoV in μm for the analysis, the dwell time per pixel in ms, the FoV in pixels (generally we generate square images so we enter one dimension and square it, for example 128, 256, or 512), and the number of cycles. To calculate the dose, chose a new cell, convert the dwell time to seconds by dividing the entered value by 1000, multiply the squared value of FoV in pixels, and multiply the number of cycles to give the duration of the measurement in seconds. The primary ion current is converted to Amperes by dividing by 10^{12} which is divided by the elementary charge to give Cs^+/s. By multiplying by the time, we find the number of Cs^+ primary ions which is divided by the FoV in μm which has been converted to cm by dividing the input by 10^8. The equation in the cell should resemble = ((((dwell time*0,001)*(FoV pixles^2)*cycles)*((primary ion current*0,000000000001))/1,6E-19)/ FoV μm^2*0,00000001). It can also be helpful to find the time needed to make the measurement. In another cell copy the time portion of the equation and report this as hours min and seconds using the custom cell (tt:mm:ss) that was used to show the implantation time. Under the cell which contains the implantation depth, it is helpful to convert the dose calculated for the analysis to depth by multiplying the value by 1.6×10^{-15} nm/(Cs+/cm2). In the cell below it is helpful to report the sum of the implantation depth and the analysis depth, this is total depth which will be excavated during the measurement. This can then be compared to the thickness of the section to be analyzed to estimate the analysis window. The analysis window is the region between where the steady state is achieved and where sample degradation begins, as a rule of thumb, we find that leaving the last 50 nm of the section will generally allow you to avoid artifacts which occur when you "burn though" the sample. Alternatively, one could also repeat the beam stab measurement with the sample to find the analysis window. Using the spreadsheet, we find that if we implant the analysis region with 1×17 Cs/cm$_2$ and analyze using D1–2 (2 pA) with a foV of 20 μm with a 5 ms dwell time for 15 cycles at 256×256 pixels, the total depth will be approximately 200 nm with an analysis window of approximately 40 nm. The analysis will take 1 h and 22 min and will require a section that is at least 250 nm thick. The power of the spreadsheet is that it allows you to see all the important parameters. The following is a basic protocol to carry out this measurement.

Step-by-Step Protocol
1. To reduce the risk of charging, the samples should first be sputter coated with several nanometers of a conductive material.
2. Load a resin-embedded specimen which has been mounted on a 5×5 Si wafer, typically we use the "Harvard" sample holder.
3. Implant the sample to remove the gold and to reach steady state typically above 1×10^{17} Cs$^+$/cm^2, using the spreadsheet to find the appropriate implantation time for the field of view for the primary ion current measured at FcO using the implantation preset and D1–1.
4. If you have not designated a detector for ^{28}Si, it would be useful to use a space of bare silicon to find that peak.
5. Use the implanted sample to tune the mass table and select the appropriate entrance and aperture slits as well as the stage height (z), optimizing lens values (EOS and EOP), and centering the secondary ion beam (Cy, P2P3). Also, note the secondary ion current relative to the primary ion current for each Diaphragm used, considering to keep the count rate at a reasonable level (we suggest below 500,000 cps).
6. Use the spreadsheet and the preciously measured FcO values to determine the field of view that will allow a significantly high dose over a duration of 50 min. For example, if D1–2 was measured at 2 pA the spreadsheet a field of view of 3 μm will reach a dose of 3×10^{17} in 36 min.
7. Move the stage to a fresh area on the sample surface and run the beam stab function from the tuning window. Be sure to set the field of view determined and the appropriate diaphragm in the tuning window. In this example, 3 μm and D1–2.

(continued)

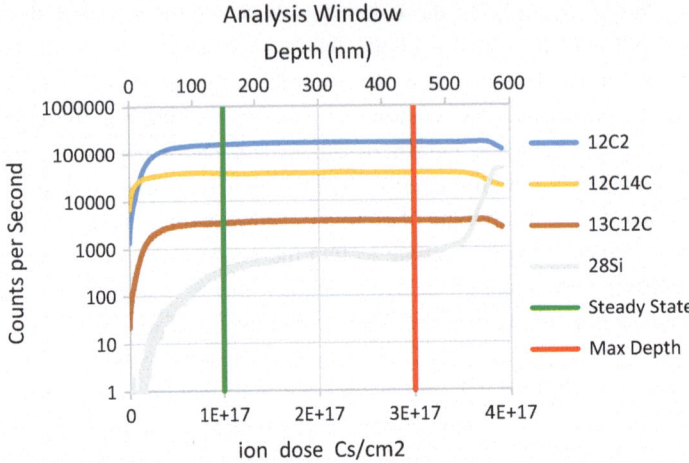

Fig. 16.4 Beam Stab plot for embedded cell. This plot shows secondary ion counts per second for 4 ions of interest $^{12}C^{12}C$, $^{12}C^{14}N$, $^{13}C^{12}C$, and ^{28}Si with respect to the primary ion dose. The apparent steady state is marked with a green line and the red line indicates where the sample burns through showing the maximum depth

8. This will generate a plot similar to the one shown in Fig. 16.4 with time in the x axis, which can be scaled to primary ion dose and depth using the spreadsheet. Here, at burn through occurred after the 36 min calculated in step 6, so we allowed the beam stab to continue the full 50 mins to observe the degradation of the sample.

16.3.5 Calibration Curve Generation

Expanding labeling strategies beyond ^{13}C labeling requires the use of a calibration standard. We make these standards by spiking a known amount of the target molecule into the epoxy used for embedding the specimen. Essentially, the calibration curve should encompass the expected concentration in the specimen. Because the addition of solvent to the epoxy will affect the ability of the material to cure, it is important to keep the spiking solution below 5% of the total volume of the epoxy. As a convention, we report the relative sensitivity factor (RSF) with respect to the concentration of the label. Only considering the label provides the opportunity to generate standards without the actual target molecule in situations where the solubility of the target molecule makes it difficult to produce a standard. In the case of amiodarone which contains 2 ^{127}I the concentrations are doubled. For the example used here, we created a serial dilution of amiodarone in DMA. Based on

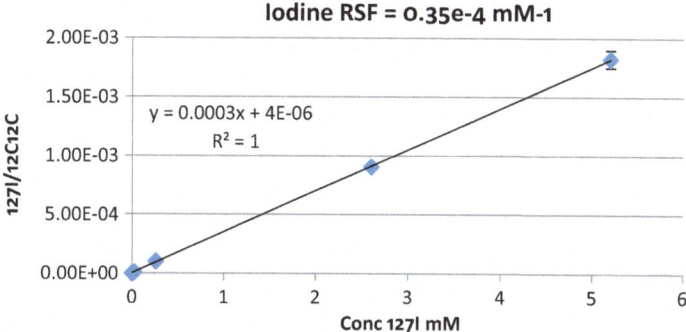

Fig. 16.5 Calibration curve for iodine generated with an Amiodarone epoxy standard series

5% V/V of the DMA solution to epoxy, 2.6, 1.3, 0.13, 0.013, and 0.0013 mM of the amiodarone are equivalent to 5.2, 2.6, 0.26, and 0.026 mM of iodine label. The slope of this line gives us the RSF which can be used directly, however because of the possibility of low levels of iodine contamination we normalize the signal to a measurement of a control cell; we term this the excess due to the drug or metabolite (Eq. 16.3). The following are basic instructions for generating a calibration curve.

Step-by-Step Protocol
1. First determine the imaging parameters which will be used in the analysis. For this reason, it could be useful to generate the calibration curve following the analysis.
2. Implant an area slightly larger than the area of a typical image, for example if typical image is 20 μm choose an area of 22 μm. Use the spreadsheet to calculate the implantation time required to reach the steady state for the implant preset.
3. After implantation, reduce the field of view, in this example 20 μm, and collect the image as would be done for the analysis.
4. Measure each standard and generate a calibration curve by plotting the ratio of the label versus the carbon signal. In general, we plot the concentration in terms of the number of labels, i.e. two ^{127}I atoms for the amiodarone in the example in Fig. 16.5.
5. Calculate the slope, this is the RSF that will be used for the calibration equation for the chosen labeling strategy.

16.3.6 Data Acquisition and Analysis

To plan data acquisition, use the spreadsheet to calculate the implantation time by entering the primary ion current used for implantation, the FoV (in principle this should be a bit larger than the FoV for analysis), and the target dose. Carry out the implantation for the given time. Now determine the imaging parameters. There is one basic idea to keep in mind, high primary current will allow you to sample more of the sample in a shorter time which is advantageous for measuring low concentrations but will be detrimental to the spatial resolution. Thus, you need to strike a balance between spatial resolution and the time it takes to make the measurement. In the case of the amiodarone, the signal was quite intense, thus a sampling depth of approximately 3 nm was deemed sufficient, where we elected to use 2 pA as primary current, a field of view of 20 μm, 1 ms dwell time, FoV pixel of 256, and 10 planes. These settings gave an analysis time of about 10 min per image. An overview of the work flow is shown in Fig. 16.6.

Figure 16.7 shows the ion images from an amiodarone treated macrophage, here we map $^{12}C^{12}C^-$, $^{12}C^{13}C^-$, $^{12}C^{14}N^-$, and $^{127}I^-$. The carbon images show that the distribution of the carbon is largely homogenous reflecting that the embedded cell and the epoxy are very well matched with regard to carbon content. This is critical as the calibration curve is generated in the pure epoxy. The $^{12}C^{14}N^-$ image gives a general overview of the cell structure, notably the nucleus is clearly observable. The iodine signal shows the location of the amiodarone. Also, shown on each image are the regions of interest selected for quantitative analysis. The RSF is calculated from the ratio of $^{127}I^-/^{12}C^{12}C^-$ thus the ROIs report this ratio. We will discuss how to apply the RSF to the ROI data below but will also take the opportunity to describe all of the quantification strategies as well.

Previously, we have shown that the isotopic enrichment in NanoSIMS images can be directly translated to biologically relevant concentrations of drugs and metabolites in subcellular structures. This relationship relies on the homogeneity and predictability of the ^{13}C concentration in resin-embedded cell material. Equation (16.1) conceptually describes the concentration of an isotopically labeled drug needed to perturb the natural isotopic ratio measured in a region of interest in a SIMS image, where $\partial^{13}C^{ROI}$ is the enrichment and $\partial^{13}C^{Control}$ is a blank measurement, both referenced to the Vienna PeeDee Belemnite $\partial^{13}C$ reference standard (VPDB), 54 M is the concentration of C in the resin-embedded cell material and $N^{13}C$ is the number of ^{13}C labels included on the drug or metabolite (Eq. 16.1).

$$[\text{drug.met}]M = \frac{\delta^{13}C^{ROI}_{VPDB} - \delta^{13}C^{control}_{VPDB}}{1000} * \frac{VPDB * 54M}{N^{13}C} \tag{16.1}$$

This equation is a straightforward conversion of the isotopic measurement to concentration. This can be done because the carbon concentration across the cell and the embedding epoxy is both known and homogenous. We have shown that the abundance of ^{13}C in our embedding material is approximately 600 mM; therefore, 10‰ (by convention we use permil to express relative deviation from VPDB reference) deviation from the natural ratio will correspond to an increase of 1% in ^{13}C concentration which is attributed to the ^{13}C labeled compound. Expressed in mM, the quantitative equation of ^{13}C takes the following form:

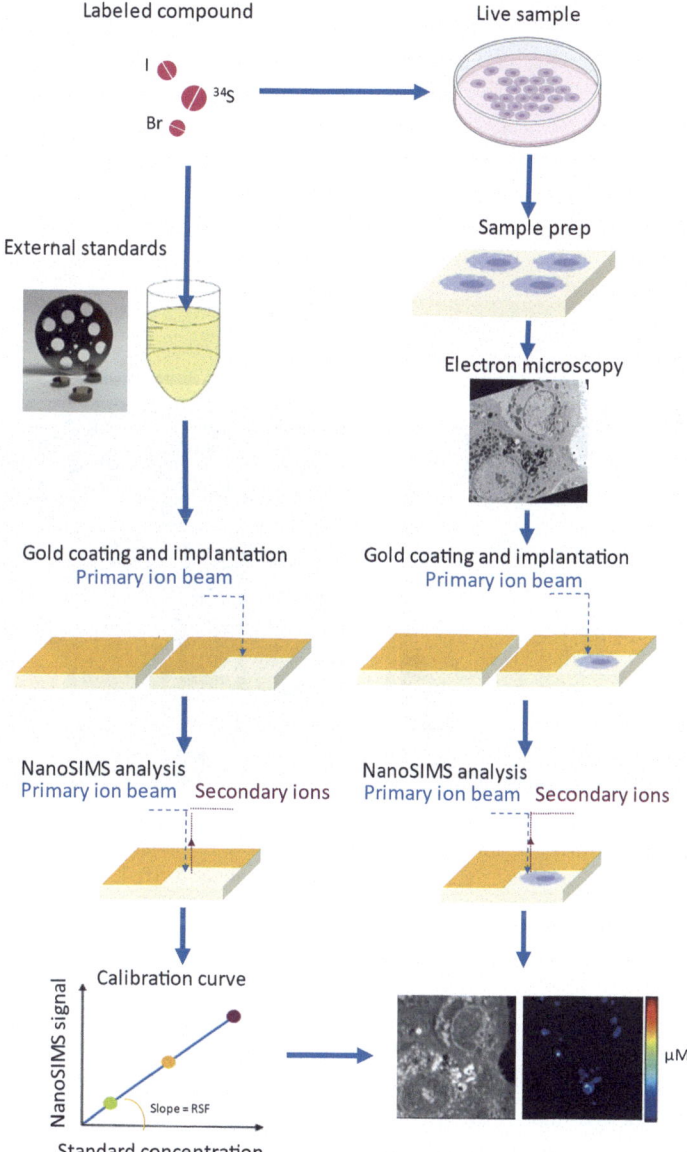

Fig. 16.6 Workflow overview. Cells are treated with the labeled compound of interest and prepared for electron microscopy imaging. The labeled compound is also used to create epoxy standards. Both the standards and the samples are sputter coated with gold and implanted to reach steady state. A calibration curve is generated which is used to translate the SIMS signal into concentration

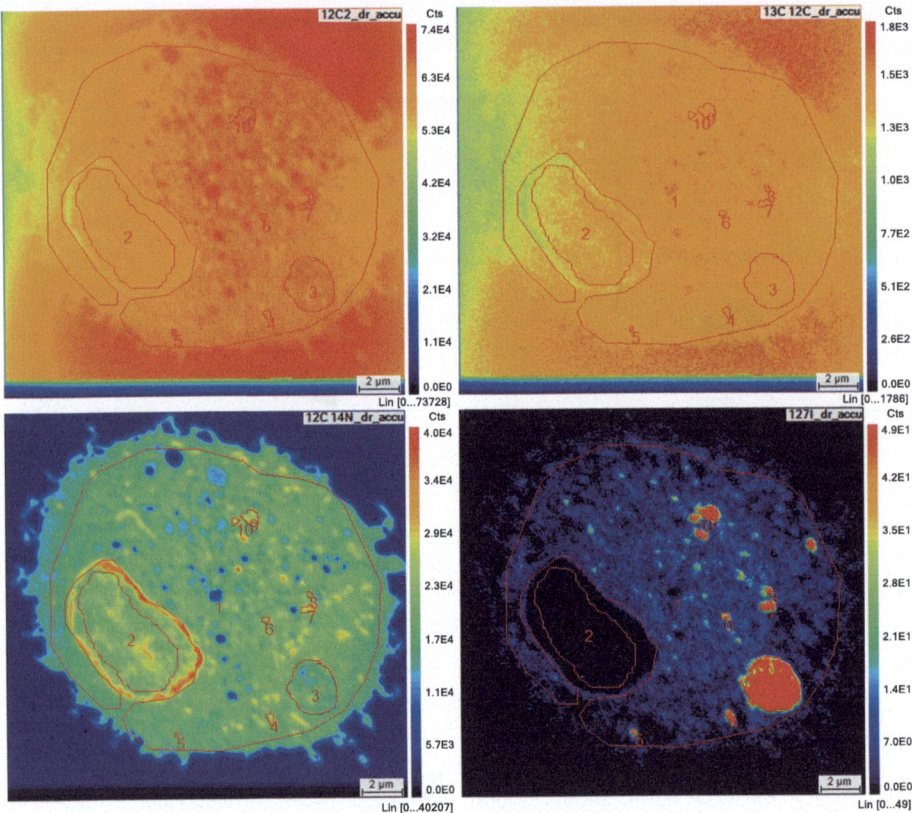

Fig. 16.7 NanoSIMS image of amiodarone treated macrophage, top left $^{12}C^{12}C$, top right $^{12}C^{13}C$, bottom left $^{12}C^{14}N$, bottom right ^{127}I, ROI 1: Cytosol, 2: Nucleus, 3–10: Lysosomes

$$[\text{drug.met}]\text{mM} = \frac{\delta^{13}C_{VPDB}^{ROI} - \delta^{13}C_{VPDB}^{control}}{1000} * \frac{600\ \text{mM} * 1000}{N^{13}C} \qquad (16.2)$$

In our experiment, we will use iodine labeling which requires the use of the RSF we have calculated from the calibration curve. However, we first adjust the $^{127}I/^{12}C^{12}C$ signal to account for background signal which may arise from biological sources or from trace contamination. We call this term drug_met excess (Eq. 16.3)

$$^{127}I\ \text{drug.met excess} = \left(\frac{^{127}I_{sample}}{^{12}C^{12}C_{sample}} - \text{Ratio}_{control}\right) \times \frac{1}{N^{127}I} \qquad (16.3)$$

Where the average $^{127}I/^{12}C^{12}C$ ratio from a control cell (Ratio$_{control}$), which serves as the background, is subtracted from the ratio measured in the region of interest and divided by the number of I labels. The drug_met excess term can be converted to the drug or

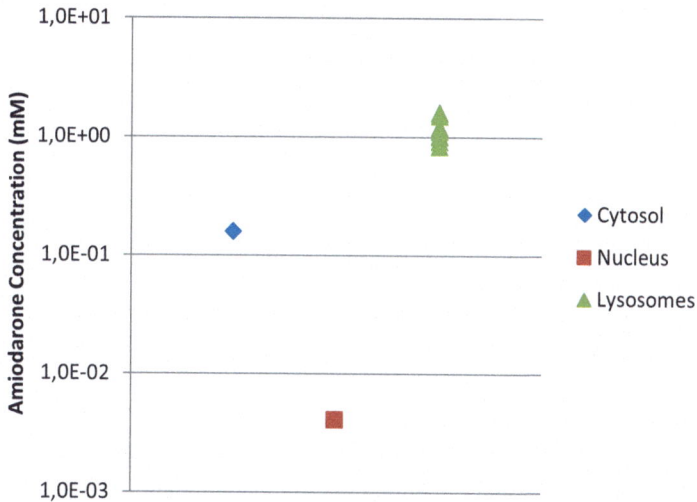

Fig. 16.8 Subcellular quantification of amiodarone in a single macrophage, region of interest analysis

metabolite concentration by applying the RSF 3.5×10^{-4} mM^{-1} found using the calibration standards (Eq. 16.4).

$$[\text{drug.met}] \text{ mM} = {}^{127}\text{I drug.met excess} \times \frac{1}{\text{RSF}} \qquad (16.4)$$

The RSF is then applied to the drug.met excess to give the concentration of the drug or the metabolite. Figure 16.8 shows a region of interest analysis of the amiodarone taken up into the NR 8383 macrophage in Fig. 16.7. As expected, we find that the concentration of the amiodarone is elevated in the in the lysosomal space and in the order of 1 mM. In fact, the trapped amiodarone is approximately 10 times more concentrated when compared to the cytosolic space. This indicates that the organelles containing the drug account for a significant proportion of the cell volume, potentially as high as 10%. This is also obvious by observing the image. Thus, it can be concluded that the macrophage's capacity to uptake this molecule relies on the ability to greatly expand the cells' lysosomal space.

To use isotopic labels other then ^{13}C that presents an heterogenous background like ^{34}S, using an RSF is not enough. First, the excess of the isotope due to the drug or metabolite need to be calculated. This is determined by using the average isotopic ratio of the control (μcontrol) measurement before applying the RSF (Eq. 16.5).

$$^{34}\text{S drug.met excess} = \frac{{}^{34}\text{S}_{\text{sample}} - ({}^{32}\text{S}_{\text{sample}} \times \mu\text{control})}{{}^{12}\text{C}^{12}\text{C}_{\text{sample}}} \times \frac{1}{\text{N}^{34}\text{S}} \qquad (16.5)$$

Equation (16.5) shows this treatment for a ^{34}S labeled molecule. Here the counts of the major isotope ^{32}S are multiplied by the average of the isotopic ratio measured in a control cell to determine the counts of ^{34}S expected to be contributed to the ROI by the endogenous sulfur in the specimen. This expected value is subtracted from the measured value of ^{34}S and divided by the number of ^{34}S labels. The value can then be entered into Eq. (16.4) along with the RSF for sulfur to give the drug or metabolite concentration.

Take-Home Message
- Be familiar with the target molecule, know the labeling opportunities as well as the fixability.
- Have a rough estimate of the expected concentration, this will give you a basic idea of how far you will need to push your parameters toward sensitivity and away from spatial resolution.
- Keep track of the ion dose, this is important for consistency in implantation but also to monitor the volume that is sampled during the measurement.
- Know when to use an RSF and which equation go with each labeling strategy.
- Get familiar with the subcellular environment, understand the ways in which molecules become sequestered in subcellular compartments.

References

1. Hann MM, Simpson GL. Intracellular drug concentration and disposition–the missing link? Elsevier; 2014. p. 283–5.
2. Mateus A, Gordon LJ, Wayne GJ, Almqvist H, Axelsson H, Seashore-Ludlow B, et al. Prediction of intracellular exposure bridges the gap between target- and cell-based drug discovery. Proc Natl Acad Sci. 2017;114(30):E6231–9.
3. Fridén M, Bergström F, Wan H, Rehngren M, Ahlin G, Hammarlund-Udenaes M, et al. Measurement of unbound drug exposure in brain: modeling of pH partitioning explains diverging results between the brain slice and brain homogenate methods. Drug Metab Dispos. 2011;39(3): 353–62.
4. Friden M, Ducrozet F, Middleton B, Antonsson M, Bredberg U, Hammarlund-Udenaes M. Development of a high-throughput brain slice method for studying drug distribution in the central nervous system. Drug Metab Dispos. 2009;37(6):1226–33.
5. Monro A. Interspecies comparisons in toxicology: the utility and futility of plasma concentrations of the test substance. Regul Toxicol Pharmacol. 1990;12(2):137–60.
6. Nilsson A, Goodwin RJ, Shariatgorji M, Vallianatou T, Webborn PJ, Andrén PE. Mass spectrometry imaging in drug development. Anal Chem. 2015;87(3):1437–55.
7. Dollery C. Intracellular drug concentrations. Clin Pharmacol Therapeut. 2013;93(3):263–6.
8. Legin AA, Schintlmeister A, Jakupec MA, Galanski MS, Lichtscheidl I, Wagner M, et al. NanoSIMS combined with fluorescence microscopy as a tool for subcellular imaging of isotopically labeled platinum-based anticancer drugs. Chem Sci. 2014;5(8):3135–43.
9. Lee RF, Escrig S, Croisier M, Clerc-Rosset S, Knott GW, Meibom A, et al. NanoSIMS analysis of an isotopically labelled organometallic ruthenium (II) drug to probe its distribution and state in vitro. Chem Commun. 2015;51(92):16486–9.

10. Wedlock LE, Kilburn MR, Liu R, Shaw JA, Berners-Price SJ, Farrell NP. NanoSIMS multi-element imaging reveals internalisation and nucleolar targeting for a highly-charged polynuclear platinum compound. Chem Commun. 2013;49(62):6944–6.

11. Kay E, Stulz R, Becquart C, Lovric J, Tängemo C, Thomen A, et al. NanoSIMS imaging reveals the impact of ligand-ASO conjugate stability on ASO subcellular distribution. Pharmaceutics. 2022;14(2):463.

12. He C, Migawa MT, Chen K, Weston TA, Tanowitz M, Song W, et al. High-resolution visualization and quantification of nucleic acid-based therapeutics in cells and tissues using Nanoscale secondary ion mass spectrometry (NanoSIMS). Nucleic Acids Res. 2021;49(1):1–14.

13. Lovrić J, Dunevall J, Larsson A, Ren L, Andersson S, Meibom A, et al. Nano secondary ion mass spectrometry imaging of dopamine distribution across nanometer vesicles. ACS Nano. 2017;11 (4):3446–55.

14. Jiang H, Passarelli MK, Munro PM, Kilburn MR, West A, Dollery CT, et al. High-resolution sub-cellular imaging by correlative NanoSIMS and electron microscopy of amiodarone internalisation by lung macrophages as evidence for drug-induced phospholipidosis. Chem Commun. 2017;53(9):1506–9.

15. Kleinfeld AM, Kampf JP, Lechene C. Transport of 13C-oleate in adipocytes measured using multi imaging mass spectrometry. J Am Soc Mass Spectrom. 2004;15(11):1572–80.

16. He C, Fong LG, Young SG, Jiang H. NanoSIMS imaging: an approach for visualizing and quantifying lipids in cells and tissues. J Invest Med. 2017;65(3):669–72.

17. He C, Hu X, Jung RS, Weston TA, Sandoval NP, Tontonoz P, et al. High-resolution imaging and quantification of plasma membrane cholesterol by NanoSIMS. Proc Natl Acad Sci. 2017;114(8): 2000–5.

18. He C, Weston TA, Jung RS, Heizer P, Larsson M, Hu X, et al. NanoSIMS analysis of intravascular lipolysis and lipid movement across capillaries and into cardiomyocytes. Cell Metab. 2018;27(5):1055–66. e3

19. Thomen A, Najafinobar N, Penen F, Kay E, Upadhyay PP, Li X, et al. Subcellular mass spectrometry imaging and absolute quantitative analysis across organelles. ACS Nano. 2020;14 (4):4316–25.

20. Becquart C, Stulz R, Thomen A, Dost M, Najafinobar N, Dahlén A, et al. Intracellular absolute quantification of oligonucleotide therapeutics by NanoSIMS. Anal Chem. 2022;94(29): 10549–56.

21. Newman CF, Havelund R, Passarelli MK, Marshall PS, Francis I, West A, et al. Intracellular drug uptake – a comparison of single cell measurements using ToF-SIMS imaging and quantification from cell populations with LC/MS/MS. Anal Chem. 2017;89(22):11944–53.

Index

The manufacturer's authorised representative in the EU is Springer
Nature Customer Service Centre GmbH, Europaplatz 3, 69115 Heidelberg,
Germany. If you have any concerns regarding our products, please
contact ProductSafety@springernature.com

Printed and bound by CPI Group (UK) Ltd, Croydon, CR0 4YY
29/04/2026
02099548-0002